METHODS IN MOLECULAR BIOLOGY

Series Editor
John M. Walker
School of Life and Medical Sciences,
University of Hertfordshire,
Hatfield, Hertfordshire AL10 9AB, UK

For further volumes:
http://www.springer.com/series/7651

Microinjection

Methods and Protocols

Edited by

Chengyu Liu and Yubin Du

National Heart, Lung, and Blood Institute, National Institutes of Health, Bethesda, MD, USA

Humana Press

Editors
Chengyu Liu
National Heart, Lung, and Blood Institute
National Institutes of Health
Bethesda, MD, USA

Yubin Du
National Heart, Lung, and Blood Institute
National Institutes of Health
Bethesda, MD, USA

ISSN 1064-3745 ISSN 1940-6029 (electronic)
Methods in Molecular Biology
ISBN 978-1-4939-8830-3 ISBN 978-1-4939-8831-0 (eBook)
https://doi.org/10.1007/978-1-4939-8831-0

Library of Congress Control Number: 2018958963

This Humana Press imprint is published by the registered company Springer Science+Business Media, LLC, part of Springer Nature.
The registered company address is: 233 Spring Street, New York, NY 10013, U.S.A.

Preface

The first dedicated microinjection device was reported over a century ago. As summarized in Chap. 1 of this book, it has been successfully used for addressing questions in a broad range of research disciplines. Microinjection and related micromanipulation techniques have played crucial roles in conceptual and technological breakthroughs that led to at least five Nobel Prizes, including the 1935 prize on embryo organizer theory, the 1991 prize on patch-clamp, the 2007 prize on mouse knockout, the 2010 prize on *in vitro* fertilization (IVF), and the 2012 prize on stem cells and animal cloning. Although microinjection is a very old technique, its utilization has increased substantially in recent years thanks to breakthrough advances in genome engineering technologies. Undoubtedly, it will continuingly be used in the foreseeable future not only for generating animal models for biomedical research but also for changing economically or ecologically important species that can broadly impact our society in general.

In this volume, about half of the chapters are focused on the various methods for generating mouse models, because a great majority of genome engineering research has been done in mice and they will continue to be the preferred species for developing new technologies. We have chapters covering the classical pronuclear injection method developed at the beginning of the 1980s and the conventional blastocyst injection method which was instrumental for developing the mouse knockout technology near the end of the 1980s. We also have chapters discussing various variations and improvements on these vintage techniques, as well as related methods, including somatic cell nuclear transfer (animal cloning) and interspecific chimera generation. Recent improvements in *in vitro* fertilization, embryo cryopreservation, and zygotic electroporation technologies will likely boost the efficiency and productivity of animal model generation, especially when it is combined with the latest highly efficient gene editing technologies. Therefore, we have included a chapter to cover each of these topics.

During the past decade, one of the most exciting advancements in biomedical research is the emergence of the so-called nuclease-based genome engineering technologies, including ZFN (zinc finger nucleases), TALEN (transcription activator-like effector nucleases), and CRISPR (clustered regularly interspaced short palindromic repeats). We have chapters describing each of these methods. Prior to these technologies, our ability to precisely manipulate the genome of living mammals was largely confined to mice and only recently to rats. The nuclease-based methods, especially the CRISPR, have essentially eliminated the species barrier for precise genome modification. Therefore, about one half of the chapters in this volume are devoted to gene editing in other species, which include mammalian species, such as rats, rabbits, pigs, and goats, as well as non-mammals including worms, fishes, frog, and chickens. We also have a chapter on microinjecting mammalian cells, which has also been proven useful for addressing various biological questions.

It is believed by many that we are in the middle of the CRISPR revolution which will move genome engineering from laboratories into farms and clinics, which will profoundly change the world we live in. Therefore, it is timely to assemble various protocols in this extremely fast-moving field to help MIMB readers to take advantage of the breakthrough technologies in gene editing and embryo micromanipulations.

Bethesda, MD, USA

Chengyu Liu
Yubin Du

Contents

Preface . . . *v*
Contributors . . . *xi*

1 Microinjection and Micromanipulation: A Historical Perspective . . . 1
Wenhao Xu

2 Production of Transgenic Mice by Pronuclear Microinjection . . . 17
Xin-an Pu, Anthony P. Young, and H. Michael Kubisch

3 Transgene Recombineering in Bacterial Artificial Chromosomes . . . 43
Michael G. Zeidler and Thomas L. Saunders

4 Using TARGATT™ Technology to Generate Site-Specific Transgenic Mice . . . 71
Ruby Yanru Chen-Tsai

5 Generating Genetically Engineered Mice Using a Spermatogonial Stem Cell-Mediated Method . . . 87
Takuya Sato and Takehiko Ogawa

6 Chimeric Mouse Generation by ES Cell Blastocyst Microinjection and Uterine Transfer . . . 99
Yubin Du, Wen Xie, Fan Zhang, and Chengyu Liu

7 Creating Knockin Alleles in Mouse Embryonic Stem Cells by CRISPR/Cas9-Mediated Homologous Recombination Without Drug Selection . . . 115
Pinghu Liu, Yan Li, Jingqi Lei, and Lijin Dong

8 Using CRISPR/Cas9 for Gene Knockout in Immunodeficient NSG Mice . . . 139
Yubin Du, Wen Xie, Fan Zhang, Uimook Choi, Chengyu Liu, and Colin L. Sweeney

9 Generation of CRISPR-Edited Rodents Using a Piezo-Driven Zygote Injection Technique . . . 169
Melissa A. Scott and Yueh-Chiang Hu

10 Delivery of CRISPR-Cas9 into Mouse Zygotes by Electroporation . . . 179
Wenning Qin and Haoyi Wang

11 Generation of Conditional Knockout Mice by Sequential Insertion of Two loxP Sites *In Cis* Using CRISPR/Cas9 and Single-Stranded DNA Oligonucleotides . . . 191
Ye Liu, Yubin Du, Wen Xie, Fan Zhang, Douglas Forrest, and Chengyu Liu

12 Improvement of Mouse Cloning from Any Type of Cell by Nuclear Injection . . . 211
Sayaka Wakayama, Satoshi Kishigami, and Teruhiko Wakayama

13 The CARD Method for Simple Vitrification of Mouse Oocytes: Advantages and Applications 229
Naomi Nakagata, Jorge Sztein, and Toru Takeo

14 The CARD Method for Mouse Sperm Cryopreservation and In Vitro Fertilization Using Frozen-Thawed Sperm 243
Toru Takeo, Jorge Sztein, and Naomi Nakagata

15 Isolation and Analysis of a Genome-Edited Single-Hepatocyte from a Cas9 Transgenic Mouse Line 257
Takayuki Sakurai, Akiko Kamiyoshi, Masato Ohtsuka, Channabasavaiah B. Gurumurthy, Masahiro Sato, and Takayuki Shindo

16 Microinjection and Oviduct Transfer Procedures for Rat Model Generation with CRISPR-Cas9 Technology 273
Tuija M. Alcantar and Xin Y. Rairdan

17 Molecular Aspects of Zinc Finger Nucleases (ZFNs)-Mediated Gene Editing in Rat Embryos 295
Xiaoxia Cui

18 Organ Generation from Knockedout Rat Blastocysts Complemented with Pluripotent Stem Cells 313
Masumi Hirabayashi and Shinichi Hochi

19 Generation of Rabbit Models by Gene Editing Nucleases 327
Dongshan Yang, Jie Xu, and Y. Eugene Chen

20 Production of Genetically Engineered Porcine Embryos by Handmade Cloning 347
Rong Li, Jinxin Miao, and Zhongde Wang

21 Electrofusion of 2-Cell Embryos for Porcine Tetraploid Embryo Production 361
Qingran Kong and Zhonghua Liu

22 Gene Knockouts in Goats Using CRISPR/Cas9 System and Somatic Cell Nuclear Transfer 373
Zhiqiang Fan, Min Yang, Misha Regouski, and Irina A. Polejaeva

23 Generating Goat Mammary Gland Bioreactors for Producing Recombinant Proteins by Gene Targeting 391
Liyou An, Lan Yang, Yuejin Huang, Yong Cheng, and Fuliang Du

24 Production of Transgenic Chickens Using Cultured Primordial Germ Cells and Gonocytes 403
Ellen J. Collarini, Philip A. Leighton, and Marie-Cecile Van de Lavoir

25 Using Microinjection to Generate Genetically Modified *Caenorhabditis elegans* by CRISPR/Cas9 Editing 431
Jyoti Iyer, Nicole DeVaul, Tyler Hansen, and Ben Nebenfuehr

26 Microinjection in Zebrafish for Genome Editing and Functional Studies 459
Wuhong Pei and Shawn M. Burgess

27 Microinjection of Marine Fish Eggs 475
Rie Goto, Taiju Saito, Takahiro Matsubara, and Etsuro Yamaha

28 Generating Gene Knockout *Oryzias latipes* and Rice Field Eel Using TALENs Method 489
Daji Luo, Ke Feng, Zuoyan Zhu, and Wei Hu

29 Functional Studies of Transcriptional Cofactors via Microinjection-Mediated Gene Editing in *Xenopus*. 507
Yuki Shibata, Lingyu Bao, Liezhen Fu, Bingyin Shi, and Yun-Bo Shi

30 Microinjection of Live Mammalian Cells: A Delivery Method that Provides Added Versatility to the Study of Cellular Function 525
Robert F. Gahl

Index *537*

Contributors

TUIJA M. ALCANTAR • *Embryo Technology Group, gRED Animal Resources, Genentech, Inc., South San Francisco, CA, USA*
LIYOU AN • *Jiangsu Key Laboratory for Molecular and Medical Biotechnology, College of Life Sciences, Nanjing Normal University, Nanjing, China*
LINGYU BAO • *Section on Molecular Morphogenesis, Eunice Kennedy Shriver National Institute of Child Health and Human Development (NICHD), National Institutes of Health (NIH), Bethesda, MD, USA; Department of Endocrinology, The First Affiliated Hospital of Xi'an Jiaotong University School of Medicine, Xi'an, Shaanxi, People's Republic of China*
SHAWN M. BURGESS • *Translational and Functional Genomics Branch, National Human Genome Research Institute, National Institutes of Health, Bethesda, MD, USA*
Y. EUGENE CHEN • *Center for Advanced Models for Translational Sciences and Therapeutics, University of Michigan Medical Center, Ann Arbor, MI, USA*
YONG CHENG • *College of Veterinary Medicine, Yangzhou University, Yangzhou, Jiangsu, China*
RUBY YANRU CHEN-TSAI • *Applied StemCell, Inc., Milpitas, CA, USA*
UIMOOK CHOI • *Laboratory of Clinical Immunology and Microbiology, Division of Intramural Research, National Institute of Allergy and Infectious Diseases, National Institutes of Health, Bethesda, MD, USA*
ELLEN J. COLLARINI • *Ligand Pharmaceuticals, Inc., Emeryville, CA, USA*
XIAOXIA CUI • *Department of Genetics, Washington University in St. Louis, St. Louis, MO, USA*
NICOLE DEVAUL • *The Department of Anatomy and Cell Biology, The George Washington University School of Medicine and Health Sciences, Tulsa, OK, USA*
LIJIN DONG • *Genetic Engineering Core, National Eye Institute, National Institutes of Health, Bethesda, MD, USA*
FULIANG DU • *Jiangsu Key Laboratory for Molecular and Medical Biotechnology, College of Life Sciences, Nanjing Normal University, Nanjing, China; Renova Life, Inc., College Park, MD, USA*
YUBIN DU • *Transgenic Core Facility, Division of Intramural Research, National Heart, Lung, and Blood Institute, National Institutes of Health, Bethesda, MD, USA*
ZHIQIANG FAN • *Department of Animal, Dairy, and Veterinary Sciences, Utah State University, Logan, UT, USA*
KE FENG • *State Key Laboratory of Freshwater Ecology and Biotechnology, Institute of Hydrobiology, Chinese Academy of Science, Hubei, People's Republic of China*
DOUGLAS FORREST • *Laboratory of Endocrinology and Receptor Biology, National Institute of Diabetes and Digestive and Kidney Diseases, National Institutes of Health, Bethesda, MD, USA*
LIEZHEN FU • *Section on Molecular Morphogenesis, Eunice Kennedy Shriver National Institute of Child Health and Human Development (NICHD), National Institutes of Health (NIH), Bethesda, MD, USA*
ROBERT F. GAHL • *Laboratory of Molecular Biophysics, Biochemistry and Biophysics Center, National Heart, Lung and Blood Institute, National Institutes of Health, Bethesda, MD, USA*
RIE GOTO • *South Ehime Fisheries Research Center, Ehime University, Minamiuwagun, Ehime, Japan*

CHANNABASAVAIAH B. GURUMURTHY • *Mouse Genome Engineering Core Facility, University of Nebraska Medical Center, Omaha, NE, USA; Developmental Neuroscience, Munro Meyer Institute for Genetics and Rehabilitation, University of Nebraska Medical Center, Omaha, NE, USA*

TYLER HANSEN • *Department of Biochemistry, Vanderbilt University School of Medicine, Tulsa, OK, USA*

MASUMI HIRABAYASHI • *The Graduate University for Advanced Studies, Okazaki, Aichi, Japan; Section of Mammalian Transgenesis, Center for Genetic Analysis of Behavior, National Institute for Physiological Sciences, Okazaki, Aichi, Japan*

SHINICHI HOCHI • *Faculty of Textile Science and Technology, Shinshu University, Ueda, Nagano, Japan*

WEI HU • *State Key Laboratory of Freshwater Ecology and Biotechnology, Institute of Hydrobiology, Chinese Academy of Science, Hubei, People's Republic of China*

YUEH-CHIANG HU • *Division of Developmental Biology, Cincinnati Children's Hospital Medical Center, Cincinnati, OH, USA*

YUEJIN HUANG • *Shanghai Jenomed Biotech Co., Ltd., Shanghai, China*

JYOTI IYER • *Department of Chemistry and Biochemistry, The University of Tulsa, Tulsa, OK, USA*

AKIKO KAMIYOSHI • *Department of Cardiovascular Research, Graduate School of Medicine, Shinshu University, Matsumoto, Nagano, Japan; Basic Research Division for Next-Generation Disease Models and Fundamental Technology, Research Center for Next Generation Medicine, Shinshu University, Matsumoto, Nagano, Japan*

SATOSHI KISHIGAMI • *Faculty of Life and Environmental Sciences, University of Yamanashi, Yamanashi, Japan*

QINGRAN KONG • *Key Laboratory of Animal Cellular and Genetics Engineering of Heilongjiang Province, College of Life Science, Northeast Agricultural University, Harbin, China*

H. MICHAEL KUBISCH • *Tulane National Primate Research Center, Covington, LA, USA*

JINGQI LEI • *Genetic Engineering Core, National Eye Institute, National Institutes of Health, Bethesda, MD, USA*

PHILIP A. LEIGHTON • *Ligand Pharmaceuticals, Inc., Emeryville, CA, USA*

RONG LI • *Department of Animal, Dairy, and Veterinary Sciences, Utah State University, Logan, UT, USA*

YAN LI • *Genetic Engineering Core, National Eye Institute, National Institutes of Health, Bethesda, MD, USA*

CHENGYU LIU • *Transgenic Core Facility, Division of Intramural Research, National Heart, Lung, and Blood Institute, National Institutes of Health, Bethesda, MD, USA*

PINGHU LIU • *Genetic Engineering Core, National Eye Institute, National Institutes of Health, Bethesda, MD, USA*

YE LIU • *Laboratory of Endocrinology and Receptor Biology, National Institute of Diabetes and Digestive and Kidney Diseases, National Institutes of Health, Bethesda, MD, USA*

ZHONGHUA LIU • *Key Laboratory of Animal Cellular and Genetics Engineering of Heilongjiang Province, College of Life Science, Northeast Agricultural University, Harbin, China*

DAJI LUO • *State Key Laboratory of Freshwater Ecology and Biotechnology, Institute of Hydrobiology, Chinese Academy of Science, Hubei, People's Republic of China; Department of Genetics, School of Basic Medical Science, Wuhan University, Hubei, People's Republic of China*

TAKAHIRO MATSUBARA • *South Ehime Fisheries Research Center, Ehime University, Minamiuwagun, Ehime, Japan*

JINXIN MIAO • *Department of Animal, Dairy, and Veterinary Sciences, Utah State University, Logan, UT, USA; National Centre for International Research in Cell and Gene Therapy, Sino-British Research Centre, School of Basic Medical Sciences, Zhengzhou University, Zhengzhou, People's Republic of China*
NAOMI NAKAGATA • *Division of Reproductive Engineering, Center for Animal Resources and Development (CARD), Kumamoto University, Kumamoto, Japan*
BEN NEBENFUEHR • *Department of Molecular, Cellular and Developmental Biology, University of Colorado, Boulder, Tulsa, OK, USA*
TAKEHIKO OGAWA • *Laboratory of Biopharmaceutical and Regenerative Sciences, Institute of Molecular Medicine and Life Science, Yokohama City University Association of Medical Science, Yokohama, Kanagawa, Japan; Department of Urology, Yokohama City University Graduate School of Medicine, Yokohama, Kanagawa, Japan*
MASATO OHTSUKA • *Division of Basic Medical Science and Molecular Medicine, School of Medicine, Tokai University, Isehara, Kanagawa, Japan; Center for Matrix Biology and Medicine, Graduate School of Medicine, Tokai University, Isehara, Kanagawa, Japan; The Institute of Medical Sciences, Tokai University, Isehara, Kanagawa, Japan*
WUHONG PEI • *Translational and Functional Genomics Branch, National Human Genome Research Institute, National Institutes of Health, Bethesda, MD, USA*
IRINA A. POLEJAEVA • *Department of Animal, Dairy, and Veterinary Sciences, Utah State University, Logan, UT, USA*
XIN-AN PU • *Comprehensive Cancer Center (CCC), Genetically Engineered Mouse Model Core (GEMMC), The Ohio State University, Columbus, OH, USA*
WENNING QIN • *The Jackson Laboratory, Bar Harbor, ME, USA*
XIN Y. RAIRDAN • *Embryo Technology Group, gRED Animal Resources, Genentech, Inc., South San Francisco, CA, USA*
MISHA REGOUSKI • *Department of Animal, Dairy, and Veterinary Sciences, Utah State University, Logan, UT, USA*
TAIJU SAITO • *South Ehime Fisheries Research Center, Ehime University, Minamiuwagun, Ehime, Japan*
TAKAYUKI SAKURAI • *Department of Cardiovascular Research, Graduate School of Medicine, Shinshu University, Matsumoto, Nagano, Japan; Basic Research Division for Next-Generation Disease Models and Fundamental Technology, Research Center for Next Generation Medicine, Shinshu University, Matsumoto, Nagano, Japan*
MASAHIRO SATO • *Section of Gene Expression Regulation, Frontier Science Research Center, Kagoshima University, Kagoshima, Japan*
TAKUYA SATO • *Laboratory of Biopharmaceutical and Regenerative Sciences, Institute of Molecular Medicine and Life Science, Yokohama City University Association of Medical Science, Yokohama, Kanagawa, Japan*
THOMAS L. SAUNDERS • *University of Michigan Transgenic Animal Model Core, Ann Arbor, MI, USA; Division of Genetic Medicine, Department of Internal Medicine, University of Michigan, Ann Arbor, MI, USA*
MELISSA A. SCOTT • *Division of Developmental Biology, Cincinnati Children's Hospital Medical Center, Cincinnati, OH, USA*
BINGYIN SHI • *Department of Endocrinology, The First Affiliated Hospital of Xi'an Jiaotong University School of Medicine, Xi'an, Shaanxi, People's Republic of China*
YUN-BO SHI • *Section on Molecular Morphogenesis, Eunice Kennedy Shriver National Institute of Child Health and Human Development (NICHD), National Institutes of Health (NIH), Bethesda, MD, USA*

YUKI SHIBATA • *Section on Molecular Morphogenesis, Eunice Kennedy Shriver National Institute of Child Health and Human Development (NICHD), National Institutes of Health (NIH), Bethesda, MD, USA*

TAKAYUKI SHINDO • *Department of Cardiovascular Research, Graduate School of Medicine, Shinshu University, Matsumoto, Nagano, Japan; Basic Research Division for Next-Generation Disease Models and Fundamental Technology, Research Center for Next Generation Medicine, Shinshu University, Matsumoto, Nagano, Japan*

COLIN L. SWEENEY • *Laboratory of Clinical Immunology and Microbiology, Division of Intramural Research, National Institute of Allergy and Infectious Diseases, National Institutes of Health, Bethesda, MD, USA*

JORGE SZTEIN • *Division of Reproductive Engineering, Center for Animal Resources and Development (CARD), Kumamoto University, Kumamoto, Japan*

TORU TAKEO • *Division of Reproductive Engineering, Center for Animal Resources and Development (CARD), Kumamoto University, Kumamoto, Japan*

MARIE-CECILE VAN DE LAVOIR • *Ligand Pharmaceuticals, Inc., Emeryville, CA, USA*

SAYAKA WAKAYAMA • *Advanced Biotechnology Center, University of Yamanashi, Yamanashi, Japan*

TERUHIKO WAKAYAMA • *Advanced Biotechnology Center, University of Yamanashi, Yamanashi, Japan; Faculty of Life and Environmental Sciences, University of Yamanashi, Yamanashi, Japan*

HAOYI WANG • *State Key Laboratory of Stem Cell and Reproductive Biology, Institute of Zoology, Chinese Academy of Sciences, Beijing, China; Institute for Stem Cell and Regeneration, Chinese Academy of Sciences, Beijing, China*

ZHONGDE WANG • *Department of Animal, Dairy, and Veterinary Sciences, Utah State University, Logan, UT, USA*

WEN XIE • *Transgenic Core Facility, Division of Intramural Research, National Heart, Lung, and Blood Institute, National Institutes of Health, Bethesda, MD, USA*

JIE XU • *Center for Advanced Models for Translational Sciences and Therapeutics, University of Michigan Medical Center, Ann Arbor, MI, USA*

WENHAO XU • *Genetically Engineered Murine Model (GEMM) Core Facility, Department of Microbiology, Immunology and Cancer Biology, University of Virginia, Charlottesville, VA, USA*

ETSURO YAMAHA • *Nanae Fresh Water Laboratory, Field Science Center for Northern Biosphere, Hokkaido University, Kameda, Hokkaido, Japan*

DONGSHAN YANG • *Center for Advanced Models for Translational Sciences and Therapeutics, University of Michigan Medical Center, Ann Arbor, MI, USA*

LAN YANG • *Lannuo Biotechnologies Wuxi Inc., Wuxi, China*

MIN YANG • *Department of Animal, Dairy, and Veterinary Sciences, Utah State University, Logan, UT, USA*

ANTHONY P. YOUNG • *Division of Pharmacology, College of Pharmacy, The Ohio State University, Columbus, OH, USA*

MICHAEL G. ZEIDLER • *University of Michigan Transgenic Animal Model Core, Ann Arbor, MI, USA*

FAN ZHANG • *Transgenic Core Facility, Division of Intramural Research, National Heart, Lung, and Blood Institute, National Institutes of Health, Bethesda, MD, USA*

ZUOYAN ZHU • *State Key Laboratory of Freshwater Ecology and Biotechnology, Institute of Hydrobiology, Chinese Academy of Science, Hubei, People's Republic of China*

Chapter 1

Microinjection and Micromanipulation: A Historical Perspective

Wenhao Xu

Abstract

Microinjection/micromanipulation is more than 100 years old. It is a technique that is instrumental in biomedical research and healthcare. Its longevity lies in its preciseness in mechanical retrieval, or delivery of biological materials, which in some cases is simply necessary or more effective than other retrieval/delivery means. Microinjection is favored for its straightforwardness in transferring contents from micromolecules to macromolecules and from organelles to cells. Microinjection/micromanipulation has been practiced over the century like an art form. Variations in handlings and instruments can be tolerated to a surprising degree with satisfactory outcomes. Throughout the century, microinjection developed as an indispensable tool along with the evolution of biomedical fields: from transgenics to gene targeting, from animal cloning to human infertility treatment, from nuclease-guided genetic engineering to RNA-guided genome editing (Fig. 1). The birth of the CRISPRology rejuvenated microinjection. For microinjection/micromanipulation, the second century has already begun with the early arrival of computerized instrumentation and lately of the high-throughput nanomanipulators potentially operable by artificial intelligence. As we yin-yang both systemic and precision approaches in research and medicine, microinjection will no doubt continue to find its unique place in the future.

Key words Microinjection, Micromanipulation, Transgenic, Gene targeting, Knockout, ES cells, Cloning, IVF, CRISPR

1 The Dawn

Microinjection/micromanipulation generally refers to the mechanical process of using a glass micropipette to insert/retrieve substances at a microscopic or macroscopic level into/out of a single living cell or multicellular entity. During the process, a needle roughly 0.5 to 15 μm in diameter penetrates a few barriers: the extracellular matrix if present, the cell membrane or the nuclear envelope. The desired contents are then injected into or retrieved from the subcellular or intercellular compartments.

Microinjection technique is more than a century old. It was invented early last century by Dr. Marshall Barber, a bacteriologist at the University of Kansas School of Medicine. In 1902,

Chengyu Liu and Yubin Du (eds.), *Microinjection: Methods and Protocols*, Methods in Molecular Biology, vol. 1874, https://doi.org/10.1007/978-1-4939-8831-0_1, © Springer Science+Business Media, LLC, part of Springer Nature 2019

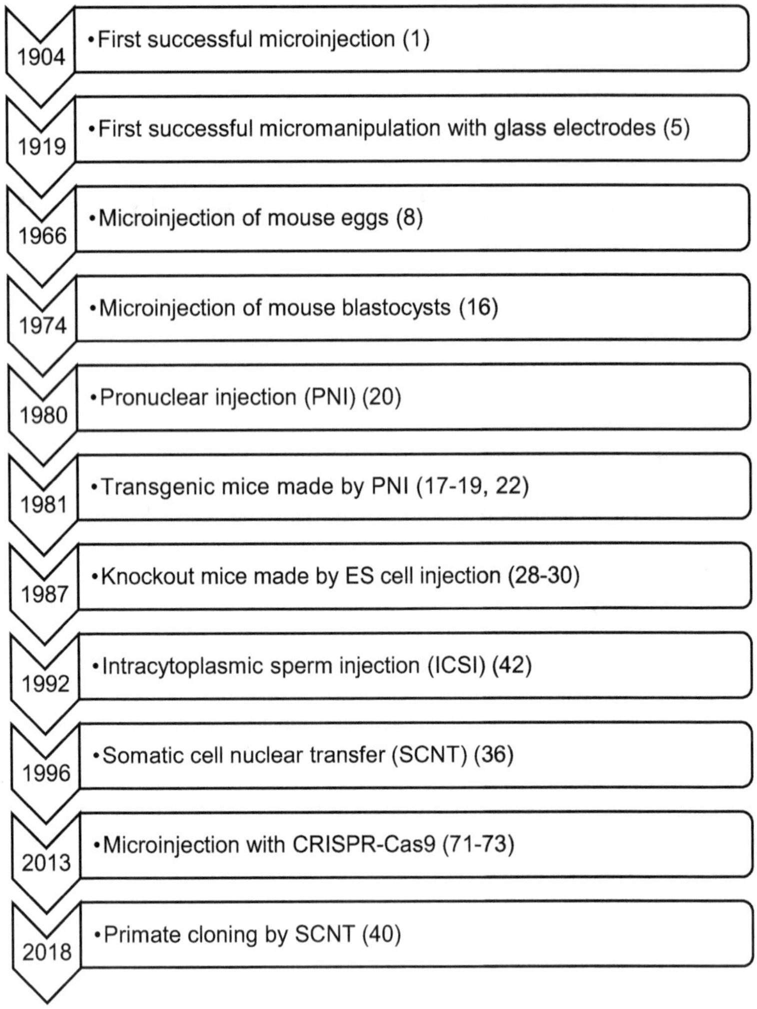

Fig. 1 Microinjection/micromanipulation Timeline. This timeline starts with the invention of microinjection (1904) and ends with the primate cloning by somatic cell nuclear transfer (2018). It covers more than a century of major microinjection/micromanipulation fanfares. References are given in the parentheses

Dr. Barber initiated the micromanipulation experiments in an effort to isolate bacterial cells. His method involved pulling glass tubes (Fig. 2) into fine tips using an ordinary Bunsen burner and fashioning them into different shapes on the pointed end (Fig. 2, *p*) and different numbers of loops (Fig. 2, *l*) on the sealed end (Fig. 2, *n*) to assume control of the movement of the mercury within.

Dr. Barber also designed the "hanging droplet" preparation in a chamber (Fig. 3, *bx*), which is still preferred today by many veteran microinjectionists. He refined his system by developing a special pipette holder—nowadays called a micromanipulator that allows three-dimensional movement of the pipette (Fig. 3, *sl* and *sv*). He also expanded the applications of his method by developing

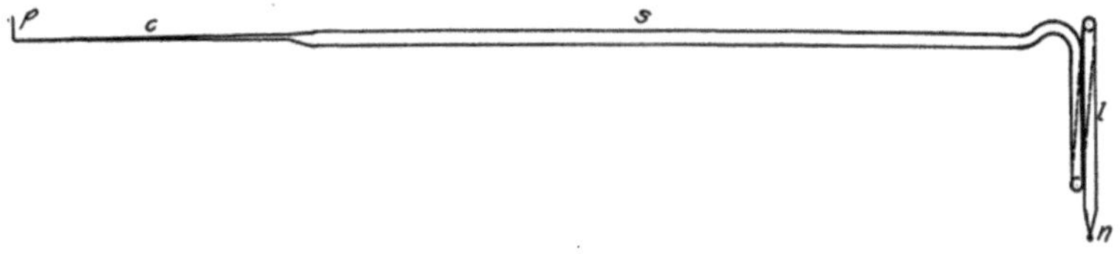

Fig. 2 The anatomy of Dr. Barber's glass capillary pipette. A 35 cm long, 4 mm diameter tubing with 0.7 mm thick wall was fabricated into 5 portions. The straight (s) portion was drawn into a capillary (c) with a bent point (p) on one end and loops (l) on the other end (n). The whole tube was nearly filled with mercury before the end n was sealed off. The greater the number of loops, the more sensitive the movement inside to changes in temperature, but the more unwieldy to handle. Reproduced with permission from Fig. 1 in ref. 2

Fig. 3 Dr. Barber's Historical Microinjection Rig. On the left, the microscope was fitted with pipette holder (ph) controlled by screws laterally (sl) or vertically (sv). Sitting on the stage was a glass box (bx) with water kept at the bottom to insure moisture and cover slip on the top with hanging drops containing cells, bacteria, or fluid. On the right, the accessories were assembled entirely for controlling the injection, the modern equivalent of a microinjector. A "magnifying" scope was repurposed to provide a rack (r) and pinion. A brass tube (t) with a cup (c) directly attached on lower end and with a thin rubber bag (b) attached through a stiff ringed wire on the upper side. The cup was filled with ice-cold water and the bag with hot water. The tube was held in the sleeve (s) attached to the arm of the scope, thus can be raised or lowered to either heat or cool the mercury inside the loop (l). A pan was placed underneath the cup to collect overflowing water. A joint (j) on the scope allowed the bag and cup to be swung aside together. Reproduced with permission from Fig. 4 in ref. 2

the first microinjector, in which movement of liquid in the capillary is controlled by expansion or contraction of mercury heated with a rubber bag filled with hot water (Fig. 3, *b*) or cooled in a cup filled with iced water (Fig. 3, *c*). With the improvement of his device, he was able to aspirate bacterial cells and inject them into various plant cells and protozoa to study the effects of implanted microorganisms on the viability of the infected organisms. However, his early attempts to make injections into smaller animal cells were less successful. He reported his method initially in the *Journal of the Kansas Medical Society* in 1904 [1], and gave more detailed accounts later [2, 3].

Dr. Barber was awarded the John Scott Medal in 1925 for "the invention of the pipette method and the apparatus for its control, used for the isolation of single micro-organisms, and the injection of substances into living cells" [4]. I myself feel humbled and honored for becoming the director who established the modern Transgenic and Gene-targeting Institutional Facility (TGIF) in 1996 at the University of Kansas School of Medicine where Dr. Barber invented microinjection.

Micromanipulation with glass microelectrodes has an almost equally long history that can be traced back to 1919 when Dr. Frederick Pratt and Dr. John Eisenberger first used a fine-pointed capillary pore electrode with an outer diameter of approximately 4–8 μm for focal stimulation of single skeletal muscle fibers [5]. However, the road to the patch-clamp technique was quite long, filled with trials and failures, and was paved by the efforts of many researchers and laboratories; numerous techniques were developed and subsequently disappeared. It was not until the late 1970s to early 1980s when the giant leap made by Dr. Ervin Neher and Dr. Bert Sakmann et al. [6, 7] to develop the patch-clamp technique revolutionized cellular electrophysiology. This involved fire-polishing the micropipette tips and a gentle negative suction applied on surface cell membranes to record single channels. It opened a way for monitoring electrical activity of virtually any cell type in situ or in vivo. Thus, the patch-clamp technique was born and quickly became the method of choice at the time for the study of excitable cells to understand the involvement of channels in the fundamental cellular processes such as nerve and muscle activity. Ervin Neher and Bert Sakmann received the Nobel Prize in Physiology or Medicine in 1991 for their discoveries concerning the function of single ion channels in cells.

2 Transgenics—Zygotic Injection

One of the major accomplishments as an extension of the early microinjection attempts was achieved by Dr. Teh P. Lin at the University of California San Francisco in 1966. He published his technique of microinjection of γ-globulin into mouse eggs in *Science* with the straightforward title "Microinjection of Mouse Eggs" [8]. Following protein injection, RNA was the next macromolecule injected into *Xenopus* oocytes to study translation of heterologous mRNA [9]. Then came DNA. In the 1970s, DNA cloning became the most popular fanfare. Genes were cloned, cut, and pasted with ease thanks to DNA recombinant technology. Invertebrate frog oocytes before mice became the most popular model system for studies of oocyte maturation and meiosis [10, 11]. Subsequently in the 1980s, the system was exploited for the purposes of expressing and identifying the function of heterologous mammalian cloned genes encoding receptors, transporters, ion channels, kinases, and oncoproteins such as the *Ras* and *Mos* [12, 13]. This was made possible due to the transcriptional and translational fidelity displayed by these frog eggs when injected with heterologous DNA and RNA, and their large size that made injection easy and allowed recovery of newly synthesized gene products. Later, mouse oocytes were also used for studying the mechanism of cytoplasmic polyadenylation of stored maternal RNAs when non-polyadenylated RNAs were injected into the cytoplasm of mouse oocytes with the aid of a picoinjector [14].

The success of gene cloning also ignited enthusiasm for gene transfer to genetically transform animals. However, the first successful transgenic mouse with Mendelian transmission was created by Dr. Rudolf Jaenisch not with microinjection, but with viral infection of embryos [15]. However, his previous effort was made by injecting SV40 virus into the blastocoel of a blastocyst, which resulted in the integration of the viral DNA in adult mice [16]. Before long, by the early 1980s, several labs successfully generated transgenic mice by pronuclear microinjection (PNI) with cloned cDNA [17–19, 22]. However, Dr. Jon Gordon and Dr. Frank Ruddle led their research team to first demonstrate the utility of PNI in inserting recombinant DNA in the embryonic genome in 1980 [20]. Though less known, Dr. Mario Capecchi did nuclear and cytoplasmic injection of *HSV-TK* gene into the cultured mouse fibroblast cell line LMTK^{-} to achieve high efficiency transformation [21]. He carried out his injection under the constant pressure to keep the flow of the fluid so as to minimize the clogging of the micropipette during continuous injections. This technique is widely adopted today. In 1981, Drs. Gordon and Ruddle first came up with the term "transgenic," which is defined as "the ability to achieve controlled genetic modifications of multicellular eukaryotic species by the insertion of genetic materials." Drs. Thomas Wagner and Peter Hoppe were granted in 1989 the US patent (#US4873191) "Genetic Transformation of Zygotes" based on their work 8 years prior [22]. Dr. Richard Palmiter, Dr. Ralph Brinster, and their research group's publication in 1982 with the headline "Gigantic Mouse Made with Rat Growth

Hormone Gene" along with the picture on the cover of *Nature* vividly illustrated the power of genetic engineering [23]. Together, they went on to publish several seminal papers from early to mid-1980s to refine the parameters for transgene microinjections including nuclear vs. cytoplasmic, mononuclear vs. binuclear, and linear vs. circular constructs [24]. Those became the gold standards still followed today by numerous microinjectionists. Microinjection became an essential technique whose use for generating transgenic animals extended beyond mouse to other species such as fly, frog, fish, rat, rabbit, dog, pig, goat, and bovine.

3 Gene Targeting—ES Cell Injection

Around the same time when transgenic mice generated by pronuclear microinjection ushered in a new era for genetic engineering, mouse embryonic stem cells (ES cells) were successfully derived from mouse embryos in 1981 by two labs at the opposite shores of the Atlantic Ocean, one in England led by Drs. Martin Evans and Matthew Kaufman [25] and another in the US led by Dr. Gail Martin [26]. The availability of pluripotent ES cells, millions in culture, is instrumental for more precise mutagenesis, based on the rare homologous recombination event, which otherwise would not be practically possible due to the limitation of the number of isolated embryos. Thanks to the increasing understanding of the mechanism of homologous recombination (HR) in somatic cells up to mid-1980s [27], subjecting mouse ES cells to HR-based gene targeting was quickly accomplished around the same time by three labs. Drs. Mario Capecchi, Martin Evans, and Oliver Smithies separately published their groundbreaking papers on introducing specific genetic modifications of the hypoxanthine phosphoribosyl transferase (HPRT) gene into mice in 1987 [28–30], for which they shared the Nobel Prize in Physiology or Medicine in 2007. Microinjection of targeted ES cells into blastocysts was their common choice of technique for generating chimeric mice. Though injection of ES cells was quite different from injecting DNA in solution, which was popular around that time, the former withstood the test of time. Ironically microinjection was originally invented for bacterial cell injection. Microinjection played a pivotal role in the initial success of generating gene-targeted mice. It became the workhorse for creating knockout and knock-in mice for many years to follow. However, some divergence from this technique such as tetraploid aggregation [31, 32] and laser-assisted 8-cell embryo injection [33] emerged to improve the contribution of the targeted ES cells to the germline lineage.

4 Animal Cloning—Somatic Cell Nuclear Transfer into Enucleated Eggs

As ES cells completely changed the landscape of genetic engineering in mice, the inability of deriving pluripotent ES cells from species other than mouse hampered the animal genetic engineering field for a long time until pluripotent rat ES cells were derived and authenticated [34, 35]. During that time, genetically altered (GA) animals had to be created by genetic modifications with somatic cells followed by somatic cell nuclear transfer (SCNT) aka cloning. In the center of this technology lies the micromanipulation technique that is applied to withdraw the nucleus of an egg and replace it with that of a somatic cell. Dolly the Sheep was the first mammal cloned, by the team led by Drs. Ian Wilmut and Keith Campbell in the Roslin Institute, from an adult somatic cell using SCNT as the micromanipulation technique [36]. Dolly was born on July 5, 1996 and died in February 2003, after she was euthanized following the discovery of a progressive lung disease. Dolly's birth was heralded as the first cloning of a mammal. However, as early as 1952, the first successful nuclear transfer was performed with a similar micromanipulation technique by Drs. Robert Briggs and Thomas King to clone a frog embryo by implanting a nucleus from a fertilized cell into an unfertilized one [37]. Several years later in 1958, Dr. John Gurdon took another step forward to transplant the nucleus of a tadpole intestinal cell into an enucleated frog egg, showing that nuclei from somatic cells in a fully developed animal could be used for cloning [38]. The difference was that Dolly was the first mammal. One year after Dolly's birth, the first two primates were cloned by Dr. Don Wolf's group using a similar micromanipulation technique, but with donor nuclei from blastomeres [39]. However, the female Rhesus monkey Neti and the male Ditto never achieved the same fame as Dolly the Sheep did. In 2018, a research group in China made another successful effort to clone Cynomolgus monkeys by essentially Dolly the Sheep technology using SCNT [40]. However, fetal fibroblasts were used in the effort whereas an adult mammary gland cell was used for cloning Dolly. Up until the birth of CRISPR technology in 2012–13, SCNT had been the go-to technique of micromanipulation for generating knock-out (KO) and knock-in (KI) animals other than the mouse and rat.

5 Infertility Treatment—Germ Cell Injection

Microinjection, though primarily invented and utilized for basic research, was also a tool for "translational medicine" to treat human infertility even before the term existed. In vitro fertilization, a process of fertilization where an egg is combined with sperm outside the body, is now a conventional clinical procedure for human infertility treatment. The first successful birth of a "test tube baby" after IVF treatment, Louise Brown, occurred in England in 1978 [41]. Dr. Robert Edwards was awarded the Nobel Prize in Physiology or Medicine in 2010 for the development of in vitro fertilization. However, when sperm fail to fertilize eggs for various reasons during in vitro incubation, a more invasive practice involving microinjection can be implemented to inject a single sperm into the cytoplasm of an egg, which is literally termed intracytoplasmic sperm injection (ICSI). The technique was developed in 1992 by Dr. Gianpiero Palermo [42]. However, before ICSI, subzonal insemination (SUZI) was actually the first technique used to subject human gametes to micromanipulation, by which a submotile sperm was injected into the perivitelline space. The first live birth by SUZI was reported in Singapore in 1989. Nevertheless, ICSI was more straightforward and soon gained widespread popularity for treating severe male infertility in males who could not be assisted by conventional IVF procedure because too few motile and morphologically normal sperm were present in the ejaculates. For Dr. Gianpiero Palermo's significant contribution in the field of assisted human reproduction, he shared the Jacob Heskel Gabbay Award for Biotechnology and Medicine with Drs. Alan Handyside and Ann Kiessling in 2009.

In the years following the first ICSI, new microinjection procedures were developed to treat more advanced male infertility even when no sperm were present in the semen. A procedure called round spermatid nuclear injection (ROSNI) or round spermatid injection (ROSI) was performed to achieve fertilization and successful birth [43, 44]. However, ROSI has not been performed as successfully as ICSI with mature sperm or elongating spermatids. The difficulty lies in the identification of the round spermatids. However, improved success was attributed to the ability of identifying round spermatids accurately before human oocyte injection in a recent study co-led by Dr. Ryuzo Yanagimachi from the University of Hawaii [45]. Dr. Yanagimachi is a pioneer in this area of research using microinjection as a tool and the mouse as a model system. His research lab successfully generated normal mice from oocytes injected with epididymal sperm [46], testicular spermatozoa, round spermatids [47], and even secondary spermatocytes [48]. In 1998, the Yana Lab was also the first to publish work on cloning mice from adult cells [49]. The first cloned female mouse was named Cumulina, which was the third cloned species after Dolly the Sheep (1996), and Neti and Ditto the Monkeys (1997). The Yana Lab first demonstrated successful use of Piezo-actuated micromanipulation in mouse [46] and human [50] ICSI. Later, most SCNT clones were generated via piezo-actuated delivery of donor nuclei into recipient oocytes to minimize the damage to the oolemma. Piezo-actuated micromanipulation was developed by

Atsushi Mimatsu and colleagues. Its utilization as a microinjection accessory tool has been extended beyond ICSI and SCNT in other forms of micromanipulation of embryos of various species.

It is interesting to note that, in Dr. Barber's earliest version of the injector, the movement of the liquid was controlled by the expansion or contraction of mercury when heated or cooled. Coincidentally, it is again mercury that was used in the first-generation piezo-assisted microinjector to reduce the amplitude of the injector pipette oscillation and significantly improve the success rate. However, due to the toxic effects of mercury, it is now replaced by the Fluorinert FC77 or 770 in the next-generation piezo-assisted microinjectors.

6 Nuclease-Mediated DNA Double-Strand Break—Microinjection Renaissance

The history of microinjection and micromanipulation intertwines with the development of scientific fields. The microinjection renaissance finally arrived after the rise of nuclease-based genome editing at the beginning of the 21st century. For a long time it has been known that gene targeting happens in a low frequency which is manageable in ES cells, but difficult for most other somatic cells. However, DNA double-strand breaks (DSBs) significantly lower the threshold for the DNA repair mechanism in cells to kick in. As early as 1994, Dr. Maria Jasin used a rare cutting endonuclease I-SceI to introduce DNA DSBs within the I-SceI recognition site either present extrachromosomally in COS1 cells [51] or inserted into the genome of 3T3 cells [52] and mouse ES cells [53]. Their studies demonstrated that I-SceI-induced DSBs stimulated homologous recombination in mammalian cells. I-SceI is encoded by an intron in the mitochondrial rRNA gene of the ω+ yeast strain *Saccharomyces cerevisiae.* I-SceI is involved in intron homing, hence the name homing endonuclease or meganuclease (MN) referring to its larger recognition site than those of restriction enzymes. I-SceI recognizes an 18-base pair sequence. Statistically, a frequency of 1 in 4^{18} could occur on average once in every 7×10^{10} base pairs, which is about 25 times the size of the mouse genome (2.8×10^{9} base pairs). The MN system depends on the prior introduction of an I-SceI binding site into the target gene and cannot be used for endogenous genes. Dr. Jasin summarized the potential and challenge of genetic manipulation of genomes with rare-cutting endonucleases [54]. In order to make I-SceI a tool for site-specific genome engineering, another approach was to modify its recognition specificity [55, 56]. This proved successful in some cases, but the intimate connection between the recognition and cleavage domains in the protein structure made it challenging to alter one without affecting the other. In other words, MN is not programmable-friendly. The search for alternative nucleases yielded an attractive candidate: the chimeric nuclease—a fusion between zinc finger DNA binding domain and the endonuclease domain of Fok I–zinc finger nuclease (ZFN). The fingers are comprised of approximately 30 amino acid modules that interact with nucleotide triplets. Because ZFN works as a dimer, one can create a pair of ZFNs that specifically recognizes any sequence of DNA triplets by constructing different zinc finger moieties. The first success with a ZFN pair designed de novo for a genomic target occurred in *Drosophila* [57] and later in human cell lines [58]. Since then, ZFN pairs have been designed, constructed, and used successfully for individual genes in a variety of organisms including nematodes, zebrafish, frogs, mice, rats, hamsters, and pigs. While the frequencies of target modification vary, yields in the vicinity of 10% of all targets are quite common. In *Drosophila*, early experiments relied on genomic integration of ZFN-coding sequences and donor DNA via P-element-mediated transformation [57, 59]. Key to these successes was the method for the delivery of ZFNs and donor DNA when desired. A welcome breakthrough occurred when it was demonstrated that excellent efficiencies of both homologous and nonhomologous events could be obtained by injecting ZFN mRNAs and donor DNA into *Drosophila* embryos [60], and later mouse and rat embryos [61, 62]. On the heel of the ZFN was TALEN (Transcription Activator-Like Effector Nuclease). The transcription activator-like effectors (TALEs) derived from the bacterial plant pathogen *Xanthomonas* can be customized in a ZFN-like fashion to target genes [63]. The difference is that interactions between the TALE-derived DNA binding domains and their targets are less complex than those between ZFNs and their targets. The repeat variable diresidue (RVD)—the two specific amino acids #12/13 of the 34 amino acid repeat—recognizes one specific nucleotide. Therefore, designing TALENs is generally more straightforward than designing ZFNs. From the perspective of microinjection, an early report compared the efficiency of microinjection routes, delivering formats, and concentrations on disruption of the rat IgM locus via TALEN [64]. TALENs were injected into the pronucleus as DNA (PNI) or into the cytoplasm (IC) as mRNA at three different concentrations each. Survival and founder statistics from injections were compared and the conclusion was drawn that the intracytoplasmic injection (IC) of mRNA was most efficient. This report exerted a significant influence on later microinjection endeavors with CRISPR/Cas9.

7 CRISPR—Breaking the Microinjection Barriers

CRISPR stands for "Clustered Regularly Interspaced Short Palindromic Repeats." It describes a DNA sequence pattern that was first reported in *E. coli* in 1987 [65], but the acronym was not introduced until 2002 along with the Cas (CRISPR-associated) protein [66]. It took two decades to verify the function of CRISPR/Cas as the acquired immunity against viruses in prokaryotes [67]. The system was reconstructed in a test tube as a tool for programmable genome editing in a landmark publication in 2012 [68]. Quickly, the system was shown to work effectively in both human cell lines [69, 70] and mice [71–73]. In cases of creating genome-edited mice by CRISPR/Cas9, microinjection is not only instrumental, but also versatile with various combinations of CRISPR reagents and delivery routes. PNI and IC were shown to be equally effective when Cas9 mRNA and sgRNA were used [73] though the reagent concentrations for PNI were commonly about ten-fold lower than those for IC. Regardless, the overall concentrations greatly exceed those for traditional PNI especially when multiplexing is employed. After recombinant Cas9 protein became available, PNI of Cas9 nuclease/gRNA complex (Cas9 ribonucleoprotein) became quite effective and popular. Nowadays, DNA, RNA, and protein are mixed and delivered together, which represents a significant milestone for microinjection in its long history, during which only single macromolecules were often delivered to one pronucleus. Deliveries of CRISPR/Cas9 reagents by microinjection have been shown not only efficient with several common mouse strains (B6F1 hybrids, C57BL/6, FVB, BALB/c) but also successful with extremely immunodeficient strains such as the NSG™ [74]. With CRISPR/Cas9, mosaicism is commonly observed in the F_0 founder animals. However, in a recent study, mosaic mice were purposely created by microinjecting CRISPR/Cas9 reagents into one of the two blastomeres, coupled with switching on the expression of a fluorescent reporter gene different from the one expressed in the neighboring blastomere, to reveal a novel *Tet3* function in regulating neocortical development [75].

Mastering microinjection requires specialized training and years of experience. The proliferative demands for microinjection of CRISPR reagents led to the development of a different means of delivery—the square-wave electroporation, which was demonstrated as a high-throughput alternative in a series of publications [76–79] using BEX Genome Editor™, BTX ECM830 Electro Square Porator, or Bio-Rad Gene Pulser Xcell™ Electroporation System. Efficient base editing (C to T conversion) with Cas9 nickase fused with cytidine deaminase was also demonstrated by microinjection as well as electroporation for the dystrophin (Dmd) and tyrosinase (Tyr) genes [80]. The editing efficiencies are very similar at two different loci regardless of whether microinjection or electroporation is utilized. In either cases, micromanipulation skill is still required to accomplish the tasks of embryo handling and implantation.

8 The Future

After more than a century, microinjection as a technique is still very much the workhorse in many research and clinical settings, and seen in increasing demand ever since the birth of CRISPR/Cas9 technology. Traditionally, microinjection and micromanipulation are performed by a microinjectionist under a specialized microscope setup equipped with micromanipulators and injectors. The micropipettes are either hand-pulled or machine-fabricated. The horizontal pipette pullers have been updated through several generations, while the vertical pullers ceased to be manufactured a while ago. The micropipette grinders have mostly given way to premade or made-to-order micropipette manufacturing businesses. Over the years, more sophisticated equipment has been manufactured to improve the micromanipulators and injectors by incorporating automation and nifty accessories. The microinjectors evolved from simple 50 cm^3 syringes to picoinjectors and femtojets. Solid mechanical micromanipulators are more often equipped with motorized outfits. The substances that can be injected have also grown over time to an impressive variety from micromolecules to macromolecules and from organelles to cells. The recipients can range from single-cell organisms (amebae, paramecia, protozoa) to cells/embryos from multicellular organisms. A century of microinjection/micromanipulation has witnessed the improvement of instrumentation from simple mechanical devices to motorized digital rigs and from glass capillaries to electrically charged lances in a prototypical microelectromechanical system (MEMS) nanoinjector [81]. Despite technological advancement, microinjection is still very much an art practiced personally by individual microinjectionists with their available or favorable tools. What will happen to microinjection in another century? Will it be cornered by competitive technologies? Will micoinjction itself be automated enough so that only little interventions from microinjectionists are required? Will it be solely operated by an inorganic algorithm? No matter where the answers lie, microinjection has made a substantial impact on biomedical research and healthcare. It will no doubt continue to grow and shape the foreseeable future of the world around us.

I apologize to those whose exciting work could not be mentioned or cited, and for any errors of historical facts I may have inadvertently made in this brief chapter. Many details and additional references can be found in the subsequent chapters. I look forward to new history to be made in the realm of microinjection and micromanipulation.

References

1. Barber M (1904) A new method of isolating microorganisms. J Kans Med Soc 4:489–494
2. Barber MA (1911) A technic for the inoculation of bacteria and other substances into living cells. J Infect Dis 8(3):348–360
3. Barber MA (1914) The pipette method in the isolation of single micro-organisms and in the inoculation of substances into living cells: with a technique for dissection, staining, and other processes carried out under the higher powers of the microscope. Philippine J Sci B 9 (4):307–360
4. Terreros DA, Grantham JJ (1982) Marshall barber and the origins of micropipette methods. Am J Phys 242(3):F293–F296
5. Pratt FH, Eisenberger JP (1919) The quantal phenomena in muscle: methods, with further evidence of the all-or-none principle for the skeletal fiber. Am J Phys 49(1):1–54
6. Hamill OP, Marty A, Neher E, Sakmann B, Sigworth FJ (1981) Improved patch-clamp techniques for high resolution current recording from cells and cell-free membrane patches. Pflugers Arch 391(2):85–100
7. Sakmann B, Neher E (1984) Patch clamp techniques for studying ionic channels in excitable membranes. Annu Rev Physiol 46:455–472
8. Lin TP (1996) Microinjection of mouse eggs. Science 151(3708):333–337
9. Gurdon JB, Lane CD, Woodland HR, Marbaix G (1971) Use of frog eggs and oocytes for the study of messenger RNA and its translation in living cells. Nature 233(5316):177–182
10. Wasserman WJ, Masui Y (1976) A cytoplasmic factor promoting oocyte maturation: its extraction and preliminary characterization. Science 191(4233):1266–1268
11. Heidemann SR, Kirschner MW (1975) Aster formation in eggs of Xenopus laevis. Induction by isolated basal bodies. J Cell Biol 67 (1):105–117
12. Birchmeier C, Broek D, Wigler M (1985) Ras proteins can induce meiosis in Xenopus oocytes. Cell 43(3 Pt 2):615–621
13. Sagata N, Oskarsson M, Copeland T, Brumbaugh J, Vande Woude GF (1988) Function of c-Mos proto-oncogene product in meiotic maturation in Xenopus oocytes. Nature 335(6190):519–525
14. Gebauer F, Xu W, Cooper GM, Richter JD (1994) Translational control by cytoplasmic polyadenylation of c-Mos mRNA is necessary for oocyte maturation in the mouse. EMBO J 13(23):5712–5720
15. Jaenisch R (1976) Germ line integration and Mendelian transmission of the exogenous Moloney leukemia virus. Proc Natl Acad Sci U S A 73(4):1260–1264
16. Jaenisch R, Mintz B (1974) Simian virus 40 DNA sequences in DNA of healthy adult mice derived from preimplantation blastocysts injected with viral DNA. Proc Natl Acad Sci U S A 71(4):1250–1254
17. Gordon J, Ruddle F (1981) Integration and stable germ line transmission of genes injected into mouse pronuclei. Science 214 (4526):1244–1246
18. Costantini F, Lacy E (1981) Introduction of a rabbit β-globin gene into the mouse germ line. Nature 294(5836):92–94
19. Brinster R, Chen HY, Trumbauer M, Senear AW, Warren R, Palmiter RD (1981) Somatic expression of herpes thymidine kinase in mice following injection of a fusion gene into eggs. Cell 27(1 Pt 2):223–231
20. Gordon JW, Scangos GA, Plotkin DJ, Barbosa JA, Ruddle FH (1980) Genetic transformation of mouse embryos by microinjection of purified DNA. Proc Natl Acad Sci U S A 77 (12):7380–7384
21. Capecchi MR (1980) High efficiency transformation by direct microinjection of DNA into cultured mammalian cells. Cell 22(2 Pt 2):479–488
22. Wagner TE, Hoppe PC, Jollick JD, Scholl DR, Hodinka RL, Gault JB (1981) Microinjection of a rabbit beta-globin gene into zygotes and its subsequent expression in adult mice and their offspring. Proc Natl Acad Sci U S A 78 (10):6376–6380
23. Palmiter RD, Brinster RL, Hammer RE, Trumbauer ME, Rosenfeld MG, Birnberg NC, Evans RM (1982) Dramatic growth of mice that develop from eggs microinjected with metallothionein-growth hormone fusion genes. Nature 300(5893):611–615
24. Brinster RL, Chen HY, Trumbauer ME, Yagle MK, Palmiter RD (1983) Factors affecting the efficiency of introducing foreign DNA into mice by microinjecting eggs. Proc Natl Acad Sci U S A 82(13):4438–4442
25. Evans MJ, Kaufman MH (1981) Establishment in culture of pluripotential cells from mouse embryos. Nature 292(5819):154–156
26. Martin GR (1981) Isolation of a pluripotent cell line from early mouse embryos cultured in medium conditioned by teratocarcinoma stem cells. Proc Natl Acad Sci U S A 78 (12):7634–7638

27. Lin FL, Sperle K, Sternberg N (1985) Recombination in mouse L cells between DNA introduced into cells and homologous chromosomal sequences. Proc Natl Acad Sci U S A 82(5):1391–1395
28. Thomas KR, Capecchi MR (1987) Site-directed mutagenesis by gene targeting in mouse embryo-derived stem cells. Cell 51 (3):503–512
29. Kuehn MR, Bradley A, Robertson EJ, Evans MJ (1987) A potential animal model for Lesch-Nyhan syndrome through introduction of HPRT mutations into mice. Nature 326 (6110):295–298
30. Doetschman T, Gregg RG, Maeda N, Hooper ML, Melton DW, Thompson S, Smithies O (1987) Targetted correction of a mutant HPRT gene in mouse embryonic stem cells. Nature 330(6148):576–578
31. Nagy A, Gócza E, Diaz EM, Prideaux VR, Iványi E, Markkula M, Rossant J (1990) Embryonic stem cells alone are able to support fetal development in the mouse. Development 110(3):815–821
32. Nagy A, Rossant J, Nagy R, Abramow-Newerly W, Roder JC (1993) Derivation of completely cell culture-derived mice from early-passage embryonic stem cells. Proc Natl Acad Sci U S A 90(18):8424–8428
33. Poueymirou WT, Auerbach W, Frendewey D, Hickey JF, Escaravage JM, Esau L, Doré AT, Stevens S, Adams NC, Dominguez MG, Gale NW, Yancopoulos GD, DeChiara TM, Valenzuela DM (2007) F0 generation mice fully derived from gene-targeted embryonic stem cells allowing immediate phenotypic analyses. Nat Biotechnol 25(1):91–99
34. Buehr M, Meek S, Blair K, Yang J, Ure J, Silva J, McLay R, Hall J, Ying QL, Smith A (2008) Capture of authentic embryonic stem cells from rat blastocysts. Cell 135(7):1287–1298
35. Li P, Tong C, Mehrian-Shai R, Jia L, Wu N, Yan Y, Maxson RE, Schulze EN, Song H, Hsieh CL, Pera MF, Ying QL (2008) Germline competent embryonic stem cells derived from rat blastocysts. Cell 135(7):1299–1310
36. Wilmut I, Schnieke AE, McWhir J, Kind AJ, Campbell KHS (1997) Viable offspring derived from fetal and adult mammalian cells. Nature 385(6619):810–813
37. Briggs R, King TJ (1952) Transplantation of living nuclei from blastula cells into enucleated frogs' eggs. Proc Natl Acad Sci U S A 38 (5):455–463
38. Gurdon JB, Elsdale TR, Fischberg M (1958) Sexually mature individuals of Xenopus laevis from the transplantation of single somatic nuclei. Nature 182(4627):64–65
39. Meng L, Ely JJ, Stouffer RL, Wolf DP (1997) Rhesus monkeys produced by nuclear transfer. Biol Reprod 57(2):454–459
40. Liu Z, Cai Y, Wang Y, Nie Y, Zhang C, Xu Y, Zhang X, Lu Y, Wang Z, Poo M, Sun Q (2018) Cloning of macaque monkeys by somatic cell nuclear transfer. Cell 172(4):881–887
41. Steptoe PC, Edwards RG (1978) Birth after the reimplantation of a human embryo. Lancet 2(8085):366
42. Palermo G, Joris H, Devroey P, Van Steirteghem AC (1992) Pregnancies after intracytoplasmic injection of single spermatozoon into an oocyte. Lancet 340(8810):17–18
43. Tesarik J, Mendoza C (1996) Spermatid injection into human oocytes. I. Laboratory techniques and special features of zygote development. Hum Reprod 11(4):772–779
44. Tesarik J, Rolet F, Brami C, Sedbon E, Thorel J, Tibi C, Thébault A (1996) Spermatid injection into human oocytes. II. Clinical application in the treatment of infertility due to non-obstructive azoospermia. Hum Reprod 11(4):780–783
45. Tanaka A, Nagayoshi M, Takemoto Y, Tanaka I, Kusunoki H, Watanabe S, Kuroda K, Takeda S, Ito M, Yanagimachi R (2015) Fourteen babies born after round spermatid injection into human oocytes. Proc Natl Acad Sci U S A 112(47):14629–14634
46. Kimura Y, Yanagimachi R (1995) Intracytoplasmic sperm injection in the mouse. Biol Reprod 52(4):709–720
47. Kimura Y, Yanagimachi R (1995) Mouse oocytes injected with testicular spermatozoa or round spermatids can develop into normal offspring. Development 121(8):2397–2405
48. Kimura Y, Yanagimachi R (1995) Development of normal mice from oocytes injected with secondary spermatocyte nuclei. Biol Reprod 53(4):855–862
49. Wakayama T, Perry AC, Zuccotti M, Johnson KR, Yanagimachi R (1998) Full-term development of mice from enucleated oocytes injected with cumulus cell nuclei. Nature 394 (6691):369–374
50. Huang T, Kimura Y, Yanagimachi R (1996) The use of piezo micromanipulation for intracytoplasmic sperm injection of human oocytes. J Assist Reprod Genet 13(4):320–328
51. Rouet P, Smih F, Jasin M (1994a) Expression of a site-specific endonuclease stimulates homologous recombination in mammalian cells. Proc Natl Acad Sci U S A 91 (13):6064–6068

52. Rouet P, Smih F, Jasin M (1994b) Introduction of double-strand breaks into the genome of mouse cells by expression of a rare-cutting endonuclease. Mol Cell Biol 14 (12):8096–8106
53. Smih F, Rouet P, Romanienko PJ, Jasin M (1995) Double-strand breaks at the target locus stimulate gene targeting in embryonic stem cells. Nucleic Acids Res 23 (24):5012–5019
54. Jasin M (1996) Genetic manipulation of genomes with rare-cutting endonucleases. Trends Genet 12(6):224–228
55. Ashworth J, Havranek JJ, Duarte CM, Sussman D, Monnat RJ Jr, Stoddard BL, Baker D (2006) Computational redesign of endonuclease DNA binding and cleavage specificity. Nature 441(7093):656–659
56. Pâques F, Duchateau P (2007) Meganucleases and DNA double-strand break-induced recombination: perspectives for gene therapy. Curr Gene Ther 7(1):49–66
57. Bibikova M, Golic M, Golic KG, Carroll D (2002) Targeted chromosomal cleavage and mutagenesis in drosophila using zinc-finger nucleases. Genetics 161:1169–1175
58. Porteus MH, Baltimore D (2003) Chimeric nucleases stimulate gene targeting in human cells. Science 300(5620):763
59. Bibikova M, Beumer K, Trautman JK, Carroll D (2003) Enhancing gene targeting with designed zinc finger nucleases. Science 300 (5620):764
60. Beumer KJ, Trautman JK, Bozas A, Liu JL, Rutter J, Gall JG, Carroll D (2008) Efficient gene targeting in drosophila by direct embryo injection with zinc-finger nucleases. Proc Natl Acad Sci U S A 105(50):19821–19826
61. Carbery ID, Ji D, Harrington A, Brown V, Weinstein EJ, Liaw L, Cui X (2010) Targeted genome modification in mice using zinc-finger nucleases. Genetics 186(2):451–459
62. Cui X, Ji D, Fisher DA, Wu Y, Briner DM, Weinstein EJ (2011) Targeted integration in rat and mouse embryos with zinc-finger nucleases. Nat Biotechnol 29(1):64–67
63. Bogdanove AJ, Voytas DF (2011) TAL effectors: customizable proteins for DNA targeting. Science 333(6051):1843–1846
64. Tesson L, Usal C, Ménoret S, Leung E, Niles BJ, Remy S, Santiago Y, Vincent AI, Meng X, Zhang L, Gregory PD, Anegon I, Cost GJ (2011) Knockout rats generated by embryo microinjection of TALENs. Nat Biotechnol 29(8):695–696
65. Ishino Y, Shinagawa H, Makino K, Amemura M, Nakata A (1987) Nucleotide sequence of the iap gene, responsible for alkaline phosphatase isozyme conversion in Escherichia coli, and identification of the gene product. J Bacteriol 169(12):5429–5433
66. Jansen R, Embden JD, Gaastra W, Schouls LM (2002) Identification of genes that are associated with DNA repeats in prokaryotes. Mol Microbiol 43(6):1565–1575
67. Barrangou R, Fremaux C, Deveau H, Richards M, Boyaval P, Moineau S, Romero DA, Horvath P (2007) CRISPR provides acquired resistance against viruses in prokaryotes. Science 315(5819):1709–1712
68. Jinek M, Chylinski K, Fonfara I, Hauer M, Doudna JA, Charpentier E (2012) A programmable dual-RNA-guided DNA endonuclease in adaptive bacterial immunity. Science 337 (6096):816–821
69. Cong L, Ran FA, Cox D, Lin S, Barretto R, Habib N, Hsu PD, Wu X, Jiang W, Marraffini LA, Zhang F (2013) Multiplex genome engineering using CRISPR/Cas systems. Science 339(6121):819–823
70. Mali P, Yang L, Esvelt KM, Aach J, Guell M, DiCarlo JE, Norville JE, Church GM (2013) RNA-guided human genome engineering via Cas9. Science 339(6121):823–826
71. Shen B, Zhang J, Wu H, Wang J, Ma K, Li Z, Zhang X, Zhang P, Huang X (2013) Generation of gene-modified mice via Cas9/RNA-mediated gene targeting. Cell Res 23 (5):720–723
72. Wang H, Yang H, Shivalila CS, Dawlaty MM, Cheng AW, Zhang F, Jaenisch R (2013) One-step generation of mice carrying mutations in multiple genes by CRISPR/Cas-mediated genome engineering. Cell 153 (4):910–918
73. Yang H, Wang H, Shivalila CS, Cheng AW, Shi L, Jaenisch R (2013) One-step generation of mice carrying reporter and conditional alleles by CRISPR/Cas-mediated genome engineering. Cell 154(6):1370–1379
74. Sweeney CL, Choi U, Liu C, Koontz S, Ha SK, Malech HL (2017) CRISPR-mediated knockout of Cybb in NSG mice establishes a model of chronic granulomatous disease for human stem-cell gene therapy transplants. Hum Gene Ther. 28(7):565–575
75. Wang L, Li MY, Qu C, Miao WY, Yin Q, Liao J, Cao HT, Huang M, Wang K, Zuo E, Peng G, Zhang SX, Chen G, Li Q, Tang K, Yu Q, Li Z, Wong CC, Xu G, Jing N, Yu X, Li J (2017) CRISPR-Cas9-mediated genome editing in one blastomere of two-cell embryos reveals a novel Tet3 function in regulating neocortical development. Cell Res. 27(6):815–829

76. Hashimotoa M, Takemotob T (2015) Electroporation enables the efficient mRNA delivery into the mouse zygotes and facilitates CRISPR/Cas9-based genome editing. Sci Rep 5:11315
77. Qin W, Dion SL, Kutny PM, Zhang Y, Cheng AW, Jillette NL, Malhotra A, Geurts AM, Chen YG, Wang H (2015) Efficient CRISPR/Cas9-mediated genome editing in mice by zygote electroporation of nuclease. Genetics 200 (2):423–430
78. Wang W, Kutny PM, Byers SL, Longstaff CJ, DaCosta MJ, Pang C, Zhang Y, Taft RA, Buaas FW, Wang H (2016) Delivery of Cas9 protein into mouse zygotes through a series of electroporation dramatically increases the efficiency of model creation. J Genet Genomics 43 (5):319–327
79. Chen S, Lee B, Lee AY, Modzelewski AJ, He L (2016) Highly efficient mouse genome editing by CRISPR ribonucleoprotein electroporation of zygotes. J Biol Chem 291 (28):14457–14467
80. Kim K, Ryu SM, Kim ST, Baek G, Kim D, Lim K, Chung E, Kim S, Kim JS (2017) Highly efficient RNA-guided base editing in mouse embryos. Nat Biotechnol 35(5):435–437
81. Aten QT, Jensen BD, Burnett SH, Howell LL (2014) A self-reconfiguring metamorphic nanoinjector for injection into mouse zygotes. Rev Sci Instrum 85(5):055005

Chapter 2

Production of Transgenic Mice by Pronuclear Microinjection

Xin-an Pu, Anthony P. Young, and H. Michael Kubisch

Abstract

Pronuclear microinjection remains the most widely used method for the germline modification of mice and other species. The method is conceptually quite simple and at least in rodents can produce transgenic offspring with relatively high efficiency. Here, we describe the various components of the production of transgenic mice including a detailed list of materials and equipment. We outline in detail the preparation of animals, the retrieval, culture and transfer of embryos, the preparation of DNA, and the microinjection process. We have added a substantial collection of notes with helpful suggestions that reflect the many years of experience we have using this technology and our continuing efforts to improve animal welfare and the efficiency of producing transgenics.

Key words Microinjection, Superovulation, Pronuclear, Transgenic, Embryo transfer

1 Introduction

It has been several decades since the generation of transgenic mice by pronuclear microinjection of exogenous DNA into zygotes was first described [1–3]. Since then thousands of transgenic mouse strains have been produced, which have provided an incalculable wealth of information on virtually every aspect of gene control and function.

Throughout the years many additional developments have been reported such as the use of bacterial and yeast artificial chromosomes (BACs and YACs, respectively) for transfer of larger amounts of genetic information, the development of alternative methods of transgene delivery such as those involving viral vectors and the creation of chimeras via injection of modified embryonic stem cells [4–10]. More recently powerful new approaches to manipulate genomes have become available based on the use of nucleases such as ZINC fingers, TALEN, or CRISPR/Cas9 [11–13].

Despite these innovations and new strategies, delivery of exogenous DNA through pronuclear microinjection continues to play

Chengyu Liu and Yubin Du (eds.), *Microinjection: Methods and Protocols*, Methods in Molecular Biology, vol. 1874,
https://doi.org/10.1007/978-1-4939-8831-0_2, © Springer Science+Business Media, LLC, part of Springer Nature 2019

an essential role and the basics of the process have remained virtually unchanged. In part this likely reflects the simplicity of the technique and the relatively high efficiency with which transgenics, at least in mice, can be produced.

It is worth noting that the technique does have some limitations. First, the lack of control over the copy numbers of inserted genes and their site(s) of insertion, leads to variability in the level of transgene expression. Second, the (occasional) generation of mosaic offspring may render the transgenic founder incapable of transmitting the transgene to the next generation. Third, the overall percentage of transgenic offspring generated via use of microinjection technology is relatively low. While it is comparatively simple and inexpensive to overcome this limitation in rodents merely by increasing the number of embryos that are injected and transferred, such measures are more problematic and often cost-prohibitive in larger animal species.

The production of transgenic mice is conceptually and technically straightforward ([14] and Fig. 1). It relies on injection of an isolated fragment of DNA (the "transgene") directly into one of the pronuclei of fertilized ova. The injected embryos are then transferred into the oviducts of pseudopregnant surrogates, which carry them to term. There are various factors that can affect the efficiency of producing transgenic animals and these have been reported elsewhere [15–17]. What is important is to be cognizant of the fact that the reliable and efficient production of transgenic mice requires a substantial investment in equipment, staff, and training. Equally important are a commitment and the access to the highest quality supplies and a strict adherence to meticulous laboratory standards when preparing solutions, media, tools, and DNA for microinjection. It is far less time consuming to invest time and money in developing and following high standards than to search for causes when things go awry.

In this chapter, we describe the procedures we have developed and used for nearly 20 years to produce transgenic mice by pronuclear microinjection, from the initial setup to the delivery of

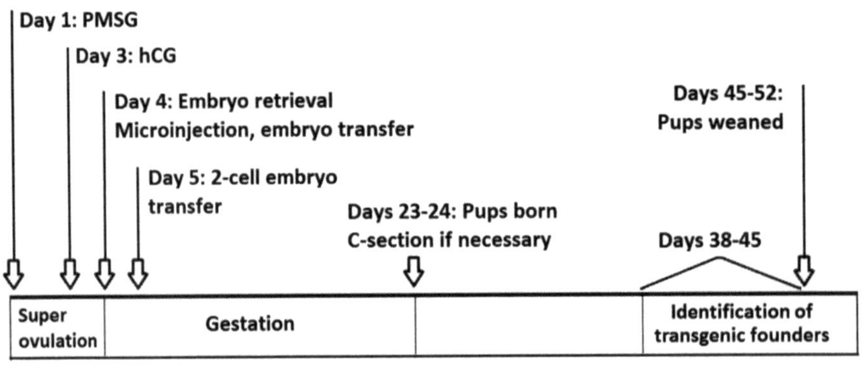

Fig. 1 Timeline for the generation of transgenic mice

progeny. We also provide some additional tips that readers may find helpful and which represent our past and ongoing efforts to make the production of transgenic mice more efficient and manageable for those involved in this process.

One cautionary note: the market for equipment and supplies in the biomedical field is constantly changing and new and better products are being released frequently. In some cases companies, whose products we have used for many years, have merged or changed names, while some products, conveniently used at one time, have been discontinued. The equipment and supplies described in this chapter are those we have found to work best in our laboratory. They are not to be interpreted as product endorsements and readers are encouraged to determine for themselves what works best in their laboratories.

2 Materials

2.1 Mice (See Table 1 and Note 1 for General Considerations)

1. Donor female mice: 20–40 juvenile females, 3–5 weeks old, from inbred or hybrid strains (*see* **Note 2**).
2. Stud males: 20–30 Stud males of the same strain as the donors, 2–10 months of age (*see* **Note 3**).
3. Vasectomized males: 10–20 vasectomized outbred males such as ICR (CD-1) or Swiss Webster. These are used to induce pseudopregnancy (*see* **Note 4**).

Table 1
Representative mouse colony for transgenic production

Mouse Gender	Mouse type	Coat color	Commonly used strains	Age (week or month)	Colony size	Housing per cage
Female: Donors	Inbred-1 Inbred-2 Hybrid	Albino Black Black	FVB C57BL/6 B6CBAF1, B6SJLF1	3–5 weeks, up to 8-week	Up to 40 per week	Up to 5
Stud male	Inbred-1 Inbred-2	Albino Black	FVB C57BL/6	2–10 months	20–30	Individual
Vasectomized (vas) male	Outbred	Albino	ICR, Swiss Webster	2–18 months	10–20	Individual
Female: Pseudo-pregnant recipients	Outbred	Albino	ICR, Swiss Webster	Optimal BW[a] 30–35 g	50–100	Up to 5
Female: Foster dams (recommended)	Outbred	Albino	ICR, Swiss Webster	2–6 months	2–5 breeder pairs	Up to 2

[a]Body weight (BW) is more important than age in this category

4. Recipients (pseudo-pregnant females): 50–100 outbred females from strains such as ICR (CD-1) or Swiss Webster, older than 6 weeks and weighing 30–35 g (*see* **Note 5**).
5. Foster dams (optional): 2–5 outbred pregnant foster dams to rear pups obtained by Cesarean section (*see* **Note 6**).

2.2 Drugs and solvent for Superovulation of Mice

1. Sterile saline.
2. Pregnant mare's serum gonadotropin (PMSG).
3. Human chorionic gonadotropin (hCG). Both PMSG and hCG are supplied as lyophilized powders. Dissolve each in sterile saline (0.9% NaCl) to prepare working solutions of 50 IU/mL. Make 1 mL aliquots and store at −80 °C.

2.3 Surgical Supplies, Tools, and Accessories

1. 70% Ethanol.
2. Isoflurane and a tank of O_2 for anesthesia (*see* **Note 7**).
3. O_2 regulator, such as the E-Z 7000 Classic System (manufactured by E-Z Systems).
4. A Research Anesthesia Machine w/O_2 Flush and accessories.
5. Buprenorphine for pain management.
6. 28G 1cc insulin syringe with 0.1 mL graduation.
7. High intensity illuminator.
8. Straight-bladed Mayo scissors for opening the skin.
9. Spring (microdissection) scissors with sharp points for opening the body wall.
10. Blunt end forceps or dissecting tweezers #5 for ova collection.
11. Fine tip forceps for embryo transfer surgery (e.g., Iris forceps or watchmaker's #5).
12. Wound clip kit: Clip applicator, clip remover, and non-sterile wound clips.
13. Serrefine clamps.
14. Ear punch.
15. Stainless steel instrument sterilization trays for storage of sterilized surgery tools.
16. Iodine Prep Pads: (Povidone-Iodine Swabsticks from Medline Industries, https://www.medline.com/product/Povidone-Iodine-Prep-Pads/Skin-Scrub/Z05-PF07445).
17. Hair clipper.
18. Surgical gloves.
19. Ultraspear PVA Classic (absorbent, Braintree Scientific, INC., SP 40815).

20. A hand held disposable surgical cautery for vasectomy (Medline Industries).
21. Heating stage and warming plate or pad: A heating stage is used for keeping the embryo handling medium warm, while the warming plate or pad is used to prevent heat loss.

2.4 Sterilization Apparatus and Accessories

1. A Steri-Dent 200 Dry Heat Sterilizer (MFI Medical Equipment Inc.) for sterilization of surgical instruments.
2. A Germinator 500 glass beads sterilizer. This is used if surgical instruments need to be re-sterilized between procedures.
3. Dry Heat Indicator Label (SPSmedical Supply Corp, DTL-125, www.SPSmedical.com) for monitoring the sterilization effectiveness of the Dry Heat Sterilizer.

2.5 Equipment, Reagents, and Media for Ova Retrieval and Embryo Culture

1. CO_2 Incubator, set at 37 °C, 5% CO_2, and humidified with a pan containing autoclaved deionized water.
2. Embryo culture-tested mineral oil.
3. Embryo-tested water.
4. Media: CZB/H as embryo handling and CZB as culture medium; both can be prepared in-house ([14], *see* **Note 8**). Other suitable media combination such as M2 and M16 are commercially available.
5. Hyaluronidase: prepared as a 20 mg/mL stock solution stored at −20 °C in aliquots of 50 μL per tube. Use one aliquot per dish of 2.5 mL CZB/H.

2.6 Stereomicroscopes for Embryo Collection and Embryo Transfer

Stereomicroscope equipped with both top and bottom light sources, such as Zeiss Stemi 2000, Zeiss Stemi 2000C or similar models from other manufacturers are suitable.

2.7 Reagents and Supplies for Transgenc Preparation

1. Spin column-based nucleic acid purification kit (commercially available).
2. Low melting agarose (LMA).
3. TAE buffer: Tris base, glacial acetic acid, EDTA (pH 8.0).
4. Ethidium bromide at 10 mg/mL.
5. 3 M Sodium acetate (NaOAc).
6. 70% and 95% Ethanol.
7. 10× Loading dye of bromophenol blue and xylene cyanol FF.
8. Microinjection buffer: 7.5 mM Tris–HCl, pH 7.4, 0.2 mM EDTA.
9. Deionized water.
10. Electroporation apparatus.

11. Gel power supply and Mini Gel System Accessories.
12. UV spectrophotometer (NanoDrop 1000 or equivalent).
13. Transilluminator, long wave UV 366 nm.
14. UV-protected goggles.
15. Microcentrifuge.
16. Sterile microcentrifuge tubes (clear and multiple colors): 0.5 mL, 1 mL.
17. Sterile 0.22 μm syringe filter units.
18. Sterile 1 cc disposable syringes.
19. Sterile disposable serological pipets: 5 mL and 10 mL.
20. Standard Pipettes: P20, P200, and P1000.
21. Sterile disposable pipette tips: 10 μL, 200 μL, and 1000 μL.
22. Sterile plastic centrifuge tubes for aliquots of media and solutions: 15 mL, 50 mL.

2.8 Microinjection Station and Setup (See Fig. 2)

1. One inverted microscope with phase contrast (DIC or Hoffman): Zeiss Axiovert 135 M (or similar microscopes from Leica, Nikon or Olympus) with a 10× ocular and 5×, 10×, 20×, and 40× objective lenses.
2. One pair of micromanipulators: Leica or Eppendorf TransferMan NK-2.
3. Microinjector: Narishige Microinjector IM-300 (programmable) or an equivalent product from Eppendorf, Leica, Nikon, or Olympus.
4. Microinjection holding pipette system: Narishige or Eppendorf CellTram Vario.
5. Microinjection chamber (*see* Fig. 3).
6. Anti-vibration airtable: TMC Pneumatic Table or Vibraplane Airtable (RS Kinetic Systems).
7. A gas cylinder with compressed nitrogen (N_2) gas for the microinjector.
8. A regulator for adjusting the pressure on the N_2 gas cylinder and a chain to secure the cylinder.
9. Tygothane tubing to connect the microinjector to the N_2 gas cylinder.
10. A digital video camera and computer system for teaching or demonstration (optional).

2.9 Equipment and Supplies to Produce Glass Pipettes and Microtools

1. A horizontal pipette puller: Sutter P-97 or P87 (Sutter Instrument, https://www.sutter.com/MICROPIPET/p-97.html).
2. Microforge: Narishige Microforge MF900 for manipulation of glass pipettes, such as breaking and polishing.

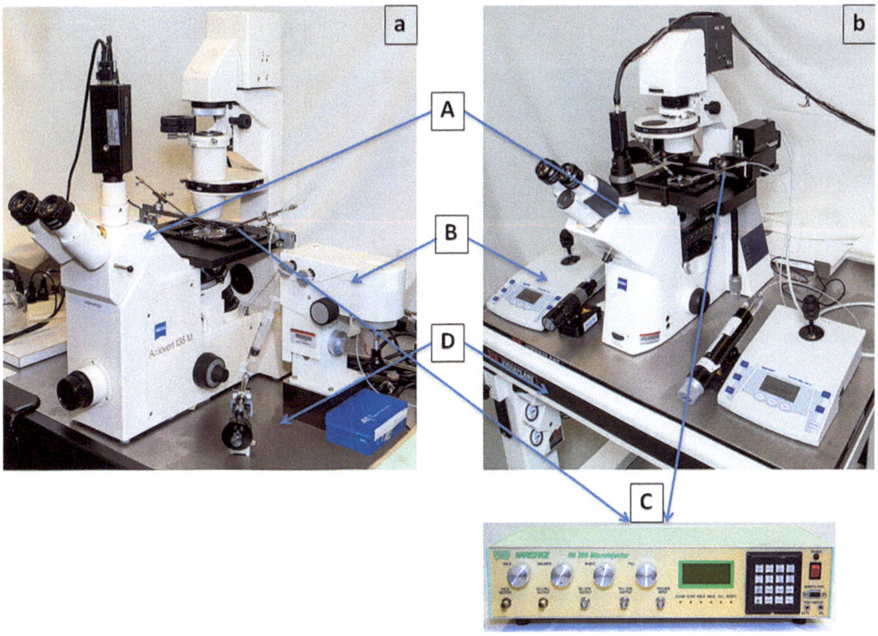

Fig. 2 Microinjection station setup. Inverted microscope (A) with Micromanipulators (B) manufactured by Leica (in **a**) or Eppendorf (in **b**) connected to a Narishige Microinjector (C) on an anti-vibration airtable (D)

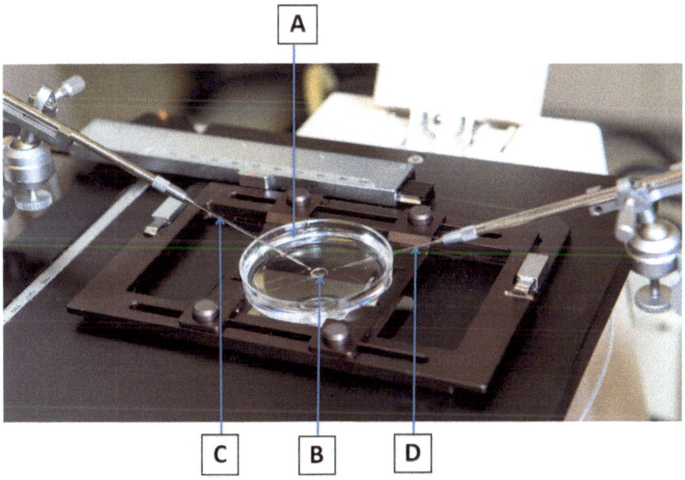

Fig. 3 Microinjection chamber. A microinjection chamber (A) containing a 25 μL drop of CZB/H (B) in the center of a 60 × 15 mm Petri dish, which is covered with mineral oil. The embryos are placed into the drop, individually positioned with a holding pipette (C) and injected with an injection pipette (D)

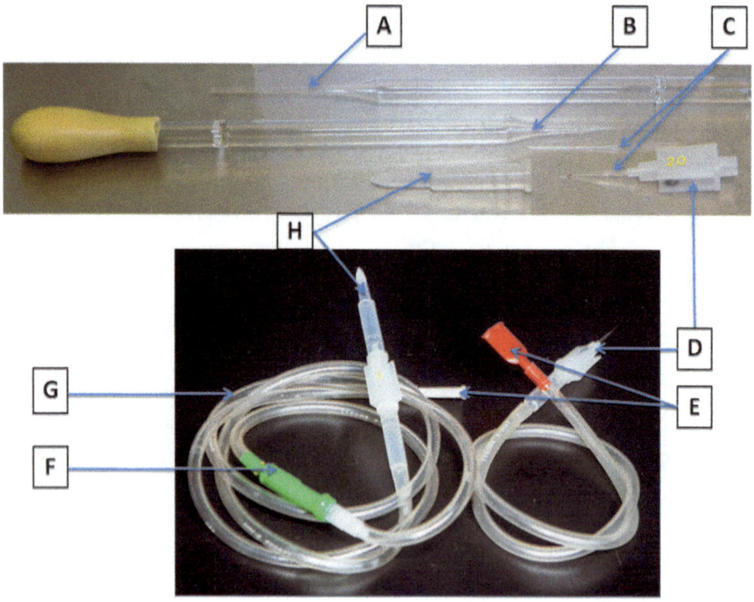

Fig. 4 Embryo handling and embryo transfer pipette assembly. Both the embryo handling pipette and the embryos transfer pipette can be prepared from a regular Pasteur pipette (A). The Pasteur pipette is drawn out and broken into two pieces (B and C). The larger segment (B) can be used for embryo handling while the small segment (C) becomes part of ET surgery pipette assembly by mounting it onto a micropipette holder (D). The holder is then attached to one end of a plastic tube (G) while the other end of the tube is connected to a mouthpiece (E). A syringe filter (F) is affixed to the plastic tube (G) and the transfer pipette is protected with a cap (H) when not in use

3. Embryo holding pipette: Either made from Borosilicate glass capillary or purchased from commercial (e.g., Eppendorf Vacu-Tip, Cat: 5175 108.000).
4. Microinjection pipette (needle): Pulled from Thin Wall Borosilicate Glass with filament (Sutter Instrument, BF100-78-15).
5. Embryo handling pipette: Made from a drawn-out Pasteur pipette (*see* Fig. 4).
6. Embryo transfer pipette: These can be either prepared as described below (*see* Fig. 4) or purchased from commercial (Biomedical Instruments http://www.biomedical-instruments.com/index.php/products-side-menu/52-pipets/pipets-for-animal-research/retransfer-pipets/8-holders-for-standard-retransfer-handling-pipets).
7. Glass test tubes capped with stainless steel closures.
8. Aspirator mouthpieces.
9. Plastic tubing for transfer pipette assembly.
10. 1 mL pipette rubber bulbs.

2.10 Miscellaneous Items

1. 4 °C Refrigerator.
2. −20 °C Freezer.
3. −80 °C Freezer.
4. Precision balances for weighing.
5. Water bath.
6. Magnetic stir plate.
7. Bunsen burner or microburner.
8. A diamond pencil and rubber bulbs.
9. Glassware for media preparation, i.e., beakers and cylinders.
10. Magnetic stir bars.
11. Spoons and weighing papers (boats) for media preparation.
12. pH strips.
13. Sterile filter units with 0.22-μm filters and 150 mL, 500 mL and 1000 mL volume capacities.
14. Sterile Disposable plastic transfer pipettes.
15. Ice and ice bucket.
16. Paper towels.
17. Sharps disposal containers.

3 Methods

3.1 Preparation of Transgene (See Note 9)

1. Extract the transgene using a spin column-based nucleic acid purification kit per manufacturer's instructions.
2. Concentrate the transgene by 3 M Sodium acetate and 95% ethanol precipitation once the DNA fragment is eluted out of the spin column.
3. Dissolve the transgene pellet in 20 μL of micro-injection buffer, usually overnight in the refrigerator (4 °C).
4. Determine the integrity and concentration of the purified transgene by testing a small aliquot on a 1% agarose mini-gel and comparing it to standard DNA size markers.
5. Filter-dilute the transgene solution through a 0.22 μm syringe filter with micro-injection buffer to a working concentration of 5 ng/μL.
6. Prepare 10 μL aliquots in microcentrifuge tubes and store at −80 °C.
7. On the day of microinjection, thaw one or two tubes of the DNA stock and make serial dilutions with DNA concentrations in the range of 1 ng/μL to 5 ng/μL (*see* **Note 10**). Each dilution should consist of at least 20 μL that is generally

sufficient for a typical microinjection session. The diluted preps are tested by gelelectrophoresis before and after microinjection to confirm DNA integrity and concentration.

3.2 Preparation of Glass Microtools (See Note 11)

1. Run a Ramp Test for each new batch of glass capillaries on a horizontal pipette puller (e.g., Sutter P-97 or P87) following the manufacturer's instruction.
2. Prepare holding pipettes: These pipettes can be either purchased or easily prepared in-house. To pull the glass pipette on the puller, we use these settings: Heat = 750, Pull = 10, Vel = 45, Time = 250, Pressure = 500. To break the pipette (at about 80 μm outer diameter) and polish the tip, we use Narishige Microforge (MF-900) and follow the manufacturer's instructions.
3. Prepare injection pipettes: These pipettes are prepared freshly prior to a microinjection session with the following parameters: Heat = 715, Pull = 90, Vel = 70, Time = 150, Pressure = 500. When the room is cold, we use alternate settings: Heat = 700, Pull = 120, Vel = 70, Time = 150, Pressure = 500. Note that the tip of the pipette is sealed at this point; it will be opened in the microinjection chamber just before embryos are to be injected (*see* Subheading 3.4 below and **Note 12**).
4. Prepare embryo handling and transfer pipettes: These pipettes can be easily produced from drawn-out Pasteur pipettes (Fig. 4a). Hold the large end of the pipette with your fingers and use a pair of forceps to grasp the smaller end. Rotate the pipette's shoulder area above the tip of the flame of a Bunsen burner until the glass begins to soften. Move the pipette out of the flame (important!) and gently pull in opposite directions. Score the glass at about 4 cm away from the shoulder with a diamond pencil, and then break the pipette into two pieces. The larger piece, which is usually smooth enough and does not need to be polished, is used as a handling pipette for ova collection and transfer between dishes (Fig. 4b). Bake-sterilize the handling pipettes in a glass test tube covered with a metal lid (or piece of double-layered heavy duty aluminum foil).
5. The smaller piece of the drawn-out Pasteur pipette can be used as a part of the embryo transfer pipette assembly (Fig. 4c). The inner and outer diameters (ID/OD) of the embryo transfer pipette can be adjusted to within a range of 110/200 μM using a Microforge. To do this, mount the pipette on a pipette holder (can be obtained from the Microforge vendor), and align the tip with the heating filament. While moving the tip of the pipette closer to the filament (do not touch!), activate the heat by stepping on the foot pedal; the heat emanating from the filament will cause the pipette opening to shrink and

become "polished." Once the ID/OD of the tip reaches around 150 μM, stop the heat and remove the pipette from the pipette holder. Finally, connect the transfer pipette to a pipette holder (Fig. 4d) and sterilize it in the laminar hood under UV light for 30–60 min. After sterilization, the tip of the pipette is capped with the cover of a pipette holder (Fig. 4h). Embryo transfer surgery pipettes can be prepared in batches and be stored in a clean Petri dish until use.

6. To assemble the final transfer tool, the pipette holder is inserted into one end of a piece of plastic tubing while a mouthpiece is attached to the other and a syringe filter is inserted somewhere inline (Fig. 4e–g). A complete embryo transfer tool is shown in Fig. 4.

3.3 Preparation of Mice and Embryos

3.3.1 Vasectomies

1. Sterilize all surgical instruments by baking in a dry sterilizer at 160 °C for a minimum of 2 h. A bead sterilizer is prepared for sterilizing the tools between surgeries if needed.
2. Anesthetize animals: The recommended method of anesthesia is the use of isoflurane, which is an inhalation anesthetic used in conjunction with O_2 (*see* **Note 7**). The mouse is placed into an induction chamber that is flooded with 4% isoflurane. Once an appropriate plane of anesthesia has been reached (which is verified by toe pinching to assess pedal reflex or by the loss of the palpebral reflex), the animal is removed from the induction chamber, placed on a suitable surface and fitted with a nose cone through which 2% (1–3% is acceptable) isoflurane/O_2 is delivered to maintain an adequate plane of anesthesia.
3. Place the anesthetized mouse on a disinfected warming plate (~34 °C) where he should remain throughout the surgery until full recovery. Shave the lower abdomen with an electric hair clipper, and swab the skin with surgical scrub solution 3 times followed by 70% ethanol.
4. Cover the disinfected surgical area with a piece of sterile surgical cloth with an opening that exposes the surgical area. With a pair of dissecting scissors, make a single mid-ventral incision (~1 cm) into the skin and then cut the body wall using a pair of spring scissors. Using a pair of watchmaker's#5 forceps, pull out one testis. The vas deferens can be easily identified by the blood vessel that is attached to it.
5. Cauterize the vas deferens using a hand held disposable surgical cautery or flame-sterilized hot forceps. Place the exposed tissue back into the body cavity and repeat the procedure on the other vas deferens.
6. Once both vasa deferentia have been cauterized, close the skin incision with sterile stainless wound clips.

7. Administer Buprenorphine subcutaneously (s.c.; 0.05–0.1 mg/kg) for pain relief.
8. Repeat **steps 2** to 7 until all the mice are vasectomized (usually 5 to 10 males per session).
9. Keep the mice in cages on a warming pad until recovery from anesthesia. The wound clips are removed about 10–14 days after the surgery.
10. Confirm the infertility of a vasectomized male at least once by mating with fertile females.

3.3.2 Superovulation of Donors (See Fig. 1 for Timeline and ***Note 2****)*

1. On Day 1, PMSG injection: thaw aliquot(s) and warm to room temperature. Inject each donor s.c. with 0.1 mL (i.e., 5 IU).
2. On Day 3, hCG injection: 46–48 h after PMSG, thaw aliquot (s) and warm to room temperature. Inject each donor s.c. with 0.1 mL (i.e., 5 IU). Following hCG injection, place the female into a stud male's cage, one female per male.
3. On Day 4, euthanize the female donors and collect ova.

3.3.3 Collection of Ova from Superovulated Donors

1. Set up media dishes for ova collection: In the morning prior to ova collection, prepare dishes of CZB culture medium and CZB/H handling medium under a laminar hood. Plate ~2.5 mL CZB/dish into four 35 × 10 Petri dishes, two for harvesting ova, and two for later embryo transfer (*see* **Note 13**).
2. Plate six drops of 25 μL of CZB into each of two 35 × 10 Petri dishes (one dish for the ova before microinjection, the other for holding them afterward), and cover the drops with mineral oil (*see* **Note 14**).
3. Place these six CZB dishes on a secondary holder such as a 100 mm Petri dish to prevent accidental spills (2 dishes/holder), and keep them in a 5% CO_2 incubator at 37 °C.
4. Plate CZB/H ~2.5 mL/dish into six to eight 10 × 35 mm Petri dishes.
5. Add one thawed aliquot of hyaluronidase to each of at least two 2.5 mL CZB/H dishes. The number of hyaluronidase-containing dishes will depend on the number of donors used. Usually one dish can hold up to 10 oviducts.
6. On the lid of a 60 × 15 mm Petri dish, plate a drop of 25 μL of CZB/H medium at the center of the lid, and cover the drop with mineral oil (~10 mL mineral oil/dish). This will be used as the microinjection chamber (*see* Fig. 3).
7. Keep CZB/H dishes on a 37 °C heating stage.
8. Prepare paper towels on a working bench, a plastic bag to hold the animal carcasses and pre-sterilized dissecting tools

consisting of two pairs of scissors and four pairs of forceps (two pairs for retrieving oviducts, and two pairs for releasing ova).

9. Check females' plugs and mark any plugged mice on their tails with a permanent ink pen to distinguish them from those not plugged. Ova from plugged and unplugged donors should be retrieved and kept separately. While unplugged females occasionally do produce fertilized ova, most do not and keeping ova separate saves time and effort otherwise needed to sort them out later.
10. Euthanize donor mice with an approved method of euthanasia, such as CO_2 gas inhalation (*see* the American Veterinary Medical Association's 2013 Guidelines for the Euthanasia of Animals https://www.avma.org/KB/Policies/Documents/euthanasia.pdf).
11. Make incisions through skin and body wall using appropriate scissors, collect oviducts, and transfer them to warm CZB/H dishes containing hyaluronidase.
12. Keep the CZB/H dishes containing oviducts on a heating stage.
13. Harvest ova under stereomicroscope: When an animal has responded to hormonal stimulation, part of the oviductal ampulla appears swollen and transparent. Hold the oviduct with one pair of dissecting forceps, using another pair of forceps stab a hole into the ampulla and apply pressure to "squeeze out" the ova. At this point the ova will be surrounded by a mass of cumulus cells. The hyaluronidase will digest the cumulus mass and the ova should then become clearly visible.
14. Rinse the ova through three dishes of CZB/H until all tissue and cellular debris has been removed. Rinse the ova once in a dish of CZB and place them into CZB microdrops, approximately 100 ova per drop. Place the dish into a 37 °C, 5% CO_2 incubator for at least 1 h prior to pronuclear microinjection.

3.3.4 Preparation of Pseudopregnant Recipients (Surrogates)

1. Mate the females with a vasectomized male the night before a scheduled embryo transfer; one male can be housed with up to ten females. Check for vaginal plugs in the morning, and check again in the afternoon (*see* **Note 5**). It is advisable to produce an excess of recipients in case an embryo transfer goes awry. Females that are not used will resume their regular estrus cycles within around 2 weeks.
2. If a large number of recipients is needed it can be helpful to synchronize their estrus cycles by the Whitten effect ([19], *see* **Notes 4** and **5**).

3.4 Pronuclear Microinjection

1. Retrieve an aliquot of DNA, set it on ice and place the blunt end of the injection pipette into the DNA solution. The internal filament will cause the pipette to become back-loaded through capillary action. After a few seconds, a tiny bubble should be visible at the shoulder of the pipette. If you don't see this bubble after a few minutes, discard and use a fresh pipette.
2. Retrieve around 100 embryos from a CZB microdrop, rinse them once in CZB/H, and place them into the CZB/H drop in the microinjection chamber.
3. Turn on the microinjector and the N_2 gas supply, set the injector balance to a continuous pressure of 0.3 psi to prevent CZB/H medium from being drawn back into the pipette by capillary action (On Narishige Microinjector this is controlled by the "Bal" button). The injection pressure can be set at a range of 1–100 psi and may have to be adjusted depending on the size of the pipette opening (*see* below **step** 7 and **Note 12**).
4. Working under the inverted microscope at a 100× magnification, lower the holding pipette into the CZB/H drop.
5. Adjust the microscope until most of the pronuclei are in focus at 200× magnification. Using the holding pipette, separate good embryos from those of poor quality.
6. Mount the DNA containing injection pipette onto the pipette holder and at 100× magnification lower it into the CZB/H drop until it is level with the holding pipette. Note that at this point, the tip of the injection pipette is still sealed.
7. Using a 400× magnification, push the injection pipette tip against the holding pipette to break off the tip and create an opening. If possible, assess the size of the opening and if it appears too large the pipette should be replaced (*see* **Note 12**).
8. At 400× magnification, using the holding pipette capture an embryo, rotate to bring one of its pronuclei to approximately the center or far side of the embryo, and focus by adjusting the micro-focus knob. Position the injection pipette tip roughly at 3 o'clock, and align it with the embryo so that its tip is next to the pronucleus and both the mid-plane of the pronucleus and the pipette tip are in focus (*see* Fig. 5 and **Note 12**).
9. Inject the embryo by inserting the injection pipette into the pronucleus. Once the pipette tip is inside the pronucleus, either push the INJ button on the microinjector or step on the foot pedal to expel the DNA solution. A slight expansion of the pronucleus should be visible indicating that the DNA injection has succeeded. Withdraw the injection pipette out of the embryo gently and swiftly.

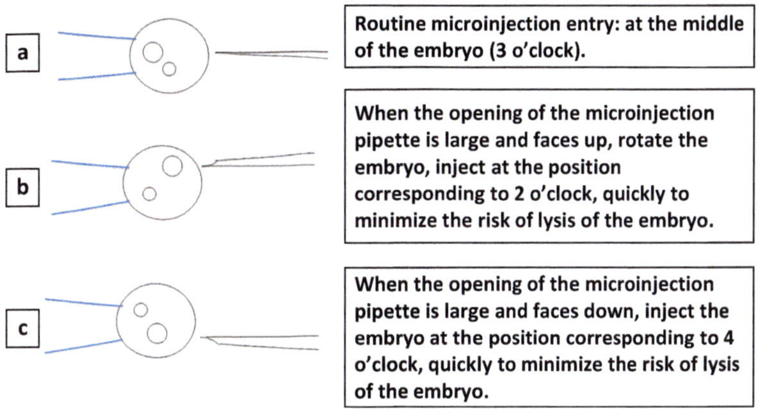

Fig. 5 Microinjection pipette entry positions. (**a**) Routine microinjection entry: at the middle of the embryo (3 o'clock). (**b**) When the opening of the microinjection pipette is large and faces up, rotate the embryo, inject at the position corresponding to 2 o'clock, quickly to minimize the risk of lysis of the embryo. (**c**) When the opening of the microinjection pipette is large and faces down, inject the embryo at the position corresponding to 4 o'clock, quickly to minimize the risk of lysis of the embryo

10. Move the injected embryo to a different area in the CZB/H drop so that injected and non-injected embryos are grouped separately.
11. Repeat **steps 8** to **10** until all embryos in the CZB/H drop have been injected, and transfer them into a fresh CZB/H dish.
12. Wash those having survived the injection procedure once in a dish of CZB medium (*see* Subheading 3.3.3) and place in groups of around 25 into drops of equilibrated CZB under mineral oil. The number of embryos in each group will depend on the number of available recipients on a given day, but each group of embryos constitutes a cohort for transfer into a recipient.
13. Culture injected embryos in a humidified 37 °C, 5% CO_2 incubator for about 30 min prior to transferring them into recipients. This allows sufficient time to assess whether embryos have survived the injection process and can be transferred. In surviving embryos, the cell membrane will be clearly demarcated from the zona pellucida.

3.5 Embryo Transfer Surgery Via Oviduct Transfer

1. Have all surgery tools and embryo transfer pipettes ready and sterilized 1 day in advance (*see* **Note 15**).
2. Identify pseudopregnant recipients: do this in the morning (0.5 dpc) and again in the afternoon if more recipients are needed (0.0 dpc).

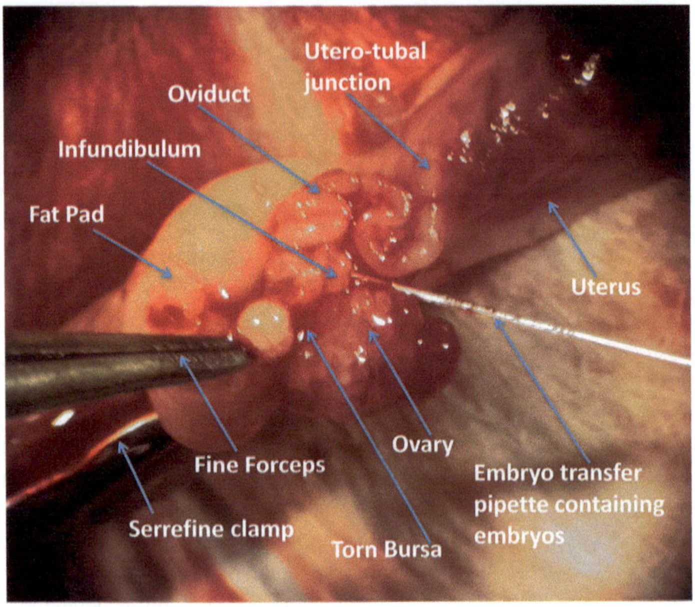

Fig. 6 Embryo transfer into the oviduct. The embryo transfer pipette containing the embryos is inserted into the infundibulum of the oviduct and the embryos are gently expelled

3. Anesthetize the animals by isoflurane inhalation (*see* Subheading 3.3.1). Shave each mouse along the right flank of the body, and swab the skin with surgical scrub solution 3 times followed by 70% ethanol.
4. Place the mouse on a piece of sterile surgical cloth (or sterile paper towel), using a pair of dissecting scissors make a single incision (~1.0 cm) through the skin and musculature of the right (or left) flank just caudal to the last rib and slightly lateral to the spine. Dissect the exposed peritoneum with a pair of spring scissors to open the peritoneal cavity. Using a pair of fine-tip forceps, lift up the body wall, with another pair of forceps, pull out the ovary and oviduct (*see* Fig. 6 and **Note 16**). Stabilize the oviduct with a Serrefine clamp attached to the fat pad.
5. Move the mouse to a warming plate under the stereoscope, which is equipped with a High Intensity Illuminator. Make sure the exteriorized tissue is properly illuminated and in focus.
6. Load 20–25 embryos directly from the microdrop into the embryo transfer pipette (*see* **Note 17**), and place the assembly over the oculars of the stereomicroscope so that both ends are suspended in the air (not touching any other object).
7. Tear the bursa over the ampulla using two pairs of sharp forceps. With a pair of forceps in one hand, grasp the tissue adjacent to the infundibulum; with the other hand, take the

transfer pipette assembly and gently insert the pipette tip into the infundibulum. Stabilize the mouthpiece of the transfer pipette between your teeth, and gently expel the embryos until air bubbles can be seen in the oviduct.

8. Detach the Serrefine clamp from the fat pad, and reposition the exteriorized tissues carefully back into the peritoneal cavity.
9. Finally, close the skin with one or two sterile stainless steel wound clips.
10. Apply an analgesic buprenorphine s.c. (0.05–0.1 mg/kg) for pain relief.
11. Repeat **steps 3** to **10** until all injected embryos have been transferred into recipients (*see* **Note 18**).
12. Place the mouse in a clean cage and keep the cage on a warming plate (or warm pad) until she has recovered from the anesthesia. Any animal having undergone a surgical procedure should be closely observed for 5 days to ensure there are no complications (*see* **Note 19**). If a transfer results in a pregnancy, the wound clips should be removed two weeks *after* the delivery of pups; some strains are sensitive to stress and removal of clips during the pregnancy may cause fetal loss in some females.

3.6 Miscellaneous

3.6.1 Cesarean Section (C-Section) and Fostering (See Note 20)

1. Identify foster dam(s): choose either a dam that is already nursing pups from injection of the same construct (same project) or a dam that has pups with a different coat color.
2. Euthanize any dam that exhibits signs of dystocia by CO_2 inhalation and immediately transfer it to a laminar flow hood.
3. Quickly open the abdomen with sterile surgery tools, remove and place the uterus containing the pups on a sterile paper towel.
4. Gently dissect the uterus and fetal membranes to retrieve the pups, remove the yolk sac and amnion, and carefully cut the umbilical cord.
5. Move the pups onto another piece of sterile paper towel on a warming plate, and clean up their bodies by wiping away the amniotic fluid and secretions using Ultraspear PVA wetted with warm (37 °C) sterile saline.
6. Pinch the pups gently to stimulate breathing until the pups start breathing and turn to a pinkish color.
7. Expose these pups to dirty bedding from the prospective foster dam, mix thoroughly, and then place the pups into her cage. Observe the acceptance of the pups by the foster dam up to 48 h.

3.6.2 Animal Identification

We use an ear punch for marking the ear right after the embryo transfer surgery to identify each animal as a part of our documentation and to keep track of the success of surgery, pregnancy, transgenic production associated with DNA concentrations, embryos stages, surgery conditions, etc. We follow the ear punch schemes as illustrated elsewhere [14].

4 Notes

1. A typical mouse colony consists of embryo donors and recipients, as well as breeding (intact) and vasectomized males (*see* Table 1). The mice are housed in shoebox cages on individually ventilated cage (IVC) racks or on conventional static racks. There are numerous mouse strains available from various vendors and the choice of which one to use will often be determined by the investigator's needs. The question of whether to obtain mice from a supplier or produce them in-house will often be driven by economic considerations. Widely used mouse vendors include (1) Charles River Laboratory (CRL), http://www.criver.com, (2) The Jackson Laboratory (JAX), https://www.jax.org, (3) Taconic Farms (TF), https://www.taconic.com, and (4) Envigo (ENV) (Formerly Harlan Labs), http://www.envigo.com.
2. Donor colony: We maintain a mouse colony with both albino and black mice (other dark-pigmented mice could be equally suitable). The advantage of using different coat colors is that orphaned progeny can be fostered into existing litters and be distinguished from natal pups by their coat color. The number of donors needed for a microinjection session is determined primarily by the yield of ova after superovulation. Three to 5 week old juveniles respond best to hormone priming, although in some strains, such as FVB/N or C57B/L6, females as old as 8 weeks of age can still produce acceptable numbers of ova. In general, outbred or hybrid mouse strains, such as B6CBAF1, B6SJLF1, B6DBAF1, produce more ova than inbred strains, such as C57BL/6, BALB/c, 129/Sv, and FVB/N [16]. Likewise, body weight and dosage of PMSG are factors which can determine superovulation outcomes [18]. We have found that injecting both hormones either s.c. or at a very shallow angle into the intraperitoneal (IP) space yields better superovulation results than traditional IP injections. In general, we obtain approximately 150 injectable embryos per session by superovulating eight C57BL/6 or 10 FVB females.
3. It is important to keep in mind that matings using virgin males tend to produce fewer fertilized ova (embryos). Based on our

records the number of embryos after a second mating is often double or triple the number we retrieve after a first mating. We therefore usually "prime" a virgin male by placing him with a female prior to using him for the production of embryos for microinjection. We also established that one-on-one matings produce better results. Our attempts (eight random trials) to place one male with two females resulted invariably in only one female producing fertilized ova. The males' performance may become less proficient after about 6 months of age, and the plug-producing frequency and fertilization rates should be routinely recorded for each male. Males should be replaced when performance declines markedly.

4. These males can be purchased commercially, although performing vasectomies in-house is feasible provided staff is properly trained. In contrast to stud males, vasectomized males can be used in matings with several females. In fact, we have occasionally placed a vasectomized male with as many as ten females whose estrus cycles had been naturally synchronized and have found as many as five of these with plugs. Synchronization can easily be accomplished by exposing females to males' dirty bedding, which works best when the bedding comes from more than one male. Alternatively, several females can be placed with a poorly performing vasectomized male which will have the same effect. Vasectomized males may last for up to 18 months and are only replaced when they fail to mate with estrous females three times in a row.
5. Pseudopregnant embryo recipients should come from outbred strains such as ICR (CD-1) or Swiss Webster, as theses strains are generally docile and possess good mothering behavior. To induce pseudopregnancy, estrous mature females, at least 6 weeks of age and weighing more than 28 g (ideally 30–35 g), are mated with vasectomized males. Pseudopregnancy is confirmed the following day by the presence of a vaginal plug. Occasionally a female that has no plug in the morning, but is left with a male, may present a plug in the afternoon (0.0 dpc). These females can still be used and the efficiency of producing pregnancies as well as transgenic offspring does not appear to be affected by the timing of the mating. In contrast, we have found that when vaginal plugs, which are clearly visible in the morning, have disappeared by the time of embryo transfer (mostly late afternoon), the pregnancy rate decreases by about 35% with a concomitant reduction in the production of transgenic pups. The conclusion is that recipients which are placed with vasectomized males in the morning and are plugged in the afternoon (0.0 dpc) can be used without reservation, while recipients that have lost their

vaginal plugs by the time of the embryo transfer after an overnight mating should be used with caution [17].

6. Foster dams are females that are used to rear pups derived by Cesarean section (C-section) from another (generally euthanized) female. Any female having given birth within 3 days prior to a C-section can serve as foster dam. If a difficult delivery is anticipated, a female can be placed with a breeding male in time to ensure her pups are born just prior to or coinciding with derivation of the C-sectioned pups.
7. In the past, mice were commonly anesthetized using 2-2-2-tribromoethanol (TBE, also known as Avertin). However, the use of TBE is now widely discouraged because it is not available as a pharmaceutical grade reagent. Isoflurane, which is an inhalant, induces anesthesia quickly and has the added advantage of facilitating a quick recovery following surgery. However, its use requires special equipment and precautions should be taken to avoid exposure of staff. The combination of ketamine and xylazine is a reliable alternative for anesthesia in mice, but ketamine is a controlled substance that requires a secured location for storage, a detailed log sheet to track usage, and the proper disposal of unused portions. Note that comparisons of the effects of various methods of anesthesia on embryo transfer outcomes suggest that there is no difference in pregnancy rates and pup production [20].
8. We prepare a liter of CZB or CZB/H at a time, filter-sterilize it, and make 40 mL aliquots in 50 mL sterile Falcone tubes. CZB/H can be stored at 4 °C, for about 6 months. CZB should always be tested by culturing wild-type embryos prior to storing aliquots. Once the quality is confirmed (at least 80% of embryos should develop into blastocysts), CZB aliquots are stored at −80 °C. We thaw 4–6 aliquots each time, and again test every aliquot prior to use by culturing embryos. We have found that frozen-thawed CZB works as well as freshly prepared medium even after 5 years of cryostorage. When freezing medium, no more than 40 mL of medium should be added to each tube to accommodate expansion during freezing.
9. We separate the DNA fragments by low melting agarose gel electrophoresis. In our experience reducing the time of exposure of DNA to ethidium bromide during electrophoresis (from overnight of 15 h to less than 7 h) substantially increases the efficiency of producing transgenic offspring (from less than 10% to up to 35%). Additionally, we always concentrate the DNA fragment by sodium acetate and ethanol precipitation, and dissolve the DNA pellet in 20 μL of microinjection buffer to obtain a concentrated transgene solution. We have found

that if the initial DNA concentration of this solution is at least 100 ng/μL, we experience less clogging of the injection pipette and better overall results in terms of embryo survival and transgenic pups.

10. The factors contributing to the fate of DNA injected into mouse embryos have been well documented [15]. Integration efficiency of transgenes improves when the DNA concentration in the injection buffer reaches the optimal level and there is a threshold after which embryo survival is markedly compromised. During our early attempts, we injected multiple DNA concentrations for each project, and found the optimal DNA concentration appears to be somewhere between 1.5 and 2.5 ng/μL. We therefore routinely make serial dilutions of our DNA stock at concentrations between 1 and 5 ng/μL, and then start to inject embryos at the highest concentration. If the injection at that concentration results in an unacceptably high rate of embryo lysis or the recurring clogging of the injection pipette we switch to the next lower DNA concentration and repeat the process. Preparing serial dilutions will not only guard against possible dilution errors, but also compensate for the effects of variations in DNA fragment size that exist between projects and which may affect injection outcomes (e.g., pipette clogging, embryo lysis, etc.).

11. Four types of glass microtools are needed for handling, microinjection, and embryo transfers: (1) an embryo handling pipette that can easily be produced by drawing out a Pasteur pipette over a flame, (2) an embryo holding and (3) microinjection pipette, both of which are used during pronuclear microinjection, and (4) an embryo transfer pipette that is needed to transfer embryos into recipients. Most of these pipettes can be either purchased or, with a little training and the appropriate equipment, be produced in-house. These glass pipettes are sharp and should be properly stored and handled with caution. We keep embryo handling pipettes and embryo transfer pipettes in glass test tubes capped with stainless steel closures, sterilize them by baking in a dry-heat oven, and store them at room temperature. For embryo holding and microinjection pipettes, pretreatment is not necessary. They can be stored at room temperature in the original package from the vendor. When preparing the embryo holding and microinjection pipettes with the pipette puller (e.g., Sutter P-97 or P87), we program it with multiple programs with closely matching settings. This is useful when we experience minor environmental fluctuations, which can affect the pipette pulling process (e.g., change in room temperature or humidity). An alternate program can then be used that may compensate for these changes. It will obviously take some time to set up several alternate

programs, but to do so ahead of time is far less disruptive than having to reprogram the puller during a microinjection session. Also note that the pipette puller may have to be reprogrammed if someone has accidentally touched the heating filament.

12. To test whether the size of the injection pipette opening is appropriate, the pipette tip should be moved close to an embryo, and the microinjector (CLR button on Narishige Microinjector) should be activated with the maximum pressure setting. If the embryo remains in place, the pipette tip is likely either not opened at all or has become clogged in which case another attempt should be made to break the tip or the pipette should be replaced. If, on the other hand, the embryo is pushed away too forcefully the opening is likely too large, which might cause lysis of embryos during injection. It will require some practice to become efficient at gauging the correct size of the opening of an injection pipette. Occasionally breaking a pipette may result in a jagged edge (Fig. 5b, c), which may necessitate changing the injecting position to 2 or 4 o'clock instead of the usual 3 o'clock position. The injection pressure should then be set as low as 1–2 psi, and the pipette should be moved swiftly in and out of the embryo to prevent damage to the membrane and lysis. A larger opening or a tip with a jagged edge can sometimes be useful for co-injection of several constructs (e.g., more than two transgenes or CRISPR constructs) as long as the incidence of embryo lysis remains acceptable. The drawback of larger pipette openings is that the tip is often not sharp enough to penetrate the pronuclear membrane. In this case, the pipette may have to be pushed through the entire pronucleus before it penetrates the membrane. If, on the other hand, the pipette tip is so blunt that it cannot even penetrate the plasma membrane, then the pipette obviously needs to be replaced.
13. We use CZB medium to load the embryos during the embryo transfer mainly because our CZB has been embryo-tested (*see* **Note 7** above). Prior to loading the embryos into the transfer pipette, the pipette is rinsed in a 2.5 mL CZB dish to make sure the liquid moves freely. If the transfer pipette becomes clogged with blood or debris during the transfer it may need to be flushed out or be replaced. In either case the embryos are expelled into CZB/H and kept on the warming stage. Before reloading, the embryos are then washed once in CZB, transferred back into a CZB drop under mineral oil, and reloaded into a fresh transfer pipette.
14. When plating CZB drops for embryo culture, care should be taken to ensure that plates as well as the micro-drops are numbered (from the bottom!) to help the tracking and identification of embryos. The culture medium should be

equilibrated before use for at least 1 h in the CO_2 incubator. It can be plated a day before (e.g., afternoon), then equilibrated overnight in the incubator.

15. We sterilize all metal surgery tools by baking them in a STERIDENT Standard 300 dry heat sterilizer. Dry sterilization has enabled us to maintain our surgery tools longer by avoiding corrosion. The duration of dry sterilization follows manufacturer's recommendation; we bake our metal tools for a minimum of 2 h. The sterilizer function is monitored weekly using Sterident Spore. Following surgery tools are immediately cleaned with 70% ethanol to remove all bloodstains and sterilized.
16. Locating the infundibulum is the most challenging step during an oviductal transfer. We choose to transfer embryos into only one oviduct to minimize the trauma to the recipient. However, for training purposes, we ask staff to practice the transfer on both oviducts and to eventually choose the one with which they have been most successful.
17. Traditionally, loading embryos for embryo transfer surgery entailed adding a few air bubbles in the medium flanking the embryos for better control and confirmation of their movement into the oviduct. We have omitted the bubbles and have found that this has no effect on the outcome.
18. If there are not enough recipients for all injected embryos, the remainder can be cultured overnight in CZB microdrops covered with mineral oil in a humidified 37 °C, 5% CO_2 incubator. The resulting 2-cell embryos can be transferred into 0.5 dpc or 0.0 dpc during the following day. In our experience even culturing embryos for 2 days and transfer at the morula stage can still result in pregnancies, although the number of pups born and therefore the production of transgenic offspring are noticeably reduced.
19. Following any surgery, animals should be closely observed for up to 5 days to make sure there are no complications and the wound clips have not fallen off. The due date of a recipient is 19–20 days after receiving embryos. To prevent cannibalism, nesting material should be provided about 2 days in advance, and changing cages should be avoided for at least a week after delivery.
20. We routinely have foster dams available in our mouse colony by setting up natural breeding pairs (or trios) a few days prior to embryo transfer surgery. This is an important precaution because occasionally embryo recipients are unable to deliver pups on their own. This is more likely to happen when only one or two embryos have survived in utero which may have developed into unusually large fetuses. In those cases a C-section

will have to be performed to retrieve the pups. Sometimes females refuse or are unable to nurse their pups and the earliest indication that something is amiss is the failure of the dam to build a nest or the absence of milk in the pups' bellies (the skin of newborn pups is translucent enough to see this). In either case, the availability of foster dams is essential if the pups are to be rescued.

Acknowledgment

The authors wish to thank Dr. Joshua Taylor for his critical review of the manuscript.

References

1. Gordon JW, Scangos GA, Plotkin DJ, Barbosa JA, Ruddle FH (1980) Genetic transformation of mouse embryos by microinjection of purified DNA. Proc Natl Acad Sci U S A 77:7380–7384 [PubMed: 6261253]
2. Brinster RL, Chen HY, Trumbauer M, Senear AW, Warren R, Palmiter RD (1981) Somatic expression of herpes thymidine kinase in mice following injection of a fusion gene into eggs. Cell 27:223–231 [PubMed: 6276022]
3. Costantini F, Lacy E (1981) Introduction of a rabbit beta-globin gene into the mouse germ line. Nature 294:92–94 [PubMed: 6945481]
4. Giraldo P, Montoliu L (2001) Size matters: use of YACs, BACs and PACs in transgenic animals. Transgenic Res 10:83–103
5. Van Keuren ML, Gavrilina GB, Filipiak WE, Zeidler MG, Saunders TL (2009) Generating transgenic mice from bacterial artificial chromosomes: transgenesis efficiency, integration and expression outcomes. Transgenic Res 18 (5):769–785 PMCID: PMC3016422
6. Botteri FM, van der Putten H, Miller AD, Fan H, Verma IM (1986) Recombinant retroviruses in transgenic mice. Ann N Y Acad Sci 478:255–268
7. Miller AD, Bender MA, Harris EA, Kaleko M, Gelinas RE (1988) Design of retrovirus vectors for transfer and expression of the human beta-globin gene. J Virol 62:4337–4345
8. Lois C, Hong EJ, Pease S, Brown EJ, Baltimore D (2002) Germline transmission and tissue-specific expression of transgenes delivered by lentiviral vectors. Science 295:868–872 [PubMed: 11786607]
9. Robertson E, Bradley A, Kuehn M, Evans M (1986) Germ-line transmission of genes introduced into cultured pluripotential cells by retroviral vector. Nature 323:445–448
10. Shen S, van den Brink CE, Kruijer W, van der Saag PT (1992) Embryonic stem cells stably transfected with mRAR beta 2-lacZ exhibit specific expression in chimeric embryos. Int J Dev Biol 36:465–476
11. Brandl C, Ortiz O, Röttig B, Wefers B, Wurst W, Kühn R (2014) Creation of targeted genomic deletions using TALEN or CRISPR/Cas nuclease pairs in one-cell mouse embryos. FEBS Open Bio 5:26–35
12. Mashiko D, Young SA, Muto M, Kato H, Nozawa K, Ogawa M, Noda T, Kim YJ, Satouh Y, Fujihara Y, Ikawa M (2014) Feasibility for a large scale mouse mutagenesis by injecting CRISPR/Cas plasmid into zygotes. Develop Growth Differ 56:122–129
13. Hsu PD, Lander ES and Zhang F (2014) Development and applications of CRISPR-Cas9 for genome engineering. Cell 157:1262–1278
14. Nagy A, Gertsenstein M, Vintersten K, Behringer R (2003) Manipulating the mouse embryo: a laboratory manual. In: Cold Spring Harbor. Cold Spring Harbor Laboratory Press, NY
15. Brinster RL, Chen HY, Trumbauer ME, Yagle MK, Palmiter RD (1985) Factors affecting the efficiency of introducing foreign DNA into mice by microinjecting eggs. Proc Natl Acad Sci U S A 82:4438–4442
16. Auerbach AB, Norinsky R, Ho W, Losos K, Guo Q, Chatterjee S, Joyner AL (2003) Strain-dependent differences in the efficiency of transgenic mouse production. Transgenic Res 12:59–69
17. Pu X, Lu M, Brower J, Kubisch HM, Young AP (2008) The effects of recipient status on

transgenic mouse production. Transgenic Res 17(5):1018–1019

18. Luo C, Juliana Zuñiga J, Edison E, Palla S, Dong W, Parker-Thornburg J (2011) Superovulation strategies for 6 commonly used mouse strains. J Am Assoc Lab Anim Sci 50:471–478
19. Whitten WK (1958) Modification of the oestrous cycle of the mouse by external stimuli associated with the male; changes in the oestrous cycle determined by vaginal smears. J Endocrinol 17(3):307–313 [PMID: 13587836]
20. Norton WB, Scavizzi F, Smith CN, Dong W, Raspa M, Parker-Thornburg JV (2016) Refinements for embryo implantation surgery in the mouse: comparison of injectable and inhalant anesthesias – tribromoethanol, ketamine and isoflurane – on pregnancy and pup survival. Lab Anim 50(5):335–343

Chapter 3

Transgene Recombineering in Bacterial Artificial Chromosomes

Michael G. Zeidler and Thomas L. Saunders

Abstract

Bacterial Artificial Chromosome (BAC) libraries are a valuable research resource. Any one of the clones in these libraries can carry hundreds of thousands of base pairs of genetic information. Often the entire coding sequence and significant upstream and downstream regions, including regulatory elements, can be found in a single BAC clone. BACs can be put to many uses, such as to study the function of human genes in knockout mice, to drive reporter gene expression in transgenic animals, and for gene discovery. In order to use BACs for experimental purposes it is often desirable to genetically modify them by introducing reporter elements or heterologous cDNA sequences. It is not feasible to use conventional DNA cloning approaches to modify BACs due to their size and complexity, thus a specialized field "recombineering" has developed to modify BAC clones through the use of homologous recombination in bacteria with short homology regions. Genetically engineered BACs can then be used in cell culture, mouse, or rat models to study cancer, neurology, and genetics.

Key words BAC clone, BAC transgenic, BAC transgenesis, Recombineering, Transgenic mice

1 Introduction

The value of bacterial artificial chromosomes (BAC) and other large transgenes is that they include all of the genetic information necessary to recapitulate endogenous expression patterns of genes contained within them [1, 2]. Genomic DNA contained within BAC clones is used as a large promoter to drive expression of the reporter of interest, most often a fluorescent protein or recombinase, in a tissue-specific fashion. The average size of genomic inserts in human and mouse BAC libraries ranges up to 200 kb [3, 4], although longer clones up to and beyond 300 kb can be identified [5]. BAC clones used to generate transgenic mice contain distant regulatory elements required for expression simply because they include more DNA sequence than phage or cosmid library clones [6–8]. Unmodified BACs have been used as transgenes to discover genes responsible for spontaneous mouse mutations and to correct

Chengyu Liu and Yubin Du (eds.), *Microinjection: Methods and Protocols*, Methods in Molecular Biology, vol. 1874, https://doi.org/10.1007/978-1-4939-8831-0_3, © Springer Science+Business Media, LLC, part of Springer Nature 2019

mutations [9–12]. The earliest uses of BAC transgenes were complementation studies to map genes in spontaneous mouse mutants. BAC transgenic mice successfully rescued circadian rhythm and deafness mutations and accelerated the discovery of the genes responsible for these phenotypes [12, 13].

A powerful application of BAC transgenesis is the combination of knockout mice with human BAC transgenes that permit the study and function of human gene expression in animal models that lack the endogenous gene. In these "humanized" transgenic mice human gene expression replaces and compensates for the lack of mouse gene expression [14–16]. Prior to the use of BAC transgenes, the production of mouse models demonstrating neuron-specific transgene expression in the central nervous system was problematic. Access to BAC transgenes and the transgenic mouse and rat models generated from them made it possible to document gene expression patterns in the central nervous system when conventional plasmid transgenes based on limited genomic information did not confer endogenous expression patterns of reporter genes [17–21]. Other studies have further demonstrated that BAC transgenes can control cell specific and temporal regulation of genes by introducing DNA cassettes that place fluorescent proteins, recombinases, or genetic elements into the genomic context contained in a BAC DNA molecule [9, 22–26]. BAC transgenes can also be used to develop cell culture models for study [27, 28]. On those occasions when the gene of interest is so large that it cannot be contained within a single BAC, alternative strategies include the use of genomic clones from yeast artificial chromosome libraries with average sizes of 430 kb [29] to produce transgenic mice [30]. Alternatively BAC can be linked to produce larger genomic clones for transgenesis [31, 32].

We present a streamlined approach to produce recombineered BAC clones for use as transgenes. Because of their large size, it is not feasible to introduce genetic modifications in BACs by conventional methods such as restriction enzymes and ligase. DNA purified from BAC clones grown in bacteria is sensitive to mechanical shearing in vitro, thus it is important to handle BAC DNA carefully. Specialized buffers have been employed to maintain the integrity of purified BAC DNA [5, 33, 34]. Current methods to modify BACs by "recombineering" techniques are based on the *E. coli* bacteriophage lambda Red system [35–37]. Recombineering makes it possible to make precise integrations of sequences such as fluorescent proteins, recombinases, and exogenous genetic regulatory elements, or other reporter genes. The transgenes derived from such genetically engineered BACs have been used to generate transgenic animal models to advance research in a number of fields. Perhaps the greatest impact of BAC recombineering in the field of transgenic science has been its application in functional genomics. Recombineering has been used to produce all the gene-targeting vectors designed to mutate every gene in the mouse genome for the International Knockout Mouse Consortium library of mouse embryonic stem (ES) cell clones [38–42].

The use of pronuclear microinjection to produce mouse models is direct and rapid. When the goal is to study the interaction of genes carried by the BAC with existing mutant genes or other genetic modifications in mouse strains, progress can be accelerated by preparing fertilized eggs for microinjection from mice that carry the mutant or targeted gene [12, 13, 43]. BAC transgenesis amplifies the power of transgenic technology for the identification of disease causing genes and generation of animal models of disease for biomedical research. We describe here a straightforward approach to BAC recombineering for the production of BAC transgenes in the production of transgenic mice or rats. The first step is to identify a BAC clone for the gene of interest. This information is readily available in genome browsers such as Ensembl.org. BAC clones are available from public repositories and commercial sources. After a BAC has been obtained, genomic DNA from the BAC is restriction mapped and compared to its expected map then the cassettes for insertion into the BAC are amplified by PCR for insertion. Drug selection cassettes are removed when BACs are used as transgenes in pronuclear microinjection or can remain in the BAC for use in cell culture models. After BACs pass quality control tests for correct insertion of cassettes they are ready to use as transgenes. Key parameters for successful transgenic production with recombineered BAC DNA include the purity and integrity, optimal microinjection DNA concentration, and robust detection of transgene integration.

2 Materials

2.1 Equipment

1. Computer with internet connection.
2. DNASTAR Lasergene or other DNA sequence analysis software.
3. Refrigerator (4 °C).
4. Freezer (−80 °C).
5. Laboratory Balance such as Mettler Toledo, Thermo Fisher.
6. Bacteria incubator (set to 37 °C).
7. Nanodrop 2000 Spectrophotometer.
8. Shaking incubator for flasks (set to 37 °C).
9. Shaking incubator for tubes such as Eppendorf Thermomixer.
10. Refrigerated floor centrifuge such as Sorvall RC 6 Plus (4 °C).
11. Bio-Rad CHEF-DR III Pulsed Field Gel Electrophoresis (PFGE) system.
12. Electroporator such as Bio-Rad Gene Pulser XCell.

2.2 Reagents

1. Anotop 10 Syringe Filters, sterile 0.02 μm pore size.
2. Nucleobond BAC DNA purification kit.
3. Nucleobond BAC DNA buffer kit (optional: Buffers S1, S2, S3).
4. 24 cm Fluted Qualitative Filter Paper Circles.
5. QIAquick PCR purification kit.
6. QIAquick gel extraction kit.
7. Proofreading Taq DNA polymerase such as Expand™ High Fidelity PCR System.
8. Plasmid pE-FLP (Addgene.com, plasmid 45978).
9. Plasmid pRed (Genebridges.com).
10. PIR1 *E. coli* (*see* **Note 1**).
11. Bromphenol blue.
12. Mouse embryo tested water (*see* **Note 1**).
13. EDTA, 0.5 M solution.
14. Ethanol, anhydrous.
15. Ethidium bromide.
16. Glycerol.
17. Isopropanol.
18. Sucrose.
19. NaCl, 5 M solution.
20. 0.5 M EDTA, 0.5 M, solution.
21. Ice-cold sterile H_2O.
22. Spermine tetrahydrochloride.
23. Spermidine trihydrocholoride.
24. Tris–HCl, pH 7.5, 1 M solution.
25. 10× Tris Borate EDTA Buffer (TBE), 1 M tris base, 1 M boric acid, 0.02 M EDTA.
26. MidRange Pulsed Field Gel (PFG) Marker.
27. Lennox L Broth Base (LB broth base).
28. Lennox L agar powder (LB agar powder).
29. Sterile pipet tips for microliter pipetters.
30. 50 ml polypropylene conical tubes, sterile.
31. 14 ml Falcon polypropylene round bottom tubes, sterile.
32. 1.5 ml microcentrifuge tubes, sterile.
33. 0.2 cm gap electroporation cuvettes, sterile.
34. 2.0 ml internal threaded cryogenic vials with silicone washer caps, sterile.

35. Thin-wall single-barrel standard borosilicate glass capillaries with internal glass filament, 4″ long (*see* **Note 1**).
36. 10% arabinose in water.
37. Chloramphenicol (cam), 2000× stock solution, 30 mg/ml dissolved in ethanol.
38. Chloramphenicol Working solution, 15 mg/ml dissolved in ethanol.
39. Kanamycin (kan).
40. Tetracycline (tet).
41. Ampicillin (amp).
42. LB-cam, 15 g cam/ml LB.
43. LB-cam, 15 g cam/ml agar plates.
44. LB-cam/tet, 15 g cam/ml, and 3 μg tet/ml agar plates.
45. LB-cam/kan, 15 μg cam/ml, and 25 μg kan/ml agar plates.
46. LB-cam/kan/amp: 15 μg cam /ml, 25 μg kan/ml, and 50 μg amp/ml agar plates.

2.3 Buffers and Other Solutions

1. 0.5× TBE: add 100 ml 10× TBE to 1900 ml H_2O.
2. 10% glycerol: Mix 10 ml glycerol and 90 ml H_2O. Autoclave. Store at room temperature.
3. 6× DNA loading buffer: 40% sucrose, 0.25% Bromphenol blue, dissolve 25 mg Bromphenol blue and 4 g sucrose in 8 ml water, adjust final volume to 10 ml with water.
4. 70% ethanol: mix 30 ml H_2O and 70 ml ethanol. Store at room temperature.

   10,000× ethidium bromide: dissolve 10 mg EtBr/ml in H_2O, store at room temperature, in glass bottle protected from light.
5. 2000× spermine stock solution: Dissolve 1 g spermine in 47.9 ml mouse embryo tested water (60 μM).
6. 2000× spermidine stock solution: dissolve 1 g spermidine in 28.1 ml mouse embryo tested water (140 μM).
7. 1000× PA Stock solution: Mix together 25 ml of spermine and 25 ml of spermidine to produce 50 ml of 1000× PA Stock, 30 μM spermine, 70 μM spermidine. Filter sterilize (Anotop 0.02 μM filters) and store 1 ml aliquots in sterile tubes at −80 °C (*see* **Note 2**).
8. PA Microinjection Buffer: 10 mM Tris–HCl, pH 7.5, 0.1 mM EDTA, pH 8.0, 100 mM NaCl, 30 mM spermine, 70 mM spermidine. To prepare 10 ml of polyamine microinjection buffer: mix together 0.1 ml of 1 M Tris–HCl, pH 7.5, 0.02 ml of 0.5 M EDTA, 0.2 ml of 5 M NaCl, and 0.01 ml of 1000× PA stock. Filter sterilize with 0.02 μM Anotop syringe filter and store at 4 °C.

9. 5× combinatorial PCR enhancer solution (CES). Mix together 540 μl 2.7 M betaine, 200 μl 6.7 mM dithiothreitol, 1 μl 55 μg BSA/ml, 67 μl 6.7% dimethyl sulfoxide, and 192 μl distilled H_2O.

3 Methods

3.1 Identify a BAC with the Gene of Interest

1. Open the Ensembl Genome Browser: http://useast.ensembl.org/index.html in your web browser. Set browser preferences to allow popup windows (*see* **Notes 3** and **4**).
2. Click on "Human" or in the pull down menu of the search box, choose the desired genome database, e.g., "Human." To search for your gene of interest enter the name of your gene in the search box and click "go". (Fig. 1).
3. If your gene is not found by Ensembl, it may be that the search term is an unrecognized synonym for the gene of interest. If you are not certain of the correct gene nomenclature, use the HUGO Gene Nomenclature Committee website to identify the correct gene name http://www.genenames.org (*see* **Note 5**).
4. The next web page shows the Results Summary (Fig. 2). Click on the gene of interest shown on the fifth line of results: Amelx (Mouse Gene, Strain: reference (C57BL/6).
5. Next, a summary of your gene of interest is displayed. To find a BAC covering that gene, click on the blue tab "Location". A figure of the genomic contig centered on your gene of interest will appear. At the top of the following page, you'll see a diagram of the chromosome. (Fig. 3) Your gene of interest is indicated within a red box in the panel labeled "Region in detail."
6. On the left panel, select "Configure this Page." In the pop-up window under the "Sequence and assembly," choose "Clones & misc. regions," select a BAC library of your choice, usually "RP23 mouse clones," "RP24 mouse clones," and "Tilepath." Click on the "Save and Close" checkmark at the top right of the popup window (Fig. 4, *see* **Note 6**).
7. After the page refreshes BACs will be listed beneath the "Region in Detail." It may be necessary to zoom out the display to view BACs from 5′ to 3′ ends in relation to the contig. Depending on the size of the gene of interest, it may be completely contained within a BAC (Fig. 5).
8. Choose a BAC for use as a transgene. Ideally, the BAC will contain the gene of interest in the center with as much 5′ and 3′ genomic sequence as possible. This will increase the probability that the gene's endogenous expression pattern and developmental timing will be reproduced in transgenic mice.

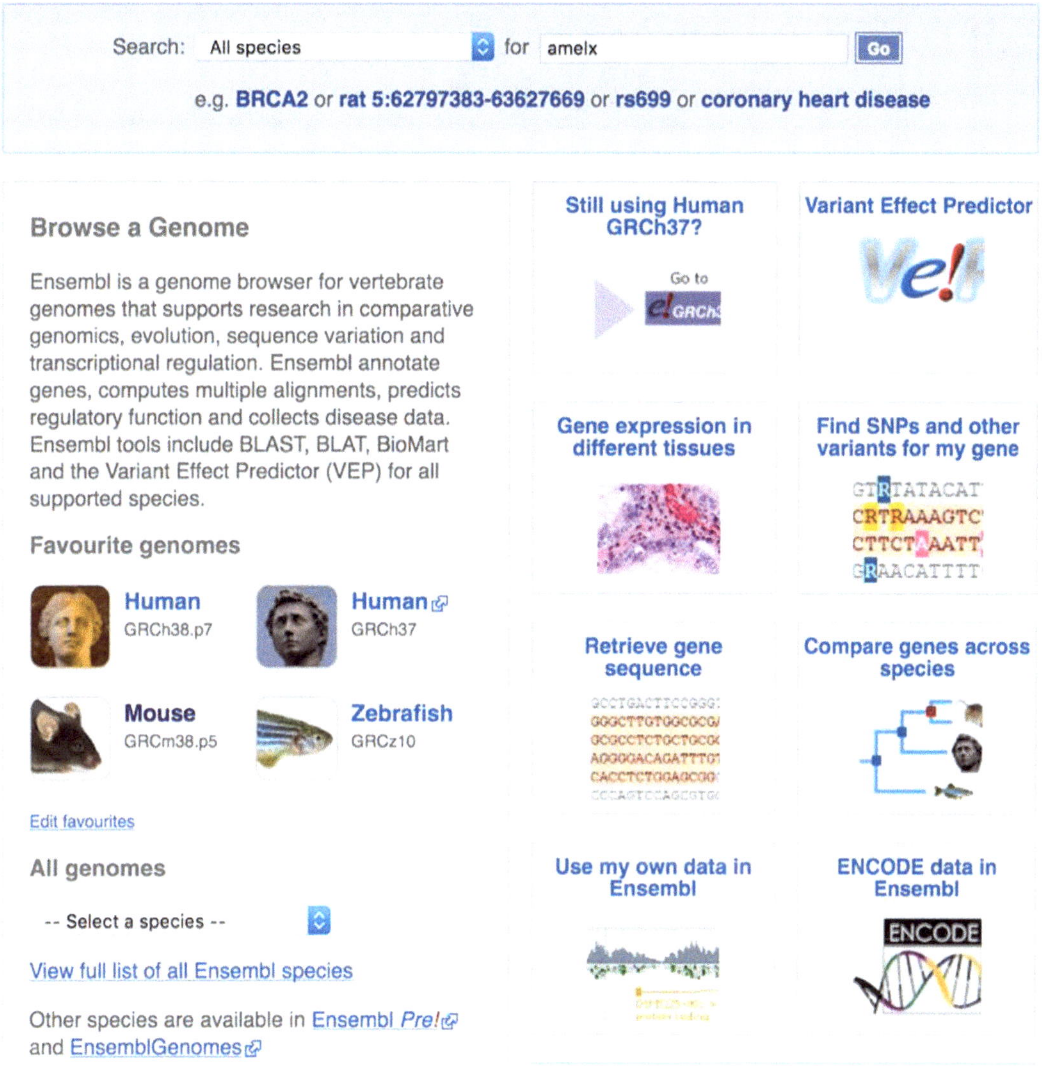

Fig. 1 Ensembl's homepage at http:/useast.ensembl.org/index.html

3.2 Importing BAC DNA Sequence Data

1. Continue from Subheading 3.1, **step 8**. Note the chromosome number that carries the gene of interest. Click on your BAC of interest. The name, genome coordinates of the BAC, length and the strain that codes for the gene will show up in a popup window (e.g., RP23-334F21: 169163162-169258987, 95826 bps, RP23-334F21).
2. Click on "Center on clone." When the window refreshes the display will correspond to the DNA sequence in the BAC.
3. Select "export data" on the left side of the screen. Ensembl exports BAC DNA sequences in various formats that include data sets such as gene information, repeat features, marker features, etc. DNA analysis software such as SeqBuilder

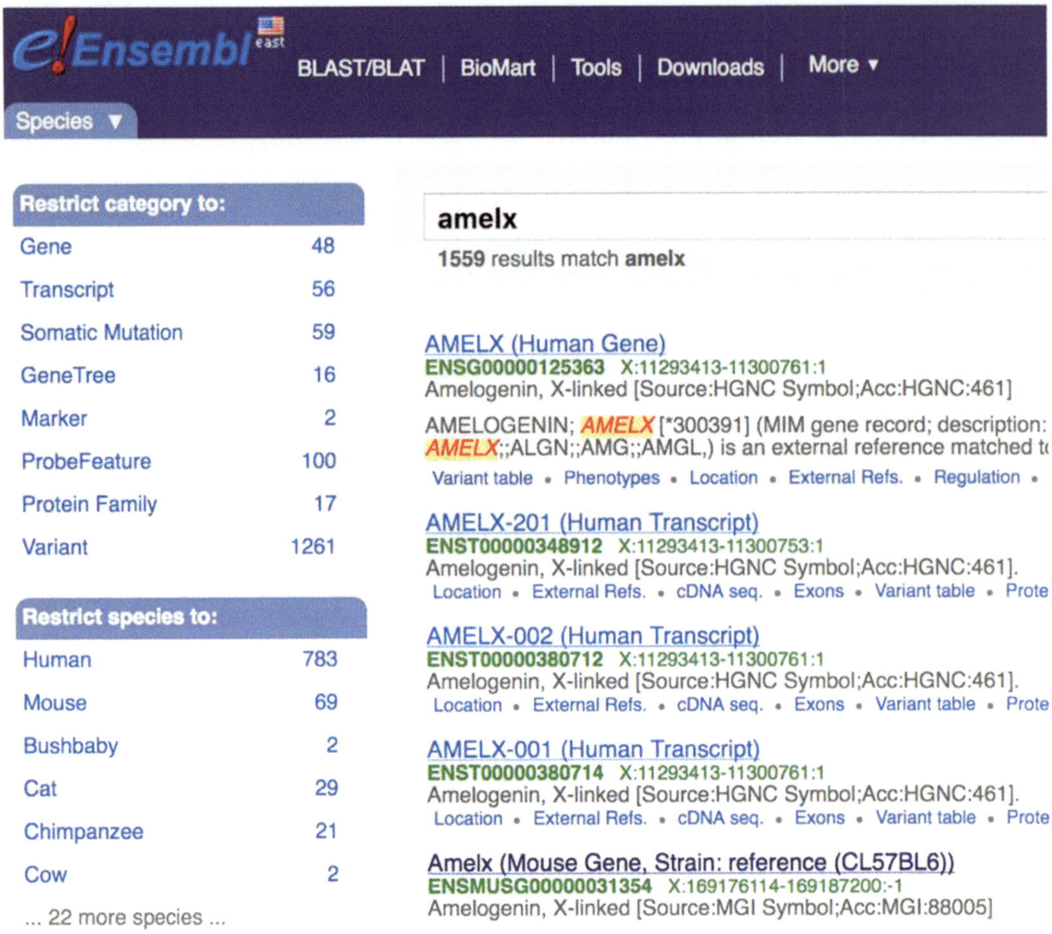

Fig. 2 Example of an Ensemble results page. Note that the mouse gene for Amelx is the fifth item on the list of results

interprets data sets and displays them as the sequence is viewed and manipulated.

4. In the popup window, decide on a format and on which features to include. To work with SeqBuilder, choose EMBL. Use the "Output:" pulldown menu to select EMBL flat file format for the output. Decide on which options you would like to have displayed in your software then click the "Next" button. The next window will allow you to choose HTML, text, or compressed text output files. Click on the "Text" hyperlink (*see* **Note** 7).
5. A new window opens with the exported file in text format. Use the browser to "Select all," paste the selection into a new document in a text editor, and save the document as an unformatted text file.

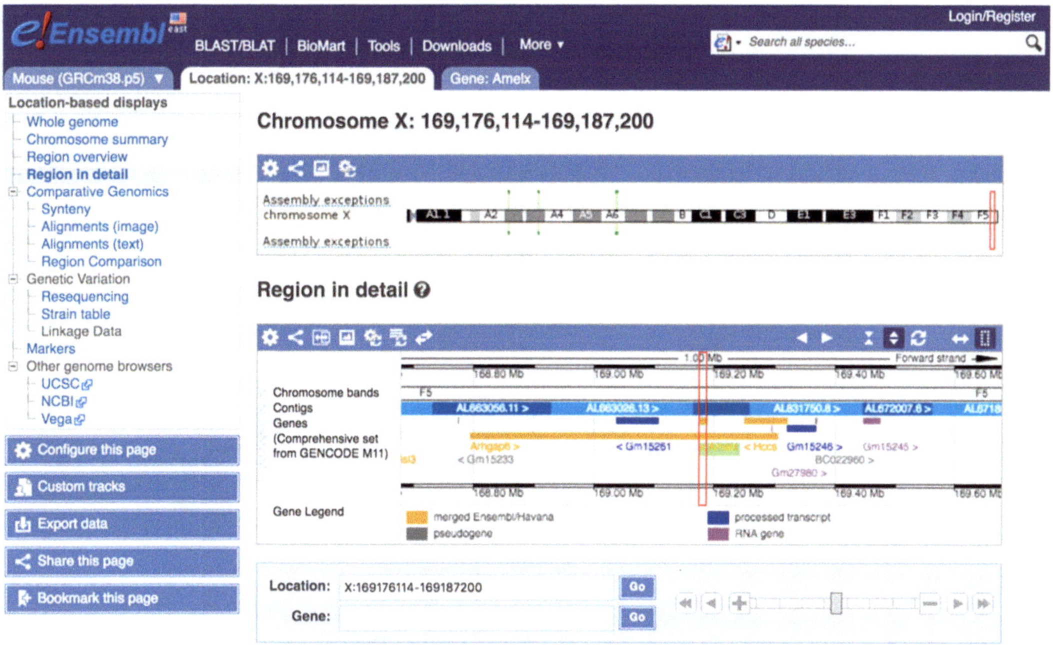

Fig. 3 Ensembl display of the chromosome and the region of detail of the chosen gene (red vertical box)

6. In SeqBuilder or other DNA sequence editing software package, use the "Open" command in the "File" menu to import the text file.
7. Verify the presence of the gene of interest in the sequence to confirm that it is contained in the selected BAC.
8. Add the sequence of the pBACe3.6 vector (NCBI Genbank accession U80929.2) to RPCI-23 library clones to complete the sequence file. In the RPCI-23 library genomic DNA was cloned between the Eco RI sites at positions 10 and 2801 (*see* **Note 8**).
9. Add the sequence of the pTARBAC1 vector (NCBI accession AY487252) to RPCI-24 library clones to complete the sequence file. In the RPCI-24 library genomic DNA was cloned between the Bam HI sites at position 1 and 2810.
10. To preserve changes or edits, save the file as a SeqBuilder file.

3.3 Order the BAC Clone

1. Go to https://bacpacresources.org (*see* **Note 9**).
2. Enter the name of the BAC of choice into the search box. Click the "Search" button.
3. If the BAC clone identity is verified the "Clone Ordering Page" will open. Follow the instructions to order the clone containing the gene of interest.

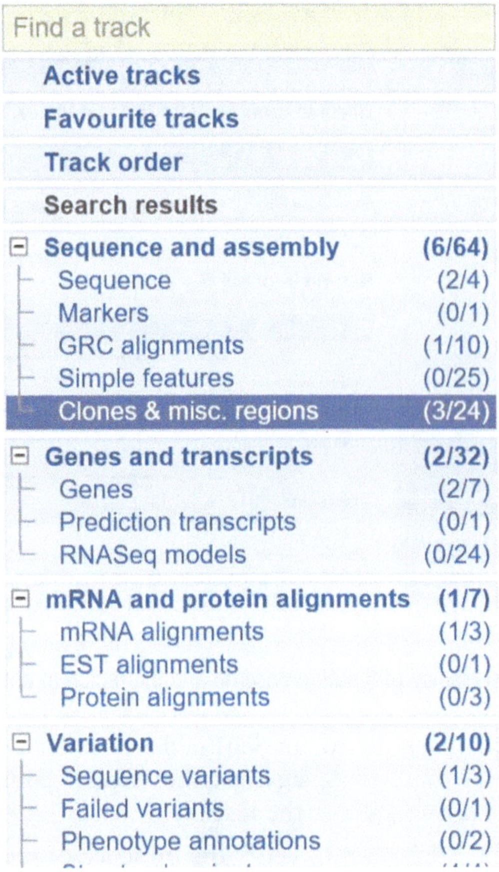

Fig. 4 Customize tracks for display by Ensembl. Choose "Clones & misc regions" then select BAC libraries to display BAC clones mapped to the genome

4. The BAC arrives in DH10B *E. coli* as a stab in LB-agar, which can be stored for a few weeks at 4 °C.
5. The first thing to do after the LB stab arrives is to freeze down an aliquot of an overnight culture in 20% glycerol at −80 °C.

3.4 BAC DNA Purification

BACs are large circular DNA molecules (average sizes exceed 150 kb) and are prone to fragmentation if handled improperly. BAC cloning vector backbones confer chloramphenicol resistance. They replicate as single copies in bacteria, thus BAC DNA yields are always lower than yields of high copy plasmids from equal volumes of bacterial culture. Always store BAC DNA in polyamine (PA) buffer at 4 °C. To prevent sheared DNA never freeze, boil, or vortex BAC DNA. Use only wide bore pipet tips and pipet gently. Kits for BAC DNA purification are available from Qiagen, Macherey-Nagel (distributed by Clontech), and Marligen. Each of the kits follows a similar protocol and gives similar results. We obtain better DNA yields from Nucleobond kits with the modified protocol below.

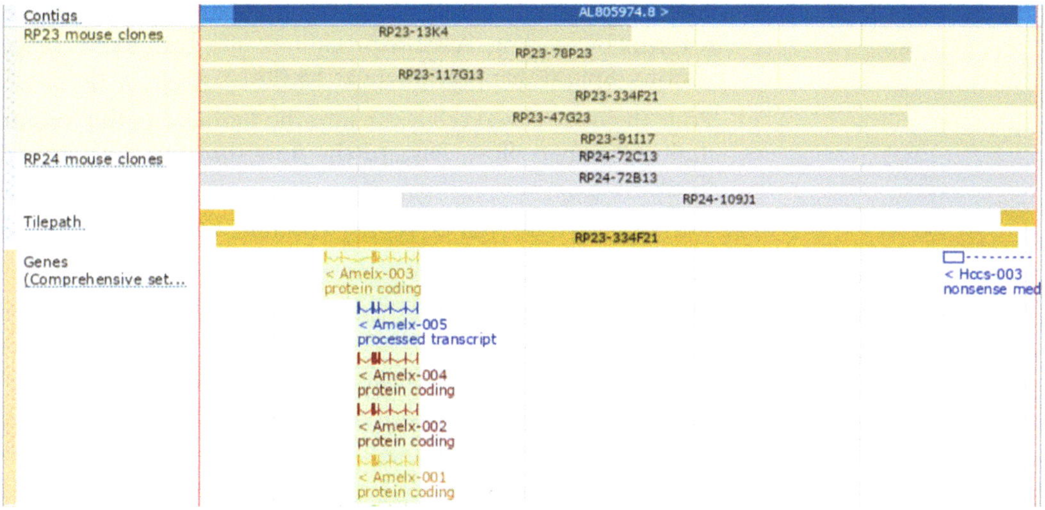

Fig. 5 Ensembl display showing the gene of interest mapped to a tilepath BAC (gold bar) and other BACs that map to the genomic region

1. Grow BAC overnight in 250 ml LB containing chloramphenicol (15 μg/ml) at 37 °C in a shaking incubator (250 rpm).
2. Cool centrifuges and rotors to 4 °C.
3. Add RNAse (kit reagent) to bacterial Cell Suspension Buffer (kit reagent).
4. Place Whatman filter into small funnel. Insert small funnel in Nucleobond BAC 100 column. Equilibrate filter and column by adding 10 ml equilibration buffer. Let it drain by gravity flow. Macherey-Nagel does not supply the Whatman filter. The filter is used to remove precipitates from the bacterial lysate and prevents columns from becoming clogged.
5. Take a 300 μl aliquot of bacteria culture, add 300 μl of 40% glycerol and store the aliquot at −80 °C.
6. Spin down the remaining bacteria for 10 min at 6000 × *g* at 4 °C. Discard the supernatant.
7. Resuspend cell pellet in 30 ml cell suspension buffer per 250 ml of cells grown, by gently swirling the centrifuge bottle. If larger volumes of bacterial culture have been used, increase buffer proportionately.
8. Be generous with kit buffer volumes. Additional buffers can be purchased from Macherey-Nagel. Marligen, Qiagen, and Nucleobond buffers are interchangeable. If necessary, matching buffers from the Qiagen Plasmid Maxi Kit can be substituted for Macherey-Nagel buffers.
9. Add an equal amount of cell lysis solution. Mix gently. Do not vortex. Keep at room temperature for 5 min.

10. Add an equal amount of neutralization solution. Do not vortex. Mix by swirling. A white precipitate will form. It should be viscous. Spin down precipitate at 15,000 × *g* for 10 min.
11. Pour the neutralized cell lysate into the funnel. Let the column drain by gravity flow. Discard the flow through. Discard the filters.
12. Add 30 ml wash buffer to the columns and let it drain by gravity flow.
13. Add 15 ml elution buffer to the column and let it drain by gravity flow.
14. Add 11 ml Isopropanol to eluate.
15. Spin 40 min at 10,000 × *g* at 4 °C (50 ml Falcon tubes collapse above 10,000 × *g*).
16. A fragile glossy to white pellet will form. It will be small, sometimes hard to see. Be careful when handling the tube, since the BAC detaches easily from the tube wall. If you see a big white pellet, something went wrong with the purification (e.g., RNAse was not added to the cell suspension buffer or has been degraded).
17. Resuspend the pellet in 3 ml 70% ethanol and divide it into two microcentrifuge tubes. Spin 10 min at high speed at 4 °C.
18. Dry the pellets 3 min. Do not let them dry completely or shearing will occur.
19. Pre-filter 1 ml of PA buffer with a 0.02 μM Anotop syringe filter. Do not use these filters with resuspended BAC DNA, the BACs will be trapped in the filters and removed from solution.
20. Resuspend the BAC DNA in 300 μl pre-filtered PA buffer. BACs go easily into solution with PA buffer. If some precipitate won't dissolve, it most likely is not DNA but waste material carried over from an improperly executed BAC DNA purification procedure.
21. Quantitate BAC DNA. For pronuclear microinjection adjust the concentration with PA buffer to 0.5–1.0 ng/μl [5].

3.5 BAC DNA Quantitation

Measure BAC DNA concentration with a NanoDrop spectrophotometer. BAC DNA concentrations are typically 50 μg-200 μg/μl. Significantly higher concentrations typically result from co-purification of RNA when RNAse is omitted from the cell suspension buffer, or has lost activity.

1. Open NanoDrop software and choose DNA measurement.
2. Raise the NanoDrop lever arm and pipet 1 μl water onto the pedestal in response to the dialogue box. Click "OK" to initiate calibration.

3. Clean both optical surfaces of the pedestal with a KimWipe.
4. Pipet 1 μl of PA buffer onto the pedestal and click "Blank" to zero the instrument.
5. Clean the pedestal and add 1 μl of BAC DNA solution and click "Measure."
6. Clean the pedestal with a KimWipe between samples and pipet 1 μl water as the last sample.
7. Save data in computer, clean the pedestal, and quit the software program.

3.6 BAC DNA Restriction Mapping

Before working with a BAC it is mandatory to check its integrity on a pulsed field gel. When a BAC shears, its products will still be tens of thousands of base pairs long. Standard agarose gel electrophoresis conditions do not have the necessary resolving power to evaluate BAC DNA integrity.

1. Use the *in silico* BAC sequence file assembled in Subheading 3.2 to choose enzymes that cut two to four times in the BAC. The cuts should result in fragments of different lengths not smaller than 8 kb.
2. Set up restriction enzyme digests with 250–500 ng BAC DNA.
3. Set up the PFGE apparatus in a cold room or attach a circulating cooling bath.
4. Pour a 0.8% agarose gel, 0.5× TBE. Make sure no bubbles disturb the gel. If imperfections are present in the gel, do not use it. Every bubble will disturb the migration of BAC DNA. Do not add ethidium bromide to the gel. Ethidium bromide will diffuse into the buffer and need to be disposed of as hazardous waste.
5. Fill the gel apparatus with 2000 ml 0.5× TBE.
6. The Pulsed Field Standards (BioRad) come as agarose blocks. Use a razor blade to cut a thin slice off the marker and load it onto the gel.
7. There is no need to prepare agarose blocks to run your sample. Load samples with 6× DNA loading dye as you would for a standard horizontal agarose gel electrophoresis. Loading liquid BAC DNA samples in 6× DNA loading buffer is as effective as loading BAC DNA in agarose blocks.
8. Load between 250 ng and 500 ng restriction enzyme digested BAC DNA per lane into the gel wells.
9. Load one lane with undigested BAC DNA as a control.
10. Enter settings for the pulsed field gel: initial switch time: 1 s, final switch time: 25 s, run time: 15 h, 6 V, angle: 120°.

11. The actual current should be around 135 A–150 A when the gel run begins. It will go up overnight. If it is much higher, check the TBE buffer concentration. If the current is too high smeared bands will result.
12. After the run, soak gel in an ethidium bromide bath for 30 min. Visualize bands with UV light. Soaking the gel in an ethidium bromide bath after electrophoresis reduces waste. The bath can be reused, which leaves the gel as the only hazardous waste product of the procedure. Ethidium bromide is a known mutagen. Observe safe handling and disposal procedures of ethidium bromide waste.
13. The markers should produce discrete bands of predicted molecular weight. Overloaded marker DNA will produce smeared bands in marker lanes.
14. Load different amounts of BAC DNA to make sure that bands will be clear. For instance, in case of RNA contamination the NanoDrop reading will show a high DNA concentration. In this case not all of the sample consists of BAC DNA and DNA band intensities will appear low. Contaminating RNA will migrate off the gel and won't interfere with visualization of BAC DNA fragments.
15. Compare the expected restriction enzyme pattern to the in silico restriction analysis. If they do not match, either the wrong BAC was received or the Ensembl BAC sequence is incorrect. BPRC will send the same BAC again for free if the restriction map is wrong. It is recommended that a different BAC be ordered if the in-silico map does not match the actual map.
16. BAC DNA samples can be stored for 2 years and longer at 4 °C in PA buffer without degradation [5].

3.7 Recombination Competent BAC Clones in *DH10B* E. coli

BAC-Recombination relies on homologous recombination in vivo in *E. coli*. Cells that are rendered transiently recombination competent are used to mediate integration of a PCR product into a BAC through homologous recombination. Recombination competency is conferred by transiently expressing the recombineering proteins (either RecE/RecT or Redα/Redβ) from a plasmid [40] under an arabinose inducible promoter. A second method is to express recombineering proteins from cells that are engineered to carry them on their bacterial chromosome under the control of a heat sensitive promoter [44]. We prefer the first method as it keeps the BAC in the same host throughout the procedure. The second method requires transfer of the BAC into different host cells after the procedure to prevent future recombination events.

Recombination competency is conferred through a plasmid carrying the Redα and Redβ genes (pRed). RepA protein encoded by pRed is needed for plasmid propagation. RepA is temperature-sensitive and lost above 30 °C. By culturing cells at 37 °C, cells cease to express pRed encoded proteins. Redα and Redβ are under control of an arabinose inducible promoter (pBAD). The plasmid confers tetracycline resistance.

1. Grow a starter culture of the BAC clone to be recombineered in 5 ml LB-cam overnight.
2. The next day, inoculate 10 ml of LB-cam with starter culture. Grow at 37 °C to an OD600 of 0.1.
3. Cool down the centrifuge. Cool H_2O on ice. Cool 20% glycerol on ice.
4. When OD600 is ~0.5, spin down culture 5 min @ 5000 rcf.
5. Wash pellet in ice-cold H_2O.
6. Spin down 5 min @ 5000 rcf.
7. Wash pellet in ice-cold H_2O.
8. Spin down 5 min @ 5000 rcf.
9. Wash pellet in ice-cold 20% glycerol.
10. Spin down 5 min @ 5000 rcf.
11. Take up electrocompetent cells in 200 μl ice-cold 20% glycerol. Distribute into 4× 50 μl aliquots. Use 1 aliquot for the transformation, freeze the rest at −80°C.
12. Mix 10 ng of pRed into 1 aliquot of electrocompetent cells.
13. Pipet mixture into an ice-cold BioRad cuvette (2 mm gap between electrodes).
14. Electroporate (BioRad Gene Pulser XCell, 2.5 kV, 25 μF, 200 Ω).
15. Use 1 ml LB medium to immediately flush the cells out of the cuvette.
16. Grow cells in LB 1 h @ 37°C.
17. Plate 50ul of electroporated cells on LB-cam/tet agar plates.
18. Grow @ 30 °C until colonies are visible (usually 30–48 h).
19. Pick 1 colony into 5 ml LB-cam/tet.
20. Grow overnight @ 30 °C.
21. The next day, inoculate two 20 ml LB-cam with overnight culture (OD_{600} ~0.1) Label one tube "+" and the other one "−."
22. Grow at 30 °C until $OD_{600} = 0.3$.
23. Add 750 μl 10% arabinose to "+" culture. Leave "−" culture alone. This will be your negative control.
24. Increase temperature to 37 °C. Grow cells 1 h.

25. Spin down culture 5 min @ 5000 rcf.
26. Wash pellet in ice-cold H_2O.
27. Spin down 5 min @ 5000 rcf.
28. Wash pellet in ice-cold H_2O.
29. Spin down 5 min @ 5000 rcf.
30. Wash pellet in ice-cold 20% glycerol.
31. Spin down 5 min @ 5000 rcf.
32. Pick up pellet in 200 μl ice-cold 20% glycerol.
33. Distribute into 4× 50 μl aliquots.
34. Freeze at −80 °C until recombination.

3.8 General Considerations for PCR Amplification of Recombination Donor DNA

The recombination donor DNA can be produced through various methods: it can be excised from a plasmid, it can be synthesized, or it can be PCR-amplified from a plasmid. We prefer the PCR-based approach (Fig. 6). PCR products can be easily purified and contaminating plasmid PCR template can be easily separated from the amplicon. Ideal characteristics for a recombineering cassette are (1) short length, (2) plasmid with R6K *ori*, and (3) a drug selection cassette that can be removed by recombinase. Reporters are subcloned into empty vectors that include an R6K ori, and FRT flanked

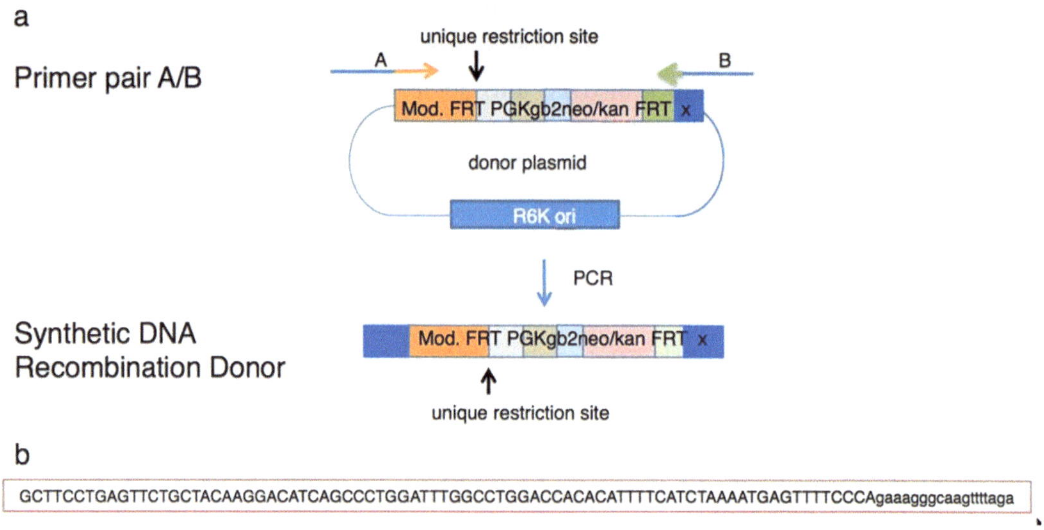

Fig. 6 Essential design elements required for the recombination donor PCR. (**a**) Amplification of the recombination donor for homologous recombination with the BAC. orange: desired modification (e.g., GFP, ERT2iCre, point mutation). FRT-PGK-gb2-neo/kan-FRT: selection cassette to determine which BAC underwent recombination. X: vector region for primer alignment. A/B: primer pair. Orange/green: this part aligns to the recombineering cassette on the donor plasmid. Blue: this part is homologous to the BAC. unique restriction site: a restriction endonuclease recognition site not present on the BAC. This information can also be included when the primers are synthesized. (**b**) Example of a recombination primer. Lower case: this region aligns to the donor plasmid. Upper case: this region is homologous to the BAC clone

kan^r cassette. EGFP is preferred to the use of lacZ since the former is less than 1 kb including a polyadenylation signal such as the bovine growth hormone termination sequence, whereas lacZ is more than 3 kb in length. The shorter EGFP reporter has clear advantages in the ease of PCR reactions to produce a recombineering product. It also affords the possibility of using a single PCR reaction with both a downstream and an upstream primer that lie outside of the homology regions for DNA sequence confirmation whereas the confirmation of a correct lacZ insertion would require three or more primer pairs.

Plasmids with an R6K origin of replication as PCR template can only be propagated in PIR1 *E. coli* cells. Carry over of an R6K *ori* plasmid template from a PCR recombination product will not confer antibiotic sensitivity to DH10B *E. coli*. Background colonies are dramatically reduced after transformation and plating of electrocompetent bacteria (*see* **Note 10**). The use of FRT flanked drug selection cassettes and the subsequent excision of the cassette by FLP recombinase is important to prevent the possible interference with expression of the reporter introduced by recombineering. Ideally, the donor plasmid features a unique restriction site, which can be used later to identify correctly targeted clones on a pulsed field gel. We usually use NotI for this purpose. The cloning vectors used for RP23 and RP24 BACs feature 2 NotI sites. The inclusion of a third NotI site in the recombineering cassette or in one of the primers used to amplify the recombination cassette will allow the easy identification of modified and original BAC clones by restriction enzyme mapping with NotI.

The recombineering cassette is produced with primers between 60 and 120 bp in length. Each primer features a region of 20 bp at the 5′ end, which anneals to the donor plasmid. The 40–100 bp sequence at the 3′ end is homologous to genomic DNA in the BAC clone and does not anneal to the donor plasmid. The donor plasmid features a FRT-kan-FRT selection cassette. It is important that the primer does not anneal with the FRT cassette to prevent multiple primer binding events on the template. Designing the primer to align 20 bp 3′ to the FRT site will result in clean PCR product (Fig. 6). If the donor plasmid to be used as the PCR template does not feature a unique restriction enzyme site, this can be added to the primer and included during primer synthesis (*see* **Notes 11** and **12**). The PCR product is either gel purified (QIAquick gel extraction kit) or column purified (QIAquick PCR purification kit). Between 200 ng and 600 ng of the PCR product is used for recombination.

3.9 Recombination

1. On ice, add 200–600 ng of recombination PCR product to an aliquot of recombination competent cells. Add the same amount to the control (un-induced) cells (*see* **Note 13**).

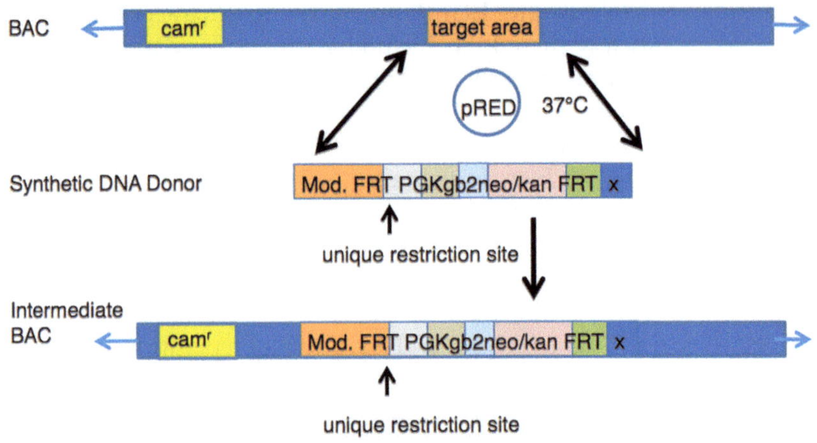

Fig. 7 The BAC clone and a synthetic donor DNA are combined in recombineering competent bacteria. Correct introduction into the BAC results in kanamycin drug resistance. DNA from kan^r BACs are analyzed to identify correctly modified BACs. The kan^r cassette has both prokaryotic and eukaryotic (mouse Pgk1) promoters. Double headed arrows indicate regions of recombination between mouse genomic DNA in the BAC DNA and mouse sequences in the synthetic DNA donor

2. Add the cell-DNA mixture to a 2 mm electroporation cuvette.
3. Electroporate the cell-DNA mixture (BioRad Gene Pulser XCell, 2.5 kV, 25 μF, 200 Ω). Transformation by heat shock is not advisable in BAC recombineering due to its low efficiency (Fig. 7).
4. Quickly resuspend the cells in LB medium.
5. Incubate 2–4 h @ 37°C.
6. Plate the whole reaction on LB-cam/kan agar plates.
7. Incubate overnight at 37 °C.
8. Compare colony count on induced and un-induced plates. The ratio should be at least 10:1. Colonies present on un-induced plates signal contamination of donor plasmid in the recombination reaction.
9. Pick 2 clones and follow Subheadings 3.4–3.6. Digest original and intermediate BAC clones with the same enzymes as in Subheading 3.6. Include the diagnostic enzyme chosen in Subheading 3.6. On the pulsed field gel, load digests of original BAC clones next to digests of intermediate BAC clones. Both should show distinct patterns.

3.10 Excision of the Selection Cassette

1. Inoculate 5 ml of LB-cam/kan with culture from the intermediate BAC.
2. Incubate overnight at 37°C.
3. Follow **steps 1–11** of Subheading 3.7.

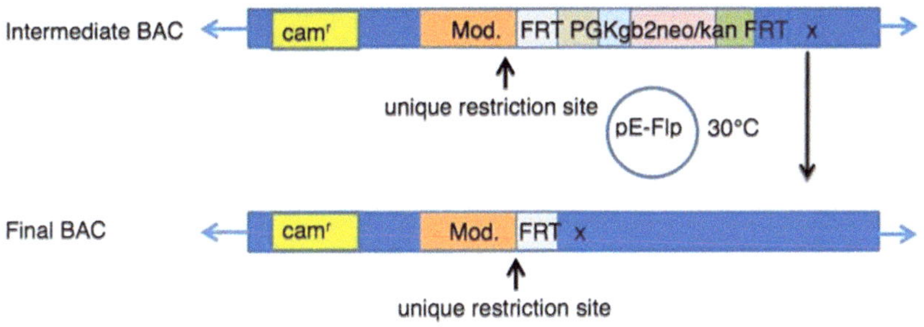

Fig. 8 Excise the selection cassette with FLP recombinase. The kan^r cassette is removed by induction of FLP recombinase in the BAC. DNA from kanamycin sensitive BACs are analyzed to identify correctly modified BACs

4. Electroporate cells with 10 ng of pE-FLP following **steps 12–15** of Subheading 3.7. The selection cassette is excised with FLP-recombinase (Fig. 8, *see* **Note 14**).
5. Incubate cells in a shaking incubator for 1 h at 30 °C.
6. Plate 50 μl of suspension on LB-cam/kan/amp agar plates.
7. Incubate overnight at 30 °C.
8. The next morning, duplicate-pick 20 colonies onto a grid on LB-cam and LB-cam/kan plates.
9. Incubate overnight at 37 °C.
10. Some of the picked colonies will be cam-resistant but kan-sensitive. These are the clones, which have lost the selection cassette and carry the final BAC (in rare events, these may be the original BAC clone that have traveled along). Pick two kan-sensitive clones from the LB-cam plate in 250 ml LB-cam and follow Subheading 3.4, **step 1**. Make sure to store purified BAC DNA in PA-buffer @ 4 °C.
11. Load DNA from the original and final BAC clones next to each other on the pulsed field gel. Follow steps described in Subheading 3.6 (restriction mapping).
12. Compare the restriction patterns of the original and the final BAC clones. If no gross rearrangements have happened, continue with sequencing (Fig. 9).

3.11 Sequencing

1. Highlight the modified region of your BAC including 500 bp 5′ and 3′ of the homology regions used in recombineering with sequence "X" between them where X is the number of bp of the recombination cassette (which includes the homology regions, the leftover FRT site, the primer alignment sequence of the donor plasmid) (Fig. 10). For long insertions it may be necessary to place primers inside the cassette (*see* **Note 15**).
2. Copy and paste sequence in search window.

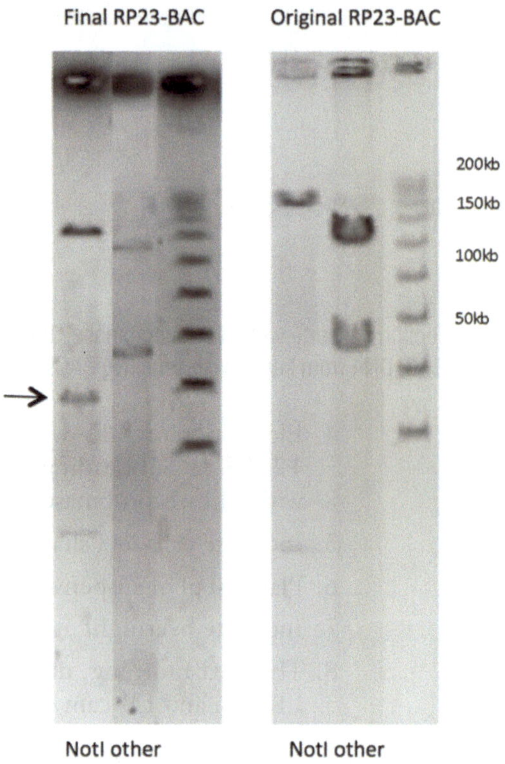

Fig. 9 Pulsed Field Gel comparison of restriction maps of original and final BAC clones. Arrow indicates band produced by the introduction of a diagnostic NotI restriction site in the final BAC. The "other" restriction enzyme introduced into the BAC causes a small band shift with respect to the entire BAC that cannot be cannot be observed in the PFG but that was confirmed by DNA sequence analysis

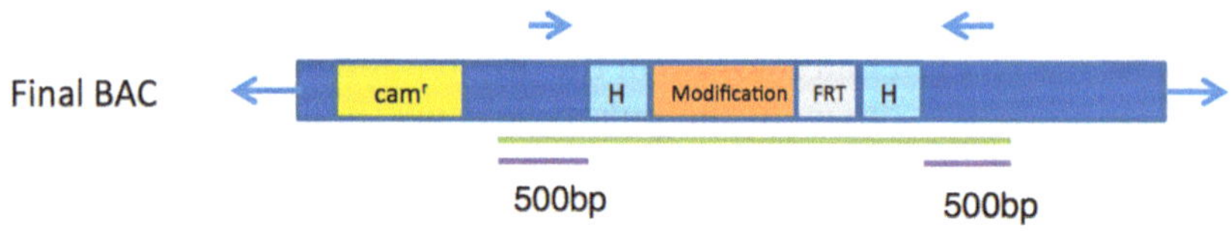

Fig. 10 DNA sequencing strategy to confirm correct integration of the desired recombineering cassettes. Sequencing primers are designed to align outside of the homology regions (H) used to amplify the recombineering cassette. Green line: sequence submitted for primer search. Purple lines: 500 bp extension on either side of the modification added to the search query

3. Set the range for the forward primer from 1 to 500 in the 5′ sequence outside of the homology region. Place the reverse primer in the 500 bp on the 3′ side of the inserted recombination donor (*see* **Note 16**).
4. Enter product size. Min: X Opt: X+200 Max: X+1000.
5. Click "Pick Primers."

6. The next page will list suggested primer pairs and annealing temperatures. Usually, the first suggestion is the most fitting for the desired parameters. Double check that the primers align outside the homology region used in recombineering.
7. PCR amplify the region of interest. Settings vary depending on your template and your primers. Use a proofreading polymerase for amplification of DNA to be sequenced.
8. The PCR product is either gel purified (QIAquick gel extraction kit, Qiagen 28704) or column purified (QIAquick PCR purification kit, Qiagen 28104) and submitted for sequencing to an in-house facility or external company.
9. Validate the sequence chromatogram. The peaks should be distinct and no Ns should be observed. The acquired sequence should match the *in-silico* design 100%.

3.12 Preparation of BAC Transgene Fragments

1. Purify BAC clone DNA as described in Subheading 3.4.
2. Digest the BAC in the presence of PA buffer with appropriate restriction enzyme(s). PA buffer includes 100 mM NaCl, thus enzymes such as NotI, that are active in 100 mM NaCl will cut to completion (*see* **Note 17**). Prepare a Sepharose 4B-CL column equilibrated in 0.1 M NaCl, 10 mM Tris (pH 7.5), 0.25 mM EDTA, in a sterile 5 ml pipet. The barrel of a 30 ml disposable syringe makes a convenient buffer reservoir.
3. After the digest is complete, apply it directly to the Sepharose 4B-CL column. Collect 0.3 ml fractions in sterile microtubes. Each microtube contains 3ul 3 mM spermine/7 mM spermidine to protect the BAC DNA after elution. Store the BAC DNA at 4 °C.
4. Use PFG to identify fractions containing the desired DNA fragment. Load 35 μl of each fraction to the wells of the PFG gel.
5. Small fragments will elute early and large fragments later. Quantitate the fraction(s) that contain the fragment of interest. For pronuclear microinjection adjust the concentration with PA buffer to 0.5–1.0 ng/μl [5].

3.13 Generation of BAC Transgenic Mice and Rats

After BAC clones are verified at the DNA sequence level to ensure the correct insertion of the desired cassette (fluorescent protein, recombinase, cDNA, genetic element, etc.) in the desired location then the purified BAC DNA can be microinjected into the pronuclei of fertilized mouse eggs for the production of transgenic mice or rats using established methods [5, 45, 46]. Primers used to verify the correct insertion of the recombineering cassette can be used to develop transgene specific PCR assays to identify transgenic founders and their offspring.

4 Notes

1. PIR1 *E. coli* bacteria are available from ThermoFisher (cat. no. C1010-10), mouse embryo tested water is available from Sigma-Aldrich (catalog no. W1503). Glass capillary tubes are available from World Precision Instruments (cat. no. TW100F-4).
2. Polyamines are very hygroscopic. We find that it is most efficient to order small quantities of high quality reagents (1 g) and then all of it should be prepared at once to produce a 1000× stock solution.
3. There are multiple public online tools to find a BAC clone that contains the gene of interest, including National Center for Biotechnology Information (NCBI) Map Viewer, Ensembl Genome Browser, and University of California Santa Cruz (UCSC) Genome Browser. The name of a BAC is composed of an abbreviation of the library and its coordinates on a microtiter plate. For instance: RP24-338R12 is a BAC clone of library RPCI-24 and located on microtiter plate 338, row R, column 12. The predominant mouse BAC libraries are RPCI-23, average insert size 197 kb [4] and RPCI-24, average insert size 155 kb. The RPCI-23 library was derived from brain and kidney tissue of 5-week old female C57BL/6J mice using pBACe3.6 as cloning vector. The RPCI-24 library was derived from male C57BL/6J mice using pTARBAC1 as cloning vector (*see* **Note 3**). It is often the case that genetically engineered BACs are used to generate transgenic mice. If the BAC in question has been modified, then the laboratory that recombineered the BAC should provide the DNA sequence of the BAC. With the DNA sequence in hand the next step is to purify BAC DNA and match the restriction map of the BAC with the in silico map (*see* Subheading 3.4 and beyond). Detailed information describing BAC library cloning vectors and library construction is available at http://bacpacresources.org/resources.htm.
4. The correct nomenclature for human genes for use as Ensembl search terms can be found at the HUGO Gene Nomenclature Committee website (www.genenames.org).
5. The correct nomenclature for mouse genes for use as Ensembl search terms can be found at the Mouse Genome Informatics website (www.informatics.jax.org).
6. The tilepath BAC clones show the clones that were completely sequenced for the assembly of the mouse genome. Their genomic location and DNA sequence are the most reliable. If a suitable BAC clone is not identified in the RP23 and RP24 libraries, additional libraries can be added to the search.

7. We find that the EMBL flat file format sequence downloads with all feature fields checked for inclusion in the file provides the most complete information on exported DNA sequences.
8. Genbank sequence files can be obtained from https://www.ncbi.nlm.nih.gov/genbank. Simply enter the accession number or name of the plasmid vector. DNA sequence files can then be copied and saved as text files for viewing in DNA editing software packages.
9. Clones can also be ordered from the BACPAC Resources Center (BPRC), imaGenes GmbH (Germany), or Geneservice Limited (United Kingdom), or other companies. The BPRC process will be described in detail. Note that the BPRC does not guarantee that clones will match predicted DNA sequence downloaded from genome browsers. It is therefore important to map BACs after they are received with restriction enzymes to validate each clone.
10. After transformation there may be background bacterial colonies that grow from transformed bacteria at low levels. The use of recombineering donor plasmid constructs that are based on the R6K origin of replication will reduce background. If background colonies interfere with progress it is advisable to do all bacterial culture (plate streaking, colony picking, broth inoculation, etc.) in a laminar flow hood such as a biosafety cabinet or a clean air PCR workstation.
11. Only use a proofreading polymerase like KOD (Novagen) or Expand High Fidelity Taq Polymerase (Sigma-Aldrich) to avoid DNA sequence errors caused by Taq polymerase DNA amplification.
12. Annealing and extension times will vary when using long primers. The template to be amplified also affects PCR efficiency. It may be necessary to test several annealing temperatures to identify conditions that will yield sufficient product for recombineering. We have found that the inclusion of the 5× CES can improve specificity and sensitivity of PCR reactions [47].
13. Recombination occurs in recombineering competent bacteria produced in Subheading 3.8 after the introduction of the PCR amplified recombineering cassette produced in Subheading 3.9. The amount of DNA needed depends on the project. Usually, the longer the PCR product, the more DNA is needed. Incubation time depends on the BAC and can be as short as 2 h or as long as 4 h. Electroporated cells take up a higher amount of DNA than heat shocked cells.
14. FLP-recombinase is expressed from plasmid pE-FLP ([48], pE-FLP was a gift from Drew Endy & Keith Shearwin (Addgene plasmid # 45978)). This plasmid has a temperature sensitive promoter for amp resistance so that it is stable at 30 °C and is lost at 37 °C.

15. The final step is verification of the engineered BAC by DNA sequencing. Primers are selected so that one primer lies outside of the homology region and a second primer is selected that lies within the cassette introduced into the BAC clone (Fig. 10). The PCR product is submitted for DNA sequencing and the results are compared to the desired result. Primer Blast is our preferred tool for primer design: http://ncbi.nlm.nih.gov/tools/primer-blast [49]. It may be necessary to test more than one primer pair and to adjust PCR conditions to find a high efficiency PCR reaction. To evaluate the sequence, a sequence chromatogram viewer is needed. We use 4Peaks software for this purpose, it can be downloaded from nucleobytes.com/4peaks.
16. We modify the default settings in the primer picking software so that longer primers with higher annealing temperatures are used to PCR amplify the 5′ and 3′ regions flanking the cassette inserted in the BAC clone [50]. Adjust Primer Parameter default settings in the following ways: (1) Set Minimum primer melting temperature to 60 °C. (2) Set Optimal primer melting temperature to 63 °C. (3) Set Maximum primer melting temperature to 66 °C. (4) Set Minimum primer melting temperature difference to 1 °C. (5) In the Specificity Checking Parameters click the box to turn on "Enable search for primer pairs specific to the intended PCR template." (6) Set Search Mode to "Automatic." (7) Set Database to "Genome (reference assembly from selected organisms)" Set Organism to "*Mus musculus* (taxid: 10090)" for mouse or to *Homo sapiens*" for human genomic DNA. (8) Click on "Advanced Parameters" then (9) set Primer Size Min to 27. (10) Set Primer Size Opt to 29. (11) Set Primer Size Max to 31.
17. For some BAC transgenes it may be preferable to isolate a fragment of the BAC clone for use as a transgene. For example, if there are neighboring genes on the BAC clone that are predicted to interfere with the interpretation of the consequences of gene expression then it may be desirable to separate those genes from the recombineered gene. When introducing loxP sites in BACs it is important to recognize that the pBACe3.6 and pTARBAC cloning vectors present in the BACs include the loxP and lox511 sites. If a loxP flanked cassette is to be introduced it is important to remove the loxP site in the cloning backbone with a plasmid such as pTAMP that can confers amp resistance on the BAC when the loxP site is excised [51]. Separation of the BAC cloning vector from the genomic sequence in the BAC can be achieved by size exclusion chromatography to eliminate both the loxP and lox511 sites from the BAC transgene prior to microinjection (*see* Subheading 4.7). In some cases it is desirable to introduce cassettes

that will be manipulated by both Cre and FLP recombinases in an animal model, such as the FLEX cassette for conditional alleles that can be reversed [52]. In these cases, we have found that the roxP flanked kan^r and Dre recombinase plasmids [53] are effective substitutes for the FRT flanked kan^r cassette.

References

1. Giraldo P, Montoliu L (2001) Size matters: use of YACs, BACs and PACs in transgenic animals. Transgenic Res 10:83–103
2. Heaney JD, Bronson SK (2006) Artificial chromosome-based transgenes in the study of genome function. Mamm Genome 17:791–807
3. Osoegawa K, Mammoser AG, Wu C, Frengen E, Zeng C, Catanese JJ, de Jong PJ (2001) A bacterial artificial chromosome library for sequencing the complete human genome. Genome Res 11:483–496
4. Osoegawa K, Tateno M, Woon PY, Frengen E, Mammoser AG, Catanese JJ, Hayashizaki Y, de Jong PJ (2000) Bacterial artificial chromosome libraries for mouse sequencing and functional analysis. Genome Res 10:116–128
5. Van Keuren ML, Gavrilina GB, Filipiak WE, Zeidler MG, Saunders TL (2009) Generating transgenic mice from bacterial artificial chromosomes: transgenesis efficiency, integration and expression outcomes. Transgenic Res 18:769–785
6. Alzhanov D, Rotwein P (2016) Characterizing a distal muscle enhancer in the mouse Igf2 locus. Physiol Genomics 48:167–172
7. Deal KK, Cantrell VA, Chandler RL, Saunders TL, Mortlock DP, Southard-Smith EM (2006) Distant regulatory elements in a Sox10-betaGEO BAC transgene are required for expression of Sox10 in the enteric nervous system and other neural crest-derived tissues. Dev Dyn 235:1413–1432
8. Dunnick WA, Shi J, Fontaine C, Collins JT (2013) Transgenes of the mouse immunoglobulin heavy chain locus, lacking distal elements in the 3′ regulatory region, are impaired for class switch recombination. PLoS One 8: e55842
9. Davis SW, Keisler JL, Pérez-Millán MI, Schade V, Camper SA (2016) All hormone-producing cell types of the pituitary intermediate and anterior lobes derive from Prop1 expressing progenitors. Endocrinology 157:1385–1396
10. Jones JM, Datta P, Srinivasula SM, Ji W, Gupta S, Zhang Z, Davies E, Hajnoczky G, Saunders TL, Van Keuren ML, Fernandes-Alnemri T, Meisler MH, Alnemri ES (2003) Loss of Omi mitochondrial protease activity causes the neuromuscular disorder of mnd2 mutant mice. Nature 425:721–727
11. Khoriaty R, Everett L, Chase J, Zhu G, Hoenerhoff M, McKnight B, Vasievich MP, Zhang B, Tomberg K, Williams J, Maillard I, Ginsburg D (2016) Pancreatic SEC23B deficiency is sufficient to explain the perinatal lethality of germline SEC23B deficiency in mice. Sci Rep 6:27802
12. Probst FJ, Fridell RA, Raphael Y, Wang A, Liang Y, Morell RJ, Touchman JW, Lyons RH, Noben-Trauth K, Friedman TB, Camper SA (1998) Correction of deafness in shaker-2 mice by an unconventional myosin in a BAC transgene. Science 280:1444–1447
13. Antoch MP, Song EJ, Chang AM, Vitaterna MH, Zhao Y, Wilsbacher LD, Sangoram AM, King DP, Pinto LH, Takahashi JS (1997) Functional identification of the mouse circadian clock gene by transgenic BAC rescue. Cell 89:655–667
14. Hu Y, Smith DE (2016) Species differences in the pharmacokinetics of cefadroxil as determined in wildtype and humanized PepT1 mice. Biochem Pharmacol 107:81–90
15. Mensah-Osman E, Labut E, Zavros Y, El-Zaatari M, Law DJ, Merchant JL (2008) Regulated expression of the human gastrin gene in mice. Regul Pept 151:115–122
16. Sarsero JP, Holloway TP, Li L, Finkelstein DI, Ioannou PA (2014) Rescue of the Friedreich ataxia knockout mutation in transgenic mice containing an FXN-EGFP genomic reporter. PLoS One 9:e93307
17. Gong S, Zheng C, Doughty ML, Losos K, Didkovsky N, Schambra UB, Nowak NJ, Joyner A, Leblanc G, Hatten ME, Heintz N (2003) A gene expression atlas of the central nervous system based on bacterial artificial chromosomes. Nature 425:917–925
18. Valjent E, Bertran-Gonzalez J, Hervé D, Fisone G, Girault JA (2009) Looking BAC at striatal signaling: cell-specific analysis in new transgenic mice. Trends Neurosci 32:538–547

19. Witten IB, Steinberg EE, Lee SY, Davidson TJ, Zalocusky KA, Brodsky M, Yizhar O, Cho SL, Gong S, Ramakrishnan C, Stuber GD, Tye KM, Janak PH, Deisseroth K (2011) Recombinase-driver rat lines: tools, techniques, and optogenetic application to dopamine-mediated reinforcement. Neuron 72:721–733
20. Keegan CE, Karolyi IJ, Knapp LT, Bourbonais FJ, Camper SA, Seasholtz AF (1994) Expression of corticotropin-releasing hormone transgenes in neurons of adult and developing mice. Mol Cell Neurosci 5:505–514
21. Alon T, Zhou L, Pérez CA, Garfield AS, Friedman JM, Heisler LK (2009) Transgenic mice expressing green fluorescent protein under the control of the corticotropin-releasing hormone promoter. Endocrinology 150:5626–5632
22. Deal KK, Cantrell VA, Chandler RL, Saunders TL, Mortlock DP, Southard-Smith EM (2006) Distant regulatory elements in a Sox10-beta GEO BAC transgene are required for expression of Sox10 in the enteric nervous system and other neural crest-derived tissues. Dev Dyn 235:1413–2432
23. Dunnick WA, Shi J, Holden V, Fontaine C, Collins JT (2011) The role of germline promoters and I exons in cytokine-induced gene-specific class switch recombination. J Immunol 186:350–358
24. Lehoczky JA, Innis JW (2008) BAC transgenic analysis reveals enhancers sufficient for Hoxa13 and neighborhood gene expression in mouse embryonic distal limbs and genital bud. Evol Dev 10:421–432
25. Pérez-Millán MI, Zeidler MG, Saunders TL, Camper SA, Davis SW (2013) Efficient, specific, developmentally appropriate cre-mediated recombination in anterior pituitary gonadotropes and thyrotropes. Genesis 51:785–792
26. Yu S, Zhou X, Hsiao JJ, Yu D, Saunders TL, Xue HH (2012) Fidelity of a BAC-EGFP transgene in reporting dynamic expression of IL-7Rα in T cells. Transgenic Res 21:201–215
27. Lehoczky JA, Thomas PE, Patrie KM, Owens KM, Villarreal LM, Galbraith K, Washburn J, Johnson CN, Gavino B, Borowsky AD, Millen KJ, Wakenight P, Law W, Van Keuren ML, Gavrilina G, Hughes ED, Saunders TL, Brihn L, Nadeau JH, Innis JW (2013) A novel intergenic ETnII-β insertion mutation causes multiple malformations in polypodia mice. PLoS Genet 9:e1003967
28. Watson CL, Mahe MM, Múnera J, Howell JC, Sundaram N, Poling HM, Schweitzer JI, Vallance JE, Mayhew CN, Sun Y, Grabowski G, Finkbeiner SR, Spence JR, Shroyer NF, Wells JM, Helmrath MA (2014) An in vivo model of human small intestine using pluripotent stem cells. Nat Med 20:1310–1314
29. Albertsen HM, Abderrahim H, Cann HM, Dausset J, Le Paslier D, Cohen D (1990) Construction and characterization of a yeast artificial chromosome library containing seven haploid human genome equivalents. Proc Natl Acad Sci U S A 87:4256–4260
30. Moreira PN, Pozueta J, Pérez-Crespo M, Valdivieso F, Gutiérrez-Adán A, Montoliu L (2007) Improving the generation of genomic-type transgenic mice by ICSI. Transgenic Res 16:163–168
31. Brandt W, Khandekar M, Suzuki N, Yamamoto M, Lim KC, Engel JD (2008) Defining the functional boundaries of the Gata2 locus by rescue with a linked bacterial artificial chromosome transgene. J Biol Chem 283:8976–8983
32. Sopher BL, La Spada AR (2006) Efficient recombination-based methods for bacterial artificial chromosome fusion and mutagenesis. Gene 12(371):136–143
33. Montoliu L, Bock CT, Schutz G, Zentgraf H (1995) Visualization of large DNA molecules by electron microscopy with polyamines: application to the analysis of yeast endogenous and artificial chromosomes. J Mol Biol 246:486–492
34. Montigny WJ, Phelps SF, Illenye S, Heintz NH (2003) Parameters influencing high-efficiency transfection of bacterial artificial chromosomes into cultured mammalian cells. BioTechniques 35:796–807
35. Copeland NG, Jenkins NA, Court DL (2001) Recombineering: a powerful new tool for mouse functional genomics. Nat Rev Genet 2:769–779
36. Court DL, Sawitzke JA, Thomason LC (2002) Genetic engineering using homologous recombination. Annu Rev Genet 36:361–388
37. Zhang Y, Muyrers JP, Testa G, Stewart AF (2000) DNA cloning by homologous recombination in *Escherichia coli*. Nat Biotechnol 18:1314–1317
38. Auwerx J, Avner P, Baldock R, Ballabio A, Balling R, Barbacid M, Berns A, Bradley A, Brown S, Carmeliet P, Chambon P, Cox R, Davidson D, Davies K, Duboule D, Forejt J, Granucci F, Hastie N, de Angelis MH, Jackson I, Kioussis D, Kollias G, Lathrop M, Lendahl U, Malumbres M, von Melchner H, Müller W, Partanen J, Ricciardi-Castagnoli P, Rigby P, Rosen B, Rosenthal N, Skarnes B, Stewart AF, Thornton J, Tocchini-Valentini G, Wagner E, Wahli W, Wurst W (2004) The

European dimension for the mouse genome mutagenesis program. Nat Genet 36:925–927
39. International Mouse Knockout Consortium, Collins FS, Rossant J, Wurst W (2007) A mouse for all reasons. Cell 128:9–13
40. Skarnes WC, Rosen B, West AP, Koutsourakis M, Bushell W, Iyer V, Mujica AO, Thomas M, Harrow J, Cox T, Jackson D, Severin J, Biggs P, Fu J, Nefedov M, de Jong PJ, Stewart AF, Bradley A (2011) A conditional knockout resource for the genome-wide study of mouse gene function. Nature 474:337–342
41. Testa G, Zhang Y, Vintersten K, Benes V, Pijnappel WW, Chambers I, Smith AJ, Smith AG, Stewart AF (2003) Engineering the mouse genome with bacterial artificial chromosomes to create multipurpose alleles. Nat Biotechnol 21:443–447
42. Valenzuela DM, Murphy AJ, Frendewey D, Gale NW, Economides AN, Auerbach W, Poueymirou WT, Adams NC, Rojas J, Yasenchak J, Chernomorsky R, Boucher M, Elsasser AL, Esau L, Zheng J, Griffiths JA, Wang X, Su H, Xue Y, Dominguez MG, Noguera I, Torres R, Macdonald LE, Stewart AF, DeChiara TM, Yancopoulos GD (2003) High-throughput engineering of the mouse genome coupled with high-resolution expression analysis. Nat Biotechnol 21:652–659
43. Hu Y, Xie Y, Wang Y, Chen X, Smith DE (2014) Development and characterization of a novel mouse line humanized for the intestinal peptide transporter PEPT1. Mol Pharm 11:3737–3746
44. Warming S, Costantino N, Court DL, Jenkins NA, Copeland NG (2005) Simple and highly efficient BAC recombineering using galK selection. Nucleic Acids Res 33:e36
45. Filipiak WE, Saunders TL (2006) Advances in transgenic rat production. Transgenic Res 15:673–686
46. Becker K, Jerchow B (2011) Generation of transgenic mice by pronuclear microinjection. In: Pease S, Saunders TL (eds) Advanced protocols for animal Transgenesis: an ISTT manual. Springer-Verlag, Berlin
47. Ralser M, Querfurth R, Warnatz HJ, Lehrach H, Yaspo ML, Krobitsch S (2006) An efficient and economic enhancer mix for PCR. Biochem Biophys Res Commun 347:747–751
48. St-Pierre F, Cui L, Priest DG, Endy D, Dodd IB, Shearwin KE (2013) One-step cloning and chromosomal integration of DNA. ACS Synth Biol 2:537–541
49. Ye J, Coulouris G, Zaretskaya I, Cutcutache I, Rozen S, Madden T (2012) Primer-BLAST: a tool to design target-specific primers for polymerase chain reaction. BMC Bioinformatics 13:134
50. Stratman JL, Barnes WM, Simon TC (2003) Universal PCR genotyping assay that achieves single copy sensitivity with any primer pair. Transgenic Res 12:521–552
51. Lee EC, Yu D, Martinez de Velasco J, Tessarollo L, Swing DA, Court DL, Jenkins NA, Copeland NG (2001) A highly efficient *Escherichia coli*-based chromosome engineering system adapted for recombinogenic targeting and subcloning of BAC DNA. Genomics 73:56–65
52. Schnütgen F, Ghyselinck NB (2007) Adopting the good reFLEXes when generating conditional alterations in the mouse genome. Transgenic Res 16:405–413
53. Chuang K, Nguyen E, Sergeev Y, Badea TC (2016) Novel heterotypic Rox sites for combinatorial dre recombination strategies. G3 (Bethesda) 6:559–571

Chapter 4

Using TARGATT™ Technology to Generate Site-Specific Transgenic Mice

Ruby Yanru Chen-Tsai

Abstract

The discovery of new gene editing tools in the past several years has moved the transgenic field to a new level. The traditional random transgenesis method by pronuclear microinjection has been largely replaced by targeted or site-specific transgenic technologies without the need of homologous recombination in embryonic stem (ES) cells. In this chapter, I describe detailed protocols of an integrase-based approach, trademarked as "TARGATT™" (***targ***et ***att****P*), to produce site-specific transgenic mice via pronuclear microinjection, whereby an intact single-copy transgene can be inserted into a predetermined chromosomal locus with high efficiency (up to 40%), and faithfully transmitted through generations. This system allows high-level global transgene expression or tissue-specific expression depending on the promoter used, or inducible expression such as induced by tetracycline or doxycycline. Using this approach, site-specific transgenic mice can be generated as fast as in 3 months. The technique presented here greatly facilitates murine transgenesis and precise structure/function dissection of mammalian gene function and regulation in vivo.

Key words Pronuclear microinjection, PhiC31 integrase, TARGATT™, Site-specific transgenic, *Rosa26* knockin, *H11* locus, *attP*, *attB*

1 Introduction

Sequence-specific recombinases such as phage integrases have rapidly gained their prominent positions as genetic tools for a wide range of applications from basic cloning to genome editing [1, 2]. Phage integrases are enzymes that mediate unidirectional site-specific recombination between their DNA recognition sequences, including the phage attachment site, *attP*, and the bacterial attachment site, *attB*, and function in mammalian cells [3–5]. As illustrated in Fig. 1, integrases catalyze recombination between appropriate *attB* and *attP*, resulting in an integration of the DNA sequence flanking the *attB* site. At the same time, recombination between *attP* and *attB* results in two new hybrid sites, *attL* and *attR*. These two new sites can no longer be recognized by

Chengyu Liu and Yubin Du (eds.), *Microinjection: Methods and Protocols*, Methods in Molecular Biology, vol. 1874, https://doi.org/10.1007/978-1-4939-8831-0_4, © Springer Science+Business Media, LLC, part of Springer Nature 2019

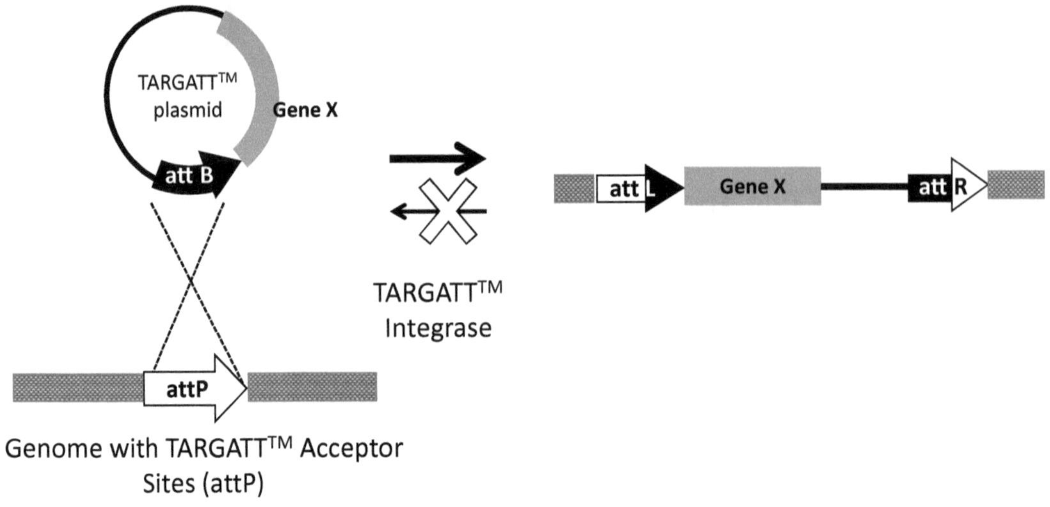

Fig. 1 Site-specific transgene integration mediated by phage integrases. Integrases such as phiC31 integrase mediate sequence-specific recombination between two largely different sites, *attP* (open arrow) and *attB* (black arrow). The sequences of the *attP* and *attB* are unrelated except for the three nucleotides in the middle where the crossover happens. Recombination between *attP* and *attB* results in two new hybrid sites *attL* and *attR*, which can no longer be recognized by integrase. So integrase is unidirectional, catalyzing only integration but no reverse excision reaction

integrases. This irreversible feature, along with the lack of a corresponding excisionase enzyme, makes the recombination reaction unidirectional, ensuring that constructs integrated into the genome do not act as substrates for the reverse reaction. The result is an improvement in integration efficiency compared with random integration or other bi-directional sequence-specific recombinases. The integrase recognition sites, *attP* or *attB*, do not naturally exist in mammalian genome. Instead, sites with similar sequences so called "pseudo-*attP*" sites exist naturally in mammalian genome and can also be recognized by integrases such as phiC31 integrase from a Streptomyces phage. In fact, phiC31 was the first phage integrase to be developed for use in mammalian cells [6]. The efficiency of recombination between an *attB* and a pseudo-*attP* site is low compared to its wild-type cognate *attP* site. PhiC31 integrase has been used to catalyze integration of circular DNA into pseudo-*attP* sites in the mouse genome for gene therapy [7], for low efficiency transgenesis in mice [8], and for integrating DNA in bovine cells [5].

The TARGATT™ transgenic technology was developed to generate site-specific transgenic animals mediated by phiC31 integrase [9]. PhiC31 *attP* sites, served as "landing pad," were pre-engineered and placed in a safe genomic harbor locus, either at the *Rosa26* or *H11* locus, of the mouse genome. These *attP* landing pad containing mice or embryos are called TARGATT™ mice or embryos. Plasmid DNA containing a transgene and an *attB*

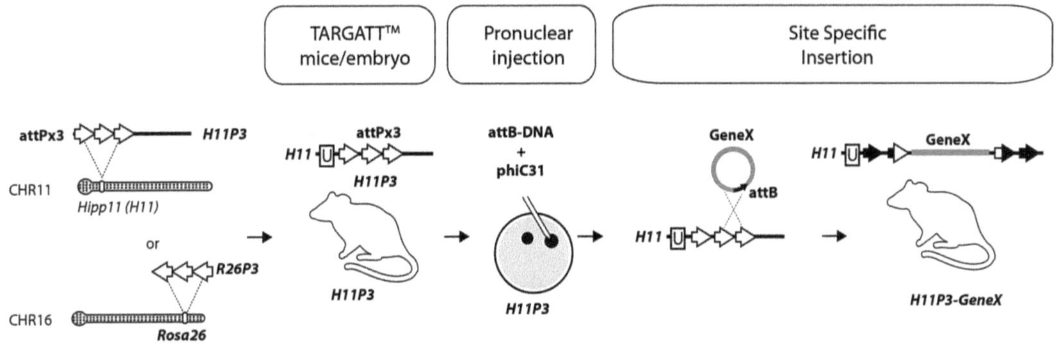

Fig. 2 Generation of site-specific transgenic mice using TARGATT™. One-cell fertilized embryos (zygotes) are harvested from TARGATT mice that contain the *attP* site at either Rosa 26 or *H11* locus. TARGATT donor vector containing the *attB* site and the gene of interest (e.g., Gene X) are microinjected into the pronucleus together with phiC31 integrase mRNA. Injected embryos are implanted into the oviduct of pseudo-pregnant females to carry the embryos to term. Founder transgenic animals will contain transgene insertion (Gene X) at the H11 locus (or Rosa26 locus)

site are co-microinjected with phiC31 integrase mRNA or protein into the pronucleus of the TARGATT™ one-cell embryos, resulting in site-specific transgene integration at the landing pad through recombination mediated by phiC31 integrase (Fig. 2). Between the two safe harbor loci, *Rosa26* has been used widely as a site for gene knockin since 1999 [10]. The *H11* locus is located in between two highly expressed genes and in an intergenic region on mouse chromosome 11. The *H11* locus was originally described by Hippenmeyer et al. in 2010 [11]. Further in vitro and in vivo experiments showed that this locus supports robust, high-level gene expression in transgenic mice [12–15], human iPS/ES cells [16], CHO cells (unpublished data), and in transgenic rats (unpublished data) and pigs [17].

For generating site-specific transgenic mice, TARGATT™ mice containing the *attP* docking sites at the *H11* or *Rosa26* locus are used as zygote donors. A mixture of a transgene vector containing a compatible site, *attB*, and in vitro transcribed phiC31 mRNA or purified phiC31 protein is microinjected into a pronucleus of each zygote. Figure 2 illustrates how this integrase-based TARGATT™ method works for generating site-specific transgenic mice. PhiC31 catalyzes recombination between *attP* and *attB*, resulting in integration of the gene of interest.

2 Materials

2.1 TARGATT™ DNA Plasmid

Several selections of the TARGATT™ plasmid can be purchased from Applied StemCell, Inc. (Milpitas, CA). Each plasmid contains a unique combination of a promoter, loxP-stop-loxP (LSL) cassette, multiple cloning site (MCS), and a polyA sequence to meet each project needs. A list of the plasmids includes:

1. TARGATT™ 2 (pCAG-MCS-polyA).
2. TARGATT™ 3 (no promoter +MSC).
3. TARGATT™ 6.1 (pCAG-LSL-MCS-polyA).
4. TARGATT™ 7 (pPGK-MCS-polyA).
5. TARGATT™ 8 (pCA-MCS-polyA).
6. TARGATT™ 9.1 (pCA-LSL-MCS-polyA).

2.2 Plasmid Preparation and Purification Reagents

1. RNase, DNase-free water.
2. Qiaprep Spin Mini Prep Kit (250) and Reagents.
3. NEB® 5-alpha Competent *E. coli* (High Efficiency).
4. LB Broth (Miller).
5. LB Broth with agar (Miller).
6. SOC Medium.
7. Agarose tablets RNase, DNase free.
8. 40× TAE Buffer.
9. RNase Zap.
10. Bleach.
11. Ethyl alcohol (Ethanol, EtOH) 100%, pure, 200 proof, mol. Boil. Grade.
12. Phenol/Chloroform/Isoamyl Alcohol (25:24:1 Mixture, pH 6.7/8.0, Liq.), store at 4C (*see* **Note 1**).
13. Chloroform (*see* **Note 1**).

Note 1: Read material safety data sheet before working with the reagent.

2.3 Mice

2.3.1 TARGATT™ Mice/Embryos

Ten TARGATT™ mice are needed for each round of pronuclear microinjection experiment.

1. Strain code 537: TARGATT™ mice, with *attP* sites at the *H11* locus, FVB strain, order from Applied StemCell, Inc.
2. Strain code 549: TARGATT™ mice, female, 3–4 weeks old, with *attP* sites at the *H11* locus, C57BL6 strain, order from Applied StemCell, Inc., shipped from Charles River Laboratory (CRL).
3. AST-0013: Cryopreserved embryos with *attP* sites at the *Rosa26* locus, C57BL6 strain from Applied StemCell, Inc. (ASC).
4. AST-0004: Cryopreserved embryos with *attP* site at the *Rosa26* locus, FVB strain from ASC.
5. Wild-type C57BL6 male mice, 7–8 weeks old as stud males to breed with TARGATT™ female mice and obtain embryos for microinjection. These mice can be obtained from CRL or other commercial vendors.

2.3.2 Vasectomized Male Mice and CD1 Female Mice

Ten vasectomized male mice and 20 CD1 mature female mice can be purchased from CRL or other animal vendors. After mating, the females are used as pseudo-pregnant recipient females for embryo implantation.

2.4 TARGATT™ Transgenic Kits

These kits can be purchased from Applied StemCell, Inc. Each kit contains:

1. TARGATT™ integrase mRNA (transgene integration tested) (20 μl at 100 ng/μl per vial, 2 or 5 vials based on the size of the kit). Each vial is good for one round of microinjection.
2. Control mRNA (20 μl) (*see* **Note 2**).
3. Embryo qualified microinjection TE (MiTE) buffer (1 ml).

2.5 TARGATT™ Genotyping Kits

The genotyping kits for detecting site-specific transgene insertion can be purchased from Applied StemCell, Inc. Each kit contains:

1. Tail tissue lysis buffer.
2. Proteinase K (20 mg/ml).
3. PCR primer sets for detecting site-specific insertion at 5′ and 3′ junctions, including H11P3, Rosa26P3, SSL, SSR.
4. Nuclease-free water.
5. Positive control DNA.

3 Methods

3.1 Construction of TARGATT™ DNA Insertion Vector

The TARGATT™ plasmids contain multiple cloning site (MCS) which can be used to insert the transgene (or gene of interest) by one-step cloning (Fig. 3).

1. Cut and purify the insertion transgene DNA from another vector, or have the transgene DNA fragment synthesized.
2. Linearize TARGATT™ plasmid with restriction enzyme (s) in MCS.
3. Ligate TARGATT™ plasmid backbone with the transgene DNA fragment.

3.2 Preparation of Plasmid DNA for Microinjection

3.2.1 Transformation

Use NEB® 5-alpha Competent *E. coli* (High Efficiency) NEB #C2987H and follow the protocol provided by the manufacturer.

1. For C2987H, thaw a tube of NEB 5-alpha Competent *E. coli* cells on ice for 10 min.
2. Add 1–5 μl containing 1 pg to 100 ng of plasmid DNA to the cell mixture. Carefully flick the tube 4–5 times to mix cells and DNA. Do not vortex.
3. Place the mixture on ice for 30 min. Do not mix.

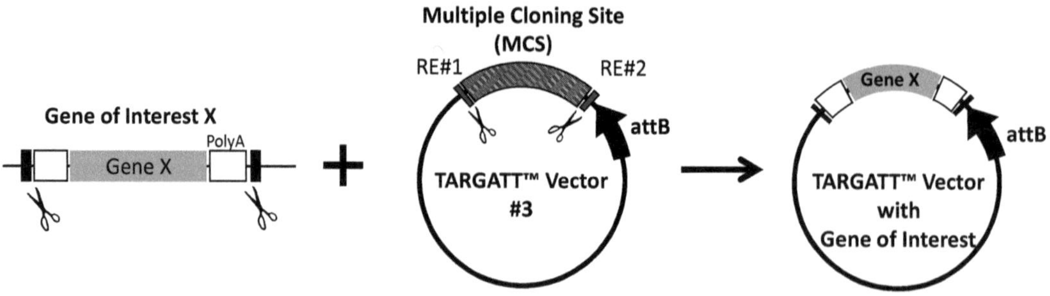

Fig. 3 Construction of TARGATT™ donor vector. The donor vector can usually be made by one-step cloning. Transgene (insertion DNA fragment) can either be synthesized or cut out from another vector and subcloned into a TARGATT™ vector (available from Applied StemCell, Inc.). There are several versions of TARGATT vectors including choice of promoter and stop cassette for conditional expression, but all containing the *attB* site and multiple cloning site (MCS) for easy subcloning

4. Heat shock at exactly 42 °C for exactly 30 s. Do not mix.
5. Place on ice for 5 min. Do not mix.
6. Pipette 950 μl of room temperature SOC into the mixture.
7. Place at 37 °C for 60 min. Shake vigorously (250 rpm) or rotate.
8. Warm LB plates with drug selection antibiotics to 37 °C.
9. Mix the cells thoroughly by flicking the tube and inverting, then perform several ten-fold serial dilutions in SOC. (20× or 50× dilution works very well).
10. Spread 50 μl of the diluted culture onto a selection plate and incubate overnight at 37 °C.

3.2.2 Bacterial Culture and Plasmid Mini-Prep Using Qiaprep Spin Mini Prep Kit

1. Start from 4 ml of DH5α bacterial culture grown in LB broth (with antibiotic) for no more than 14 h at 37 °C.
2. Collect the bacteria in 2 ml tubes by spinning 2 ml of culture twice in the same tube at 8000 rpm (6800 × *g*), for 3 min at room temperature.
3. Resuspend pelleted bacterial cells in 500 μl Buffer P1 (with RNaseA added, stored at 4 °C).
4. Add 500 μl Buffer P2 and mix thoroughly but gently by inverting the tubes 6–8 times (or until the solution becomes clear). Do not allow the reaction to proceed for more than 5 min. Do not shake or vortex.
5. Add 700 μl Buffer N3 and mix immediately and thoroughly by inverting the tubes 6–8 times.
6. Centrifuge for 10 min at 13,000 rpm (~17,900 × *g*) in a table top microcentrifuge.
7. Apply the supernatant to pre-labeled QIAprep spin columns. Discard tubes with pellets.

8. Centrifuge for 30–60 s at 13,000 rpm, discard the flow-through.
9. Wash twice with 500 µl Buffer PB. Centrifuge for 30–60 s at 13,000 rpm, discard the flow-through (*see* **Note 3**).
10. Wash twice with 750 µl Buffer PE. Centrifuge for 30–60 s at 13,000 rpm, discard the flow-through.
11. Centrifuge for 60 s at 13,000 rpm to remove residual buffer.
12. Place the QIAprep columns in a clean 1.5 microcentrifuge tubes. To elute DNA, add 100 µl of three-fold diluted Buffer EB (*see* **Note 4**).
13. Check DNA concentration on NanoDrop, use 3× diluted EB as a blank (*see* **Note 5**).

3.2.3 Phenol/Chloroform Extraction of Plasmid DNA

1. Add an equal volume of TE-saturated phenol-chloroform to the aqueous DNA sample (*see* **Note 1**).
2. Mix well by hand for 30–60 s.
3. Centrifuge the tube for 5 min at room temperature on the highest setting. You should see clearly separated layers: top phase (aqueous DNA), interphase (a white layer may appear, consisting of precipitated protein particles), and bottom phase (organic protein phase).
4. Collect the aqueous DNA layer only and place it in a new microfuge tube. Avoid aspirating of the interphase.
5. Repeat phenol-chloroform extraction.
6. Add equal volume of chloroform to the recovered aqueous DNA layer.
7. Mix well by hand for 30–60 s.
8. Centrifuge the tube for 5 min at room temperature on the highest speed setting.
9. From this point on, take all precautions to avoid RNase contamination. Work in RNAsae-free hood. Use all RNase-free solutions and equipment.
10. Pipette the aqueous DNA layer and place it in a new 2 ml microfuge tube.
11. Repeat chloroform extraction.
12. Concentrate DNA by ethanol precipitation.

3.2.4 Ethanol Precipitation and Filtration

1. Measure the volume of collected DNA solution. If needed transfer to 2 ml RNase-free, DNase-free microfuge tube.
2. Add 1/10 volume of 3 M Na-acetate (pH 5.2). Mix by tapping the tube and inverting.

3. Add 2.7 volumes of 100% EtOH. Mix by tapping the tube and inverting (2.5 volumes will work as well).
4. Place the tube either at −20 °C overnight or −80 °C for 30 min or on dry ice for 5 min.
5. Centrifuge at highest speed at 4 °C for 30 min to pellet. Pay attention to the tube orientation in the centrifuge. The pellet may not be visible.
6. Remove supernatant by pipetting. Avoid aspirating the DNA pellet.
7. Wash the pellet with 1 ml of 70% EtOH.
8. Centrifuge at highest speed for 5 min at room temperature.
9. Repeat the wash if possible.
10. Remove the wash by pipetting. Quickly spin at room temperature. Gently pipette off the rest of the wash.
11. Air dry for 5 min. Resuspend the pellet in a small volume of RNase-free microinjection TE buffer (miTE: 0.1 mM EDTA, 10 mM Tris pH 7.5) or 10 mM Tris–HCl pH 7.5. If the DNA was extracted from 5 × 4 ml cultures and was pooled, use 50 μl of resuspension buffer (*see* **Note 6**).
12. Measure DNA concentration on NanoDrop. Adjust the volume to 30 ng/μl with the appropriate buffer.
13. Use sterile 0.2 μm filter.
14. Place the filter in the collection tube and add 60 μl RNase-free water.
15. Spin at 10,000 rpm for 30 s at room temperature. Save flow-through for RNase analysis.
16. Add 60 μl of the DNA sample, spin at 10,000 rpm for 30 s at room temperature. Repeat if working with larger volumes using the same filter.
17. Measure the DNA concentration after filtration.
18. Prepare 20 μl aliquots. Use RNase-free tubes and tips. Store frozen at −80 °C until use.

3.2.5 Quality Control

1. Run 300 ng of uncut and linear (single cut) DNA on 1% agarose TAE gel with wide wells. Include three other restriction digests to confirm plasmid integrity. Make sure that the equal volumes are loaded to each well. Keep the final volume low at 15 μl or less.
2. Resolve samples using electrophoresis at 75–100 V until all three fractions (open circle (oc), linear (l), and supercoiled (sc)) are clearly separated. The supercoiled fraction should be the predominant fraction (>70%) in uncut plasmid prep in order to achieve a high success rate of site-specific insertion.

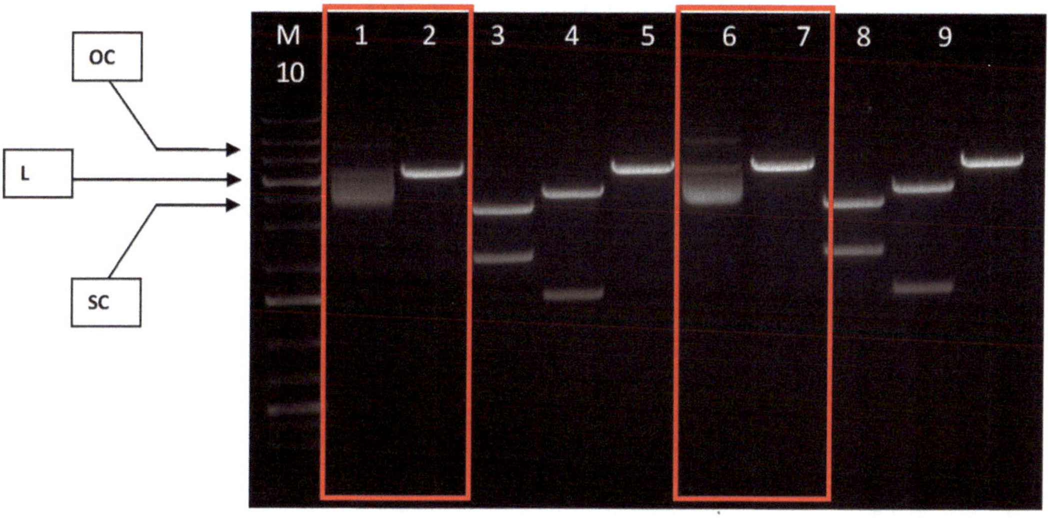

Fig. 4 Gel electrophoresis to check the quality of TARGATT™ insertion plasmid. 1% agarose gel was electrophoresed at 75-100 volts. Lanes 1 and 6 contain two different plasmid preps of an uncut TARGATT™ plasmid, respectively. Lanes 2 and 7 depict the same plasmid cut with single cutter Not I. Supercoiled form (SC) runs the fastest on the gel, whereas linearized plasmid (L) runs as a single band equivalent to the size of the plasmid. Other forms such as open circle (OC) can also present in the prep. Both preps contain over 70% supercoiled form (lower band) and passed quality control

If the linear or open circle fractions are above 30%, do not use the prep and start over. Use the gel picture in Fig. 4 for reference. Lanes 1 and 6 contain an uncut TARGATT™ plasmid. Lanes 2 and 7 depict the same plasmid cut with single cutter Not I. Both preps contain over 70% supercoiled form (lower band) and passed quality control.

3. Test for absence of RNase by incubating 200 ng plasmid DNA with 200 ng control RNA at 37 °C for 1 h, and then heat inactivate at 85–90 °C for 5 min, and store on ice for 2 min. Run the mixture on a 1% agarose gel. Gel box should be kept cold during the run. There should be no RNA degradation shown as a smear on a gel.
4. Dilute the DNA with microinjection TE buffer (MiTE) supplied in the TARGATT™ transgenic kit to a final concentration of 6 ng/μl, and store it at −80 °C till microinjection. If the DNA prep will be used to be co-injected with PhiC31 integrase protein, use Tris buffer without EDTA (10 mM Tris–HCl pH 7.5).
5. Mix equal volume (10 μl each) of the plasmid DNA (6 ng/μl) and TARGATT™ integrase mRNA supplied in the TARGATT™ transgenic kit. The mixture should be made fresh on the day of microinjection. The 20 μl of mixture should be sufficient for injecting 200 embryos.

3.3 Pronuclear Microinjection and Oviduct Embryo Transfer

The mixture of insertion plasmid DNA and integrase mRNA is microinjected into the pronuclei of one-cell embryos (zygotes) with a pause in the cytoplasm using a constant flow mode of the microinjector. As a result, the mixture is distributed into both the pronuclei and cytoplasm. If the embryos are homozygous for TARGATT *attP* sites, either of the two pronuclei can be injected. But if the embryos are from TARGATT *attP* male bred with wild-type females, the injection should be into the male pronuclei. In the case when it is not easy to tell the male pronucleus from the female one, inject both pronuclei if possible to ensure integrase is present in the *attP* containing nucleus.

Upon completion of microinjection, embryos are implanted into the oviduct of pseudo-pregnant females at about 20–25 embryos per female to carry the embryos to term.

3.4 Identifying Founders by Genotyping

Upon weaning the pups (2–3 weeks after birth), tail snips can be obtained from each pup for genotyping using a TARGATT™ Mouse Genotyping Kit.

3.4.1 Genomic DNA Extraction from Tail Samples

1. Cut approximately 2 mm tail snip and place in an Eppendorf tube.
2. Add 200 μl tail tissue lysis buffer and Proteinase K, and lyse at 55 °C for 4 h to overnight.
3. Heat the samples at 90 °C for 5 min to inactivate proteinase K.
4. Proceed to PCR reaction or store at −20 °C.

3.4.2 Genotyping by PCR

1. Set up PCR reactions on ice according to the instruction of the Taq polymerase manufacturer. Use the table below for Qiagen Tag polymerase. Use "Primer Set H11P3" (AST-2005) or "Primer Set Rosa26P3" (AST-2006) to examine whether the starting mouse/embryo has attP insertion at the correct locus (H11 or Rosa26). Use "Primer Set SSL" and "Primer Set SSR" to determine site-specific transgene integration.

Component	Amount
DNA-Tail lysis	1 μl
10× PCR buffer	2 μl
dNTPs (10 mM)	0.4 μl
Primer set (10 μM)	1 μl
Taq Polymerase (5 U/μl)	0.2 μl
Nuclease-free water	15.4 μl
Total volume	20 μl

2. Perform PCR amplification using the following program.

Step	Number of cycles	Temperature	Time
1	1	94 °C	3 min
2	35	94 °C	30 s
3		58 °C	25 s
4		72 °C	45 s
5	1	72 °C	5 min

3. Run PCR products on a gel to analyze.

3.4.3 PCR Data Analysis

Figure 5 depicts examples of TARGATT™ genotyping scheme on identifying site-specific transgenic founders. Primer sets H11P3.F (forward primer)/H11P3.R (reverse primer) and R26P3.F/R26P3.R are used to detect the *attP*x3 at the *H11* locus or *Rosa26* locus, respectively, prior to transgene insertion. Here is a list of PCR products based on the genotype at the locus.

1. 690 bp PCR product for *attP*x3 modified *H11* allele.

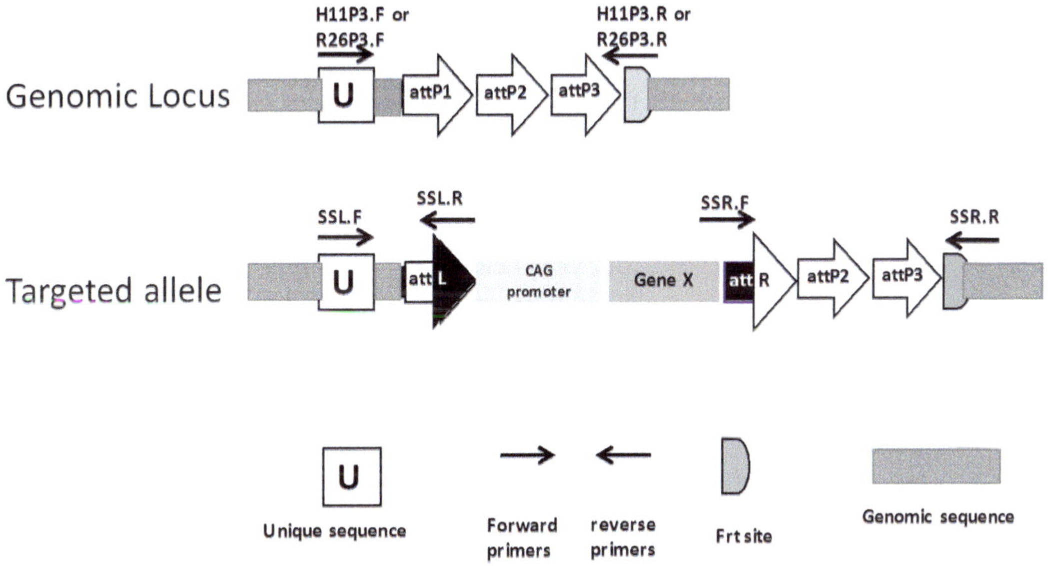

Fig. 5 TARGATT™ genotyping data analysis. The TARGATT™ mice contain three copies of the *attP* site (*attP*x3) in tandem at either the *H11* or *Rosa26* locus. Site-specific recombination between an *attP* site on the genomic DNA and the *attB* site on the TARGATT™ insertion vector results in transgene integration. Depending on which one of the three *attP* sites is recombined with the *attB* site, PCR fragments of different sizes are expected. The example shown here is a result of recombination of the first *attP* (*attP1*) site with the *attB* site. Using the SSL.F/SSL.R primer set, the expected PCR fragment size is 136 bp. Using the SSR.F/SSR.R primer set, the expected PCR fragment size is 225 bp. *Note*: indels are sometimes observed during recombination at either the 5′ or 3′ sites, in which cases, primers farther away from the junctions are used for PCR amplification

2. 364 bp PCR product for wild type *H11* allele.
3. *H11* heterozygotes show both the 690 bp and 364 bp bands.
4. 663 bp PCR product for *attP*x3 modified *Rosa26* allele.
5. 278 bp PCR product for wild type *Rosa26* allele.
6. *Rosa26* heterozygotes show both the 663 bp and 278 bp bands.

To detect site-specific transgenic founders at both the 5′ and 3′ junctions, two sets of PCR reactions are carried out; one with primer set SSL.F/SSL.R for 5′ and one with SSR.F/SSR.R for 3′ junction fragment (Fig. 5). Anticipated size of PCR fragments using primer set SSL are:

1. 136 bp PCR product for 5′-insertion at *attP* site 1.
2. 206 bp PCR product for 5′-insertion at *attP* site 2.

Anticipated size of PCR fragments using primer set SSR are:

1. 225 bp PCR product for insertion at *attP* site 2.
2. 155 bp PCR product for insertion at *attP* site 3.

To detect transgene insertion, primers specific to the gene of interests need to be used for PCR. All PCR fragments should be sequenced to confirm correct gene insertions (*see* **Note 7**).

3.5 Discussion

3.5.1 Key Points for Successful TARGATT

Since the first paper on using PhiC31 to generate site-specific transgenic mice was published in 2011 [9], we have successfully generated site-specific transgenic mouse models for over eighty DNA constructs using the TARGATT™ technology. Besides the models described in the publications of which ASC employees are co-authors, examples of TARGATT models generated by Applied StemCell, Inc. for its customers are published in refs. 1, 12–15, 18–21. A data summary of twenty-five TARGATT™ projects indicates that many factors contribute to the outcome of the experiments and some of the general guidelines are detailed as the following.

1. Strain background: The TARGATT™ mice are in two strain background, C57BL6/J and FVB. There is no statistic difference between the two strains in terms of efficiency.
2. The efficiency at the *Rosa26* locus appears to be slightly higher than the *H11* locus with an average efficiency of 10% at *Rosa26* versus 8% at *H11*. The overall efficiency varied, ranging from 0.3% to 40%.
3. Insertion size: There is a general trend that the bigger the insertion is, the lower the efficiency tends to be. With this method, the largest fragment we successfully inserted is 22 kb.

4. Mini-circle DNA versus plasmid DNA: There is no statistic difference in insertion efficiency between using a mini-circle vector with one *attB* site compared to a plasmid DNA with two *attB* sites.
5. The quality of the insertion plasmid prep is crucial to the success of obtaining site-specific transgenic mice. The plasmid prep needs to be RNase-free to ensure that integrase mRNA is active in the injection mix. In addition, the insertion plasmid in the injection mix needs to be in a supercoiled form in order to increase the rate of site-specific insertion. Random insertion events increase with decreasing fraction of supercoiled form.

3.5.2 CRISPR/Cas9 and TARGATT™ are Two Complementary Technologies

The discovery of CRISPR/Cas9 technology [22–24] has provided an unprecedented, revolutionary, and relatively easy method for genome editing in both cells and animals. The Cas9 protein makes site-specific double-stranded DNA breaks, and then uses the host cell repair system including non-homologous end joining (NHEJ) for error-prone repair or homology-directed repair (HDR) for precise DNA insertion. NHEJ is a major pathway which generates indels, resulting in frameshift. This feature is used as a strategy for generating knockout mice. However, HDR happens at a low rate in a host cell. As a result, DNA insertion rate is low using CRISPR/Cas9 when ssODNs (single-stranded oligodeoxynucleotides <200 bp) are used as donor DNA, although a recent paper by Quadros et al. [25] described a more efficient gene knockin (up to 100%) using long single-stranded DNA (ssDNA) donors with pre-assembled crRNA + tracrRNA + Cas9 ribonucleoprotein complex [25]. However, in this study, the longest ssDNA donor is 1.5 kb. Another limitation of CRISPR/HDR is that this method does not work in host cells that have inactive homologous recombination pathways such as nondividing or slow-dividing cells.

In comparison, phiC31 integrase is self-autonomous such that the enzyme itself contains all the functional components for recombination including DNA cutting, strand exchange, and re-ligation, and is therefore independent of the host cell recombination pathways. As a result, TARGATT™ works in nondividing cells. Another advantage of TARGATT is its higher efficiency in large DNA fragment insertion in transgenic mice with up to 40% live born animals being positive founders. DNA fragments up to 22 kb have been successfully inserted. Table 1 is a comparison of TARGATT™ with CRISPR/Cas9 and recommends which technology to use for specific transgenic models. More importantly, combination of CRISPR and TARGATT can be used to do gene replacement directly in embryos, bypassing ES cell homologous recombination step. For example, in the case of making a humanized mouse model (replacing a mouse gene with a human gene), the first step is to knock-in an *attP* site (~50-70 bp) using CRISPR at a genomic site

Table 1
CRISPR/Cas9 and TARGATT™ are two complementary technologies

Project purpose	CRISPR/Cas9	TARGATT™
Knockout (KO)	Efficient	
Point mutation	Yes	
Knock-in (200 nucleotides –2 kb)	Yes (<2 kb)	
Knock-in transgene in a safe genomic locus		Yes (up to 20 kb)
Knock-in gene replacement (two step approach)	First step Knock-in *attP* site	Second step Insert transgene

and at the same time knocking out the endogenous mouse gene. The second step is to insert transgene at the *attP* site using TARGATT.

In summary, CRISPR/Cas9 and TARGATT™ are two complementary technologies, and together provide a set of genetic tools for making diverse genetic modifications in mice.

4 Notes

1. Phenol/Chloroform/Isoamyl Alcohol and Chloroform are health hazard level. Read MSDS data sheet first.
2. Purified PhiC31 integrase protein will be available soon.
3. The PB washes diminish but do not abolish RNase contamination.
4. Make sure that pH of the three-fold diluted EB buffer is in a range 7–8.5. The lower pH will diminish DNA yield.
5. The plasmid DNA yield from a single prep should be in the range of 7.5 to 25 μg. To obtain more DNA, several preps can be performed at the same time and pooled; e.g., if you start with 5× 4 ml preps, you should finish with 5× ~100 μl DNA preps.

   Pool the DNA preps together, adjust the final volume to 500 μl with three-fold diluted EB buffer, and proceed to phenol/chloroform extraction.
6. If the DNA prep will be used to be co-injected with PhiC31 integrase protein, make sure that the resuspension solution does not contain EDTA. DNA resuspension can take time; therefore, it is a good practice to let it sit (with the lid closed tight) for several hours to overnight at room temperature before quantifying and use.

7. Indels at the 5′ and 3′ insertion sites are often observed, in which cases, can result in no PCR fragment amplification, thus causing false negative calling of the animals. To avoid losing of a good founder, primers further away from the 5′ and 3′ junction site can be used to PCR amplify the junction fragments.

Acknowledgment

The author would like to thank all Applied StemCell, Inc. employees who have contributed to improving the TARGATT™ technologies. Thanks also go to our customers who entrusted us on generating TARGATT™ models for their research.

References

1. Fogg PCM, Colloms S, Rosser S, Stark M, Smith MCM (2014) New applications for phage integrases. J Mol Biol 426 (15):2703–2716
2. Calos MP (2016) Phage integrases for genome editing. Genome Editing Part of the series. In: Advances in experimental medicine and biology. Springer, New York, pp 81–91 Date: 04 March 2016. Series ISSN 0065-2598
3. Groth AC, Olivares EC, Thyagarajan B, Calos MP (2000) A phage integrase directs efficient site-specific integration in human cells. Proc Natl Acad Sci U S A 97:5995–6000
4. Keravala A, Groth AC, Jarrahian S, Thyagarajan B, Hoyt JJ, Kirby PJ, Calos MP (2006) A diversity of serine phage integrases mediate site-specific recombination in mammalian cells. Mol Gen Genomics 276:135–146
5. Ma QW, Sheng HQ, Yan JB, Cheng S, Huang Y, Chen-Tsai Y, Ren ZR, Huang SZ, Zeng YT (2006) Identification of pseudo attP sites for phage phiC31 integrase in bovine genome. Biochem Biophys Res Commun 345:984–988
6. Thorpe HM, Smith MC (1998) In vitro site-specific integration of bacteriophage DNA catalyzed by a recombinase of the resolvase/invertase family. Proc Natl Acad Sci U S A 95:5505–5510
7. Olivares EC, Hollis RP, Chalberg TW, Meuse L, Kay MA, Calos MP (2002) Site-specific genomic integration produces therapeutic Factor IX levels in mice. Nat Biotechnol 20:1124–1128
8. Hollis RP, Stoll SM, Sclimenti CR, Lin J, Chen-Tsai Y, Calos MP (2003) Phage integrases for the construction and manipulation of transgenic mammals. Reprod Biol Endocrinol 1:79
9. Tasic B, Hippenmeyer S, Wang C, Gamboa M, Zong H, Chen-Tsai Y, Luo L (2011) Site-specific integrase-mediated transgenesis in mice via pronuclear injection. Proc Natl Acad Sci U S A 108:7902–7907
10. Soriano P (1999) Generalized lacZ expression with the ROSA26 Cre reporter strain. Nat Genet 21:70–71
11. Hippenmeyer S, Youn YH, Moon HM, Miyamichi K, Zong H, Wynshaw-Boris A, Luo L (2010) Genetic mosaic dissection of Lis1 and Ndel1 in neuronal migration. Neuron 68:695–709
12. Fan X, Petitt M, Gamboa M Huang M, Dhal S, Druzin ML, Wu JC, Chen-Tsai Y, Nayak NR (2012) Transient, inducible, placenta-specific gene expression in mice. Endocrinology 153:5637–5644
13. Guenther CA, Tasic B, Luo L, Bedell MA (2014) Kingsley DM (2014) A molecular basis for classic blond hair color in Europeans. Nat Genet 46(7):748–752
14. Feng D, Dai S, Liu F, Ohtake Y, Zhou Z, Wang H et al (2016) Cre-inducible human CD59 mediates rapid cell ablation after intermedilysin administration. J Clin Invest 126 (6):2321–2333
15. Li K, Wang F, Cao WB, Lv XX, Hua F, Cui B, Zhang XW, Shang S et al (2017) TRIB3 promotes APL progression through stabilization of the oncoprotein PML-RARa ad inhibition of p53-mediated senescence. Cancer Cell 31 (5):697–710

16. Zhu F, Gamboa M, Farruggio AP, Hippenmeyer S, Tasic B, Shule B, Chen-Tsai Y, Calos MP (2014) Dice, an efficient system for iterative genomic editing in human pluripotent stem cells. Nucleic Acids Res 42 (5):e34
17. Ruan J, Li H, Xu K, Wu T, Wei J, Zhou R, Liu Z, Mu Y, Yang S-l, Ouyang H, Chen-Tsai RY, Li K (2015) Highly efficient CRISPR/Cas9-mediated transgene knockin at the H11 locus in pigs. Sci Rep 5:14253
18. Booze ML, Hansen JM, Vitiello PF (2016) A novel mouse model for the identification of thioredoxin-1 protein interactions. Free Radic Biol Med 99:533–543
19. Sun N, YunJ LJ, Malide D, Liu C, Rovira L et al (2015) Measuring in vivo mitophage. Mol Cell 60(4):685–696
20. Devine WP, Wythe JD, George M, Koshiba-Takeuchi K, Bruneau BG (2014) Early patterning and specification of cardiac progenitors in gastrulating mesoderm. elife 3:e03848
21. Villamizar CA (2014) Characterization of the vascular pathology in the ACTA2 R258C mouse model and cerebrovascular characterization of the ACTA2 null mouse. UT GSBS dissertations and thesis (open acess). Paper 508
22. Cong L, Ran FA, Cox D, Lin S, Barretto R, Habib N, Hsu PD, Wu X, Jiang W, Marraffini LA, Zhang F (2013) Multiplex genome engineering using CRISPR/Cas systems. Science 339:819–823
23. Mali P, Yang L, Esvelt KM, Aach J, Guell M, DiCarlo JE, Norville JE, Church GM (2013) RNA-guided human genome engineering via Cas9. Science 339:823–826
24. Wang H, Yang H, Shivalila CS, Cheng AW, Shi L, Jaenisch R (2013) One-step generation of mice carrying mutations in multiple genes by CRISPR/Cas-mediated genome engineering. Cell 153(4):910–918
25. Quadros RM, Miura H, Harms DW, Akatuka H, Sato T, Aida T, Redder R, Richardson GP et al (2017) Easi-CRISPR: a robust method for one-step generation of mice carrying conditional and insertion alleles using long ssDNA donors and CRISPR ribonucleoproteins. Genome Biol 18:92

Chapter 5

Generating Genetically Engineered Mice Using a Spermatogonial Stem Cell-Mediated Method

Takuya Sato and Takehiko Ogawa

Abstract

Mouse spermatogonial stem cells (SSCs) can be grown in culture for long periods. Cultured SSCs, also called germline stem (GS) cells, maintain themselves by self-renewing proliferation while retaining the ability to differentiate into sperm. Thus, when transplanted into the seminiferous tubules of a host mouse testis, they settle in the basal compartment of the tubules and establish spermatogenenic colonies. The sperm produced in the host are competent to produce offspring. This can be exploited for the generation of genetically modified mice, through the transplantation of genetically modified GS cells. In this section, we describe a method of genome editing-mediated GS cell modification and transplantation.

Key words Spermatogonial stem cell, GS cell, Cell transplantation, Intratubular injection, CRISPR/Cas9, Genome editing, Gene targeting

1 Introduction

The culture method of mouse spermatogonial stem cells (SSCs) was first established in 2003 and the cell line was termed germline stem (GS) cells [1]. They proliferate in culture without losing the potential for either self-renewal or spermatogenesis. When GS cells are transplanted by intra-tubular injection into a host mouse whose testes are devoid of germ cells, they form spermatogenic colonies in the seminiferous tubules [1, 2]. In most cases, sperm can be retrieved from the host testes for microinsemination, and in the most successful cases, the host mice become fertile and can produce offspring through natural mating [1]. Since the number of colonies formed by transplantation is proportional to the number of transplanted SSCs, the transplantation method can be used as a quantitative SSC functional assay. Thus, studies using intratubular injection of GS cells have been used to clarify the details of SSC dynamics [3]. Combining this method with the genetic modification of GS cells will be indispensable for further research in this area.

Chengyu Liu and Yubin Du (eds.), *Microinjection: Methods and Protocols*, Methods in Molecular Biology, vol. 1874, https://doi.org/10.1007/978-1-4939-8831-0_5,

Transgenes can be introduced and established in GS cells through conventional gene transfer techniques [4, 5]. Gene targeting is possible by conventional homologous recombination, as performed in embryonic stem (ES) cells [4, 5]. However, the genetic modification of GS cells has proved more difficult than ES cells, mainly due to low gene transfer and genome targeting efficiencies in GS cells.

The rapid development of genome editing technologies, such as those meditated by transcription activator-like effector nucleases (TALEN) and clustered regularly interspaced short palindromic repeat (CRISPR)/Cas9, has had dramatic effects on biological experimentation in general, and in particular, on the genetic modification of GS cells. Wu et al. succeeded in repairing a one base mutation in the genome of GS cells derived from a mutant mouse. They introduced a single-stranded 89 bp oligodeoxynucleotide, along with CRISPR components, into GS cells, correcting the mutated sequence by homologous recombination [6]. In addition, we have demonstrated the high efficiency and accuracy knock-in of a longer 2.8–4.7 kbp transgene into GS cells using either CRISPR/Cas9 or TALEN [7]. The efficacy of CRISPR has also been shown in rat GS cells [8]. In this chapter, the entire procedure is described, from the genome modification of mouse GS cells to the production of offspring carrying the modification, using an example of a transgene knock-in at the Rosa26 locus.

2 Materials

2.1 JK1 Feeder Cells

1. JK1 cells (Cell Biolabs).
2. DMEM+10% FBS: Dulbecco's modified Eagle's medium, 1.54 g/l sodium bicarbonate (v/v), 10% heat-inactivated fetal bovine serum (v/v), and 1× Antibiotic–Antimycotic.
3. Mitomycin C solution: Dissolve in phosphate-buffered saline (PBS) at a concentration of 0.5 mg/ml.
4. Gelatin-coated plates: Dissolve gelatin purified from porcine skin in double-deionized water, 0.2% (w/v), and autoclave. To coat a 12-well culture plate, pour 0.5 ml of gelatin solution into each well, and incubate the wells at room temperature (22–25 °C) for 1 h. Remove excess gelatin solution by aspiration before use.
5. Tissue culture dishes, 100 mm.
6. Cell culture plates, 12 well.
7. Trypsin solution (0.25%).
8. Water bath (37 °C).
9. Centrifuge (swing-out rotor).
10. CO_2 incubator (CO_2 at 5%, humidified, 37 °C).

2.2 GS Cells

GS cells such as GS-DG1 (#AES0002) and Mvh-Venus GS (#AES0196) are available from the RIKEN BioResource Center. To establish a new GS cell line from mice, refer to a previous publication [9]. Since the C57BL/6 strain is relatively resistant to the establishment of GS cell lines, it is recommended to use DBA/2, ICR, or an F1 cross between C57BL/6 and those strains instead.

2.3 GS Cell Culture Medium

GS medium (previously described in [9]) consists of StemPro-34 SFM, supplemented with the compounds listed in Table 1.

2.4 Plasmids

For plasmid preparation, use a transfection-grade plasmid DNA isolation kit (e.g., QIAGEN EndoFree Plasmid Maxi Kit). Dissolve the purified plasmid in Tris-EDTA buffer at 2 μg/μl.

1. CRISPR/Cas9 plasmid vector: Prepare oligonucleotides of Rosa26-left-gRNA and Rosa26-right-gRNA (sequences shown below), which target the first intron of the Rosa26

Table 1
GS medium components

Component	Final concentration
Insulin	25 μg/ml
Transferrin	100 μg/ml
Putrescine	60 μM
D-(+)-Glucose	6 mg/ml
Pyruvic acid	30 μg/ml
DL-Lactic acid	0.1% (v/v)
Bovine serum albumin	5 mg/ml
L-Glutamine	2 mM
MEM vitamin (100×)	1×
MEM non-essential amino acids (100×)	1×
β-Mercaptoethanol	55 μM
Ascorbic acid	100 nM
D-Biotin	10 μg/ml
Heat-inactivated FBS (ES cell-qualified)	1% (v/v)
Mouse EGF	20 ng/ml
Human bFGF	10 ng/ml
Human GDNF	10 ng/ml
Sodium bicarbonate	1.4 g/l

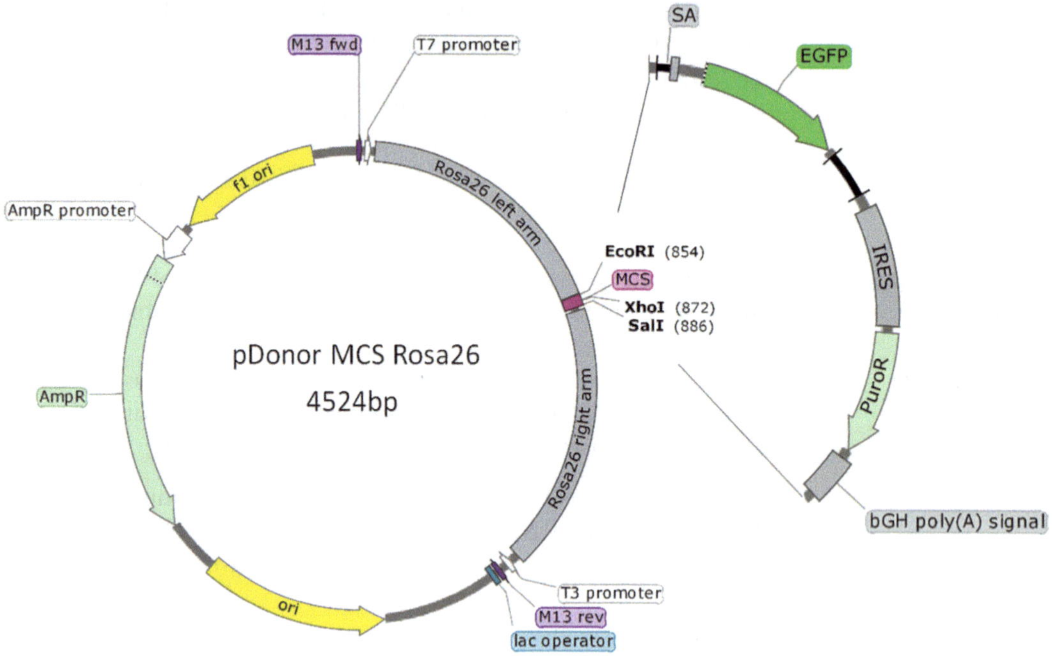

Fig. 1 Construction of the targeting vector for the Rosa26 locus. The SA-GFP-IRES-Puro fragment was inserted into the multiple cloning site (MCS) of the pDonor MCS Rosa26 plasmid. SA, splicing acceptor; IRES, internal ribosomal entry site; PuroR, puromycin-resistance gene

locus. Insert each into the gRNA cloning site of nickase-type pX335 vector (Addgene #42335) to produce two CRISPR/Cas9 plasmid vectors (*see* **Note 1**). This plasmid preparation is described in detail by Zhang et al. in *Nature Protocols* [10].

Rosa26-left-gRNA: 5'-TGATCTGCAACTCCAGT-3'.
Rosa26-right-gRNA: 5'-TGCCCAGAAGACTCCCGC-3'

2. Donor plasmid vector: pDonor MCS Rosa26 (Addgene # 37200) was used as a donor vector for the Rosa26 locus. It consists of a multiple cloning site (MCS) flanked by homology arms of 800 bp, each corresponding to the first intron of the Rosa26 locus (Fig. 1) [11]. We cloned the cassette sequence, which contains splicing acceptor (SA), enhanced green fluorescent protein (EGFP), internal ribosome entry site (IRES), and puromycin resistance (PuroR) gene sequences, into the MCS of the donor vector (Fig. 1).

2.5 Electroporation

1. Lonza Nucleofector 2b.
2. Lonza Cell Line Nucleofector Kit T.
3. Puromycin.
4. PCR primers: Forward Rosa26: 5'-CTACTGTGTTGGCGGA CTGG-3', Reverse Rosa26: 5'-AAGACAACAAACACCTGAA CTTTG-3' and Transgene reverse 5'-CCGGATCCACTA

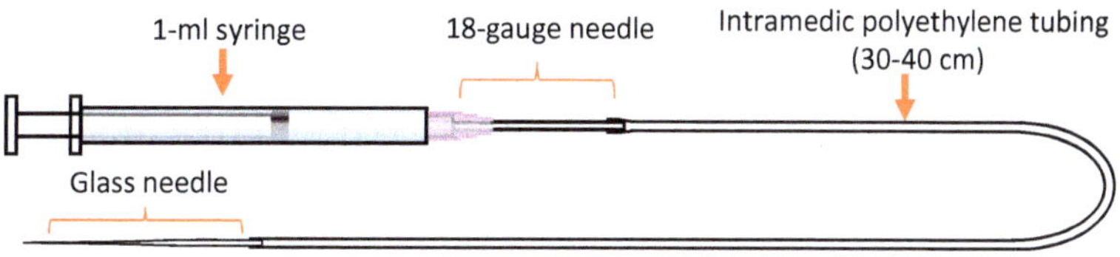

Fig. 2 Setup of the injection system. The injection tool is composed of a 1 ml syringe, an 18-gauge needle, Intramedic polyethylene tubing, and a glass needle, connected in order

GTTCTAGA-3′. Expected sizes of the PCR products are 2552 bp in wild-type alleles and 1678 bp in targeted alleles.

5. PCR reagents: KOD-Plus-NEO.
6. Thermal cycler.

2.6 Spermatogonial Transplantation

1. Cell strainer (40 μm).
2. Glass capillary (outer diameter 1.2 mm).
3. Micropipette puller.
4. Microinjection needles: To make glass needles for cell injection, pull a glass capillary with a micropipette puller. Make the diameter of the needle tip 50–80 μm by snapping off the tip with forceps at an appropriate site.
5. Intramedic polyethylene tubing (inner diameter 1.14 mm, outer diameter 1.57 mm).
6. Disposable Syringe (1 ml).
7. Injection device: Connect an 18-gauge needle and Intramedic polyethylene tube to a 1 ml syringe (Fig. 2). The tip of the 18-gauge needle should be rolled up with a file or similar tool to avoid damaging the Intramedic polyethylene tubing. Connect the glass tube to the polyethylene tube just before use.
8. Trypan blue solution (0.4%).
9. Recipient male mouse: WBB6F1 W/W^v (6–10 weeks old).

3 Methods

3.1 Preparation of JK1 Feeder Cells

1. Basic handling, thawing, and culturing of cryopreserved JK1 cells is described in the attached manual from suppliers.
2. Seed the JK1 cells at approximately 1–10 × 10^4 cells/ml in 100 mm dishes with DMEM +10% FBS, and culture them to become confluent, which usually takes 7 days. Then, remove the medium, rinse with 10 ml PBS, and add 5 ml DMEM +10% FBS. Then, add 100 μl mitomycin C and incubate for 3 h in a 5% CO_2-humidified incubator.

3. Wash 3 times with 10 ml PBS, add 1 ml 0.25% trypsin, and incubate for 10 min.
4. Add 10 ml DMEM +10% FBS to stop the trypsin reaction, collect the suspended cells, and transfer into a 50 ml tube for centrifugation at 300 × *g* for 5 min at room temperature.
5. Remove the supernatant by aspiration and suspend the pelleted cells in 5 ml DMEM +10% FBS. Count the cells using a small aliquot of the suspension and resuspend them at a concentration of 2.5×10^5 cells/ml with DMEM +10% FBS.
6. Seed 1 ml of the cell suspension into each well of a 12-well plate and incubate in a CO_2 incubator at 37 °C until use. We recommend that these feeder cells be used within 1–2 weeks.

3.2 Maintenance of GS Cells

1. Quickly thaw freeze-stocked GS cells in a water bath at 37 °C, and transfer them to 10 ml GS medium. Centrifuge at 300 × *g* for 5 min at room temperature.
2. Remove the supernatant by aspiration and add 1 ml GS medium to suspend the cells.
3. Remove medium from a plate well seeded with JK1 feeder cells. Transfer the suspended GS cells to the well. Incubate at 37 °C, 5% CO_2, replacing the medium every 3–4 days. Continue culturing for 7–10 days until GS cell colonies become large and dense (*see* **Note 2**).
4. To passage the cells, remove the medium and rinse the cells with PBS once. Add 200–300 μl 0.25% trypsin and incubate at room temperature for 5 min.
5. After confirming cell detachment with an inverted microscope, add 1 ml GS medium and dissociate the cells to a single-cell suspension by pipetting. Centrifuge and remove the supernatant. Add 3 ml GS medium and resuspend the cells. Seed 1 ml each in JK1-seeded 12-well plate and incubate in a CO_2 incubator. Repeat **steps 4–5** to maintain GS cells.

3.3 Electroporation

1. Prepare a plate well with JK1 feeder cells for seeding of transfected GS cells. Replace the DMEM +10% FBS with 1 ml GS medium and incubate in a CO_2 incubator at 37 °C until use.
2. Prepare 2–3 wells of adequately-grown GS cells. Aspirate the medium, rinse once with PBS, and add 200–300 μl 0.25% trypsin. Incubate the plate at room temperature for 5 min to dissociate the cells.
3. After confirming cell detachment with an inverted microscope, add 1 ml GS medium and dissociate cells to a single-cell suspension by pipetting. Count the cell concentration using a small aliquot. Take a volume of the suspension containing $1\text{–}2 \times 10^6$ cells and centrifuge at 300 × *g* for 5 min at room temperature.

4. During centrifugation, prepare the Electroporation Reagent by mixing 82 μl Nucleofector Solution and 18 μl Supplement; both components of Cell Line Nucleofector Kit T. To this Electroporation Reagent, add 2 μg of each CRISPR/Cas9 vector and 6 μg of targeting vector and mix by gentle shaking or tapping.
5. After cell centrifugation, carefully remove the supernatant, add the Electroporation Reagent from **step 4**, and gently resuspend the cell pellet. Transfer the cell suspension to an electroporation cuvette attached to a Cell Line Nucleofector. Insert the cuvette into the Nucleofector Cuvette Holder of Nucleofector 2b. Select program A-023 and carry out electroporation.
6. After electroporation, immediately remove the cuvette from the Nucleofector and gently transfer the cell suspension to the pre-warmed plate prepared in **step 1**, using the disposable pipette included in the Nucleofector kit.
7. After 2 days of culturing, passage the cells at a 1:3 ratio.
8. Once GS cell proliferation is observed (7–10 days after transfection), exchange the medium with GS medium containing 0.15 μg/ml puromycin to select for resistant clones. JK1 cells can survive at this concentration of puromycin.
9. Replace the medium with fresh GS medium containing puromycin after 2 days.
10. Replace the medium with GS medium without puromycin after 5–7 days. Change the medium every 2–3 days, until the GS cell colonies grow to approximately 300 μm in diameter.

3.4 Colony Isolation and Genotyping

1. Under observation with an inverted microscope, pick up individual colonies using a P200 micropipette and transfer them to individual microtubes. We usually transfer 10–12 clones in a single experiment.
2. Add 200 μl 0.25% trypsin into each microtube and incubate for 5 min at room temperature.
3. Add 200 μl GS medium to stop trypsinization. Dissociate cells to single-cell suspension by pipetting. Centrifuge at 300 × *g* for 5 min at room temperature.
4. After centrifugation, aspirate the supernatant and resuspend the cells in 1 ml GS medium. Aspirate the medium from a JK1-seeded plate, and transfer the GS cell suspension in the JK1-seeded plate.
5. Change the culture medium every 3 days. Once the GS cells have proliferated to an appropriate density (*see* **Note 2**), passage them at a 1:3 ratio, usually every 7–10 days. To select clones of knocked-in GS cells, genomic PCR is performed. To harvest GS cells for genomic PCR (*see* **Note 3**), one well of a

12-well plate is necessary. Remove the medium and rinse the cells with PBS, then add 200–300 μl 0.25% trypsin for incubation for 5 min at room temperature.

6. After confirming cell detachment with an inverted microscope, add 1 ml GS medium and dissociate cells to a single-cell suspension by pipetting.
7. Centrifuge the cell suspension at 300 × *g* for 5 min at room temperature and aspirate the supernatant. Extract genomic DNA from the harvested GS cells using a genomic DNA extraction kit (such as the DNeasy kit, QIAGEN), according to the manufacturer's instructions.
8. Perform PCR using the genomic DNA as a template, with the appropriate primers and according to the instructions provided with PCR reaction components.
9. Perform electrophoresis with the PCR products along with controls and size markers on a 1% (w/v) agarose gel. Confirm that the positions of PCR products on the gel are compatible with the expected sizes.

3.5 Spermatogonial Transplantation

1. Prepare 2–3 wells of adequately grown GS cells. Remove the medium by aspiration and add 1 ml PBS for gentle rinsing once.
2. Add 200–300 μl 0.25% trypsin solution and incubate at room temperature for 5 min.
3. After confirming cell detachment with an inverted microscope, add 1 ml GS medium and pipette repeatedly to dissociate cells.
4. Filter the cell suspension through a cell strainer (40-μm mesh size) into a 50 ml centrifuge tube. Count the cells using a small aliquot. Centrifuge at 300 × *g* for 5 min at room temperature, and resuspend the cell pellet with an appropriate volume of GS medium, usually 20–40 μl, for a concentration of 5–20 × 10^6 cells/ml.
5. Add 0.4% Trypan blue solution to the cell suspension to a final concentration of 0.08% and mix well. Transfer the cell suspension to a 1.5 ml tube and place it on ice.
6. Use Avertin or other anesthetizing reagents for intraperitoneal injection into the recipient mouse. Under adequate anesthesia, lay the mouse down on his back and spray 70% EtOH over the lower abdomen. Open the lower abdomen by midline dissection to reach the abdominal cavity. Hold the epididymal fat with forceps and pull the testis out of the groin. Then, place the mouse under a stereomicroscope, with head far and tail close to the practitioner. Recognize the axes of the testis: longitudinal (head-tail), dorsoventral (dorsal-abdominal), and right-left (lateral-central). Place the testis head-side to the practitioner,

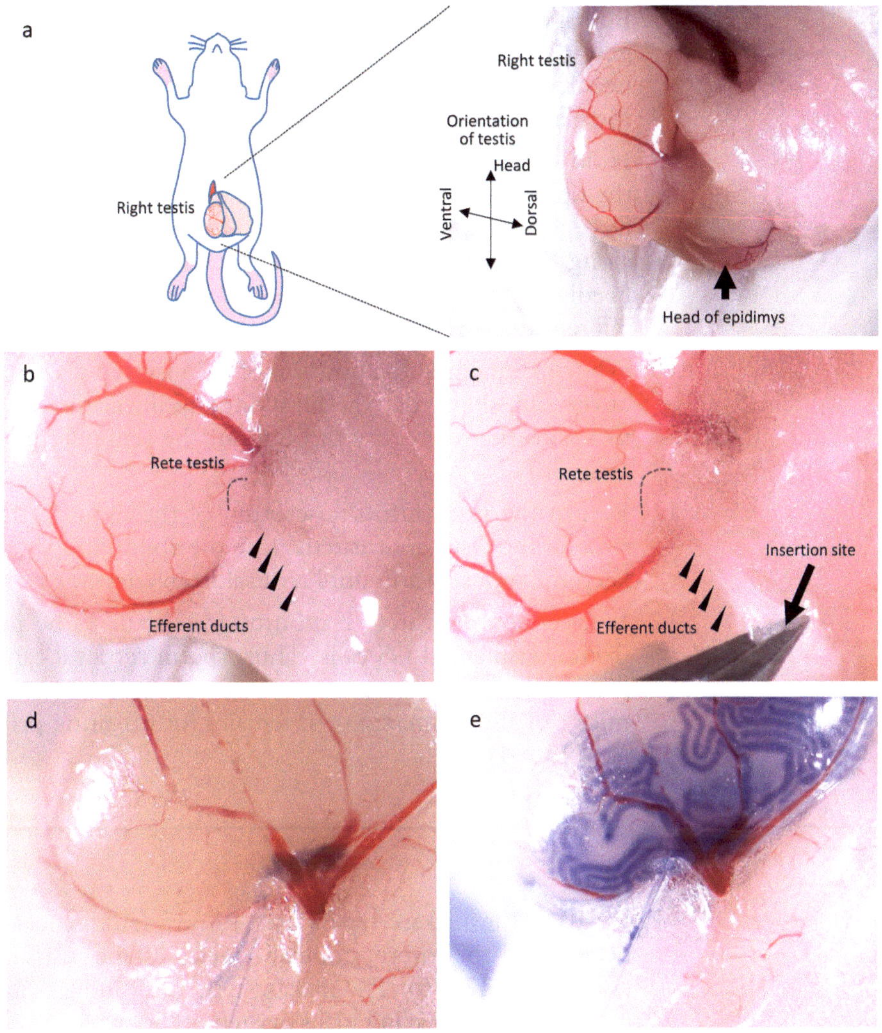

Fig. 3 Process of spermatogonial injection. (**a**) Positioning of the right testis for injection. (**b**) Position of efferent ducts and rete testis. (**c**) Efferent ducts were maintained as a bundle while exposed. The efferent ducts and rete testis are clearly visible. (**d**) Insertion of the glass needle into the bundle to insert the tip into the rete testis. (**e**) Injection of a GS cell suspension into the rete testis followed by spreading to the seminiferous tubules

dorsal-side up (Fig. 3a). With this placement, the rete testis and its outlet, which leads to the efferent ducts, are facing upward, close to the practitioner.

7. Push the epididymal fat to the left to reach the right testis, and vice versa, and identify the efferent ducts, which are covered by a thin membrane (Fig. 3b). Hold the ducts as a bundle with the forceps and gently pull it away from the outlet of the rete testis (Fig. 3c). Using additional forceps, free the bundle from the adhering tissue, the tunica albuginea of the testis. The bundle will be free except for the connection to the outlet of the rete testis.

8. Resuspend the cells in the GS cell suspension tube again by gentle pipetting. Load 10 μl of the cell suspension into a glass needle from the rear end using a micropipette. Then, connect the glass needle to the injection device.
9. Hold the distal portion of the bundle with forceps and insert the tip of the glass needle into the bundle at a site close to the forceps, on the side of the testis (Fig. 3c). Advance the needle along the efferent ducts toward the testis until the tip of glass needle reaches the junction where the rete testis connects with efferent ducts (Fig. 3d).
10. While maintaining the position of the inserted glass needle tip at the site, slowly deploy the syringe to inject the cell suspension into the rete cavity throughout the seminiferous tubules (Fig. 3e).
11. Return the testis into the scrotum of the mouse. After both sides of the testis are treated, close the abdominal incision and keep the mouse warm until arousal from anesthesia.
12. It takes approximately 2 months for sperm to be produced from transplanted GS cells. Thus, sperm retrieval can be performed at 2 months or later for microinsemination. The host mice can also be mated with females for offspring production (*see* **Note 4**).

4 Notes

1. We used the nickase-type Cas9: PX335 (Addgene #42335), instead of the more popular nuclease-type Cas9: PX330 (Addgene #42230). The nickase-type Cas9 induces a double-strand break by making nicks in each of the two close opposing strands of the target sequence with the paired guide-RNA [12]. Thus, the double-nicking Cas9 system is more specific and accurate than the original nuclease-type Cas9.
2. Timing of passaging is important to maintain the quality of GS cells. Avoid overgrowth of GS cells, which results in loss of spermatogenic ability. Ideally, GS cells should be passaged when colonies contain approximately 100 cells, and have grape-like appearance (Fig. 4).
3. For genomic PCR of GS cells, contamination from JK1 cells could make the results unclear, especially when distinguishing between homo- or heterozygous insertion of the transgene. To remove JK1 cells, seed the suspension of GS cells in a feeder-free 6-well plate, and incubate for 2–3 h in a 5% CO_2-humidified incubator. Then, collect GS cells, which will be floating or weakly adhered weakly to the dish, by pipetting the medium gently several times. The JK1 cells will remain attached to the dish.

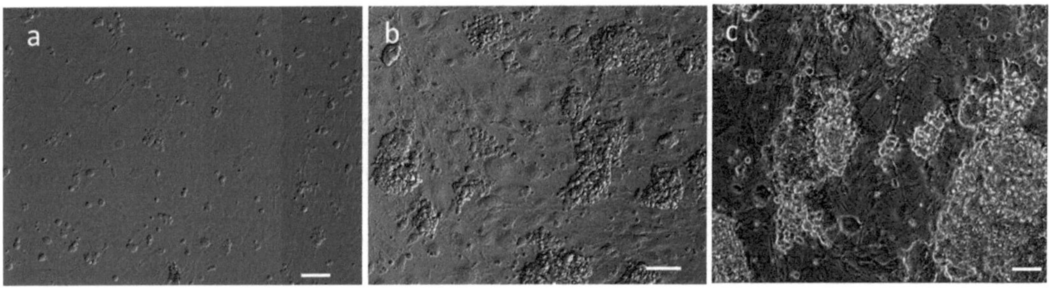

Fig. 4 Timing of GS cell passaging. (**a**) Low cell density GS cells shortly after seeding. (**b**) GS cells at a density suitable for passaging, with many colonies of approximately 100 cells. (**c**) Overgrown GS cells. Most colonies have grown too large. Scale, 100 μm

4. It is possible to obtain offspring by natural mating of the host males with females. However, their fertility largely depends on the extent of GS cell colonization in the testis. In general, the rate of offspring production is not high and it takes time even when successful. Therefore, microinsemination is recommended to obtain offspring, by retrieving sperm from the host testis 2 months after transplantation or later.

Acknowledgments

We thank Anri Miyamoto and Kumiko Katagiri for technical assistance. This study was funded by a Grant-in-Aid for Young Scientists (A) 26713012 (to T. S.); The Uehara Memorial Foundation (to T. S.); and Innovative Areas, "Mechanisms regulating gamete formation in animals" 25114007 (to T.O.).

References

1. Kanatsu-Shinohara M, Ogonuki N, Inoue K, Miki H, Ogura A, Toyokuni S, Shinohara T (2003) Long-term proliferation in culture and germline transmission of mouse male germline stem cells. Biol Reprod 69(2):612–616. https://doi.org/10.1095/biolreprod.103.017012
2. Ogawa T, Ohmura M, Tamura Y, Kita K, Ohbo K, Suda T, Kubota Y (2004) Derivation and morphological characterization of mouse spermatogonial stem cell lines. Arch Histol Cytol 67(4):297–306. https://doi.org/10.1679/aohc.67.297
3. Kanatsu-Shinohara M, Shinohara T (2013) Spermatogonial stem cell self-renewal and development. Annu Rev Cell Dev Biol 29:163–187. https://doi.org/10.1146/annurev-cellbio-101512-122353
4. Kanatsu-Shinohara M, Toyokuni S, Shinohara T (2005) Genetic selection of mouse male germline stem cells in vitro: offspring from single stem cells. Biol Reprod 72(1):236–240. https://doi.org/10.1095/biolreprod.104.035659
5. Kanatsu-Shinohara M, Ikawa M, Takehashi M, Ogonuki N, Miki H, Inoue K, Kazuki Y, Lee J, Toyokuni S, Oshimura M, Ogura A, Shinohara T (2006) Production of knockout mice by random or targeted mutagenesis in spermatogonial stem cells. Proc Natl Acad Sci U S A 103(21):8018–8023. https://doi.org/10.1073/pnas.0601139103
6. Wu Y, Zhou H, Fan X, Zhang Y, Zhang M, Wang Y, Xie Z, Bai M, Yin Q, Liang D, Tang W, Liao J, Zhou C, Liu W, Zhu P, Guo H, Pan H, Wu C, Shi H, Wu L, Tang F, Li J (2015) Correction of a genetic disease by

CRISPR-Cas9-mediated gene editing in mouse spermatogonial stem cells. Cell Res 25 (1):67–79. https://doi.org/10.1038/cr.2014.160

7. Sato T, Sakuma T, Yokonishi T, Katagiri K, Kamimura S, Ogonuki N, Ogura A, Yamamoto T, Ogawa T (2015) Genome editing in mouse Spermatogonial stem cell lines using TALEN and double-nicking CRISPR/Cas9. Stem Cell Rep. https://doi.org/10.1016/j.stemcr.2015.05.011
8. Chapman KM, Medrano GA, Jaichander P, Chaudhary J, Waits AE, Nobrega MA, Hotaling JM, Ober C, Hamra FK (2015) Targeted Germline modifications in rats using CRISPR/Cas9 and Spermatogonial stem cells. Cell Rep 10(11):1828–1835. https://doi.org/10.1016/j.celrep.2015.02.040
9. Sato T, Katagiri K, Kubota Y, Ogawa T (2013) In vitro sperm production from mouse spermatogonial stem cell lines using an organ culture method. Nat Protoc 8 (11):2098–2104. https://doi.org/10.1038/nprot.2013.138
10. Ran FA, Hsu PD, Wright J, Agarwala V, Scott DA, Zhang F (2013) Genome engineering using the CRISPR-Cas9 system. Nat Protoc 8 (11):2281–2308. https://doi.org/10.1038/nprot.2013.143
11. Perez-Pinera P, Ousterout DG, Brown MT, Gersbach CA (2012) Gene targeting to the ROSA26 locus directed by engineered zinc finger nucleases. Nucleic Acids Res 40 (8):3741–3752. https://doi.org/10.1093/nar/gkr1214
12. Ran FA, Hsu PD, Lin CY, Gootenberg JS, Konermann S, Trevino AE, Scott DA, Inoue A, Matoba S, Zhang Y, Zhang F (2013) Double nicking by RNA-guided CRISPR Cas9 for enhanced genome editing specificity. Cell 154(6):1380–1389. https://doi.org/10.1016/j.cell.2013.08.021

Chapter 6

Chimeric Mouse Generation by ES Cell Blastocyst Microinjection and Uterine Transfer

Yubin Du, Wen Xie, Fan Zhang, and Chengyu Liu

Abstract

The ability to generate chimeric mice through microinjecting embryonic stem (ES) cells into blastocysts is a critical step for the conventional ES cell-mediated knockout technology. In recent years, designer nuclease-based methods, especially the CRISPR technology, have substantially decreased the needs for blastocyst microinjection. However, this method has still remained as a valuable technique for generating sophisticated genetic models as well as for stem cell research. In this chapter, we describe the detailed procedures used in our laboratory on how to use ES cells to produce chimeric mice, including derivation and inactivation of MEF feeder cells, culturing and handling of mouse ES cells, collection and microinjection of blastocysts, and finally implantation of injected blastocysts into the uteri of pseudopregnant surrogate mothers.

Key words Blastocyst microinjection, Embryo transfer, Mouse ES cells, Knockout, Chimeric mice, MEF feeders

1 Introduction

Two decades of research converged on a method that matured in the late 1980s, and won a Nobel Prize yet another two decades later in 2007 for developing the revolutionary mouse knockout technology. DNA homologous recombination technology started in the 1960s by Oliver Smithies [1], combined with mouse embryonic stem (ES) cell derived from embryos by Martin Evans [2] made it possible to modify genes in the ES cells [3, 4]. With embryo manipulation work developed earlier [5], a general strategy for knocking out mouse gene was developed by Mario Capecchi and colleagues [6]. Several years later, conditional knockout mouse using cre-lox method was developed [7, 8].

Using these methods, a great majority of mouse genes have been conditionally modified in mouse ES cells through internationally coordinated consortia [9], and these ES cell lines are available from publicly funded repositories for reasonable fees (https://www.

Chengyu Liu and Yubin Du (eds.), *Microinjection: Methods and Protocols*, Methods in Molecular Biology, vol. 1874, https://doi.org/10.1007/978-1-4939-8831-0_6, © Springer Science+Business Media, LLC, part of Springer Nature 2019

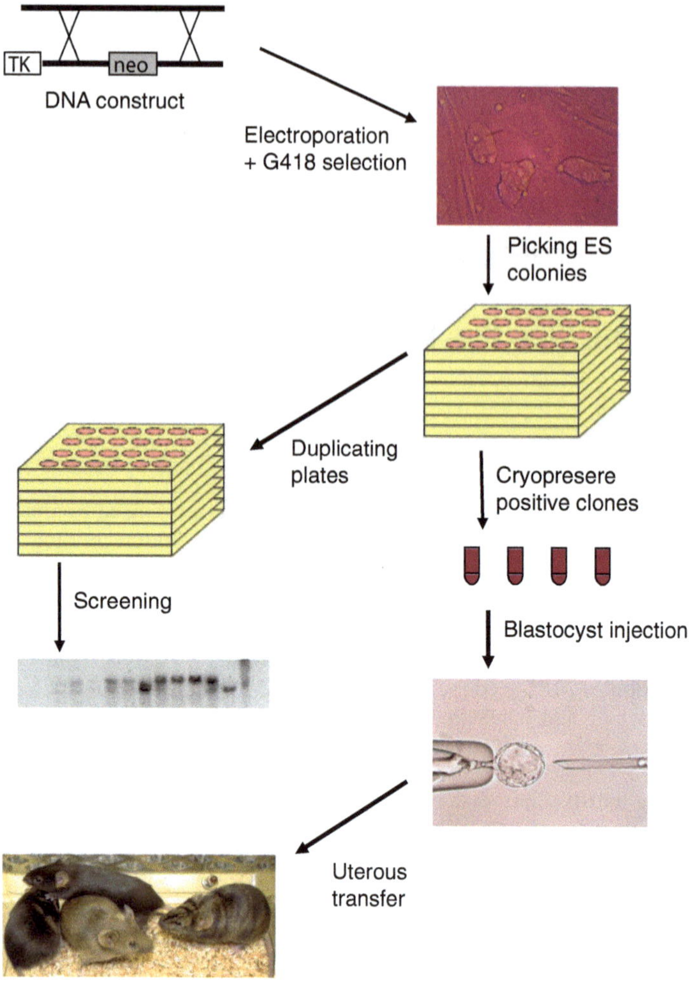

Fig. 1 Outlining of the conventional ES cell-mediated gene knockout procedures. A targeting construct containing a selection marker gene (neo) and several Kb of homologous arms is electroporated into a germline-competent ES cell line. After antibiotic selection, ES colonies are picked and grown up in multi-well plates. A duplicate set of plates are used to extract genomic DNA for genotyping, while the original set of plates are cryopreserved. After genotyping, positive ES clones are thawed from the cryopreserved plates and expanded. These ES clones are microinjected into blastocyst stage embryos to generate chimeric mice, which are used to breed with wild-type mice for passing the ES cell-derived gametes to the next generation

komp.org, http://www.mousephenotype.org, https://www.mmrrc.org). Converting these valuable ES cell lines into live mice requires the generation of chimeric mice by mixing ES cells with early embryos. Although a few alternative methods for generating chimeric mice have been reported, such as ES cell-embryo aggregation [10] and morula injection methods [11], the most widely used method is blastocyst microinjection as outlined in Fig. 1. In this method, ES cells are injected into blastocyst stage embryos using a pulled micro glass needle, and then the injected blastocysts are surgically implanted into the uteri of pseudopregnant surrogate mothers. If the injected ES cells are pluripotent, they will contribute to various lineages of the embryos, including the germline (usually sperm but can sometimes be eggs), which can pass the modified gene to the next generation.

In the past decade, designer nuclease-based technologies, including ZFN (zinc finger nucleases) ([12] and Chapter 17 of this book), TALEN (transcription activator-like effector nucleases) [13], and CRISPR (clustered regularly interspaced short palindromic repeats) [14, 15], have been developed and increasingly used for precisely manipulating the genome of mice and a variety of other species. The CRISPR technology has emerged as the predominant method for engineering mouse lines due to its simplicity and extraordinary efficiency. It has also dramatically shortened the timespan for gene knockout and small knockins because the CRISPR reagents can be directly injected into zygotes without using ES cells and chimera generation. However, the conventional blastocyst microinjection method is still a valuable tool because: (1) at this time the efficiency for using CRISPR to generate conditional knockouts and large knockins are still low and unpredictable, so a significant number of facilities still use the conventional methods for complicated projects; (2) floxed ES clones are already available for a majority of mouse genes, which can be obtained and used for generating conditional knockout mice through blastocyst injection; (3) besides generating gene-targeted mice, blastocyst microinjection is a valuable method for other areas of biomedical research, such as testing pluripotency of stem cells, tetraploid complementation studies [16], interspecific chimera study ([17] and Chapter 18 of this book); (4) at the present time, there are still concerns about CRISPR's nonspecific activities at off-target sites, although most off-target mutations, if any, can be eliminated through breeding. Therefore, it is unlikely that the blastocyst microinjection method will become obsolete anytime soon.

2 Materials

2.1 Targeted Mouse ES Cells and Cell Culture Reagents

1. Germline-competent mouse ES cells: mouse ES clones with desired genetic modifications can be obtained from various repositories, such as https://www.komp.org, http://www.mousephenotype.org, and https://www.mmrrc.org, or engineered in individual laboratories.
2. Inactivated mouse embryonic fibroblast (MEF): MEF cells can be obtained from commercial sources or prepared from E13.5 embryos as described in Subheading 3.1.

3. Tissue culture hood.
4. Water bath.
5. Centrifuge with rotor for spinning down cells.
6. Inverted microscope for cell culture.
7. Tissue culture dishes (35 mm, 60 mm, 100 mm, and 150 mm).
8. Gelatin solution (0.1% in distilled water) for coating dishes.
9. PBS without Ca^{2+} and Mg^{2+}.
10. DMEM medium for culturing MEF.
11. Knockout DMEM medium for culturing ES cells.
12. ES cell qualified fetal bovine serum (FBS).
13. 1 M HEPES solution (pH 7.4).
14. Tissue culture grade β-mercaptoethanol (1000×).
15. 100× pen-strep solution.
16. 100× nonessential amino acids (AA).
17. 200 mM L-glutamine solution (100×).
18. 0.25% Trypsin-EDTA.
19. MEF culture medium: DMEM with 10% FBS, 1× pen-strep, 1× L-glutamine.
20. ES cell culture medium: Knockout DMEM with 15% FBS, 1× nonessential AA, 1× pen-strep, 1× L-glutamine, 10 mM HEPES (pH 7.4), 0.1 mM β-mercaptoethanol, 1000 IU/mL leukemia inhibitory factor (LIF), and optional G418 (200 μg/mL).
21. Serological pipets.

2.2 Mice

1. Twenty C57BL/6 stud males: 2–12 months old; individually caged (*see* **Note 1** for choosing recipient mouse strains).
2. Fifty to 100 C57BL/6 females: 4–12-week-old; up to five mice per cage.
3. Twenty vasectomized CD-1 male mice: 2–12 months old; individually caged. Vasectomy procedure can be performed in house or by the animal supplier for a reasonable fee.
4. Fifty to 100 CD-1 females: 6–16-week-old; up to five mice per cage.

2.3 Equipment

1. CO_2 incubator: 37 °C, 6% CO_2, and humidified by a pan containing distilled water.
2. Benchtop sterilizer for autoclaving surgical instruments.
3. Hot bead sterilizer for sterilizing surgical instruments between animals when more than one animal is used during a session.
4. Heating water blanket for keeping the cage warm for aiding post-surgical recovery.

5. Stereo microscope with both top and bottom light sources for embryo collection and embryo transfer procedures.
6. Inverted microscope with 5x, 10x, 20x, and 40x objectives for blastocyst microinjection.
7. Two Eppendorf TransferMan 4r micromanipulators.
8. Two Eppendorf CellTram vario.
9. Air table for reducing vibration of the microinjection setup.
10. Optional digital video camera and computer system for training and recording.

2.4 Tools and Supplies

1. Microinjection chamber: custom made by using rubber cement to glue a standard histological glass slide to a 2 mm thick aluminum frame. Lids of tissue culture dishes (60–100 mm) can also be used, but it requires using much more mineral oil in order to fully immerse the microinjection drop to prevent evaporation.
2. Holding pipette and microinjection needles: both micropipettes are commercially available (Eppendorf VacuTip FCH, holding capillary, 25° tip angle, 60 μm inner diameter, 0.5 mm flange, sterile (5175240006); and TransferTip (ES), for ES cell transfer, 20° tip angle, 15 μm inner diameter, 1 mm flange, sterile (5175107004)). Alternatively, microinjection needles and holding pipette can be made by using a pipette puller and microforge following procedures described by Colin Stewart [18].
3. Embryo transfer pipette (Fig. 3a): the embryo transfer pipette is made by heating a long-tipped Pasteur pipette using a flame, and then pull the softened glass to a thin tube (150–200 μm in diameter) outside of the flame. The pipette is connected to a plastic mouth piece and rubber aspirator tubing (Sigma, Cat: A5177-5EA) with a disposable syringe filter to prevent accidently sucking medium into the mouth or dripping saliva into the medium.
4. Small hair clipper for removing hair from the surgical sites of mice.
5. Betadine (Povidone-Iodine Swabsticks).
6. Alcohol swab.
7. 29G x ½″ 1-CC syringe and 3 mL disposable plastic syringe .
8. 18G and 25G hypodermic needles.
9. Sterile surgical gloves.
10. 5–0 Vicryl absorbable suture.
11. Surgical instrument sterilization tray.
12. Micro dissecting scissors.

13. Small iris forceps.
14. Micro dissecting tweezers (#5).
15. Dieffenbach micro clamp.
16. Wound clip applier.
17. Suturing needle holder.
18. Pipette tips.

2.5 Embryo Culturing Media and Chemicals

1. M2 embryo culture medium.
2. ES cell microinjection medium: Knockout-DMEM with 10% FBS, 1% nonessential AA, 1x L-glutamine, 25 mm HEPES (pH 7.4).
3. Embryo culture-tested mineral oil.
4. Ketamine/Xylazine anesthetics: prepared fresh by mixing 1.7 mL 0.9% NaCl with 0.2 mL Ketamine (100 mg/mL) and 0.1 mL Xylazine (20 mg/mL) (*see* **Note 2** for proper storage and required record keeping).
5. Bupivacaine (0.25% injectable solution).
6. Meloxicam (5 mg/mL injectable solution).

3 Methods

3.1 Preparation of MEF Cells

MEF cells are usually derived from the carcass of E13.5 embryos. Depending on needs, MEFs can be derived from various mouse strains, such as CD-1 outbred mice for economy, or strains with single (neomycin, hygromycin, or puromycin) or multi (DR4) antibiotic resistance for culturing ES cells that need selection. Ready-to-use MEFs are also available from commercial sources.

1. Set up mating of desired mouse strain for timed pregnancy.
2. Sacrifice females that are visually pregnant 13 days after checked positive for mating.
3. Dissect E13.5 embryos out of uteri and their yolk sacs, and wash in PBS three times.
4. Under dissecting microscope, using sharp tweezers pinch off and discard the heads. Tear open abdominal cavity and remove liver (red color) and other internal organs.
5. Use a pair of tweezers to place the embryo carcasses into the barrel of a 3 mL disposable plastic syringe with an attached 18G needle. Add 2 mL MEF medium into the syringe, and then push the plunger to force the carcasses through the 18G needle to break them into small pieces. Further triturate the pieces by drawing in and pushing out the tissues 4–5 times through the 18G needle. Up to 10 embryos can be processed in one

100 mm culture dish. Add 15 mL MEF medium to each 100 mm dish and shake to distribute the tissue pieces evenly throughout the dish.

6. Place the dishes in the CO_2 incubator undisturbed for 48 h to allow the tissue clumps to attach and fibroblasts to migrate out.
7. Aspirate off the culture medium and wash the dish with PBS twice. Add 3 mL of Trypsin-EDTA and incubate for 5 min at 37 °C to lift off attached cells and tissue clumps. Add 15 mL MEF medium and split the content into two 150 mm culture dishes. Add 20 mL MEF medium to each dish and culture for two more days in the CO_2 incubator.
8. Trypsinize the cells using Trypsin-EDTA, and add MEF medium to inactivate trypsin. Centrifuge at 270 × *g* for 5 min to pellet the cells and then resuspend in MEF medium at ~2×10^6 cells/mL.
9. Add equal volume of 2x cryopreservation medium (50% FBS, 30% MEF medium, and 20% DMSO). Aliquot 1 mL cell suspension (~1×10^6 cells) into each cryopreservation vial and label the vial as P1 MEFs. Freeze the cells at −80 °C overnight before transferring into liquid nitrogen for long term storage.
10. Prior to using, P1 MEFs can be thawed and further expanded for 3–4 additional passages (1 to 4 expansion for each passage). For inactivation, the MEFs are harvested and resuspended in MEF medium in a 50 mL sterile tube. Place the tube inside the chamber of a γ-irradiator and irradiate for 3000 RADS (exposure time is dependent on the source of the irradiation) (*see* **Note 3** for MEF inactivation using Mitomycin C).
11. After inactivation, MEF cells can be used directly or cryopreserved at 2×10^6 per vial for later use.

3.2 Preparing ES Cells for Microinjection

ES cells are normally stored in liquid nitrogen but can be temporarily stored at −80 °C. We usually thaw ES cells 2 days before injecting them, but culturing dishes need to be coated with gelatin and plated with MEF feeders prior to thawing.

1. Coat each 60 mm dish with 2 mL 0.1% gelatin at room temperature overnight or at 4 °C for at least 2 h.
2. Aspirate off gelatin solution. Plate one vial of MEFs (~2×10^6 cells) in 5 mL of MEF medium and culture until the cells attach (*see* **Note 4** for attaching time) at 37 °C in a CO_2 incubator.
3. Take ES cell frozen vials out of the liquid nitrogen freezer and quickly immerse them in a 37 °C water bath till majority of ice in the vial melt. Wipe the tube clean and disinfect outside with 70% alcohol. Transfer the content into a sterile 15 mL tube, and then gradually dilute out the cryoprotectant (DMSO) by slowly adding ES cell medium while shaking. If the cryopreservation medium is diluted by 10-folds or more (<1% DMSO),

ES cells can be directly plated without centrifugation. Otherwise, *see* **step 4** below.

4. (Optional) Centrifuge the cells for 5 min at 270 x *g*. Aspirate off the supernatant and suspend the ES cells in 7 mL ES cell medium.
5. Aspirate off medium from MEF dish prepared in **step 2** above, and gently add resuspended ES cells into the dish. Culture in a CO_2 incubator at 37 °C.
6. Change medium and check ES cells under the inverted microscope every day to make sure cells do not become over confluent (<80% confluency) and ES colonies are morphologically normal (spindle shaped colonies with smooth and shining surface, and borders for individual cells are hard to distinguish).
7. If necessary, split the cells by first washing with PBS, and then digest with 1.5 mL 0.25% trypsin-EDTA for 5 min at 37 °C. Add 5 mL ES medium and pipet up and down ~10 times to break up colonies into small clumps. We usually make a 1:4 to 1:10 split depending on when the cells will be subsequently used.
8. On the day prior to the scheduled blastocyst microinjection, we trypsinize ES cells for ensuring that majority of cells are in log growing phase.

3.3 Blastocyst Collection

1. Four days prior to the scheduled microinjection, select estrus C57BL/6 female mice and mate with stud males in the afternoon (*see* **Note 5**). The next morning, check the females for vaginal plugs. Plugged females are group caged until needed for blastocyst collection.
2. On the day of microinjection, the females are euthanized by CO_2 exposure. The abdominal cavity of each mouse is cut open using a pair of scissors, and the entire uterus together with oviducts and ovaries are removed and placed in a 60 mm culture dish containing 3 mL M2 medium.
3. Clean the reproductive organs by dissecting out associated fat, and place the cleaned organs into a new 60 mm dish. Separate the two uterine horns by cutting with a pair of small scissors.
4. Fill a 3 mL disposable plastic syringe (with 25G needle) with M2 medium. Bend the needle to ~45° using a pair of forceps. Under a stereo microscope, insert the needle into the lumen of each uterine horn near the uterine-oviduct junction. Flush each uterine horn with 1 mL medium. Up to 10 uterine horns can be flushed in the same 60 mm dish.
5. After all uterine horns are flushed, gently swirl the dish for 30–60 s to gather blastocysts to the center area of the dish. Pick up blastocysts and morulae (some morulae will develop into usable blastocysts a few hours later) using an embryo

transfer pipette and wash them through two drops of M2 medium. Culture the embryos in a M2 drop in the CO_2 incubator until needed for microinjection in the afternoon.

3.4 Blastocyst Injection

1. After finishing collecting blastocysts, trypsinize ES cells and replate them onto a gelatin-coated dish (without MEF). Culture in a CO_2 incubator for 1.5–2 h, which is enough for most feeders to reattach but most ES cells only loosely adhere to the MEFs.
2. Aspirate off the medium with unattached cells and debris. Use a 5 mL serological pipet containing 3 mL injection medium (*see* **Note 6**), gently rinse off the ES cells adhering to the MEF feeders. Place the collected ES cells into a 15 mL sterile tube and store on ice for the entire duration of microinjection.
3. Add a drop of injection medium into the center of the injection chamber, and immediately cover the drop with mineral oil. Transfer a few microliters of ES cell suspension into the drop (*see* **Note 7**). Place the injection chamber onto the microscope stage.
4. Attach the holding pipette and blastocyst injection needle into the capillary holders, which are connected to the CellTram vario through Teflon tubes filled with mineral oil (*see* **Note 8**).
5. Lower the tip of both the holding pipette and injection needle into the drop containing ES cells. Focus the microscope to make sure the ES cells are in focus, and adjust micromanipulators to make sure the holding pipette and injection needle are in the center of the view field and nearly touch the bottom of the injection chamber.
6. Draw healthy looking ES cells into the injection needle by turning the CellTram vario (*see* **Note 9** for selecting healthy ES cells).
7. Transfer about 30 blastocysts into the medium drop containing ES cells. Move the tip of the holding pipette to the vicinity of a blastocyst, and turn the CellTram vario to firmly grasp the blastocyst. Use the Eppendorf 4r micromanipulator to align the microinjection needle so that it is in the same straight line and focus plane as the holding pipette and blastocyst (Fig. 2a). Quickly insert to injection needle into the blastocyst, ideally near the boundary of inner cell mass (ICM) and trophoblasts (Fig. 2b). Turn the CellTram vario to slowly expel 5–15 ES cells into the cavity, and then withdraw the needle (Fig. 2c). Move the injected blastocyst to a different area of the medium drop.
8. After all blastocysts are injected, take them out of the injection chamber using a mouth pipette and culture at 37 °C in a CO_2 incubator until uterine transfer (see below).

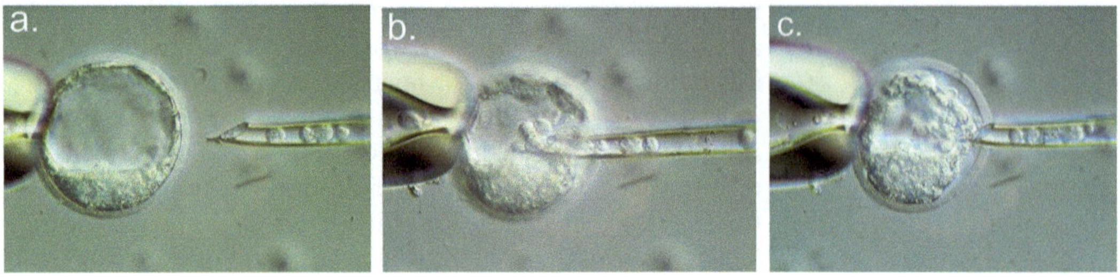

Fig. 2 Blastocyst microinjection procedures. (**a**) A blastocyst is firmly grabbed by the holding pipette, and the microinjection needle loaded with ES cells is brought to the vicinity of the blastocyst. (**b**) With a sudden and forceful move of the micromanipulator, the needle is inserted into the blastocyst cavity, slowly turn CellTram vario to expel 5–15 ES cells out of the injection needle. (**c**) Slowly withdraw the needle while leaving all or most of injected ES cells inside the blastocoel

3.5 Uterine Transfer

1. Three days prior to the scheduled blastocyst injection (1 day after mating the blastocyst donor mice), select estrus CD-1 females and set up mating with vasectomized male mice to prepare foster mothers. The next morning, check the mice for vaginal plugs, and house plugged females in groups until needed for uterine transfer (2 days after plug). As a general rule, we prefer to obtain one plugged foster mother for every 3 plugged blastocyst donor mice (*see* **Note 10** for timing of embryo transfer and alternative methods).
2. After finish blastocyst injection (Subheading 3.4, **step 8**), anesthetize recipient foster mother by injecting (i.p.) Ketamine/Xylazine mixture at 10 μL/g of body weight.
3. Remove hair from the dorsal lumbar area of the back using a small hair clipper. Disinfect the clipped area using alcohol swab followed by Betadine.
4. Use a pair of dissecting scissors making a small dorsal midline incision in the disinfected skin area, and then tear by spreading the back of the scissors to ~1 cm long. Separate skin and body wall using the back of the scissors.
5. Put the animal on its left side, and use iris forceps to move the skin opening to the right ovary/uterus area (the ovary associated fat pad can be seen through the muscle body wall as a slightly paler area). Cut a small incision in the body wall and expand the opening to 0.5–1 cm long with the back of scissors. Avoid cutting major blood vessels.
6. Use a pair of small iris forceps grab the fat pad and gently pull the fat and associated reproductive organs out of abdominal cavity. Hold the fat pad in place with a Dieffenbach clip (Fig. 3c).
7. Load the embryo transfer pipette with some M2 medium and a small air bubble near the tip. Then, suck the mouth piece to draw in 7 injected blastocysts followed by another small air bubble (*see* Fig. 3b). The two air bubbles mark the boundaries of the suspended embryos, which are useful for ensuring

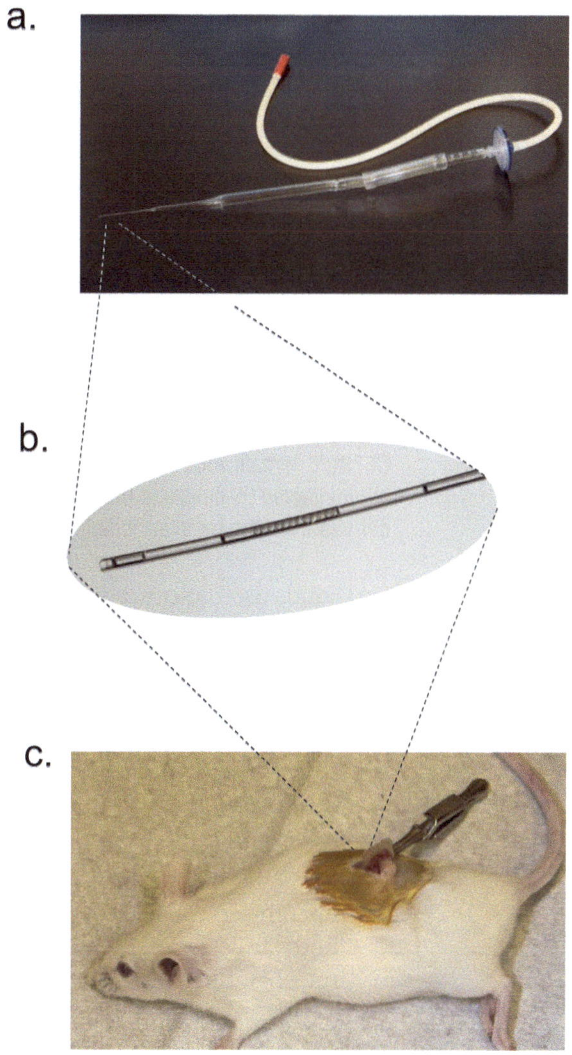

Fig. 3 Implantation of blastocysts into the uterus of a surrogate mother. (**a**) A photograph of a typical embryo transfer pipette, which consists of a pulled Pasteur pipet connected with a plastic mouth piece through tubing and a disk filter. (**b**) A magnified view of the tip of the transfer pipette showing embryos sandwiched between two air bubbles. (**c**) An anesthetized surrogate mother with her reproductive organs (including ovary, oviduct, and a segment of the uterine horn) exteriorized through a small incision on the body wall and skin. A Dieffenbach clip is clamped to the fat pad associated with the ovary for holding the organs in place. Then, the tip of the transfer pipette is inserted into the lumen of the uterine horn through a hole punched by a sterile needle

successful expelling of all embryos with minimum volume of M2 medium.

8. Under the dissecting microscope, use a 25G needle punching a hole on the uterine wall near the oviduct junction. While using a pair of fine tweezers loosely holding the uterus in place with one

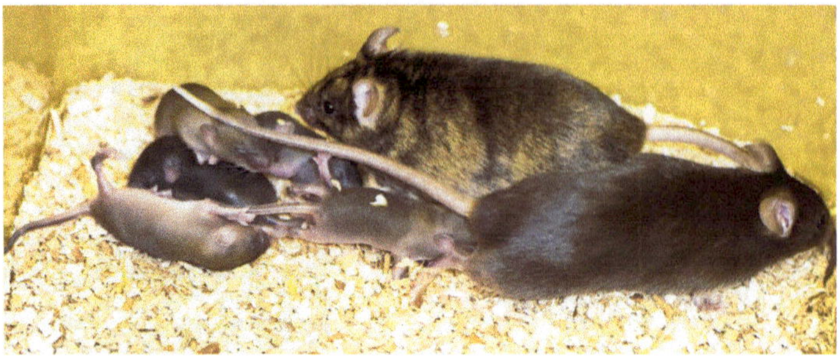

Fig. 4 A photograph showing germline pups born to a mating pair (C57BL/6 mother x chimeric father). The male chimera (agouti and black mix color) was produced by injecting ES cells derived from 129 mouse strain into a recipient blastocyst collected from C57BL/6 strain. After breeding the chimera with a C57BL/6 female mouse (black), those pups with agouti color are derived from microinjected ES cells and may carry the targeted gene, while the black pups are derived from the recipient blastocysts

hand, use the other hand carefully insert the tip of the transfer pipette into the hole punched by the needle. Gently blow into the mouthpiece until the two air bubbles enter the uterus.

9. Remove the Dieffenbach clip and carefully push the uterus, ovary, and fat pad back into the abdominal cavity. Using absorbable 5–0 Vicryl suture to sew the body wall incision, and drop one drop of 0.25% Bupivacaine solution, which serve as a long-lasting local anesthetic.
10. Turn the mouse on its right side, and move the midline skin opening to the left ovary/uterus area and cut open the muscle body wall as described in **step 5**, and repeat **step 6–9** for transplanting 7 blastocysts into the left uterine horn.
11. Close the skin incision with stainless steel surgical wound clips. Inject (i.p.) the mouse with diluted Meloxicam solution (2–5 mg/kg of body weight) for reducing pain. Place the mouse in a warm cage, which is placed on top of a circulating warm water blanket.
12. After mice awake (usually 30–60 min post-surgery), the cage is returned to the animal room. Wound clips are removed 10–14 days after surgery.
13. Pups are born 16–17 days after the embryo transfer procedure. At postnatal day 5, skin patches with various darkness indicate chimerism. At 7–8 day, the coat color of the pups becomes clear, which indicates the degree of chimerism. The offspring are weaned when they are 19–21 days old.
14. Usually male chimeras with high percentage of coat color derived from ES cells are preferable for mating with wild-type C57BL/6 females for achieving germline transmission. As shown in Fig. 4, pups bear the color of the ES cells are germline pups, which may carry the gene altered in the ES cells.

4 Notes

1. Most earlier ES cell lines were derived from 129 mouse strain, which has coat colors ranging from agouti, chinchilla, and albino depending on the substrains. When these ES cells are injected into C57BL/6 (B6) blastocysts, chimeras are easily recognizable. After breeding with B6 mice, germline transmission can also be recognized through coat color because the 129 color is dominant. Due to increased robustness, ES cells derived from F1 hybrid strain (often 129 and B6 mix) are also popular for gene-targeting experiments. Various ES cell lines have also been derived from B6 mice. For easy visualization of chimerism and germline transmission, some of these ES cell lines were intentionally derived from B6 mice with the c-2J albino mutation [19], or with restored agouti locus [20]. For blastocyst injection, it is recommended to use B6 variants with a different coat color as blastocyst donors, such as albino B6 ES cells can be injected into black B6 blastocysts or vice versa, and agouti-restored B6 ES cells can be injected into black B6 blastocysts. While chimerism can always be recognized if the ES cells and recipient blastocysts are in different color, some germline transmitted pups may not be visualized if the ES cell line was derived from mice that are heterozygous on the color gene, such as the F1 hybrid ES cells and agouti-restored B6 ES cells. In these cases, half of the germline pups will show the dominant color (usually agouti) while the other half of germline pups may have the recessive color, which is usually the same color as the blastocyst donor mice. Therefore, if only a few germline offspring with the desired color are obtained, genotyping all offspring regardless of coat color can increase the number of germline offspring.
2. Ketamine is a controlled substance regulated by the US Drug Enforcement Administration (DEA). A license is required for purchasing and storing this anesthetic. A lock box is needed to store these chemicals. A log book is needed to record the usage until it is depleted or expired.
3. If a γ-irradiator is not available, MEFs can be inactivated by treating with Mitomycin C. Briefly, grow P3 or P4 MEFs to 70–90% confluency in MEF medium, and then change to MEF medium containing 10 μg/mL Mitomycin C and culture for 2–3 h. After washing the MEFs three times with PBS, they are trypsinized, centrifuged down, and resuspended in MEF medium at the density of 4×10^6/mL. Next, equal volume of 2x cryopreservation medium (50% FBS, 30% MEF medium, and 20% DMSO) is added, and then aliquoted into

cryopreservation vials at 1 mL per vial (2 x 10^6 cells). The vials are placed in a freezing container and store overnight at −80 °C before transferring into a liquid nitrogen freezer for long term storage.

4. Most MEF feeders will attach to gelatin-coated dishes within 2 h after plating, but we have noticed that MEFs from certain mouse strains, such as CF-1, take much longer (~6 h for CF-1 MEF) to attach. Under emergency situations, ES cells and MEFs can be plated simultaneously as long as LIF-containing ES medium is used.
5. A swollen and reddish genital area is a good indication that the female is in estrus. Estrus females can be selected from a pool of breeding age animals. On average, only ~20% of the females are in estrus, but sometimes the females in the same cage are in-sync which makes the selection process more unpredictable. An obvious alternative method is to inject the females with hormones to induce superovulation, which may or may not be better than natural mating depending on the substrains, age, and housing environments.
6. Normal ES cell medium contains 10 mM HEPES (pH 7.4), but we normally increase its concentration to 25 mM in injection medium, because the injection drop will be out of the CO_2 incubator for an extended period of time. We also lower the FBS concentration from 15% to 10% and omit LIF from the injection medium due to possible deleterious effects of LIF and other serum growth factors on blastocysts. It is imperative to omit G418, which is often present in ES culture medium for maintaining selection, from the injection medium because the recipient blastocysts are usually not neomycin-resistant.
7. It is important to load the injection chamber with an appropriate density of ES cells. Too few cells make it difficult to find healthy cells for loading the injection needle. Too many cells make it difficult to separate the good ES cells from the bad ones and selectively load the good cells. Also, too many dead cells often lead to sticky needle and make it hard to aspirate and expel cells. If a high proportion of cells are lysed, adding 300 IU of DNase I into the injection drop can help in reducing stickiness. Some people prefer to use a cooling stage to keep the injection chamber at 4 °C for reducing cell death.
8. For precisely controlling the microinjection needle, it is critical that the entire line, from CellTram vario to injection needle, is completely filled with oil. Any air bubbles inside the line will reduce the precision of moving cells in and out of the injection needle. For holding pipettes, precision is less important and therefore the line can be filled with either oil or air.

9. Healthy ES cells are small and round, while MEFs are much larger under the microinjection microscope. Healthy undifferentiated ES cells tend to stick to each other forming aggregates. Preferentially loading small aggregates (usually 2–3 cells) into the injection needle provides another level of selection for good ES cells, which should enhance germline transmission later. We usually preload a large number of ES cells into the same needle so many blastocysts can be injected, but some scientists prefer to load just enough cells (usually 10–20) for injecting each blastocyst.
10. Injected blastocysts should be implanted into the uteri of E2.5 pseudopregnant foster mothers on the same day of microinjection. However, if insufficient number of E2.5 surrogate mothers are available, E0.5 and E1.5 mother can also be used but the blastocysts should be transferred into the oviducts instead of uteri. Surgical implantation of injected embryos requires expertise and extensive experience. Transfer without surgery using ParaTechs' Non-Surgical Embryo Transfer (NSET™) Device (http://www.paratechs.com/nset/) or transfer tips [21] have also led to acceptable results.

Acknowledgements

This work was supported by the Intramural Research Program of the National Heart, Lung, and Blood Institute at the US National Institutes of Health.

References

1. Smithies O, Connell GE, Dixon GH (1962) Chromosomal rearrangements and the evolution of haptoglobin genes. Nature 196:232–236
2. Evans MJ, Kaufman MH (1981) Establishment in culture of pluripotential cells from mouse embryos. Nature 292:154–156
3. Smithies O, Gregg RG, Boggs SS, Koralewski MA, Kucherlapati RS (1985) Insertion of DNA sequences into the human chromosomal beta-globin locus by homologous recombination. Nature 317:230–234
4. Thomas KR, Capecchi MR (1986) Introduction of homologous DNA sequences into mammalian cells induces mutations in the cognate gene. Nature 324:34–38. https://doi.org/10.1038/324034a0
5. Papaioannou VE, McBurney MW, Gardner RL, Evans MJ (1975) Fate of teratocarcinoma cells injected into early mouse embryos. Nature 258:70–73
6. Mansour SL, Thomas KR, Capecchi MR (1988) Disruption of the proto-oncogene int-2 in mouse embryo-derived stem cells: a general strategy for targeting mutations to non-selectable genes. Nature 336:348–352. https://doi.org/10.1038/336348a0
7. Orban PC, Chui D, Marth JD (1992) Tissue- and site-specific DNA recombination in transgenic mice. Proc Natl Acad Sci U S A 89:6861–6865
8. Gu H, Zou YR, Rajewsky K (1993) Independent control of immunoglobulin switch recombination at individual switch regions evidenced through Cre-loxP-mediated gene targeting. Cell 73:1155–1164
9. Austin CP, Battey JF, Bradley A, Bucan M, Capecchi M, Collins FS, Dove WF, Duyk G, Dymecki S, Eppig JT, Grieder FB, Heintz N, Hicks G, Insel TR, Joyner A, Koller BH, Lloyd KC, Magnuson T, Moore MW, Nagy A, Pollock JD, Roses AD, Sands AT, Seed B, Skarnes

WC, Snoddy J, Soriano P, Stewart DJ, Stewart F, Stillman B, Varmus H, Varticovski L, Verma IM, Vogt TF, von Melchner H, Witkowski J, Woychik RP, Wurst W, Yancopoulos GD, Young SG, Zambrowicz B (2004) The knockout mouse project. Nat Genet 36:921–924. https://doi.org/10.1038/ng0904-921

10. Wood SA, Allen ND, Rossant J, Auerbach A, Nagy A (1993) Non-injection methods for the production of embryonic stem cell-embryo chimaeras. Nature 365:87–89. https://doi.org/10.1038/365087a0
11. Valenzuela DM, Murphy AJ, Frendewey D, Gale NW, Economides AN, Auerbach W, Poueymirou WT, Adams NC, Rojas J, Yasenchak J, Chernomorsky R, Boucher M, Elsasser AL, Esau L, Zheng J, Griffiths JA, Wang X, Su H, Xue Y, Dominguez MG, Noguera I, Torres R, Macdonald LE, Stewart AF, DeChiara TM, Yancopoulos GD (2003) High-throughput engineering of the mouse genome coupled with high-resolution expression analysis. Nat Biotechnol 21:652–659. https://doi.org/10.1038/nbt822
12. Geurts AM, Cost GJ, Freyvert Y, Zeitler B, Miller JC, Choi VM, Jenkins SS, Wood A, Cui X, Meng X, Vincent A, Lam S, Michalkiewicz M, Schilling R, Foeckler J, Kalloway S, Weiler H, Menoret S, Anegon I, Davis GD, Zhang L, Rebar EJ, Gregory PD, Urnov FD, Jacob HJ, Buelow R (2009) Knockout rats via embryo microinjection of zinc-finger nucleases. Science 325:433. https://doi.org/10.1126/science.1172447
13. Sung YH, Baek IJ, Kim DH, Jeon J, Lee J, Lee K, Jeong D, Kim JS, Lee HW (2013) Knockout mice created by TALEN-mediated gene targeting. Nat Biotechnol 31:23–24. https://doi.org/10.1038/nbt.2477
14. Wang H, Yang H, Shivalila CS, Dawlaty MM, Cheng AW, Zhang F, Jaenisch R (2013) One-step generation of mice carrying mutations in multiple genes by CRISPR/Cas-mediated genome engineering. Cell 153:910–918. https://doi.org/10.1016/j.cell.2013.04.025
15. Shen B, Zhang J, Wu H, Wang J, Ma K, Li Z, Zhang X, Zhang P, Huang X (2013) Generation of gene-modified mice via Cas9/RNA-mediated gene targeting. Cell Res 23:720–723. https://doi.org/10.1038/cr.2013.46
16. Eggan K, Akutsu H, Loring J, Jackson-Grusby-L, Klemm M, Rideout WM 3rd, Yanagimachi R, Jaenisch R (2001) Hybrid vigor, fetal overgrowth, and viability of mice derived by nuclear cloning and tetraploid embryo complementation. Proc Natl Acad Sci U S A 98:6209–6214. https://doi.org/10.1073/pnas.101118898
17. Kobayashi T, Yamaguchi T, Hamanaka S, Kato-Itoh M, Yamazaki Y, Ibata M, Sato H, Lee YS, Usui J, Knisely AS, Hirabayashi M, Nakauchi H (2010) Generation of rat pancreas in mouse by interspecific blastocyst injection of pluripotent stem cells. Cell 142:787–799. https://doi.org/10.1016/j.cell.2010.07.039
18. Stewart CL (1993) Production of chimeras between embryonic stem cells and embryos. Methods Enzymol 225:823–855
19. Schuster-Gossler K, Lee AW, Lerner CP, Parker HJ, Dyer VW, Scott VE, Gossler A, Conover JC (2001) Use of coisogenic host blastocysts for efficient establishment of germline chimeras with C57BL/6J ES cell lines. BioTechniques 31:1022–1024 1026
20. Pettitt SJ, Liang Q, Rairdan XY, Moran JL, Prosser HM, Beier DR, Lloyd KC, Bradley A, Skarnes WC (2009) Agouti C57BL/6N embryonic stem cells for mouse genetic resources. Nat Methods 6:493–495. https://doi.org/10.1038/nmeth.1342
21. Cui L, Zhang Z, Sun F, Duan X, Wang M, Di K, Li X (2014) Transcervical embryo transfer in mice. J Am Assoc Lab Anim Sci 53:228–231

Chapter 7

Creating Knockin Alleles in Mouse Embryonic Stem Cells by CRISPR/Cas9-Mediated Homologous Recombination Without Drug Selection

Pinghu Liu, Yan Li, Jingqi Lei, and Lijin Dong

Abstract

The rapidly evolving CRISPR/Cas9-mediated genome editing provides the convenience of genome manipulation directly in mouse zygotes for a number of genomic manipulations; but knockins of large insertions prove to be relatively inefficient at least with double-strand DNA as targeting constructs. Here, we describe an alternative approach to the direct genome editing in mouse zygotes by generating knockin alleles in mouse embryonic stem cells first with CRIPSR-mediated homologous recombination. Our results show this approach is efficient and requires no drug selection in mouse embryonic stem cells as in classic gene targeting. As the result, knockin alleles across many target loci are created in mouse embryonic stem cells and readily transmitted through germline. The knockin alleles created in ES cells can also serve as valuable tools for in vitro stem cell differentiation.

Key words CRISPR/Cas9, Genetic manipulation, Homologous recombination, Mouse embryonic stem cells, Genome editing, Reporter, Indel, Knockin

1 Introduction

Genome editing, when applied to mouse genetic engineering, offers the convenience of genome manipulation directly in zygotes [1], bypassing the need for embryonic stem (ES) cells as in classic gene targeting [2]. In its simplest terms, genome editing works by making a double-strand break(s) in genomic DNA near a target site, followed by repairs made through either homologous recombination (HR) when an externally supplied DNA template is present, or non-homologous end joining which alters the open reading frame of the target DNA by a small insertion or deletion (indel). The center piece of the technology, however, is the ability to guide a DNA endonuclease to predetermined target sites in the genome producing site-specific double-strand breaks, or to program substrate sequence specificity of the endonuclease involved. CRISPR/Cas9 system has become popular because its target

Chengyu Liu and Yubin Du (eds.), *Microinjection: Methods and Protocols*, Methods in Molecular Biology, vol. 1874,
https://doi.org/10.1007/978-1-4939-8831-0_7, © Springer Science+Business Media, LLC, part of Springer Nature 2019

programmability is the simplest to achieve among all the systems developed to date.

CRISPR/Cas9-mediated genome editing in mouse zygotes is highly efficient for an arrange of applications in mouse genomic manipulation. In our hands, for example, indel mutagenesis, genomic deletion, and oligo-mediated HR are typically achieved at high efficiencies at which 20 to 100% of F0 pups carry intended mutations. Consistent with observations of many other laboratories, however, knockin of relatively large insertions with double-strand DNA as HR template, such as knockin of the Cre recombinase or a fluorescence reporter, is still far less efficient by direct injection of mouse C57B6 zygotes. In our hands, knockin of reporter or Cre alleles into various target loci directly in zygote typically requires multiple injections and intensive screening through many F0 pups before recovering one correct HR event. In this chapter, we describe CRISPR/Cas9-mediated HR in mouse ES cells, rather than directly in mouse zygotes, for the knockin of large insertions. With this approach, knockin alleles have been produced in a more predictable manner. Integrated knockin alleles are stable in ES cells in general, and readily transmitted through germline. Knockin alleles created by the procedures described here are useful for mouse ES cell differentiation studies as well. Step by step instructions with notes detailing challenging steps are provided for design and synthesis of target-specific gRNAs by T7 in vitro transcription (T7-IVT) (*see* Subheading 3.1), validation of gRNAs experimentally (*see* Subheading 3.2), assembly of targeting constructs (*see* Subheading 3.3), mouse embryonic stem (ES) cell handling (*see* Subheading 3.4), and genome editing in mouse R1 ES cells (*see* Subheading 3.5) respectively, for successful engineering of knockin alleles. Finally, HR events can be identified by a simple screening procedure (*see* Subheading 3.6), without the need of either positive or negative selection as in classic gene targeting.

2 Materials

2.1 Equipment

1. Thermocycler.
2. Table-top refrigerated microcentrifuge.
3. Nanodrop spectrophotometer.
4. Gel electrophoresis systems (horizontal and vertical).
5. UV transilluminator.
6. Cell culture incubator.
7. Laminar flow hood.
8. Benchtop low-speed centrifuge with swing bucket rotor.
9. Bright field cell counter.
10. Liquid nitrogen storage container (LN2).

2.2 Reagents

2.2.1 Buffer and Cell Culture Media

1. Lysis buffer for genomic DNA isolation: 10 mM Tris–HCl (pH 7.5), 10 mM EDTA pH 8.0, 10 mM NaCl, 0.2% SDS, 200 μg/ml proteinase K.
2. Mouse fibroblast culture medium: DMEM with high glucose, 10% Fetal Bovine Serum, 1× MEM nonessential amino acids solution, 2 mM L-alanyl-L-glutamine, 1% penicillin/streptomycin.
3. Mouse ES cell culture medium: KnockOut™ D-MEM, 1× MEM nonessential amino acids, 10–4 M β-mercaptoethanol, 1× GlutaMAX™-I Supplement, 15% ES-qualified fetal bovine serum, 100 unit/ml Penicillin and streptomycin, 2000 U/ml ESGRO® (LIF).
4. Mouse ES cell freezing medium: 75% ES cell culture medium, 15% ES-qualified fetal bovine serum, 10% DMSO.

2.2.2 gRNA Synthesis and Purification

1. Plasmid vector carrying the sequence of an appropriate gRNA scaffold.
2. 20 mg/ml proteinase K.
3. 10% SDS.
4. Phenol:Chloroform:Isoamyl Alcohol (25:24:1, v/v).
5. T7 RNA Synthesis Kit.
6. RNA extraction and purification Kit.

2.2.3 PCR Amplification and Purification for gRNA Evaluation

1. *S. pyogenes* Cas9 Nuclease.
2. Taq polymerase 2× Master Mix.
3. DNA Gel and PCR Clean-Up kit.
4. Gibson assembly kit such as In-Fusion® HD EcoDry™ Cloning Kit from Takara.
5. 10 mg/ml RNase A.
6. 8% TBE gels (vertical).
7. 1–2% agarose gels.
8. 1×TBE.
9. 20 mg/ml Ethidium Bromide.
10. Cel-I based mutation detection kit such as Surveyor Mutation Detection Kit from Integrated DNA Technology.

2.2.4 gRNA Transfection

1. Cell transfection reagent such as Lipofectamine® 2000 Transfection Reagent.
2. 0.5% Trypsin/0.53 mM EDTA.
3. Reduced-serum medium Opti-MEM®.
4. Cas9 mRNA transcribed via mRNA in vitro transcription kit.

5. Target-specific gRNA.
6. Plasmid as HR template purified by EndoFree Plasmid Mega Kit.

2.3 Supplies

1. Sterile 1.5 ml microcentrifuge tubes.
2. 0.2 ml nuclease-free PCR tubes.
3. Multi-well tissue culture plate (6-, 12-, and 96-well plates).
4. 60 mm tissue culture plate.
5. Sterile serological pipettes.
6. Disposable, sterile 15 ml or 50 ml conical tubes.
7. Sterile cryogenic storage vials (i.e., cryovials).
8. Controlled rate freezing apparatus or isopropanol chamber.

3 Methods

3.1 Design and Synthesis of Target-Specific Guide RNAs (gRNAs)

3.1.1 gRNA Design

The crucial factor in choosing a gRNA is the distance from the site of strand cleavage to the site of integration (Fig. 1a) when a single double-strand break is involved. Based on literature and our experience, it is often necessary to keep this distance within 15 base pairs for efficient HR. If no gRNA is available within such a short range on the target DNA, we recommend using a pair of gRNAs to make two double-strand breaks across the integration site instead (Fig. 1b). With this alternative strategy, however, sequence alterations must be built into the targeting construct (*see* Subheading 3.4) so that either the PAM or the seed for both gRNAs are abolished in the recombined allele to prevent cutting from reoccurring at the recombined locus (Fig. 1b). When multiple gRNAs are available to choose from, several web-based resources are useful for selection of potentially more active ones. We recommend the webtool Crispr scan (http://www.crisprscan.org), which uses a gRNA-scoring algorithm designed to capture the sequence features affecting gRNA stability and Cas9 loading to predict overall activity [3]. One useful feature is that output sequences of ranked gRNAs have the T7 promoter sequence automatically attached to their 5′ ends with optimal number of 3′ G nucleotides present (Fig. 1c). The output sequences are, therefore, directly used as the forward primer sequences for PCR reactions generating T7-IVT templates for gRNA synthesis as discussed next in Subheading 3.1.2. gRNAs may thus carry one to several G nucleotides at their 5′ end for optimal activity (*see* **Note 1** and Fig. 1c).

3.1.2 Preparing T7-IVT Template for gRNA Synthesis

1. Assembly of the T7-IVT template: Using a plasmid vector carrying the sequence of an appropriate gRNA scaffold (e.g., the tracrRNA for SpCas9) as the template, a PCR reaction can

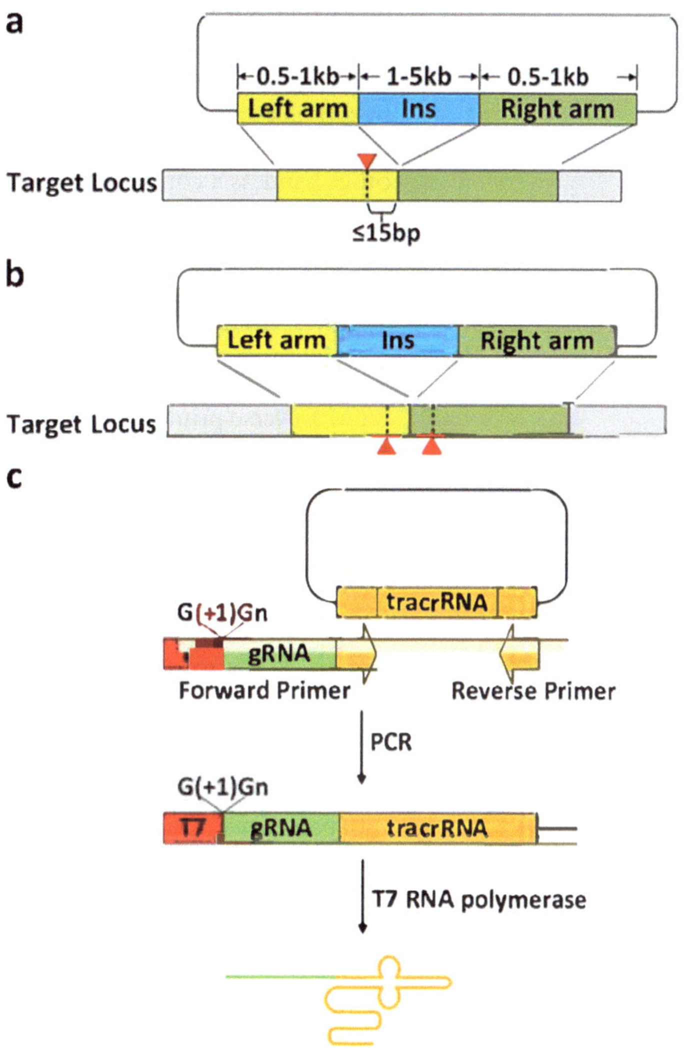

Fig. 1 Design and synthesis of sequence-specific guided RNAs. (**a**) A diagram of targeting construct and target locus. Insertion for the knockin allele (Ins) is generally 1–5 kb long, flanked by homologous arms of 0.5–1 kb. The selected gRNA should reside in the vicinity so that the distance between DNA cleavage site (red triangle, between third and fourth base pairs downstream in most cases) to the site of integration is within 15 base pairs. (**b**) Alternatively, we recommend using a pair of gRNAs to make two double-strand breaks (red triangles) across the integration site. With this alternative strategy, however, sequence alterations must be built into the targeting construct (*see* Subheading 3.4) so that either the PAM or the seed for both gRNAs are abolished in the recombined allele to prevent cutting from rcoccurring at thc recombined locus. (**c**) In vitro transcription of gRNA. A forward primer is designed and synthesized so that the sequence-specific gRNA sequence (17–20 base pairs) is attached to the T7 promoter at 3′ end. A variable number of G nucleotides follows the T7 transcription starting site (+1) for the optimal activity of the gRNA. Along with a reverse primer of the last 20 nucleotides of the tracrRNA, T7-IVT templates are generated via PCR amplification, which is then transcribed in the presence of T7 RNA polymerase

be carried out to assemble a T7-IVT template (Fig. 1c). The forward primer used in the PCR reaction, as discussed Subheading 3.1.1, consists of the T7 promoter sequence followed by a target-specific gRNA sequence and the first 20 base pairs of the tracrRNA scaffold (Fig. 1c). The reverse primer, on the other hand, is a common primer derived from the last 20 base pairs of the tracrRNA scaffold. A typical PCR reaction is assembled and run as the following:

PCR ddH_2O: 14 μl.

Template plasmid DNA: 1 μl (~30 ng).

2 μM Forward primer: 5 μl.

2 μM Reverse primer: 5 μl.

2×Taq master mix: 25 μl.

Total volume: 50 μl.

Step 1: 95 °C 30 s.

Step 2: 95 °C 20 s.

Step 3: 56 °C 30 s.

Step 4: 68 °C 30 s, back to **step 2** for 35 cycles.

Step 5: 68 °C 5 min.

2. Purification of the T7-IVT template produced by PCR: To purify the PCR product for the subsequent T7-IVT reactions, add proteinase K and SDS at final concentration 150 μg/ml and 0.5% respectively to the PCR reaction, and incubate at 50 °C for 30 min followed by Phenol/chloroform extraction once and ethanol precipitation. After resuspension in 10 μl of RNase-free water, 1 μl of the final product is loaded onto a 2% agarose gel to check for the presence of the T7-IVT template DNA with the expected size of 120 bp. Proceed to gRNA synthesis by T7-IVT reaction next.

3.1.3 gRNA Synthesis by T7-IVT

Using a commercially available T7-IVT kit such as the HiScribe™ T7 Quick High Yield RNA Synthesis kit (NEB), gRNAs can be synthesized in microgram quantities.

1. First, thaw the necessary kit components, mix and pulse-spin in a microcentrifuge to collect solutions to the bottoms of tubes. While Kept on ice, assemble the reaction at room temperature in the following order:

Nuclease-free water 17 μl.

NTP Buffer Mix 10 μl (6.7 mM each NTP final).

Template DNA 1 μl (75 ng (1 picomole)).

T7 RNA Polymerase Mix 2 μl.

Total reaction volume 30 μl.

2. Mix thoroughly, pulse-spin in a microcentrifuge, and incubate at 37 °C overnight.
3. When the reaction is completed, add 20 μl nuclease-free water and 2 μl of RNase-free DNase I to each T7-IVT reaction and continue to incubate at 37 °C for additional 30 min to remove DNA template. Proceed to gRNA purification.

3.1.4 Purification of gRNAs from T7-IVT Reactions

As the last step in gRNA production, product of T7-IVT reactions needs to be purified and concentrated. We routinely carry out the purification step by using RNA spin columns, such as the ones provided in a RNeasy kit (Qiagen).

1. The volume of each T7-IVT reaction needs to be adjusted to 100 μl with RNase-free water.
2. Add 350 μl Buffer RLT from the kit, and mix well. Proceed immediately to the next step.
3. Add 250 μl ethanol (96–100%) to the diluted RNA, and mix well by pipetting.
4. Transfer the sample (700 μl) to an RNeasy Mini spin column placed in a 2-ml collection tube. Close the lid gently, and centrifuge for 15 s at ≥8000 × *g*. Discard the flow-through (*see* **Note 2**).
5. Add 500 μl Buffer RPE to the RNeasy spin column. Close the lid gently, and centrifuge for 15 s at ≥8000 × *g* to wash the spin column membrane. Discard the flow-through (*see* **Note 3**).
6. Add 500 μl Buffer RPE to the RNeasy spin column. Close the lid gently, and centrifuge for 2 min at ≥8000 × *g* to wash the spin column membrane.
7. Place the RNeasy spin column in a new 1.5 ml collection tube (supplied). Add 30 μl RNase-free water directly to the spin column membrane. Close the lid gently, and centrifuge for 1 min at ≥8000 × *g* (≥10,000 rpm) to elute the RNA.

3.2 Validate gRNAs Experimentally

We found that gRNAs selected directly by web-based algorisms vary in their performance. The consideration to experimentally validate and compare gRNAs selected by algorisms is twofold. First, by narrowing down to one or two best performing gRNAs, it takes less effort overall in subsequent procedures. Secondly, we reasoned that by selecting gRNAs that cut the target site more efficiently, off-target binding and cleavage could be potentially reduced as various off-target sites most likely compete with the target site for Cas9-gRNA ribonuclear particles based on binding affinity.

3.2.1 Measuring DNA Cleavage Activity In Vitro

gRNA-mediated target DNA cleavage by Cas9 can be tested by in vitro target DNA cleavage assays which are similarly set up as in restriction digestions [4].

1. The target region is amplified first by conventional PCR as one to several hundred-base pair fragments with R1 ES cell genomic DNA as the template.
2. PCR products are purified with mini columns such as those with the NucleoSpin® Gel and PCR Clean-Up kit.
3. A DNA in vitro cleavage assay is then assembled as the following:

   Nuclease-free water 20 μl.

   10× Cas9 Nuclease Reaction Buffer 3 μl.

   300 nM sgRNA 3 μl.

   1 μM Cas9 (SpCas9) 1 μl.

   Preincubate for 10 min at 25 °C to allow for assembly of the Cas9/gRNA ribonuclease particles.

   Add 3 μl of the purified PCR product (~30 nm) to the ribonuclease mixture and incubate at 37 °C for 1 h.
4. Treat the reaction with 1 μg/ml RNase A and proceed to agarose gel electrophoresis for DNA fragment analysis.

3.2.2 Measuring DNA Cleavage Activity in Cell Culture

The Surveyor nuclease-based chromosomal cleavage assay, the Surveyor Assay, provides a closer approximation of gRNA activities as factors that are not testable by the in vitro cleavage assay, such as chromatin accessibility by Cas9-gRNA ribonuclease and stability of gRNAs in cells, are tested by Surveyor Assay [5]. We have routinely performed Surveyor Assays in mouse embryonic fibroblast (MEF) and found good correlations between Surveyor Assay activities and target cleavage in mouse ES cells and zygotes (data not shown). Based on this observation, we engineered a Surveyor Assay cell line, derived from immortalized MEF, in which a Cas9 expression cassette is stably integrated into the genome and its expression is under the tight control of Doxycycline (data not shown).

1. Culture of MEF: Cells are cultured in DMEM medium supplemented with 10% Fetal Bovine Serum, MEM Non-Essential Amino Acids Solution, 2 mM L-alanyl-L-glutamine, and 1% penicillin/streptomycin.
2. Seeding of MEF cells and induction of Cas9 expression: The day before transfection, 3–4 × 10^4 cells are seeded into each well on a 12-well plate, and 2 μg/ml Doxycycline (Dox) is added into the media to induce Cas9 expression.
3. On the day of transfection, media is replaced with fresh culture media without Dox prior to transfection.

4. Transfection of MEF with target-specific gRNAs: gRNAs are delivered into the cells using the lipofectamine 2000. Specifically, for each well, 100 μl OPTI-MEM media is mixed with 6 μl lipofectamine 2000 and incubated at room temperature for 5 min. Meanwhile, 2 μg of gRNA is diluted in a separate tube with 100 μl OPTI-MEM before mixing with the diluted lipofectamine. The mixture of lipofectamine and gRNA is incubated at room temperature for 20 min before applying to the cells. The cells are subject to the lipofectamine/gRNA mixture for 4–5 h, followed by media change with fresh media including 1 μg/ml Dox. The Dox concentration is increased to 2 μg/ml the following day. Cells are kept in culture for another 24–48 h before harvesting. Proceed to Surveyor assay (**steps 5–9**).
5. Genomic DNA preparation: Genomic DNA is extracted from transfected cells by classical ethanol precipitation. Specifically, cells from 12-well plate are lysed in 500 μl lysis buffer containing 200 μg/ml protease K and the lysate are incubated at 55 °C for several hours. An equal volume of isopropanol is added to each sample, mixed well. The samples are incubated at −20 °C for 1 h before centrifuged at a speed of more than 15,000 × *g* for 15 min at 4 °C. The pellet is washed with 70% ethanol, dried, and resuspended in 200–400 μl 10 mM Tris–HCl buffer (pH 8.5). The dissolved DNA is further diluted into a concentration of 100 ng/μl (*see* **Note 4**).
6. PCR amplification: For each samples DNA, a PCR reaction of 100 μl is set up using not more than 100 ng of genomic DNA. The PCR product is analyzed on an agarose gel and subsequently purified via purification column (*see* **Note 5**). The DNA fragment is eluted in 30 μl elution buffer containing 10 mM Tris–HCl (pH 8.0–8.5), 1.5 mM $MgCl_2$, and 50 mM KCl (*see* **Note 6**). Concentration of the PCR product is measured by nanodrop analyzer.
7. Heteroduplex formation: 500–600 ng PCR products is diluted in elution buffer to a total volume of 20 μl, denatured at 95 °C for 10 min, and annealed at a slow temperature decreasing rate to ensure the heteroduplex formation (*see* **Note 7**). Specifically, PCR products are placed in a thermocycler and subject to the following running program: 95 °C for 2 min; 95–85 °C (ramp rate − 2 °C/s); 85–25 °C (ramp rate − 0.3 °C/s); and 4 °C hold.
8. Surveyor Nuclease digestion: Following the slow annealing, the PCR products are split into two fractions. Half of the solution (10 μl) is treated at 42 °C for 40–45 min with 1 μl of SURVEYOR Nuclease S in the presence of 1.3 μl $MgCl_2$ and

1 μl SURVEYOR Enhancer S from SURVEYOR® Mutation Detection Kit. The digestion is terminated by addition of 1.5 μl stop solution. The other half of the annealed solution serves as the uncut control for the subsequent analysis.

9. Gel electrophoresis and mutation analysis: The cutting fragments of the PCR product from digestion are analyzed on Novex 8% TBE gel. The digested PCR products, along with the uncut controls, are mixed with 4 μl 5XTBE Hi-Density Sample Buffer and run on the 8% TBE gels for 1–2 h at 120 V. The TBE gel is then stained with 0.5 μg/ml ethidium bromide for 20 min at room temperature, and the DNA bands were then detected under the UV transilluminator and photograph. Fig. 2 shows an example of the Surveyor assay result. Cleavage product of the expected fragment is observed in the PCR product of the genomic DNA isolated from gRNA-transfected cells, but not from the untransfected cells.

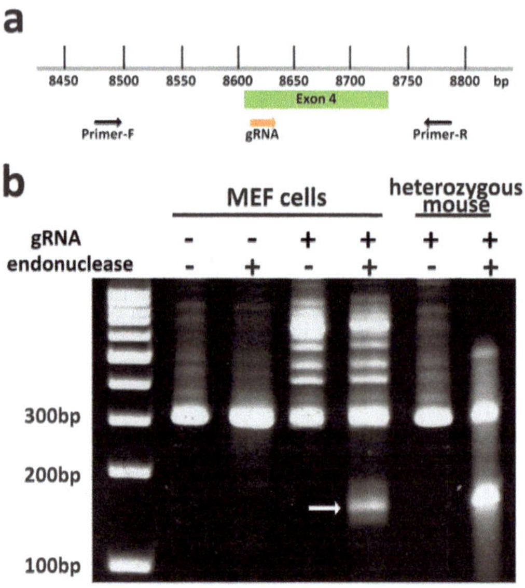

Fig. 2 Surveyor assay. (**a**) Schematics of the Surveyor assay design. The primers are designed such that a ~300–500 bp amplicon can be amplified with the DNA cleavage site situated in the center. (**b**) Gel electrophoresis showed that delivery of gRNA into cells generated a cleaved fragment (white arrows) with predicted size, indicating indel mutations. The mutagenesis rate can be roughly estimated by the ratio between the cleaved DNA fragment and the intact PCR amplicon. As shown in Figure b, the ratio between the cleaved and uncut DNA fragments in a heterozygous transgenic mouse is about 1, while the ratio in the transfected cell mixture is significantly less

3.3 Design and Assembly of Targeting Constructs

3.3.1 Targeting Construct Design

We and others have attempted to determine the optimal length of homology arms flanking an insertion cassette for CRISPR-mediated HR in mouse ES cells. Results from our experiments showed that when homology dropped below 100 base pairs on both arms, little or no CRISPR-mediated HR is detected (data not shown). Optimal length of homology is between 500 to 1000 base pairs for a flanked cassette of 1 to 5 kb. Homology arms longer than 1000 base pairs, however, did not improve HR rate (data not shown). Targeting constructs, in their final forms, are double-strand DNA cloned into a plasmid vector, which are not linearized before transfection (Fig. 1a). Importantly, with either one or two double-stand breaks at or near the target site produced by CRISPR/Cas9 (Fig. 1a, b), HR rates in the R1 ES cells are usually high enough allowing for recovering individual HR events directly without the need of drug selection at various target loci we have tested so far (Fig. 5).

3.3.2 Assembly of Targeting Constructs

It is generally efficient to assemble the targeting constructs from overlapping PCR fragments by Gibson assembly. This is particularly the case when the homology arms are short and no drug selection cassette is involved. Gibson assembly is a molecular cloning method which allows for the joining of multiple DNA fragments with overlapping ends in a given order in a single, isothermal reaction [6]. We found that 40 to 50 base pair overlaps at both ends of joining fragments are sufficient.

1. PCR amplification of homology arms: Homology arms of a targeting construct are directly amplified from R1 genomic DNA to keep isogenic to R1 ES cell (SV129S6) using Herculase II Fusion DNA Polymerases. PCR optimization may be necessary.
2. PCR amplification of the knockin cassette: The knockin cassette is amplified from an appropriate and often-existing DNA construct by using Herculase II Fusion DNA Polymerases. PCR optimization may be necessary.
3. Preparation of the receiving vector: The plasmid vector for receiving the overlapping PCR fragments can be prepared by either restriction digestions or PCR amplification of a cloning vector.
4. Gibson reaction: While Kept on ice, assemble the Gibson reaction with an In-Fusion® HD EcoDry™ Cloning Kit (Takara) in the following order:

   dH_2O up to 10 μl.

   Linearized destination vector (10–100 ng/μl) 1 μl.

   PCR insert(s) (20–200 ng/μl) 1 μl for each PCR product.

Incubate the reaction for 15 min at 37 °C, followed by 15 min at 50 °C, then place on ice.

5. Proceed to transformation with bacteria competent cells. Select no more than 6 bacteria colonies per assembly reaction from the transformation plate to analyze by plasmid mini preparation, restriction digestions and sequencing to identify correctly assembled clones.

3.4 R1 ES Cell Handling

The mouse R1 ES cell line was originally derived in Dr. Andras Nagy's laboratory in early 1990s and has served as a robust tool for germline transmission of engineered alleles in classic gene targeting since then [7]. CRISPR-mediated HR for targeted insertion of knockin alleles can be achieved efficiently in R1 ES cells without the need of drug selection. This section provides a set of basic protocols handling R1 ES cell cultures geared for genome editing by CRISPR/cas9.

3.4.1 Thawing of R1 ES Cells

1. Take one vial of R1 cells (1.5–2 × 10^6 cells per vial) from the LN_2 storage freezer and it is kept in hands with gloves for about 1 min before being placed in a 37 °C water bath with swirling (*see* **Note 8**).
2. When ice crystal in the frozen vial is about to disappear completely (usually takes no longer than 1 min), the content of the vial is quickly transferred into a 15 ml cell culture conical tube prefilled with 10 ml of the ES medium under a cell culture hood with a 1 ml transfer pipette.
3. Spin at 200 × *g* in a regular clinical centrifuge for 5 min at room temperature.
4. The supernatant is removed with vacuum suction (*see* **Note 9**), and ES cells resuspend in 1 ml of ES medium.
5. Evenly distribute the cells into each well of a 6-well tissue culture plate pre-seeded with a feeder layer of mouse embryonic fibroblast (MEF, *see* **Note 10**) in 3 ml ES cell culture medium per well.
6. Place the 6-well plate in a cell culture incubator with CO_2 level set between 5% to 7.5% at 37 °C immediately.
7. Change the medium next day and daily afterward. The cells are usually ready for split in 2 days after thawing.

3.4.2 Passage of R1 ES Cells in a 6-Well Plate

1. The R1 cells are ready to split when reaching 70–80% confluence.
2. Wash the R1 cells once with 3 ml D-PBS without Ca^{2+}/Mg^{2+}, add 0.5 ml of 0.05% trypsin-EDTA per well of a 6-well plate ES cell culture, and incubate at 37 °C in a cell culture incubator with CO_2 levels set between 5% to 7.5% for not longer than 5 min.

3. Add ES culture medium to inactivate the trypsin and carefully pipet up and down with a sterilized and plugged Pasteur pipette for about 15 to 20 times to break up cell clamps into single-cell suspension (*see* **Note 11**).
4. Transfer cells from each well to a 15 ml conical tube, count the cells, spin down and replate the cells as in Subheading 3.4.1.
5. Passage of the ES cells growing in 96-well plates or other culture plate requires adjustments from this procedure, which is detailed in **Note 12**.

3.4.3 Duplicate a 96-Well Plate of R1 ES Cell Culture

1. After removing the medium, cells are washed once with 200 μl per well D-PBS without Ca^{2+}/Mg^{2+}.
2. Add 30 μl 0.05% trypsin-EDTA to each well and incubate in the cell culture incubator for 5 min.
3. Immediately add 70 μl ES cells culture medium with serum to each well to inactivate trypsin. With a multiple channel pipettor, pipet the cell aggregates in each well up and down for 15 to 20 times to break ES colonies into single-cell suspension. Care must be taken to avoid excessive foaming.
4. Split the content in each well between two fresh 96-well plates pre-seeded with feeders in 150 μl ES cell culture medium per well.

3.4.4 Freezing of R1 ES Cells in a 96-Well Plate

1. After being washed with 200 μl D-PBS without Ca^{2+}/Mg^{2+} per well, ES cell colonies are detached in each well by adding 30 μl 0.05% trypsin-EDTA and incubating at 37 °C for 5 min.
2. Immediately after incubation, add 75 μl ES medium to each well, pipet cell aggregates up and with a multi-channel pipettor for 15–20 times, avoiding excessive foaming.
3. Transfer 75 μl of the content to a U-bottom 96-well plate pre-filled with 75 μl 2× ES cell freezing medium per well, place the plate into a styrofoam box, and move the box into a −80 °C freezer (*see* **Note 13**). The frozen plate can be kept under −80 °C for about 3 months without significant loss of cell viability.

3.4.5 Thawing of R1 ES Cells in a 96-Well Plate

1. Take the 96-well plate out of −80 °C freezer and place it in a plastic zipper bag.
2. Then put the bag in the 37 °C water bath with a weight on top. It will take about 15 min for thawing.
3. Carefully taking the plastic bag with the plate out from the water bath, dry the bag with paper towel, and remove the plate from the bag.

4. Spin the plate for 5 min at 175 × *g* in a plate centrifuge. Before opening the lid, wipe the plate with 70% alcohol under a culture hood.
5. Use the pipettor to remove most of the liquid in the wells, not to disturb the cells at the bottom of the wells (*see* **Note 14**).
6. Add 100 μl fresh ES culture medium to each well, mix thoroughly by pipetting up and down, and transfer the contents to a new 96-well plate with feeders and ES culture medium.
7. Wash the wells of the frozen plate once with 100 μl fresh ES culture medium per well and transfer the residual cells to the fresh plate.
8. Culture the cells and change medium daily afterward. It will take 2 to 3 days for ES cells to grow back to 70–80% confluency.

3.4.6 Picking Colonies of R1 ES Cells

The R1 ES cell colonies are generally ready for picking in about 5–7 days after seeding. The most suitable colonies to select are those with rounded or oval shape, with shinning edge and often a dark center under phase contrast.

1. Remove the ES cells culture medium from the culture plate, add 10 ml D-PBS without Ca^{2+} Mg^{2+} per P100 mm plate or 5 ml per P60 mm plate, and put the plate under the dissection microscope setup in the hoods.
2. With the pipettor set to a volume of 6 μl, under the microscope, circle the desired colony with the pipette tip to loosen the surrounding fibroblast layer, then scrape and aspirate the colony with the pipette tip and transfer to the empty well of a U-bottom 96-well plate. Continue picking and transferring ES cell colonies to fresh wells using a new tip each time, until a suitable number is picked (*see* **Note 15**).
3. Add 30 μl 0.05% trypsin-EDTA to each well in U bottom 96-well plate, and incubate at 37 °C for 5 min.
4. Then add 50 μl ES culture medium to inactivate the trypsin. Disperse each ES cell colony in the U-bottom 96-well plate by pipetting up and down for 15 to 20 times using a multiple channel pipette to break up each colony, avoiding excessive foaming.
5. Transfer the suspension to a flat-bottom 96-well plate with MEF prepared the day before (including foam), pre-filled with 150 μl of ES cell medium per well, using a fresh tip for each well.

6. Continue culture for 48–72 h, the new ES colonies should appear and reach 70–80% confluence.
7. Duplicate the 96-well plate as described in Subheading 3.4.3 and freeze one of the plates as described in Subheading 3.4.4, when majority of the wells become confluent, usually 24 h after.

3.5 Genome Editing of R1 ES Cells with CRISPR/Cas9

Key factors to success include the state of ES cells in culture, transfecting dosage of gRNA(s), and overall transfection efficiency. The R1 ES cells need to be at a passage lower than P15 and handled appropriately as described in Subheading 3.4. Importantly, R1 ES cell colonies cannot be large at the time of transfection to be able to reach most of the cells within individual colonies. The control over ES cell colony size is achieved by appropriate initial seeding densities with culture duration limited to less than 24 h. R1 ES cells cannot tolerate high dosage of gRNA in general. We recommend no more than 4 μg of a given gRNA for each well of a 6-well plate (see below). With conditions recommended here, transfection efficiency is typically above 70% (Fig. 3).

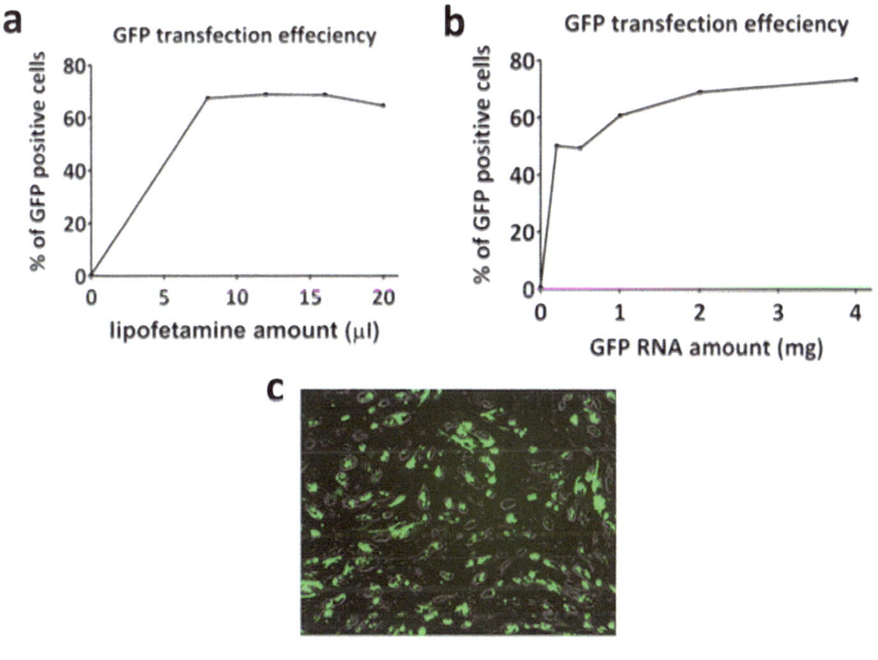

Fig. 3 Transfection efficiency of mouse ES cells. R1 cells were transfected with various amounts of lipofectamine 2000 (**a**) or different amounts GFP mRNA (**b**). 24 h after transfection GFP positive cells were counted by flow cytometry. (**c**) Fluorescence microscopy image showing that transfection efficiency via lipofectamine 2000 is typically above 70% with R1 ES cells

3.5.1 Transfection of R1 ES Cells by Lipofectamine 2000

1. 24 h before transfection, plate 2×10^5 R1 ES cells in each well of a 6-well plate pre-seeded with feeder (gamma-irradiated MEF).
2. Next morning, change medium of each well with 1.5 ml regular ES culture medium without antibiotics. Meanwhile, set two 1.5 ml micro-centrifuge tubes under the culture hood. In tube 1, mix 200 μl Opti-MEM Reduced Serum Medium with 4 μg Cas9 mRNA, 2 μg gRNA and 2 μg DNA plasmid (*see* **Note 16**). In tube 2, add 12 μl lipofectamine 2000 to 200 μl Opti-MEM Reduced Serum Medium.
3. Then mix and combine the contents of tube 1 and tube 2, tab the tube gently, and incubate at room temperature for 15–20 min.
4. Add the mixture to one well of cultured cells, gently swirl the plate to evenly distribute the transfection reagents, and return the plate to the incubator.
5. 4–6 h later, remove the medium and change to fresh ES culture medium (*see* **Note 17**).

3.5.2 The Rosa26-Cas9 R1 ES Cell Line

To simplify genomic manipulation of R1 ES cells further, a SpCas9 expression cassette was integrated into the Rosa26 locus [8] (dong et al., unpublished). Expression of Cas9 in the Rosa26-Cas9 R1 ES line is under tight control of the tet-inducible promoter. When using this line to perform genome editing, only gRNA and the targeting plasmid are transfected into the ES cells, and the rest of the procedures are the same as described in Subheading 3.5.1 except that 2 μg/ml Doxycycline is added to the culture medium at the time of seeding R1 ES cells the day before transfection. Knockin alleles using the Rosa26-Cas9 R1 ES line have been shown to transmit through germline robustly (data not shown).

3.6 Screening for HR Events

R1 ES clones with correct knockin at the target gene can be recovered in a two-step screening procedure without the need of drug selection. A flow chat of HR screening is presented in Fig. 4, and detailed procedures are described below.

3.6.1 Pooling Transfected R1 Cells into a 96-Well Plate with Feeder

1. Prepare a single-cell suspension from the transfection plate 24 h after transfection, as described in Subheading 3.4.2.
2. Resuspend ~5000 transfected R1 ES cells in 20 ml ES cell culture medium (*see* **Note 18**), mix well, then transfer 200 μl (~50 cells) into each well of a 96-well plate with feeder using a multiple channel pipettor to prepare 96 pools of transfected R1 ES cells.
3. Culture the plate until the ES colonies appear. Almost all the wells with pooled R1 ES cells will reach 70–80% confluence in 5–6 days (*see* **Note 19**).

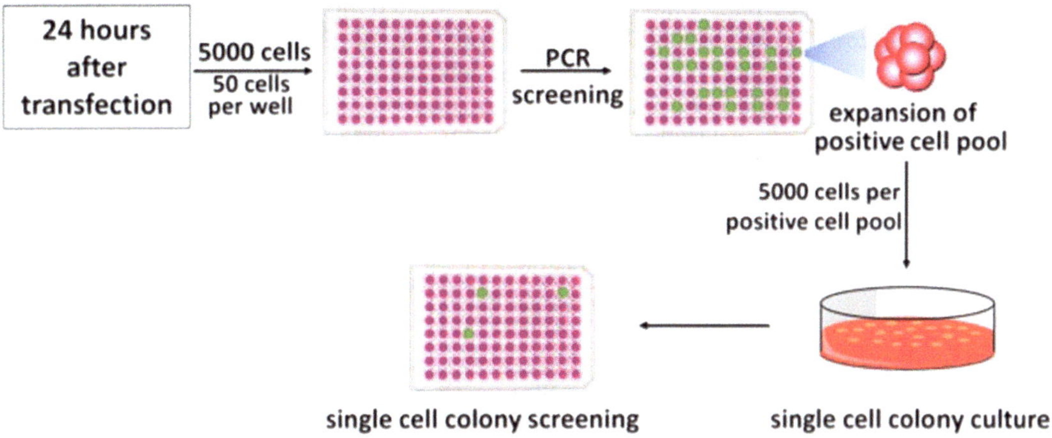

Fig. 4 Two-step screening for homologous recombination events of a knockin allele without drug selection. 24 h after transfection, ~5000 transfected R1 ES cells are resuspended and seeded into a 96-well plate at a density of ~50 cells per well. When reaching confluence, the 96-well plate is duplicated; genomic DNAs are isolated from a duplicated plate, and the 50-cell pool containing recombinant alleles are identified via PCR screening (shown in green). The positive R1 cell pools are selected from the other duplicate plate and expanded. 5000 cells from one positive pool are plated onto a P100 cell culture plate to form single-cell colonies. Each colony is transferred into a 96-well plate for continuous culture. The individual cell clones in 96-well plate are subject to a second round of PCR screening to identify the clones containing the desired knockin allele (shown in green)

3.6.2 Duplicating the 96-well Plates with Pooled R1 ES Cells as Described in Subheading 3.4.3

3.6.3 Freezing Down One of the Duplicate 96-well Plates as Described in Subheading 3.4.4

3.6.4 Prepare Genomic DNA from the Other 96-Well Plate

1. Remove the culture medium and gently wash the wells with 200 μl DPBS. Add 50 μl of lysis buffer to each well and incubate overnight at 50 °C.
2. Next morning, add 5 μl 3 M sodium acetate pH 5.2 and 150 μl 100% ethanol to each well, mix carefully with the multiple channel pipettor to avoid cross contamination, and place the plate at −20 °C for 30 min.
3. Spin the plates for 30 min at 2650 × *g* at 4 °C.
4. Quickly invert plate over a piece of paper towel to dump out the liquid.
5. Rinse the plate with 200 μl 70% ethanol each well. Spin for another 30 min at 2650 × *g* at 4 °C, and then empty wells again.
6. Invert the plate on paper towel and spin 30 s at 138 × *g* to dry the DNA samples.
7. Add 120 μl 10 mM Tris–HCl pH 8.0 with RNase A (10 μg/ml) per well immediately. The genomic DNA is mostly dissolved next day and ready for downstream applications (*see* **Note 20**).

3.6.5 PCR Screening to Identify Positive Pools

With one of the primers located upstream or downstream of a homology arm, HR events can be identified in a PCR reaction (Fig. 5). We have routinely used the 2XTaq from NEB for this purpose and achieved satisfactory results. A typical PCR reaction for HR screening can be assembled and run as the flowing (*see* **Note 21**).

Genomic DNA: 100 ng.

2 μM Forward Primer: 3 μl.

2 μM Reverse Primer: 3 μl.

Taq 2× Master Mix: 15 μl.

Adjust volume to 30 μl with Nuclease-free water.

Step 1: 95 °C 30 s.

Step 2: 95 °C 20 s.

Step 3: 58 °C 30 s.

Step 4: 68 °C 60 s per kb, back to **step 2** for 35 cycles.

Step 5: 68 °C 5 min.

3.6.6 Thawing and Expansion of the Positive Pools

1. After identifying the positive pools, thaw out the 96-well plate as described in Subheading 3.4.5. In 2 to 3 days, most wells will become confluent.
2. Transfer individual positive pools of R1 ES cells from the 96-well plate to a 24-well plate(s) with feeders and ES medium as described in Subheading 3.4.2. It will take another 48 to 72 h for cells to grow up and reach 70–80% confluence.

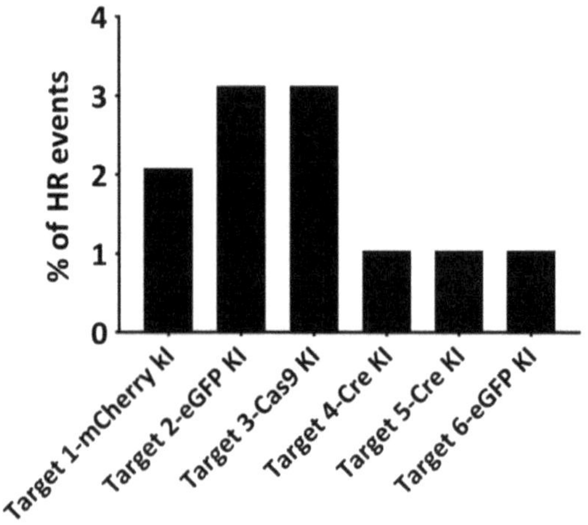

Fig. 5 Rates of HR across a few target loci. Different cassettes (mcherry, GFP, and Cre) were correctly integrated into six different loci via CRISPR-mediated homologous recombination

3.6.7 Replate R1 ES Cells from Positive Pools at Cloning Density to Identify Individual Positive Clones

1. Trypsinize Rl ES cells in the 24-well plate as described in Subheading 3.4.2 to prepare single-cell suspensions.
2. For each positive pool, resuspend ~5000 cells in 20 ml ES cell culture medium (*see* **Note 22**), mix well, and plate into one P100 mm dish with feeder and 10 ml ES medium.
3. Freeze back the rest of Rl ES cells in pool as described in Subheading 3.4.4.
4. Change medium after 48 h and daily afterward. The individual clone will appear in 3 to 4 days and be ready to pick on the 6th or 7th days.

3.6.8 Picking Colonies and Conducting a Second Round PCR Screening Identify Individual Positive Clones

1. Prepare MEF feeders in 96-well plate the day before colony picking, and proceed to picking colonies as described in Subheading 3.4.6.
2. Duplicate plates are made when majority of the 96 wells reach 70 to 80% confluence, usually taking 2 to 3 days in culture, as described in Subheading 3.4.3.
3. While one of the duplicate 96-well plates is frozen down as the source of positive clones, genomic DNA samples are prepared from the second 96-well plate as described in Subheading 3.6.4.
4. PCR screening is performed as described in Subheading 3.6.5 to identify individual positive clones.

3.6.9 Thawing Out and Expanding Positive Clones in 6-well Plates as Dscribed in Subhaediing 3.4.1

3.6.10 Freeze Down Positive Clones for Storage

1. Treat the cells with trypsin as described in Subheading 3.4.2, transfer all the content to a 15 ml conical tube, count the cells, spin down and resuspend 1–2 × 10^6 cells per ml of 1× freezing medium.
2. Freeze cells in a −80 °C freezer overnight in a Nalgene Cryo®1 °C Freezing container and then keep the vials in the vapor phase of LN_2 for longer storage.

3.6.11 Prepare Positive Clones for Microinjection (Include PCR Validation)

1. Thawing out vials of Rl ES cells for microinjection as described in Subheading 3.4. Each clone for micro-injection is validated after thawing by PCR as described below.
2. Replate ~3000 cells of each clone in a P60mm tissue culture dish with feeders and ES culture medium.
3. When individual colonies are visible (usually taking 6–7 days in culture), pick not more than 12 colonies into a 96-well plate with feeder and ES culture medium as described in Subheading 3.4.6.
4. Prepare genomic DNA samplers as described in Subheading 3.6.4, and perform PCR as described in Subheading 3.6.5 for confirmation of HR. If a positive clone had contaminations

from wild-type Rl ES cells during processing, a percentage of the subclones should be negative for HR. It is important that only use the clones with all the subclones being positive for HR signals for subsequent microinjection.

3.7 Prepare R1 ES Cells for Microinjection

1. On the day of microinjection, refresh the culture with 1.5 ml ES medium in a 6-well plate at least 1 h prior to split.
1. Prepare single-cell suspension as described in Subheading 3.4.2, transfer all the content to a 15 ml conical tube, and count the cells.
2. Spin down the cells at 200 × *g* for 5 min at room temperature, and resuspend the cells in 1 ml ES cell microinjection medium.
3. Keep the vials on ice, and transfer them to the microinjection laboratory.

4 Notes

1. T7 transcription starts at the first G (+1) at the 3′ end of T7 promoter.
2. After centrifugation, carefully remove the RNeasy spin column from the collection tube so that the column does not contact the flow-through. Be sure to empty the collection tube completely.
3. Buffer RPE is supplied as a concentrate. Ensure that ethanol is added to Buffer RPE before use.
4. Addition of RNase a to the 10 mM Tris–HCl may help avoid the stickiness of resuspended genomic DNA.
5. It is preferable to optimize the PCR conditions (including the selection of polymerase and annealing temperature) before performing Surveyor Assay to ensure the genomic DNA is amplified with high efficiency and specificity.
6. The addition of salt in the final elution buffer helps the annealing of denatured PCR products in the following step.
7. It is important not to use excess of the PCR products (less than 80 ng/μl) to ensure efficient annealing between wild type and mutant strand in this step.
8. Avoid placing LN_2-kept frozen vials containing ES cells immediately into a 37 °C water bath to reduce the incidence of frozen vial explosion due to rapid temperature increase.
9. In mouse ES cell culture, when making single-cell resuspensions from cell pellets, it is a clever idea to leave a very small amount of medium behind and tap the bottom of the conical

tube gently to have the cell pellet partially resuspended before adding additional medium to reduce cell clamping.

10. Gamma-irradiated Mouse Embryonic Fibroblast (MEF) is used as feeders. It is recommended that MEF feeder cells be plated 1 day prior to plating ES cells, which will allow the feeder layer to be approximately 80% confluent right before ES cell plating.
11. To appropriately trypsinize R1 ES cells, one may need to practice and determine the number of times pipetting up and down of the cell suspension through a plugged Pasteur pipette to achieve a single cell suspension without being too harsh on the ES cells. Cells that either stayed in trypsin too long or being pipetted through a Pasteur pipette too many times will lyse and release genomic contents turning the cell suspension viscous. Under this condition, majority of ES cells are lost.
12. Trypsinization of ES cells in 96-well plate needs 200 μl D-PBS without Ca^{2+} Mg^{2+} washing and 30 μl trypsin per well, inactivate trypsin by adding 100 μl ES culture medium, after pipet up and down transfer all the content to 24-well plate with feeders. There is no need of spinning down by centrifuge. Change fresh ES medium the second day and continue culture. For 24-well plate and 12-well plate, we use 200 μl trypsin and 300 μl trypsin, respectively, following 1 ml or 2 ml washing with D-PBS without Ca^{2+} Mg^{2+}. The cell suspension can be spun down and resuspend in fresh ES culture medium and transfer to new wells to continue culture.
13. When freezing the cells in 96-well plate, add 75 μl 2× freezing medium first to the U bottom 96-well plate and put the plates on ice before trypsinizing the cells. After making single-cell suspension, transferring 75 μl to the U bottom plate, mix gently by pipetting up and down 2–3 times. Then seal the plate with tape and put back on ice for about 20–30 min. Usually do not put the plates directly to the −80 °C freezer. It is recommended to put the plates in a styrofoam box first and then put the box into −80 °C freezer to ensure the temperature dropping not too fast.
14. Set up the pipettor at 120 μl and remove most of the supernatant of the desired well. Keep the tip at the edge of the well, and leave a very small amount of the liquid in the well to make sure no cell pellet be disturbed.
15. Keep the lid of the U bottom 96-well plate closed when picking up the colonies so the picked colonies in the well won't dry out. Don't let the colonies stay in the U bottom plate for more than 30 min before trypsinizing them as described.

16. Targeting plasmids are prepared with endotoxin-free columns (Qiagen) and after the last precipitation, DNA preparations are resuspended in sterile ddH_2O under a cell culture hood. The plasmid need not be linearized as in classic gene targeting.
17. When preparing the transfection mixture, it is recommended to mix the DNA target donor, gRNA and cas9 mRNA with Opti-MEM® Reduced Serum Medium first, and then dilute lipofectamine 2000 in 200 μl Opti-MEM® Reduced Serum Medium. Don't keep the diluted lipofectamine 2000 for more than 5 min before mix it with the tube 1.
18. 24 h after transfection, trypsinize the ES cells as described in Subheading 3.4.2; count and resuspend the ES cells at 5×10^5 cells/ml. It is recommended to do a serial dilution to reach the final 5×10^3 cells/ml suspension. Add 1 ml of the 5×10^3 cell/ml cell suspension to 20 ml ES cell medium, mix thoroughly, and distribute 200 μl to each well of a 96-well plate as quickly as possible.
19. After pooling the transfected cells into 96-well plate as described in Subheading 3.6.1, put all the rest of the mixture of the transfected cells into one well in 6-well plate and continue culture for 2–3 days. Harvest the pool of the transfected cells, extract the genomic DNA, and use this as positive control to optimize the PCR condition when screening the homology recombination in later experiments.
20. Dissolve the genomic DNA completely in the 10 mM Tris–HCl pH 8.0 before performing screening PCR, by putting the plate at room temperature overnight, or at 50 °C for 1 h.
21. Use the mixture of the genomic DNA from **Note 17** to optimize the PCR condition.
22. Prepare the P100 plate with 2.5×10^6 MEF feeders the day before subcloning. Trypsin treated the ES cells, counting the cells and resuspend the ES cells at 5×10^5 cells/ml, and it is recommended to do serial dilution to make 5×10^3 cells/ml cell suspension. Remove the medium in P100 and change to 15–20 ml ES culture medium, and add 1 ml 5×10^3 cells/ml cell suspension to the plate.

References

1. Yang H et al (2013) One-step generation of mice carrying reporter and conditional alleles by CRISPR/Cas-mediated genome engineering. Cell 154(6):1370–1379
2. Thomas KR, Capecchi MR (1987) Site-directed mutagenesis by gene targeting in mouse embryo-derived stem cells. Cell 51(3):503–512
3. Moreno-Mateos MA et al (2015) CRISPRscan: designing highly efficient sgRNAs for CRISPR-Cas9 targeting in vivo. Nat Methods 12(10):982–988
4. Jinek M et al (2012) A programmable dual-RNA-guided DNA endonuclease in adaptive bacterial immunity. Science 337(6096):816–821

5. Pilato B et al (2012) DHPLC/SURVEYOR nuclease: a sensitive, rapid and affordable method to analyze BRCA1 and BRCA2 mutations in breast cancer families. Mol Biotechnol 52(1):8–15
6. Gibson DG et al (2009) Enzymatic assembly of DNA molecules up to several hundred kilobases. Nat Methods 6(5):343–345
7. Nagy A, Rossant J, Nagy R, Abramow-Newerly-W, Roder JC (1993) Derivation of completely cell culture-derived mice from early-passage embryonic stem cells. Proc Natl Acad Sci U S A 90(18):8424–8428
8. Friedrich G, Soriano P (1991) Promoter traps in embryonic stem cells: a genetic screen to identify and mutate developmental genes in mice. Genes Dev 5(9):1513–1523

Chapter 8

Using CRISPR/Cas9 for Gene Knockout in Immunodeficient NSG Mice

Yubin Du, Wen Xie, Fan Zhang, Uimook Choi, Chengyu Liu, and Colin L. Sweeney

Abstract

NOD.Cg-*Prkdc*^{scid} *Il2rg*^{tm1Wjl}/SzJ (NSG) mice are an immunodeficient strain that enables human cell xenografts. However, NSG mice possess a complex genetic background that would complicate cross-breeding with other inbred transgenic or knockout mouse strains to establish a congenic strain with a desired genetic modification in the NSG background. Newly developed clustered regularly interspaced short palindromic repeats (CRISPR)/Cas9 technology enables modification of the mouse genome at the zygote stage without the need for extensive cross-breeding or the use of embryonic stem cells. In this chapter, we use the knockout of the X-linked *Cybb* gene as an example to describe our procedures for genetically modifying NSG mice using the CRISPR/Cas9 method. Briefly, two sgRNAs were designed and made to target exon 1 and exon 3 of the *Cybb* gene, and either sgRNA was then microinjected together with Cas9 mRNA into fertilized eggs collected from NSG mice. The injected embryos are subsequently transferred into the oviducts of pseudopregnant surrogate mothers. Offspring born to the foster mothers were genotyped by PCR and DNA sequencing. In this chapter, we describe our experiment procedures in detail and report our genotyping results for demonstrating that NSG mice can be genetically modified using the CRISPR/Cas9 technology in a highly efficient manner.

Key words CRISPR/Cas9, NSG, Microinjection, Oviduct embryo transfer, Genotyping, *Cybb*, Immunodeficient mice, Knockout

1 Introduction

Immunodeficient mouse strains, such as NOD.Cg-*Prkdc*^{scid} *Il2rg*^{tm1Wjl}/SzJ (NSG) mice [1], allow for xenotransplants of human cells. NSG mice carry two mutations in the nonobese diabetic (NOD)/ShiLtJ genetic background: knockout of the DNA repair gene *Prkdc* for severe combined immunodeficiency (scid), rendering the mice deficient in producing mature B and T lymphocytes, and a completely null allele of the IL2 receptor common gamma chain (*IL2rg*) gene, resulting in a deficiency of natural killer (NK) cell activity due to prevention of signaling through

Chengyu Liu and Yubin Du (eds.), *Microinjection: Methods and Protocols*, Methods in Molecular Biology, vol. 1874, https://doi.org/10.1007/978-1-4939-8831-0_8, © Springer Science+Business Media, LLC, part of Springer Nature 2019

multiple cytokine receptors. Together, these immune deficiencies render NSG mice defective for graft rejection. Consequently, the NSG mouse strain is useful for human cell transplants to model human hematopoiesis and immune cell function, stem cell biology, tumor engraftment, and infectious disease research.

Further genetic modifications to the NSG mouse strain can generate improved models for studying human diseases or enhancing human cell transplants, such as by expressing human hematopoietic cytokines for increased engraftment of specific human hematopoietic cell lineages [2–4]. Until recently, the introduction of new genetic modifications into the complex NSG background has relied on crossbreeding with other established knockout or knockin mouse strains, requiring a complicated and lengthy breeding program to ensure that the new congenic strain contains the NSG alleles as well as the new desired mutation, assuming that their chromosomal locations allow for co-inheritance. As an alternate approach, direct genetic modification of the NSG strain would enable a more rapid establishment of a new desired congenic strain.

Conventional gene-targeting technology for generating knockout or knockin mice involves introducing mutations by homologous recombination in mouse embryonic stem (ES) cells, followed by microinjection into wild-type blastocysts, resulting in chimeric mice with potential germline transmission [5]. This methodology requires screening of ES cell clones for rare homologous recombination events, typically following drug selection for the targeted insertion of a donor DNA sequence containing a drug-resistance gene (thereby disrupting the targeted gene in case of a knockout strategy, or inserting a desired new sequence for a knockin approach). Although initial attempts were unsuccessful at deriving ES cells capable of germline transmission from NOD and NSG mouse strains, more recently the inclusion of PD0325901 mitogen-activated protein kinase kinase inhibitor and CHIR99021 glycogen synthase kinase 3 inhibitor in serum-free derivation conditions has enabled the establishment of NSG germline-competent ES cells [6], providing a means for more rapid targeted genetic modification of NSG mice. However, due to the generally low efficiency of modifying both alleles of a target gene by homologous recombination in ES cells, additional mouse breeding is often required to establish homozygosity using this approach.

In the past few years, clustered regularly interspaced short palindromic repeats (CRISPR)/Cas9 nuclease has proved to be a more efficient method for targeted gene editing, following induction of a double-strand break by Cas9 nuclease at the target DNA sequence [7]. For gene knockout, the design is straightforward, relying on error-prone non-homologous end joining (NHEJ) or microhomology-mediated end joining (MMEJ) [8] at the Cas9 nuclease cutting site in the genome to induce knockout mutations, instead of using homologous recombination as in ES cell-based approaches. Although typically less efficient than gene knockout strategies, Cas9 nuclease can also be used for gene knockin through homology-directed repair at the genomic double-strand break by a co-administered donor plasmid or single-stranded DNA oligos possessing homology to the genomic target [7, 9].

Target specificity of Cas9 nuclease relies on a guide RNA (gRNA) containing a sequence (typically 20 bases in length) that is complementary to the target genomic DNA sequence, which must be adjacent to a protospacer adjacent motif (PAM) that is specific for each nuclease. The most commonly used Cas9 for genome engineering is derived from *Streptococcus pyogenes* (SpCas9) which utilizes a site-specific CRISPR RNA (crRNA) and a trans-activating RNA (tracrRNA) for providing a scaffold for binding to the Cas9 protein and the target DNA sequence; however, for simplicity, these two components are often fused together into a single-guide RNA (sgRNA) [10–12]. Using a traditional transgenic delivery method, a mixture of sgRNA and Cas9 (either mRNA or protein) can be microinjected into fertilized eggs [13]. One or both alleles can be modified and identified by genotyping, and homozygotes can be obtained in the first generation, bypassing the need for breeding chimeric mice for germline transmission and for crossing with mice expressing FLP or Cre recombinase to remove the selection marker (usually the neomycin-resistance gene), consequently saving two generations of mouse breeding when compared with the conventional ES cell-mediated method.

We previously reported the CRISPR/Cas9-mediated knockout of the *Cybb* gene on the X-chromosome to establish a mouse model of X-linked chronic granulomatous disease (X-CGD) in the NSG strain (*see* **Note 1**), which is useful for modeling human hematopoietic cell transplantation for developing gene therapies for the treatment of X-CGD [14]. In the current chapter, we describe our methodology in detail, including the making of CRISPR/Cas9 reagents, embryo microinjection, surgical procedures for embryo implantation, and genotyping procedures as well as detailed results, for demonstrating the feasibility of using CRISPR/Cas9 to genetically modify immunocompromised mouse strains.

2 Materials

2.1 Mice (See Note 2)

1. NSG mice as embryo donors: ten stud males (10–32 weeks old) individually caged; 30 females (6–16 weeks old) housed in groups (up to five mice per cage).

2. Vasectomized male mice: ten CD-1 outbred male mice (8–40 weeks old) individually caged. Vasectomy can be requested from the animal supplier for a reasonable fee.
3. Recipient females: 20 CD-1 females (8–12 weeks old) for mating with vasectomized males for generating pseudopregnant foster mothers.

2.2 Instruments and Supplies for Preparing CRIPSR Reagents and Genotyping

1. Bacterial culturing incubator with shaker, set to 37 °C and 200–250 rpm.
2. Heat blocks set to 50–56, 70, and 98 °C.
3. PCR thermocycler.
4. 250 or 500 mL glass Erlenmeyer flasks.
5. Microwave oven.
6. Horizontal gel casting tray and gel combs.
7. Optional: Gel electrophoresis apparatus suitable for TBE-Urea polyacrylamide gels, with power supply.
8. Gel electrophoresis apparatus suitable for agarose gels, with power supply.
9. Ultraviolet (UV) transilluminator for visualizing ethidium bromide-stained DNA, preferably with 365 nm wavelength UV for gel band excision (or an appropriate light source for visualizing alternatives to ethidium bromide).
10. NanoDrop 2000 or equivalent spectrophotometer.

2.3 Equipment and Supplies for Embryo Microinjection and Mouse Surgery

1. Stereo microscopes for embryo collection and embryo transfer procedures.
2. Sutter P-97 horizontal pipette puller for pulling microinjection needles.
3. Microinjection setup: inverted microscope, micromanipulators, Eppendorf FemtoJet, Eppendorf CellTram Vario injector for holding pipette.
4. P10 or P20 Pipetman and Eppendorf GELoader pipette tips for loading injection needles.
5. CO_2 incubators for culturing embryos.
6. Mouth pipette for transferring embryos.
7. Hair clipper.
8. Povidone-Iodine Swabsticks.
9. Alcohol prep pads (70% isopropyl alcohol).
10. Sterile surgical gloves.
11. Ethicon 5–0 Vicryl suture.
12. 29-G × ½ in. 1.0 mL insulin syringe.

13. Micro dissecting scissors.
14. Small iris forceps and No. 5 micro dissecting tweezers.
15. Small Dieffenbach micro clamp.
16. Wound clip applier.
17. Suturing needle holder.
18. Heating water blanket.
19. Benchtop Sterilizer for autoclaving surgical instruments.
20. Hot beads sterilizer for sterilizing surgical tools between animals when more than one animal is used during a session.
21. Mouse ear puncher.
22. Surgical scissors or razor blade (sterile).
23. RNase-free 1.5 mL microcentrifuge tubes.

2.4 Media, Chemicals, and Other Reagents

1. sgRNA expression plasmid for T7 in vitro transcription (DR274; Addgene plasmid # 42250) [15].
2. Cas9 expression plasmid for T7 in vitro transcription (MLM3613; Addgene plasmid # 42251) [15].
3. PCR-grade or molecular biology grade water (double-distilled, sterilized, nuclease-free).
4. Single-stranded DNA oligos for cloning sgRNAs into DR274 plasmid (*see* Subheading 3.2; examples listed are for exon 1 and exon 3 of the mouse *Cybb* gene, as described in Subheading 3.1):

   Cybb Ex-1 forward: 5′-taggTACTTACAATGACAAAGA-3′.

   Cybb Ex-1 reverse: 5′-aaacTCTTTGTCATTGTAAGTA-3′.

   Cybb Ex-3 forward: 5′-taggATTTCGACACACTGGCAGC-3′.

   Cybb Ex-3 reverse: 5′-aaacGCTGCCAGTGTGTCGAAAT-3′.

   When the forward and reverse strand oligos for each target site (Ex-1 or Ex-3) are annealed, the nucleotides shown in uppercase letters (corresponding to the guide sequence) hybridize to each other and form double-strand DNA, while the four nucleotides in lowercase letters form sticky ends which increase cloning efficiency.
5. BsaI (Eco31I) restriction endonuclease.
6. Competent *E. coli* bacteria for transformation with plasmid DNA.
7. Luria-Bertani (LB) medium.
8. Kanamycin (50 mg/mL for 1000× stock).
9. LB agar plates prepared with 50 μg/mL Kanamycin.
10. Plasmid DNA miniprep and midiprep kits.

11. Optional: Cas9 protein or Cas9 mRNA from commercial source (as an alternative to synthesizing Cas9 mRNA in-house, as described in Subheading 3.3.2).
12. DraI restriction endonuclease.
13. M13 reverse primer: 5′-CAGGAAACAGCTATGAC-3′ (for sequencing DR274 plasmid to confirm sgRNA cloning).
14. Invitrogen MEGAshortscript T7 transcription kit.
15. Invitrogen mMESSAGE mMACHINE T7 Ultra Transcription Kit.
16. PmeI restriction endonuclease.
17. 3 M sodium acetate (pH 5.2).
18. Phenol and phenol/chloroform (water-saturated, buffer-saturated, or acidic; *see* **Note 3**).
19. Chloroform.
20. 100% ethanol.
21. 70% ethanol (molecular biology grade).
22. Optional: RNA loading dye.
23. Optional: single-stranded RNA ladder.
24. Optional: 6% TBE-Urea polyacrylamide gel.
25. M2 culture medium.
26. M16 culture medium.
27. Embryo culture-tested mineral oil.
28. Hyaluronidase: concentrated solution (10×) is prepared by dissolving the entire bottle (100 mg) in 20 mL of M2 medium. Filter through a 0.22 μm filter, aliquot into microcentrifuge tubes, and store at −20 °C in the dark.
29. Ketamine/Xylazine: 0.2 mL of ketamine (100 mg/mL) and 0.1 mL of xylazine (20 mg/mL) are mixed with 1.7 mL of 0.9% saline for a 2 mL anesthetic mixture (*see* **Note 4**).
30. 0.25% Bupivacaine.
31. Meloxicam 5 mg/mL Solution for injection.
32. Kaneka Easy DNA Extraction kit (version 2) or equivalent genomic DNA isolation kit.
33. PCR primers specific for CRISPR target site in genome. Primers used for amplification of mouse *Cybb* exons 1 and 3 (depicted in Fig. 5a, b) were:

    Primer 1 (exon 1 forward): 5′-GTTGGAAGAGCCTGTGAG AAGA-3′.

    Primer 2 (exon 1 forward): 5′-GGGAACAGCCTTTCAGTT GG-3′.

Primer 3 (exon 1 reverse): 5′-ACCGAATCTACCTGCAAGC A-3′.

Primer 4 (exon 1 reverse): 5′-TGCACTCTGGTAAATGCTG G-3′.

Primer 5 (exon 3 forward): 5′-AGGATAGGAGTTCTTGCCG C-3′.

Primer 6 (exon 3 forward): 5′-AACTGTGGTCTGGGAGGT GA-3′.

Primer 7 (exon 3 reverse): 5′-TCCAGGAACTTTTGCCCTC A-3′.

34. ThermoFisher DreamTaq Green PCR Master Mix or equivalent.
35. Optional: 6× DNA loading dye (if using a different PCR buffer than DreamTaq Green).
36. 0.2 mL PCR tubes (or appropriate size tubes for PCR thermocycler to be used).
37. 100 base pair (bp) DNA ladder.
38. Agarose powder (molecular biology grade), or precast 1–3% agarose gels.
39. Ethidium bromide (*see* **Note 5** regarding safety, handling, and disposal) or an alternative less-toxic/less-mutagenic DNA stain.
40. Gel electroporation buffer: 1× Tris-acetate-EDTA (TAE) or 0.5× Tris-borate-EDTA (TBE) buffer.
41. NucleoSpin Gel and PCR Clean-Up kit (Machery-Nagel), or equivalent PCR clean-up kit or gel extraction kit.
42. Optional: TOPO-TA Cloning Kit for cloning PCR fragments for DNA sequencing.

3 Methods

3.1 General Considerations for Designing and Making CRISPR sgRNAs

When designing CRISPR gRNAs, a critical aspect of nuclease cutting activity relies on the presence of a PAM sequence adjacent to the complementary 20-nucleotide sequence in the target DNA. For *Streptococcus pyogenes* Cas9, the required PAM sequence is NGG (any nucleotide followed by two guanine nucleotides) immediately to the 3′ of the 20-nucleotide genomic target sequence, and the Cas9 nuclease cut occurs at 3–4 bp upstream of the PAM site (within the target sequence). It is important to note that this PAM sequence must be present at the genomic target site, but should not be included in the gRNA.

Although CRISPR target sites can be manually identified by searching target genomic DNA sequences for PAM sequences, easy-to-use online CRISPR gRNA design tools have been developed for ready identification of target sequences. Many such tools also perform analysis of similar DNA sequences in the target organism's genome that might serve as additional off-target sites for CRISPR cutting with a particular gRNA (based on one or more base mismatches with the target sequence), in order to rank and identify CRISPRs with higher potential specificity for a particular target. Examples of these tools include Keith Joung lab's ZiFiT Targeter (http://zifit.partners.org/ZiFiT/), Feng Zheng lab's CRISPR design tool (http://crispr.mit.edu), and Jin-Soo Kim lab's Cas-Designer (http://www.rgenome.net/cas-designer/). Use of these tools simply requires selecting the option for *Streptococcus pyogenes* Cas9 nuclease (if other options are available), pasting a target DNA sequence corresponding to the genomic region to be targeted (which can include 250 or more bases, depending on the tool), and selecting the target organism.

It should be noted that at present CRISPR design tools cannot predict the actual cutting activity of a particular CRISPR/gRNA, which may vary based on target location, sequence, chromatin accessibility, and other as yet undetermined factors. Consequently, it is highly advisable to design at least two gRNAs aimed at different sequences or regions within the target gene, to improve the likelihood that one will result in sufficient cutting activity for gene editing.

While some of these tools also allow for the design of alternative CRISPRs with different PAM and guide requirements, the protocols described here assume the use of the standard *Streptococcus pyogenes* Cas9 nuclease (SpCas9). If using an alternative nuclease, a different gRNA expression plasmid would be needed in Subheading 3.2 to provide the appropriate gRNA scaffold, and a different nuclease mRNA instead of SpCas9 would be needed in Subheadings 3.3 and 3.4.

As illustrated in greater detail in Subheading 3.7, the presence of microhomology sequences flanking the CRISPR cutting site can result in the deletion of the intervening sequences by MMEJ, thereby affecting the frequency of frameshift knockout deletions versus in-frame deletions that may not result in gene knockout [8]. Consequently, when designing CRISPRs to knock out a target gene, it may be useful to assess the suitability of a particular gRNA target using the Cas-Designer or Microhomology-Predictor tool at http://www.rgenome.net.

Once a desired guide has been designed, the appropriate guide sequence can be cloned into a plasmid vector such as DR274 (Fig. 1), which is used for in vitro transcription of an sgRNA using the T7 promoter present in this plasmid (*see* **Notes 6** and 7), as described in Subheadings 3.2 and 3.3.1. This plasmid also contains the remaining nonspecific portion of the gRNA (corresponding to the tracrRNA and a portion of the crRNA

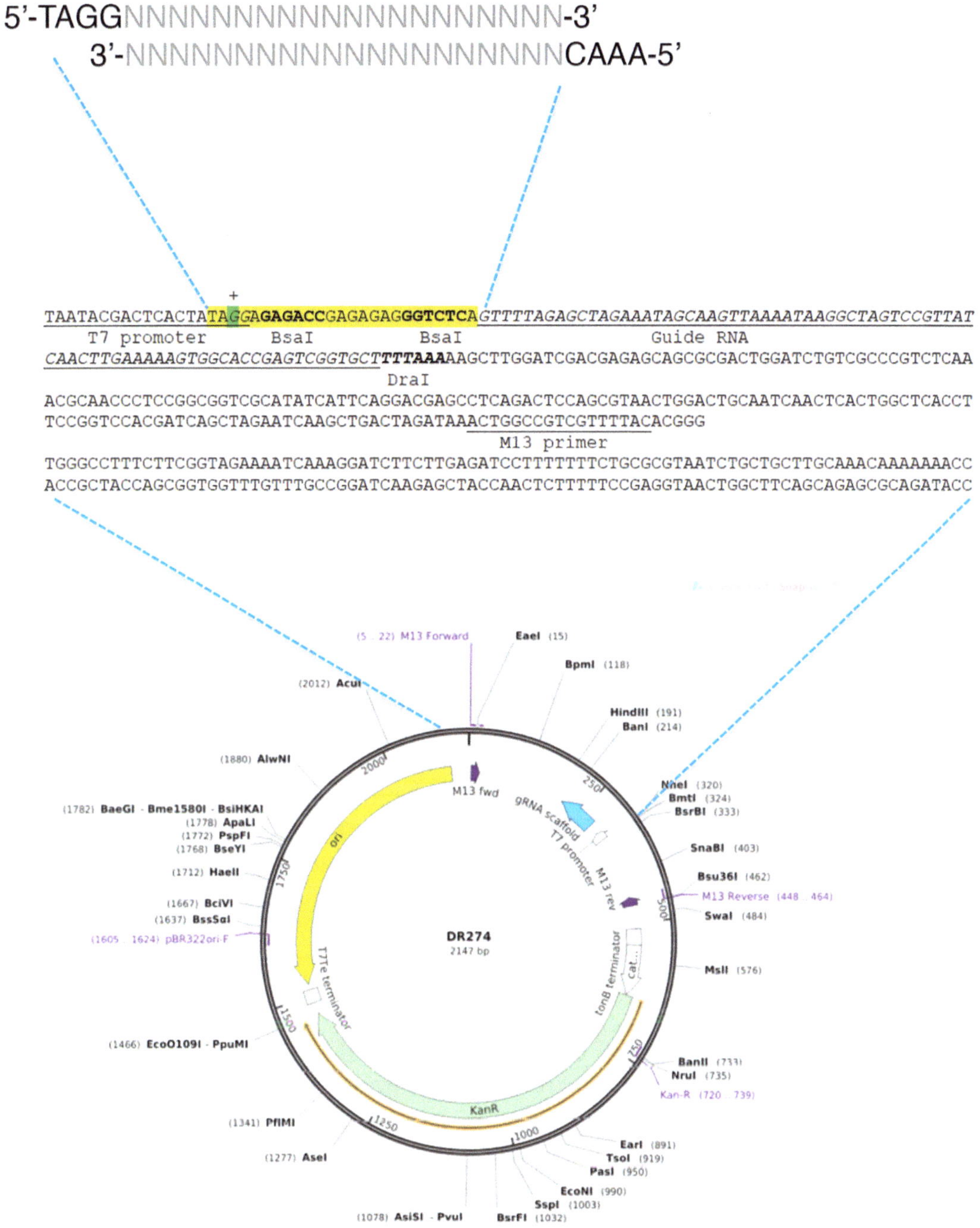

Fig. 1 Cloning of sgRNAs into the pDR274 vector. For generating sgRNAs using in vitro transcription, the pDR274 vector (https:/www.addgene.org/42250/) is first digested with BsaI restriction enzyme, which recognizes GGTCTC sequence (bold) but cuts further downstream and create a 4 bp overhang. Then, annealed forward and reverse strand oligos containing the 20 bp sgRNA sequence and matching overhangs (TAGG for forward oligos and AAAC for reverse oligos) were ligated into the digested vector, so that the sequence highlighted yellow is replaced by the 20 bp sgRNA sequence. For in vitro transcription, the completed DNA construct is digested with DraI (recognition sequence shown in bold), and the linearized plasmid is used as templated for synthesizing RNA using T7 RNA polymerase

sequence), which reconstitutes a complete sgRNA upon cloning of the 20-nucleotide guide sequence. In order to clone the guide sequence, two corresponding single-strand oligonucleotides DNAs should be synthesized (which can be done commercially, as for PCR primers) as described below, with the string of N's corresponding to the guide sequence matching the genomic target (without the PAM) in the forward strand oligonucleotide, and corresponding to the complementary sequence of this guide sequence in the reverse strand oligonucleotide (*see* **Note 8**). Note that the reverse strand oligonucleotide is listed below in 3′ to 5′ orientation (for purposes of showing DNA strand alignment with the forward strand oligonucleotides), so care should be taken to change the reverse strand oligonucleotide sequence to 5′ to 3′ orientation before synthesizing.

Forward oligonucleotide: `5′-taggNNNNNNNNNNNNNNNNNNNN-3′.`

Reverse oligonucleotide: `3′-NNNNNNNNNNNNNNNNNNNNcaaa-5′.`

The T7 promoter requires that the sgRNA sequence start with two G nucleotides for optimal expression during in vitro transcription; this "GG" will be appended to the start of the sgRNA during transcription. If your desired guide does not start with one or two G nucleotides, these G's can simply be inserted in front of the guide sequence, generally without substantially affecting CRISPR targeting. However, if the guide target sequence does start with one or two G's, you can shorten the 20-nucleotide guide sequence in the forward and reverse oligos by the same number of G's. For example, the *Cybb* exon 1 sgRNA target (GGTACTTACAATGACAAAGA) starts with GG, so we shortened the guide sequence in the exon 1 forward and reverse oligos by two bases, as follows:

Cybb Ex-1 forward oligo: `5′-taggTACTTACAATGACAAAGA-3′.`

Cybb Ex-1 reverse oligo: `3′-ATGAATGTTACTGTTTCTcaaa-5′.`

The exon 3 sgRNA target (GATTTCGACACACTGGCAGC) starts with G, so we shortened the guide sequence in the exon 3 forward and reverse oligos by one base, as follows:

Cybb Ex-3 forward oligo: `5′-taggATTTCGACACACTGGCAGC-3′.`

Cybb Ex-3 reverse oligo: `3′-TAAAGCTGTGTGACCGTCGcaaa-5′.`

For the *Cybb* exon 1 sgRNA, this will result in no changes to the overall sgRNA sequence following T7 transcription; for exon 3 sgRNA, this will result in an additional G appended to the beginning of the sgRNA sequence following T7 transcription.

3.2 Cloning sgRNAs into pDR274 Vector

For inactivating the mouse *Cybb* gene, we designed two sgRNAs to target exon 1 and exon 3 using the following protocol (*see* Fig. 1 for plasmid map and cloning site):

1. Order commercial synthesis of both the forward and reverse strand oligos (*see* Subheading 3.1 for designing sgRNAs and Subheading 2.4, **item 4** for sequences of the four oligos). Dissolve the lyophilized oligos in TE at the concentration of 100 μM.
2. Digest 10 μg of DR274 plasmid with BsaI restriction enzyme for 2 h at 37 °C. After confirming complete digestion by running a mini agarose gel (*see* Subheading 3.6 for a detailed protocol on agarose gel preparation and electrophoresis), purify the linearized plasmid DNA by adding 20 μL of 3 M sodium acetate and 200 μL of saturated phenol. Mix well and centrifuge for 1 min at 14,000 × *g*. Collect the aqueous phase and transfer it to a new tube, then add 200 μL of chloroform. Mix well, then centrifuge for 1 min at 14,000 × *g*. Transfer the aqueous phase to a new tube.
3. Precipitate linearized plasmid DNA with 400 μL of 100% ethanol at −20 °C for at least 15 min. Centrifuge at 4 °C for 10 min at 14,000 × *g* to pellet DNA. Decant the supernatant, and wash the DNA pellet with 1 mL of 70% ethanol. Centrifuge for 1 min, and decant the supernatant. Air dry DNA pellet for 10 min. Dissolve the linearized DNA in 100 μL of TE buffer and store at 4 °C until **step 5**.
4. Add 10 μL of forward and 10 μL of reverse oligos to 80 μL of 1× PCR reaction buffer in a PCR tube. Using a PCR thermocycler to heat the tube to 94 °C for 5 min and then gradually cool down to 50 °C over 15 min. Then, place the annealed oligos on ice until use.
5. Mix 1 μL of annealed oligos, 2 μL of linear DR274, 2 μL T4 DNA ligase buffer, 14 μL ddH_2O, and 1 μL T4 ligase. Incubate at 16 °C overnight.
6. Transform 1 μL of the ligation mix into *E. coli* competent cells and plate onto an LB agar plate containing 50 μg/mL Kanamycin, following the manufacturer's supplied protocols. Culture overnight at 37 °C in a bacteria incubator.
7. Pick 24 colonies and transfer each into 1–5 mL of LB medium containing 50 μg/mL Kanamycin . Culture bacteria 8 h or overnight in a bacterial shaker incubator for plasmid minipreps (performed using manufacturer's supplied protocol), setting aside a residual amount (~100 μL) of the bacterial culture for further culturing. Perform DNA sequencing on the plasmid DNA samples using M13 reverse primer, to identify clones containing the correct sgRNA sequences.
8. Inoculate 10–50 mL of LB medium containing 50 μg/mL Kanamycin with 1:1000–1:10,000 dilution of remaining bacterial culture from one of the correct bacterial clones, and culture overnight in a bacterial shaker incubator for plasmid midiprep (using manufacturer's supplied protocol) to obtain DNA template for synthesizing sgRNA (*see* Subheading 3.3.1).

3.3 Preparation of sgRNA and Cas9 mRNA

A protocol for sgRNA synthesis using the MEGAshortscript T7 Transcription kit follows (*see* **Notes 6** and 7). All reagents are RNase-free.

3.3.1 T7 Promoter Driven sgRNA Synthesis

1. Prepare DNA template for in vitro transcription by digesting DR274 (into which the sgRNA sequence has been cloned) with DraI restriction enzyme as follows: in a microfuge tube, mix 10 μg of DR274 in 180 μL of nuclease-free water, 20 μL of 10× digest buffer, and 100 Units of DraI. Incubate at 37 °C for 1 h.
2. Purify digested DNA template (or PCR product) by adding 20 μL of 3 M sodium acetate and 200 μL of phenol/chloroform. Mix well and centrifuge for 1 min at 14,000 × *g*. Collect the aqueous phase and transfer it to a new tube, then add 200 μL of chloroform. Mix well, then centrifuge for 1 min at 14,000 × *g*. Transfer the aqueous phase to a new tube.
3. Precipitate DNA with 400 μL of 100% ethanol at −20 °C for at least 15 min. Centrifuge at 4 °C for 10 min at 14,000 × *g* to pellet DNA. Decant the supernatant, and wash the DNA pellet with 1 mL of 70% ethanol. Centrifuge for 1 min, and decant the supernatant. Air dry DNA pellet for 10 min. Dissolve DNA in 10 μL of nuclease-free water. Determine DNA concentration with a NanoDrop.
4. In an RNase-free microfuge tube at room temperature, gently mix 1 μg of template DNA (PCR product or plasmid) in 8 μL of nuclease-free water with 2 μL of T7 reaction buffer, 2 μL of each NTP, and 2 μL of T7 enzyme mix (from MEGAshortscript kit). Incubate at 37 °C for 2–4 h.
5. Add 1 μL of TURBO DNase (from MEGAshortscript kit) to the reaction and continue incubation at 37 °C for 15 min.
6. Add 115 μL of nuclease-free water and 15 μL of 3 M sodium acetate to the reaction. Add 150 μL of phenol/chloroform and briefly vortex. Then centrifuge at 14,000 × *g* for 1 min. Collect the aqueous phase and transfer to a new tube.
7. Add 150 μL of chloroform and briefly vortex. Centrifuge at 14,000 × *g* for 1 min. Collect the aqueous phase and transfer to a new tube.
8. Add 300 μL of 100% ethanol and mix well, then chill for at least 15 min at −20 °C to precipitate sgRNA.
9. Centrifuge at 4 °C for 10 min at 14,000 × *g* to pellet sgRNA, and carefully remove and discard the ethanol supernatant. Wash the sgRNA pellet with 70% ethanol, and then centrifuge and discard ethanol supernatant.

10. Dry the sgRNA pellet at room temperature for 10 min, then dissolve in 100 μL of nuclease-free water. Measure the sgRNA concentration by a NanoDrop and check the RNA integrity by 6% TBE-Urea polyacrylamide gel if desired, following the manufacturer's protocol (*see* **Note 9**).
11. Aliquot sgRNA and store at −80 °C.

3.3.2 T7 Promoter Driven Cas9 mRNA Synthesis

A protocol for Cas9 mRNA synthesis using the mMESSAGE mMACHINE T7 Ultra Transcription kit follows (*see* **Note 10**). All reagents are RNase-free.

1. Linearize 12 μg of MLM3613 T7-Cas9 plasmid by digesting with 100 Units of PmeI restriction enzyme in 20 μL of 10× enzyme digestion buffer and 180 μL of nuclease-free water at 37 °C for 1 h.
2. Extract digested DNA by adding 20 μL of 3 M sodium acetate and 200 μL of saturated phenol. Mix well and centrifuge for 1 min at 14,000 × *g*. Collect the aqueous phase and transfer it to a new tube, then add 200 μL of phenol/chloroform. Mix well, then centrifuge for 1 min at 14,000 × *g*. Transfer the aqueous phase to a new tube.
3. Precipitate DNA with 400 μL of 100% ethanol at −20 °C for at least 15 min. Centrifuge at 4 °C for 10 min at 14,000 × *g* to pellet DNA. Decant the supernatant, and wash the DNA pellet with 1 mL of 70% ethanol. Centrifuge for 1 min, and decant the supernatant. Air dry DNA pellet for 10 min. Dissolve DNA in 12 μL of nuclease-free water.
4. At room temperature, add 20 μL of nuclease-free water, 40 μL of T7 2× NTP/ARCA, 8 μL of 10× T7 reaction buffer, 4 μL (1 μg/μL) of linearized MLM3613, and 8 μL of T7 enzyme mix. Incubate at 37 °C for 2 h.
5. Add 4 μL of TURBO DNase; mix well and incubate at 37 °C for 15 min.
6. Add 144 μL of nuclease free water, 80 μL of 5× E-PAP buffer, 40 μL of 25 mM $MnCl_2$, and 40 μL of ATP solution. Add 16 μL of E-PAP; incubate at 37 °C for 45 min.
7. Add 40 μL of ammonium acetate stop solution. Add 400 μL of phenol/chloroform to extract. Mix vigorously, then centrifuge for 1 min at 14,000 × *g*. Transfer the aqueous phase to a new tube. Add 400 μL of chloroform, then mix vigorously. Centrifuge for 1 min at 14,000 × *g*. Transfer the aqueous phase to a new tube.
8. Precipitate Cas9 mRNA by adding 800 μL of 100% ethanol. Keep at −20 °C for at least 15 min. Centrifuge at 4 °C for 10 min at 14,000 × *g*. Decant the supernatant without disturbing the pellet. Wash pellet by adding 1 mL of 70% ethanol,

then centrifuge for 1 min at 14,000 × *g*. Decant the supernatant and air dry for 10 min at room temperature.

9. Dissolve Cas9 mRNA in 400 μL of nuclease-free water. Measure the Cas9 mRNA concentration by a NanoDrop. The yield should be approximately 400 μg in 400 μL of nuclease-free water. Check the mRNA integrity by agarose gel (1% agarose) electrophoresis if desired (*see* **Note 11**).
10. Aliquot mRNA and store at −80 °C.

3.4 Cytoplasmic Microinjection of CRISPR Reagents (See Note 12)

1. In our experience, NSG female mice do not respond to hormone-stimulated ovulation well, so we prefer to use natural mating to produce zygotes. The day before scheduled microinjection, select females that are in estrus (Fig. 2) and mate each female with an individually caged stud male.
2. In the next morning, check the females for vaginal plugs. Euthanize plugged females with CO_2 and dissect out both oviducts from each female (part of the uterus horn and whole ovary can be dissected to avoid damage to the oviduct). Place the dissected tissues in a 35 mm culture dish containing M2 medium to rinse off blood and other loose debris (*see* Fig. 3 for an outline of the remaining steps of microinjection and embryo transfer).
3. Place 4–5 cleaned oviducts into one 50 μL M2 drop under mineral oil. Under the stereo microscope, using a 29-G insulin needle, tear open the ampulla to allow the cumulus mass to

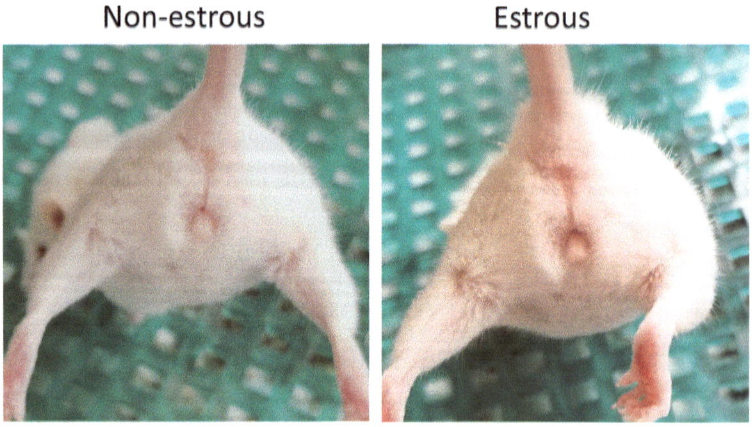

Fig. 2 Photographs of female NSG mice that are in estrous and non-estrous. Mouse's estrous cycle is typically 4–5 days, and therefore on average only 20–25% of female mice will mate on a particular day. For strains that do not respond to hormone-stimulated superovulation, it is important to select and mate the females in estrous in order to obtain enough zygotes on the microinjection day. The morphological differences between estrous and non-estrous female mice are subtle and difficult to describe using language alone. As shown in the photographs, the genital area of an estrous female swells slightly and shows a little deeper in pink color

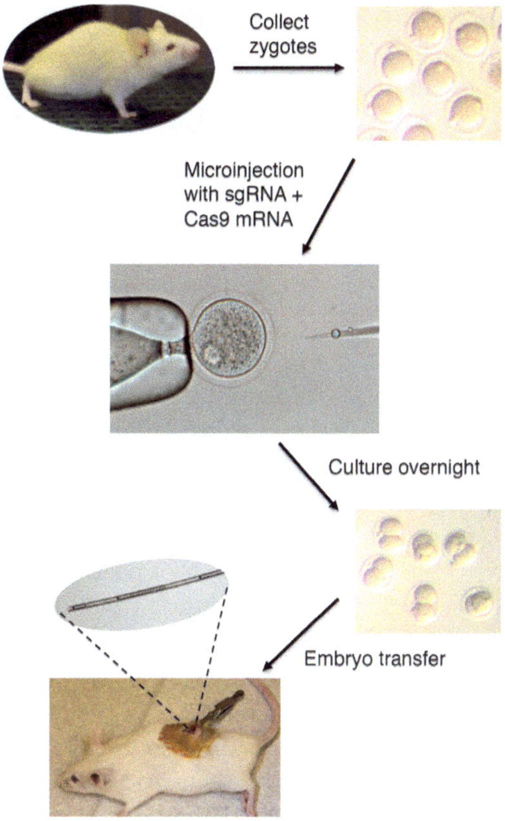

Fig. 3 Outlining of microinjection and embryo transfer procedures. Plugged female NSG mice are euthanized and fertilized eggs are collected from their oviducts. These zygotes are subsequently microinjected with Cas9 mRNA and either Cybb exon 1 or exon 3 sgRNA. After culturing injected eggs overnight, those that reach 2-cell stage of development are implanted into the oviducts of pseudopregnant surrogate mothers using a mouth-controlled glass transfer pipette

extrude spontaneously into the medium. Discard the oviduct tissue.

4. After releasing the egg clutches from all oviducts, add 5 μL of 10× concentrated hyaluronidase solution to each M2 drop. Gently swirl the dish and incubate at room temperature for a few minutes to allow hyaluronidase to release follicle cells from the eggs. Wash eggs using an embryo transfer pipette through three drops of M2 medium to remove residual hyaluronidase.
5. Transfer 30–50 zygotes into the M2 drop in the injection chamber, and place the injection chamber onto the microscope stage, then lower the tip of the holding pipette into the M2 drop.
6. Mix the desired amount of sgRNA and Cas9 mRNA (*see* **Note 13**), and dilute with microinjection buffer to the final

concentration of 100 ng/μL of Cas9 mRNA and 50 ng/μL of sgRNA (*see* **Note 14**). Then, centrifuge the mixture for 5 min at 14,000 × *g* in a microcentrifuge to pellet any possible particles that may clog the injection needle.

7. Use a P10 or P20 Pipetman and an Eppendorf microloader tip to deliver 2–3 μL of sgRNA/Cas9 solution into the barrel of a microinjection needle through the open (back) end. The filament inside the needle will facilitate transporting the solution to the needle tip through capillary action.
8. Lower the injection needle into the injection chamber. Break open the tip of the microinjection needle by hitting the needle tip against the holding pipette.
9. Use the holding pipette to hold an egg and insert the needle into its cytoplasm. Sometimes the needle has to pass through the entire cytoplasm in order to penetrate the plasma membrane. When the needle tip is inside the cytoplasm, push the injection foot paddle of the FemtoJet 4i to inject the sgRNA/Cas9 solution (*see* **Note 15**).
10. After injection, the egg is transferred to a different area using the holding pipette to separate it from the uninjected eggs. After all eggs in the chamber are injected, transfer them to an M16 drop in the CO_2 incubator and culture overnight to allow them to develop into 2-cell stage embryos (*see* **Notes 16** and **17**).

3.5 Surgical Implantation of Injected Embryos

1. On the day of microinjection, estrus CD-1 females are selected and set up with vasectomized male mice for mating. The next morning, check the mice for vaginal plugs. The plugged mice can be used as recipient mothers for embryo transfer.
2. The day after the injection, examine microinjected embryos using a stereo microscope. Count the number of embryos that have reached the 2-cell stage of development. A large number of eggs at the 1-cell stage indicate toxicity of the gRNA. Transfer the 2-cell embryos into an M2 drop.
3. Anesthetize the recipient mothers by injecting ketamine/xylazine solution (intraperitoneal, with a 29-G needle) at a dose of 0.1 mL per 10 g of body weight.
4. Remove hair from the back using a small hair clipper. Disinfect the clipped area using alternating applications of 70% isopropyl alcohol and Betadine.
5. Cut a midline incision (~1 cm) in the cleaned skin area, and place the mouse on its left side. Move the skin incision to the right side near the right ovary area, then make a small cut in the body wall and extend it to 0.5–1 cm long using the back of the scissors' blades.

6. Pull the fat pad that is associated with the ovary with a pair of iris forceps until a segment of the uterine horn appear. Clamp the fat pad with a Dieffenbach clip to hold the organs in place.
7. Load the embryo transfer pipette with 12 embryos. Because the embryos are barely visible in the transfer pipette, two air bubbles are sucked into the transfer pipette to mark the boundaries of the suspended embryos.
8. Under the dissecting microscope, find the oviduct and ampulla. Use a 29-G needle to punch a small hole in the oviduct wall one turn before the ampulla after the ovary. Carefully insert the tip of the transfer pipette into the hole with the pipette opening pointing toward the ampulla. Gently blow into the mouthpiece to expel the two air bubbles and the embryos sandwiched between them. The presence of the air bubbles in the ampulla is a good indication that all embryos have been successfully transferred.
9. Remove the Dieffenbach clip and carefully push the ovary back into the abdominal cavity. Close the abdominal wall with a cruciate suture utilizing absorbable 5–0 Vicryl suture.
10. Drop one drop of 0.25% Bupivacaine solution on the muscle at the surgical site. Bupivacaine solution is a long-lasting local anesthetic for reducing pain at the surgical site.
11. For implanting embryos into the left oviduct, turn the mouse on its right side and make a 0.5–1 cm incision in the left ovary area of the body wall. Then, repeat **steps 6–10**.
12. Close the skin incision with stainless steel surgical wound clips. Inject 0.2 mL meloxicam (intraperitoneal, with a 29-G needle).
13. Place the mice in a pre-warmed mouse cage placed on a circulating water blanket. Normally, the mice will wake 30–60 min post-operation. They can then be returned to the animal room.
14. Wound clips are removed 10–14 days after surgery. Pups developed from injected embryos will be born approximately 19 days after the embryo transfer procedure. The offspring will be weaned when they are 18–21 days old.

3.6 Mouse Tail Biopsy and Genotyping by PCR

Mouse distal tail snips may be collected from mice at up to 21 days of age without anesthesia (*see* **Note 18**). Sample collection and mouse marking may be performed concurrent with weaning of mice. **Steps 3** through **5** of the protocol described here use the Kaneka Easy DNA Extraction kit for rapid (<20 min) DNA isolation, but this may be readily replaced with any other genomic DNA extraction method appropriate for animal tissues (*see* **Note 19**). PCR analysis may be used both for initial identification of CRISPR-induced mutations in founder mice and for routine genotyping to screen for gene-modified offspring (*see* **Note 20**).

1. Mark each mouse by ear punch or other method to identify the mouse and the corresponding DNA sample.
2. Snip the distal 2–5 mm of the tail with a sterile razor blade or scissors, and transfer the tail snip to a labeled 1.5 mL microcentrifuge tube (*see* **Note 21**). Repeat for each mouse.
3. Add 100 μL of Kaneka kit Solution A to each tube containing a tail snip. Incubate at 98 °C for 10 min on a heat block. Remove tubes from heat block, and allow the sample to cool for at least 2 min.
4. Add 14 μL of Kaneka kit Solution B to each tube. Vortex.
5. Centrifuge tubes at maximum microcentrifuge speed for 1 min to pellet particulate matter and undigested tissue. Transfer liquid containing DNA to a new microcentrifuge tube, and store DNA at −20 °C until ready for PCR analysis.
6. Design PCR primers specific for the CRISPR-targeted genomic region (*see* **Note 22**). Given the potential for large deletions occurring in either direction from the CRISPR target site, initial PCR screening for unknown deletions at the target site may require designing either multiple overlapping primer pairs to cover the entire region or a single pair of primers spaced far enough apart to generate a PCR product of sufficient size to identify deletions of hundreds of bp on either side of the target, as was necessary for the identification and genotyping of the larger *Cybb* mutations (Figs. 4a–d and 5a, b; *see* **Note 23**).
7. Prepare a PCR reaction mix by combining the following reagent volumes, multiplied by (the total number of samples +1): 12.5 μL of 2× DreamTaq Green PCR Master Mix, 0.5 μL of 10 μM forward primer (200 nM final), 0.5 μL of 10 μM backward primer (200 nM final), and 9.5 μL of sterile PCR-grade water.
8. In separate PCR tubes for each sample to be genotyped, add 23 μL of PCR reaction mix to 2 μL of mouse genomic DNA (*see* **Note 24**).
9. Set up thermocycler settings (*see* **Notes 25** and **26**): 95 °C for 1 min for initial DNA denaturation; 25–35 cycles of 95 °C for 15–30 s, 55–60 °C (typically 3–5° below the lowest primer Tm) for 15–30 s, 72 °C for ~1 min per 1000-bp of PCR product size; and finally 72 °C for 7 min followed by indefinite incubation at 4 °C.
10. Prepare a 1–3% agarose gel as follows: add 1–3 g of agarose powder to 100 mL of 1× TAE or 0.5× TBE buffer (agarose percentage based on weight/volume, i.e., 1 g of agarose per percent per 100 mL for a 1% agarose gel) in an Erlenmeyer flask large enough to hold 2–4× the buffer volume (*see* **Note 27**). Microwave for 1.5–3 min (longer times for higher percentage

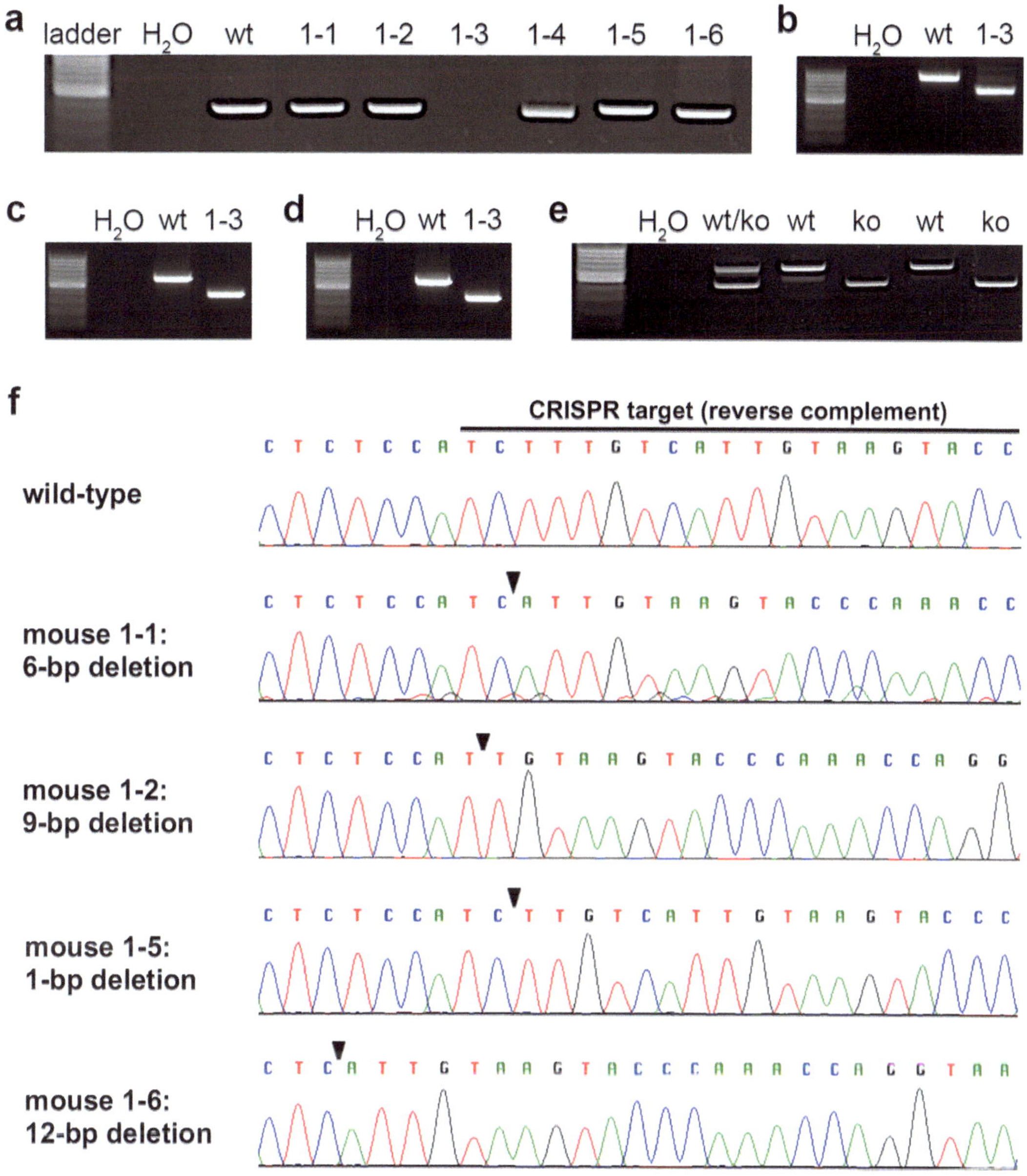

Fig. 4 PCR and sequencing analysis of CRISPR-induced mutations in *Cybb* exon 1. (**a**) Gel electrophoresis of PCR products using primers 2 and 3 for mice (numbered 1–1 through 1–6), including water (H_2O) and wild-type (wt) mouse DNA controls. (**b**) PCR analysis of mouse number 1–3 using primers 1 and 4 to amplify alleles with 235 and 240 bp deletions that appear as a single band. PCR analysis of this mouse using (**c**) primers 1 and 3 amplifies only the 235 bp deleted allele, while PCR using (**d**) primers 2 and 4 amplifies only the 240 bp deletion. (**e**) PCR genotyping of the *Cybb* exon 1 knockout mouse strain containing the 235 bp deletion established from mouse number 1–3, using primers 1 and 3 to distinguish between the 628 bp wild-type (wt) allele and the 393 bp knockout (ko) allele. (**f**) DNA sequences from PCR products for wild-type exon 1 and four mice with small deletions (at locations indicated by arrowheads)

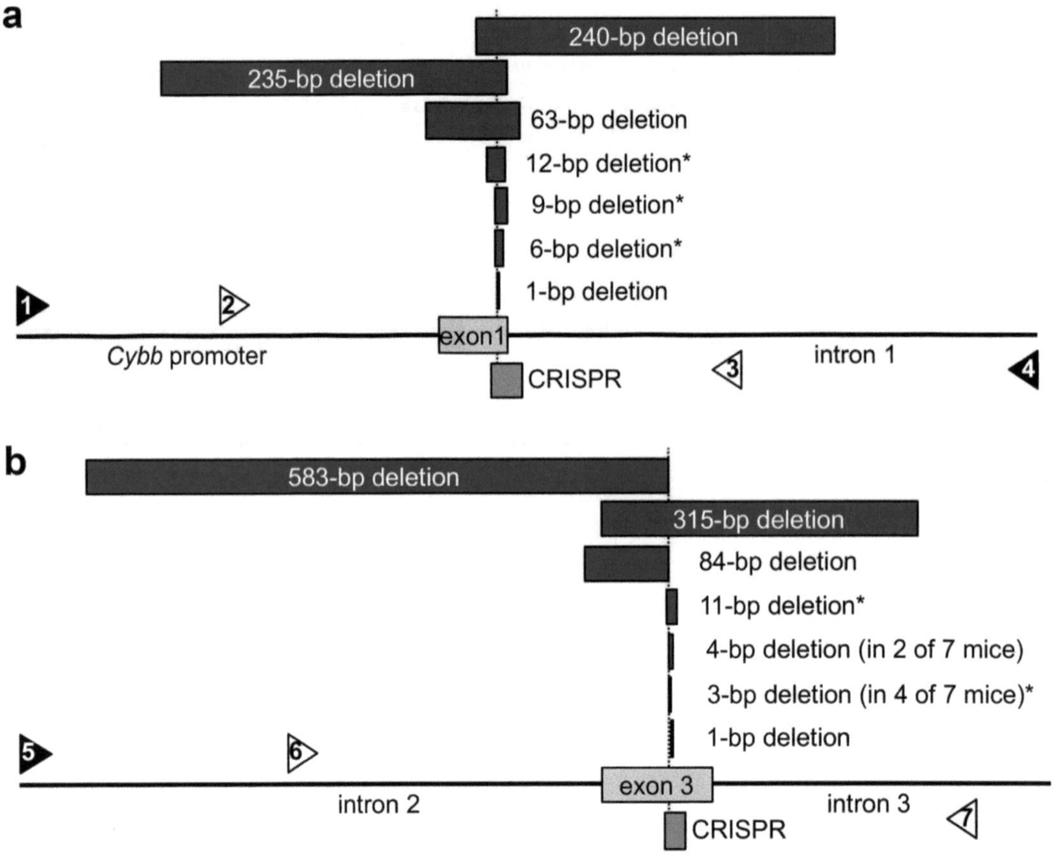

c

Sequence	Deletion (bp)	Micro-homology	# Mice
Exon 1:			
GAATGAAGGACTCTCCATCTTTGTCATTgtaagtacc			
GAATGAAGGACTCTCCATC------ATTgtaagtacc	6	TC	1/6
GAATGAAGGACTCTCCAT---------Tgtaagtacc	9	CAT	1/6
GAATGAAGGACTC------------ATTgtaagtacc	12	TC	1/6
Exon 3:			
AACTGTATGCTGATCCTGCTGCCAGTGTGTCGAAATC			
AACTGTATGCTGATCCTGC---CAGTGTGTCGAAATC	3	TGC	4/7
AACTGTATGCTGATCCTG-----------TCGAAATC	11	TG	1/7

Fig. 5 Summary of mouse *Cybb* mutations from CRISPR targeting. Shown are the CRISPR target region and cut site (dotted line) for (**a**) exon 1 or (**b**) exon 3. Deletions matching MMEJ predictions are denoted by "*". (**c**) Mutations matching predicted MMEJ-induced deletions. Wild-type sequences (bold) are listed in 5′ to 3′ orientation, with exons (uppercase), introns (lowercase), and CRISPR targets (underlined). CRISPR target and PAM sequences are shown in reverse complement, as they are on the complementary strand to the *Cybb* gene orientation

gels), until agarose melts without liquid boiling over. Remove flask from microwave, swirl gently and carefully to mix, and allow to cool at room temperature until the outside of the flask can be touched safely (approximately 5 min, or less for higher percentage gels, which solidify faster). Add ethidium bromide to 0.5 μg/mL final volume (i.e., 5 μL of 10 mg/mL stock to a 100 mL gel) and gently and carefully swirl flask to mix. Assemble a gel casting tray with a gel comb inserted and with seals in place to prevent leakage, and pour the molten agarose into the tray (using approximately 30 mL for a small 8.5 cm × 8.5 cm gel tray, or 100 mL for a larger 12 cm × 12 cm gel tray). Allow gel to solidify for 20–30 min.

11. Remove the gel comb and the seals, and transfer the gel tray containing the gel into a gel electrophoresis apparatus. Orient the gel so that the samples will migrate toward the end of the apparatus with the red electrode. Add sufficient TAE or TBE buffer (that same buffer used to prepare the gel) to completely submerge the gel.
12. Run gel as follows: load a DNA ladder into one well for product size analysis. From each PCR reaction, load up to 25 μL into a well of gel (*see* **Note 28**). Place cover on gel apparatus, and perform gel electrophoresis at 5 V/cm (referring to the distance between electrodes in the apparatus; i.e., 100 V for a 20 cm apparatus) for 30–45 min, or until the DNA loading dye migrates approximately 75% of the way across the gel.
13. Visualize the DNA bands in the gel using a UV transilluminator (*see* **Note 29**); larger insertions or deletions may be readily distinguishable based on size differences from wild-type control (*see* **Notes 30** and **31**).

3.7 Confirmation of Mutations by DNA Sequencing and Analysis of Results

Initial identification of founder mouse mutations may be performed by purifying individual PCR products from Subheading 3.6 (using a PCR clean-up kit for samples containing a single PCR product, or by gel band excision and purification of separate bands using a gel extraction kit for samples containing multiple PCR products), followed by DNA sequencing (*see* **Note 32**). A protocol for PCR clean-up or gel band excision using a dual-purpose NucleoSpin Gel and PCR Clean-up kit follows (**steps 1–5**), and a discussion of CRISPR-induced mutations (with examples from *Cybb* targeted mice) can be found at the end of this section:

1. For DNA purification by cleanup of the PCR reaction, transfer PCR reaction to a 1.5 mL tube and proceed directly to **step 2**. For DNA extraction from agarose gels following gel electrophoresis, visualize ethidium bromide-stained DNA using a UV

transilluminator in the dark (*see* **Notes 29** and **33**). Quickly cut out each desired DNA band using a clean scalpel, removing excess agarose from the excised slice. Transfer each excised gel band to a separate pre-weighed 1.5 mL tube. Reweigh the tube to determine the weight of the gel slice.

2. Add 200 μL of Buffer NTI to each 100 μL of PCR product or to each 100 mg of gel slice for gels containing ≤2% agarose (for gels containing >2% agarose, double the volume of Buffer NTI), then vortex. For PCR clean up, proceed directly to **step 3**. For gel purification, incubate tubes for 5–10 min at 50–56 °C, vortexing briefly every 2–3 min, until the gel slice is completed dissolved.
3. For each sample, place a NucleoSpin column in a collection tube, and add up to 700 μL of the sample to the column. Centrifuge for 30 s at 11,000 × *g*. Discard the flow-through from the collection tube and place the column back in the collection tube. Load any remaining volume of the sample (beyond the initial 700 μL) into the same tube and repeat centrifugation, discarding flow-through.
4. Add 700 μL of Buffer NT3 to the column. Centrifuge for 30 s at 11,000 × *g*. Discard the flow-through from the collection tube and place the column back in the collection tube. Repeat this wash step with another 700 μL of Buffer NT3, centrifuging and discarding the flow-through as above. Centrifuge for 1 min at 11,000 × *g* to remove any residual buffer.
5. Transfer the column to a new 1.5 mL microcentrifuge tube. Add 30 μL of Buffer NE or water. Incubate at room temperature for 1 min, then centrifuge for 1 min at 11,000 × *g* (*see* **Note 34**). Determine DNA concentration using a NanoDrop.
6. Sequence purified PCR product commercially using a sequencing primer (*see* **Notes 35** and **36**).
7. Once a desired mutation has been identified, the founder mouse can be bred and the resulting offspring genotyped (*see* **Note 37**) to confirm germline transmission of the mutation in order to establish the gene-modified strain (*see* Fig. 4e for PCR-based genotyping of offspring of mouse number 1–3 confirming germline transmission of the 235-bp deletion in *Cybb* exon 1 on the X-chromosome; shown are genotyping results for wild-type or knockout males and a female mouse heterozygous for one wild-type and one knockout allele).

CRISPR-induced mutations can include base deletions, additions, or substitutions at or near the target site. For CRISPR-mediated knockout of the *Cybb* gene at exon 1 or exon 3 in NSG embryos [14], all detected mutations were deletions ranging in size from 1- to >500-bp, stretching upstream and/or downstream from the genomic target site (Figs. 4f and 5a, b). Some smaller deletions may be the favored results of MMEJ rather than NHEJ at the CRISPR cut site [8], due to short regions (two or more bp) of microhomology flanking the cut site that can result in the exact deletion of the intervening sequence plus one of the microhomology repeats (Figs. 5c and 6). For CRISPR-mediated knockout of the *Cybb* gene in NSG mice, deletions matching MMEJ predictions

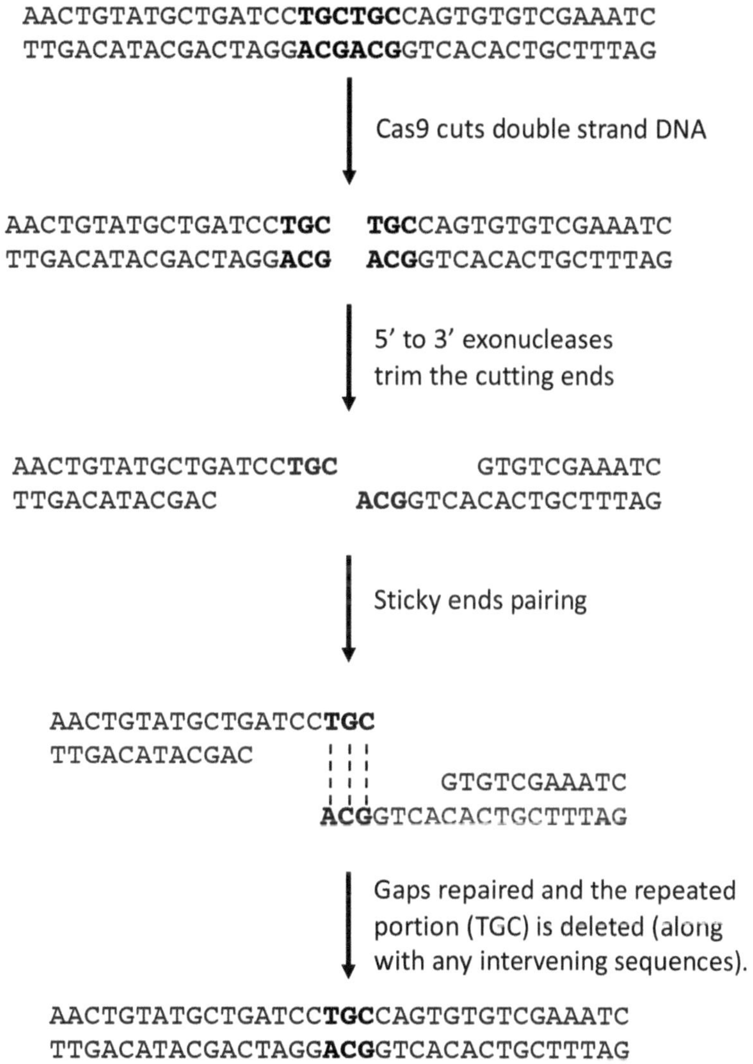

Fig. 6 Schematic interpretation of a prevalent microhomology-mediated end joining. Four out of seven offspring derived from embryos injected with Cybb exon 3 sgRNA carried exactly the same 3 bp deletion, supporting the notion that CRISPR-mediated deletions are not completely random. This diagram illustrates a plausible mechanism for interpreting biased deletion patterns mediated by short homologous sequences (in bold) near the two ends created by Cas9 cutting

occurred frequently, as they were detected in at least half of the founder mice (Fig. 5c). These deletions may be in-frame (encompassing one or more codons in multiples of 3-bp), and consequently may not result in knockout of the target gene. For this reason, when designing CRISPRs for gene knockout, it may be helpful to first identify potential microhomology sequences surrounding the CRISPR cut site in the genome, so as to choose CRISPRs that favor frameshifts rather than in-frame deletions (*see* **Note 38**).

4 Notes

1. This *Cybb* knockout NSG mouse strain (NSG.Cybb[KO]), which contains a 235 bp deletion in *Cybb* exon 1, will be available from The Jackson Laboratory as JAX stock # 030610. Due to their increased immunodeficiency beyond that of NSG mice, housing in a high barrier level specific pathogen-free facility is strongly recommended for these mice, to prevent spontaneous infection with bacterial and fungal agents that have low pathogenicity to NSG mice or that are non-pathogenic to immunocompetent mice.
2. Federally funded animal research in the United States requires review and approval from the appropriate Institutional Animal Care and Use Committee. The studies described here were approved by the National Institute of Allergy and Infectious Diseases Animal Care and Use Committee (ACUC) under animal use protocol LCIM 1E and by the National Heart, Lung, and Blood Institute ACUC under animal use protocol H-0125R2. For these studies, all mice were bred and maintained under specific pathogen-free conditions at an American Association for the Accreditation of Laboratory Animal Care accredited animal facility and housed in accordance with the procedures outlined in the Guide for the Care and Use of Laboratory Animals.
3. Acidic phenol may be used to partition DNA into the organic phase to remove DNA from RNA preparations, in place of performing DNase digestion.
4. Because ketamine (Ketaset) is a controlled substance regulated by the United States Drug Enforcement Administration, a license is required for purchasing and storing this anesthetic, and a lock box and log sheets are needed to store it and record its usage.
5. Ethidium bromide can be toxic at high concentrations, and its metabolites may be mutagenic in some species. Appropriate personal protective equipment should be used when handling;

consult your institution's safety requirements for approved handling practices and disposal procedures for ethidium bromide solutions and gels.

6. As an alternative to performing in vitro T7 transcription and purification of the sgRNA, the complete 100 bp sgRNA can be commercially synthesized (by Synthego or Origene, for example), or gRNA can be commercially synthesized as a separate target-specific crRNA and annealed to a common tracrRNA component (synthesized by Integrated DNA Technologies, for example).
7. As an alternative to cloning the sgRNA binding region into DraI-digested DR274 plasmid for using the vector's T7 promoter, a PCR product may be used as a template for in vitro transcription. In this case, the DR274 plasmid can be used as a template for amplifying the sgRNA scaffold portion (common to all sgRNAs), while the T7 promoter and the 20-nucleotide target-specific region can be added to the scaffold using a forward primer (5′-TTAATACGACTCACTATAGGN_{20}GTT TTAGAGCTAGAAATAGC-3′, where N_{20} is the 20-nucleotide target sequence of the sgRNA, the sequence before N_{20} is the T7 promoter, and the sequence after N_{20} is the first portion of the sgRNA scaffold), and the reverse primer (5′-AAAAGCACCGACTCGGTGCC-3′, which is complementary to the 3′ end of sgRNA scaffold). *See* Subheading 3.6, **steps 7–9** for a general PCR protocol.
8. The complementary sequence refers to A/T base complementarity and G/C base complementarity from the opposite DNA strand to the 20-nucleotide target sequence. The complement to the TAGG sequence from the forward oligonucleotide should not be included in reverse oligonucleotide, and the complement to the AAAC sequence from the reverse oligonucleotide should not be included in the forward oligonucleotide, as these unpaired 5′ TAGG and AAAC overhangs are needed for cloning the annealed oligonucleotides into the restriction enzyme-digested DR274 plasmid in subsequent steps.
9. Following gel electrophoresis, good quality intact RNA should appear as a distinct band without smearing. If accurate size analysis of the sgRNA is desired, use a low range single-stranded RNA ladder instead of a DNA ladder.
10. As an alternative to performing in vitro T7 transcription and purification of Cas9 mRNA, multiple vendors offer Cas9 mRNA or Cas9 protein for purchase.
11. A detailed protocol for agarose gel preparation and electrophoresis can be found in Subheading 3.6. Following gel electrophoresis, good quality intact RNA should appear as a distinct

band without smearing. Polyadenylated Cas9 mRNA is approximately 4.5 kb in size. However, in comparison with a DNA ladder, polyadenylated Cas9 mRNA appears to migrate at a size of approximately 1.5 kb on an agarose gel [16]; use a high range single-stranded RNA ladder and electrophorese under denaturing conditions if a more accurate size comparison is desired.

12. Electroporation of zygotes with CRISPR/Cas9 has been shown to be an effective alternative to microinjection [17, 18], including electroporation of zygotes while they are still in the oviducts to negate the need for embryo transfer [19]. A detailed protocol on electroporation of mouse zygotes for CRISPR/Cas9 delivery can be found in this book in Chapter 10 by Qin and Wang [20].
13. The same amount of commercially available Cas9 protein can be used instead of Cas9 mRNA to achieve genomic incision. When using Cas9 protein instead of Cas9 mRNA, premix the Cas9 protein and gRNA and incubate at room temperature for 10–20 min to form the ribonucleoprotein complex before using for microinjection.
14. A single injection needle containing 2–3 μL of CRISPR reagents can be used to inject up to 100 zygotes; however, the total volume of CRISPR reagents needed may be greater than this if additional needles are required, depending on whether the needle tip is broken correctly and whether the reagents clog the needle. Preparation of 15–20 μL total of diluted CRISPR reagents (sgRNA and Cas9 diluted in microinjection buffer) should be more than sufficient for an entire microinjection procedure; this excess volume will facilitate centrifugation of the diluted reagents to pellet any particles that may clog the needle.
15. It is very important to adjust and optimize the microscope, so that movement of cytoplasmic organelles near the needle tip is clearly visible when the fluid containing CRISPR reagent is forced out of the needle. Otherwise, it is very easy to inject each egg with too much or too little reagent. A high percentage of lysed eggs indicates that the needle opening is too big or the volume injected is too high; change needle or adjust pressure accordingly.
16. M16 medium does not contain HEPES buffer, so it needs to be equilibrated in the CO_2 incubator for several hours before using.
17. We prefer to perform oviduct transfer the day after microinjection so we can monitor the egg development. If there is a shortage of pseudopregnant foster mothers, embryos can be further cultured in KSOM medium (KSOM medium is

preferable to M16 for culturing embryos beyond the 2-cell stage of development). Embryos ranging from the 1-cell stage to the blastocyst stage can be transferred into the oviducts.

18. As an alternative to tail snips, ear punch biopsies may be used as a source of genomic DNA, and may be collected from mice older than 21 days without requiring anesthesia.
19. This extraction method yields genomic DNA containing impurities that can inhibit subsequent PCR reactions, so use of larger tail snips (>5 mm) can be problematic. Excessively large tail snips should be trimmed to reach this size limit; alternately, when proceeding with DNA extraction of larger tissue samples, dilution of the extracted DNA prior to PCR is recommended. As an alternative method, QIAGEN's DNeasy Blood & Tissue kit yields purer DNA and can accommodate larger amounts of starting tissue, but requires 6 or more hours of tissue digestion for DNA extraction.
20. Upon identification of mutations present in founder mice, phenotypic analysis of the mice may indicate which mutations have resulted in functional gene knockout, prior to further breeding of the founder mice, in order to save time and effort in breeding for mutations that do not cause the desired phenotype.
21. Mouse biopsy samples may be stored short term (a few days to weeks) at −80 °C to protect from DNA degradation.
22. Design forward and backward PCR primer pairs using primer design software such as Primer-Blast (https://www.ncbi.nlm.nih.gov/tools/primer-blast/). Primer pairs typically have melting temperatures (Tm) within 3–5 °C of each other, as computed by primer design software. For sequencing across the CRISPR cut site, primers should be designed at least 150–200 bases from the CRISPR site, because some deletions can be quite large and DNA sequencing results may be poor within the first 50–100 bases of the primer binding site.
23. Forward and backward primer pairs initially designed for genotyping at each exon are indicated by white arrowheads in Fig. 5a, b, while additional primers indicated by black arrowheads were required to identify larger deletions. Mouse number 1–3 was heterozygous for 235 and 240 bp deletions at exon 1, but no PCR products could be amplified from this mouse using primers 2 and 3 in Fig. 4a, since each of these primers binds within one of the deleted regions as depicted in Fig. 5a. The additional primers were therefore required to amplify and distinguish between these similarly-sized PCR products, as shown in Fig. 4b–d.

24. A control sample of wild-type mouse genomic DNA should be included for comparison when screening for mutations; a negative control sample containing PCR-grade water instead of DNA template should also be included to check for DNA contamination of the PCR reagents.
25. Other DNA polymerases besides Taq polymerase may require different temperatures and times for the PCR denaturation, annealing, and extension steps. Follow manufacturer's recommendations.
26. Detection of large insertions may require increasing the PCR extension time to amplify larger products.
27. Higher percentage agarose gels are better for separating smaller sizes of linear DNA. A 2% agarose gel is considered optimal for resolving linear DNA between 50 and 2000 bp in length.
28. DreamTaq Green PCR master mix already contains DNA loading dye. When using other PCR buffers that lack dye, add 6× DNA loading dye to samples prior to loading samples into gel.
29. For visualizing DNA stained with ethidium bromide, a 302/312 nm UV transilluminator gives stronger fluorescence intensity which is useful for capturing gel images, but these UV wavelengths can damage DNA in a little as 45 s, negatively affecting subsequent DNA sequencing or plasmid cloning. For visualizing DNA during gel band excision, use of a 365 nm UV transilluminator will cause less DNA damage, but produces weaker fluorescence; minimize UV exposure time during gel band excision to avoid DNA damage.
30. Similar sized PCR products within a sample may appear as a single band during gel electrophoresis, as illustrated in Fig. 4a, b.
31. Genetic mosaicism in founder animals may be evident by the presence of multiple PCR bands in a mouse genomic DNA sample (i.e., more than two bands for an autosomal gene, more than two bands for X chromosome genes in a female mouse, or more than one band for X or Y chromosome genes in male mice).
32. For PCR products containing multiple gel bands that cannot be easily separated by gel electrophoresis, identification of all mutations present at the genomic target site can be performed by TOPO-TA cloning of the mixed PCR product according to the manufacturer's protocol, followed by sequencing of DNA prepared from multiple TOPO vector clones.
33. Use of a lab coat, gloves, and a UV blocking face shield is important when performing gel band excision, to protect from UV damage to skin and eyes.

34. For recovery of DNA fragments larger than 1000 bp, heat to 70 °C for 5 min after adding Buffer NE or water, instead of incubating at room temperature.
35. Typically, either of the primers used for the initial PCR can be used for subsequent sequencing. However, DNA sequence quality typically deteriorates beyond 300–500 bases from the primer binding site, so it may be necessary to design another sequencing primer closer to the CRISPR target site to obtain good quality sequences for larger PCR products. Sequence quality may also be poor within the first 50–100 bases of the primer binding site, so primers should be designed at least that far away from the CRISPR target site for accurate sequencing of CRISPR-induced mutations.
36. DNA sequence analyses for comparisons between wild-type sequences and CRISPR-induced mutations can be performed using free online tools such as the Basic Local Alignment Search Tool (BLAST; https://blast.ncbi.nlm.nih.gov) or commercial programs such as Sequencher DNA Analysis Software.
37. In the case of small mutations that cannot be easily distinguished from wild-type based on size differences of PCR products, analyze the sequences for the presence of a unique restriction endonuclease site that can distinguish between the wild-type and the mutation sequence, which would allow for genotyping to be performed by digesting PCR products with this restriction endonuclease and analyzing digested products by gel electrophoresis.
38. Predictions of MMEJ-induced deletions at the CRISPR target site may be determined using the Cas-Designer tool or Microhomology-Predictor tool of the Center for Genome Engineering, Institute for Basic Science, Korea (http://www.rgenome.net).

Acknowledgements

This work was supported by the Intramural Research Program of the NIH, NIAID and NHLBI.

References

1. Shultz LD, Lyons BL, Burzenski LM, Gott B, Chen X, Chaleff S, Kotb M, Gillies SD, King M, Mangada J, Greiner DL, Handgretinger R (2005) Human lymphoid and myeloid cell development in NOD/LtSz-*scid* $IL2R\gamma^{null}$ mice engrafted with mobilized human hemopoietic stem cells. J Immunol 174 (10):6477–6489
2. Billerbeck E, Barry WT, Mu K, Dorner M, Rice CM, Ploss A (2011) Development of human $CD4^{+}FoxP3^{+}$ regulatory T cells in human stem cell factor-, granulocyte-macrophage colony-stimulating factor-, and interleukin-3-expressing NOD-SCID $IL2R\gamma^{null}$ humanized mice. Blood 117(11):3076–3086

3. Takagi S, Saito Y, Hijikata A, Tanaka S, Watanabe T, Hasegawa T, Mochizuki S, Kunisawa J, Kiyono H, Koseki H, Ohara O, Saito T, Taniguchi S, Shultz LD, Ishikawa F (2012) Membrane-bound human SCF/KL promotes in vivo human hematopoietic engraftment and myeloid differentiation. Blood 119(12):2768–2777
4. Brehm MA, Racki WJ, Leif J, Burzenski L, Hosur V, Wetmore A, Gott B, Herlihy M, Ignotz R, Dunn R, Shultz LD, Greiner DL (2012) Engraftment of human HSCs in nonirradiated newborn NOD-*scid* $IL2r\gamma^{null}$ mice is enhanced by transgenic expression of membrane-bound human SCF. Blood 119 (12):2778–2788
5. Capecchi MR (2005) Gene targeting in mice: functional analysis of the mammalian genome for the twenty-first century. Nat Rev Genet 6 (6):507–512
6. Landel CP, Dunlap J, Patton JB, Manser T (2013) A germline-competent embryonic stem cell line from NOD.Cg-$Prkdc^{scid}$ $Il2rg^{tm1Wjl}$/SzJ (NSG) mice. Transgenic Res 22(1):179–185
7. Wang H, Yang H, Shivalila CS, Dawlaty MM, Cheng AW, Zhang F, Jaenisch R (2013) One-step generation of mice carrying mutations in multiple genes by CRISPR/Cas-mediated genome engineering. Cell 153 (4):910–918
8. Bae S, Kweon J, Kim HS, Kim J-S (2014) Microhomology-based choice of Cas9 nuclease target sites. Nat Methods 11(7):705–706
9. Yang H, Wang H, Shivalila Chikdu S, Cheng Albert W, Shi L, Jaenisch R (2013) One-step generation of mice carrying reporter and conditional alleles by CRISPR/Cas-mediated genome engineering. Cell 154(6):1370–1379
10. Jinek M, Chylinski K, Fonfara I, Hauer M, Doudna JA, Charpentier E (2012) A programmable dual-RNA-guided DNA endonuclease in adaptive bacterial immunity. Science 337 (6096):816–821
11. Cong L, Ran FA, Cox D, Lin S, Barretto R, Habib N, Hsu PD, Wu X, Jiang W, Marraffini LA, Zhang F (2013) Multiplex genome engineering using CRISPR/Cas systems. Science 339(6121):819–823
12. Mali P, Yang L, Esvelt KM, Aach J, Guell M, DiCarlo JE, Norville JE, Church GM (2013) RNA-guided human genome engineering via Cas9. Science 339(6121):823–826
13. Sung YH, Kim JM, Kim H-T, Lee J, Jeon J, Jin Y, Choi J-H, Ban YH, Ha S-J, Kim C-H, Lee H-W, Kim J-S (2014) Highly efficient gene knockout in mice and zebrafish with RNA-guided endonucleases. Genome Res 24 (1):125–131
14. Sweeney CL, Choi U, Liu C, Koontz S, Ha S-K, Malech HL (2017) CRISPR-mediated knockout of *Cybb* in NSG mice establishes a model of chronic granulomatous disease for human stem-cell gene therapy transplants. Hum Gene Ther 28(7):565-575. https://doi.org/10.1089/hum.2017.005
15. Hwang WY, Fu Y, Reyon D, Maeder ML, Tsai SQ, Sander JD, Peterson RT, Yeh JRJ, Joung JK (2013) Efficient genome editing in zebrafish using a CRISPR-Cas system. Nat Biotechnol 31(3):227–229
16. Auer TO, Duroure K, Concordet J-P, Del Bene F (2014) CRISPR/Cas9-mediated conversion of eGFP- into Gal4-transgenic lines in zebrafish. Nat Protoc 9:2823-2840. https://doi.org/10.1038/nprot.2014.187
17. Kaneko T, Sakuma T, Yamamoto T, Mashimo T (2014) Simple knockout by electroporation of engineered endonucleases into intact rat embryos. Sci Rep 4:6382. https://doi.org/10.1038/srep06382
18. Hashimoto M, Takemoto T (2015) Electroporation enables the efficient mRNA delivery into the mouse zygotes and facilitates CRISPR/Cas9-based genome editing. Sci Rep 5:11315. https://doi.org/10.1038/srep11315
19. Takahashi G, Gurumurthy CB, Wada K, Miura H, Sato M, Ohtsuka M (2015) GONAD: genome-editing via oviductal nucleic acids delivery system: a novel microinjection independent genome engineering method in mice. Sci Rep 5:11406. https://doi.org/10.1038/srep11406
20. Qin W, Wang H (2018) Delivery of CRISPR-Cas9 into mouse zygotes by electroporation. In: Liu C, Du Y (eds) Microinjection: methods and protocols. Humana, New York, NY

Chapter 9

Generation of CRISPR-Edited Rodents Using a Piezo-Driven Zygote Injection Technique

Melissa A. Scott and Yueh-Chiang Hu

Abstract

Direct modification of the genome of the zygotes (i.e., one-cell embryos) by the CRISPR/Cas9-editing reagents, followed by embryo transfer to pseudopregnant females for live birth, has been the most effective method to generate laboratory rodent models for research. The method relies on proper delivery of the editing reagents into zygotes, which is commonly achieved by a standard or slightly modified pronuclear microinjection technique. In this chapter, we describe in detail an alternative delivery method, named piezo-driven cytoplasmic microinjection, which offers a superior embryo survival and birth rate. Because this method uses a much wider injection needle than that in pronuclear injection, it allows a larger volume of the editing materials to be transported into the zygotes, leading to an increase in the targeting efficiency. This also eliminates the clogging issues seen regularly in pronuclear injection. Moreover, Cytochalasin B that is used to soften zygotes during piezo-driven microinjection has been suggested a role in improving the knockin efficiency, which provides an additional benefit to use this injection method.

Key words Piezo, Zygote, Cytoplasmic injection, Pronuclear injection, CRISPR/Cas9, Cytochalasin B

1 Introduction

The simplicity and efficiency of the CRISPR/Cas9-mediated genome engineering system has triggered a revolution in the practice of rodent model generation in the scientific community. The genetically modified rodent models, including knockouts, knock-ins, and conditional alleles, can be generated effectively in a single step via direct modification of the genome of one-cell embryos by the CRISPR/Cas9 [1–5]. The modified embryos are subsequently transferred into pseudopregnant surrogate mothers for fetal development and birth. The success of the procedure depends on careful design and preparation of the CRISPR/Cas9 reagents and selection of the appropriate delivery method [5]. The latter affects how the reagents should be prepared, the number of embryos should be used and the overall targeting efficiency. The pronuclear microinjection technique that was established in the early 1980s remains

Chengyu Liu and Yubin Du (eds.), *Microinjection: Methods and Protocols*, Methods in Molecular Biology, vol. 1874, https://doi.org/10.1007/978-1-4939-8831-0_9,

the predominant method to deliver exogenous materials, including the CRISPR/Cas9 system, into the zygotes [6]. This technique has also been modified for cytoplasmic injection without penetrating the pronuclei, which improves embryo survival [7, 8]. Electroporation poses a new and attractive delivery method that is simple and does not require highly technical training to operate [7, 9–12]. However, its usage is limited to the generation of the animal models with gene deletions or knockin using single-strand DNA donor oligos. In our practice, we use a piezo-driven microinjection technique to deliver the editing reagents into the cytoplasm of the zygotes, the advantages of which include its superior zygote survival, birth rate, and targeting efficiency (*see* **Note 1**).

Piezo-driven microinjection technique uses a pipette holder quipped with a piezoelectric actuator that creates a quick mechanical pulse to vibrate the pipette tip and pierce the zona pellucida and oocyte membrane (oolemma). Because the size of the injection needle (outer diameter > 10 μm) is much larger than that for the pronuclear injection (usually 0.5–2 μm), this technique has been used in many specialized procedures, including intracytoplasmic sperm injection, nuclear transfer cloning, and blastocyst injection with embryonic stem cells [13–16]. Yet, it achieves a survival rate approaching 100% following injection of either unfertilized oocytes or fertilized zygotes. Jaenisch group first used the piezo-driven microinjection technique to deliver the CRISPR/Cas9 reagents to mouse zygotes and showed an exceptional targeting efficiency in generating all types of the mutant alleles [1, 2]. We modified the protocol, described in steps below, and have performed more than 200 mouse and rat projects with excellent success since early 2014. In addition, Cytochalasin B, an actin polymerization inhibitor, which is added to the incubation buffer to soften the zona pellucida and zygote membrane for accommodating the large needle, may play a role in enhancing the efficiency of large DNA insertion via homologous recombination [17]. Because our average knockin efficiency using single-strand DNA donors is 46%, such high rate may be attributed to the use of Cytochalasin B and the large injection needle that is able to deposit much more CRISPR/Cas9 reagents inside the zygotes.

2 Materials

2.1 Reagents

1. Mice or rats of any strain.
2. Pregnant mare serum gonadotropin (PMSG).
3. Human chorionic gonadotropin (hCG).
4. Injection pipette wash buffer: 10% Polyvinylpyrrolidone (PVP) in RNase-free water; keep it at 4 °C for up to 3 months and at −20 °C for long-term storage.

5. M2 medium: run through a syringe filter (pore size: 0.22 μm, diameter: 13 mm) before use.
6. 20× hyaluronidase stock, 10 mg/ml: prepare hyaluronidase in M2 medium, and then aliquot 1 ml into tubes; keep it at 4 °C for up to 1 month and at −20 °C for long-term storage.
7. M2 + hyaluronidase: add 80 μl of 20× hyaluronidase stock to 1.5 ml of M2 medium before use.
8. 100× Cytochalasin B (CB) stock, 1 mg/ml: add 1 ml of RNase-free DMSO to 1 mg of CB, and then aliquot 40 μl into tubes. Store the stock at −20 °C.
9. M2 + CB, 10 μg/ml: add 2 μl of CB stock to 198 μl of filtered M2 medium before use. Refreeze the remaining CB stock for up to 15 times without losing effect.
10. Fluorinert FC-770 (mercury substitute).
11. Mineral oil.
12. Microinjection buffer: 1 mM Tris–HCl, pH 7.4, 0.1 mM EDTA in an RNase-free condition.
13. 5× ssDNA donor stock, 500 ng/μl: resuspend in microinjection buffer. Store the stock at −20 °C. HPLC or PAGE purification is not necessary when ordering ssDNA donor from the vendor.
14. 5× circular dsDNA donor plasmid stock, 500 ng/μl: pretreat at least 20 μg of the plasmid with proteinase K (final concentration: 200 ng/μl) at 55 °C for at least 2 h to remove RNase; clean with any column-based purification method and elute with microinjection buffer.
15. Injection mix preparation, 5 μl per injection: mix Cas9 protein and sgRNA (final concentration: 100 and 50 ng/μl, respectively) in microinjection buffer and incubate at 37 °C for 10 min to form ribonucleoproteins (skip this step if using Cas9 mRNA); add ssDNA or circular dsDNA donor to the mix at a final concentration of 100 ng/μl.
16. 1 ml syringe with 25 G × 5/8″ needle.
17. 100 mm petri dishes: bottoms are suitable for zygote wash/collection and lids are suitable for micromanipulation.
18. Transfer mouth pipette: pull 5¾″ Pasteur pipettes to make thin transfer pipettes over a flame and polish the opening; attach it to the mouthpiece sets for embryo manipulation.

2.2 Equipment

1. Inverted microscope equipped with micromanipulators and microinjectors (Fig. 1). We use a pneumatic microinjector for holding zygotes and a hydraulic microinjector filled with RNase-free water for injection.
2. PIEZO Impact Drive.

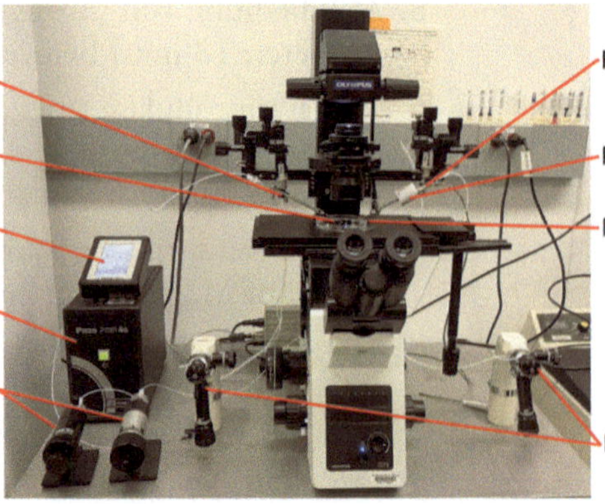

Fig. 1 The setup of piezo-driven microinjection station. The system is comprised of a standard inverted microscope equipped with micromanipulators and microinjectors, and a Piezo Impact Drive that pulsates the injection pipette tip through an actuator on the pipette holder

3. Injection pipette: blunt-end glass pipette (8 μm inner diameter and bent at a 15° angle; commercially available). The choice of the bent angel depends on the micromanipulator setup.
4. Zygote holding pipette: blunt-end glass pipette with fire-polished end (15–20 μm inner diameter, 65–95 μm outer diameter, and bent at a 15° angle). The choice of the bent angel depends on the micromanipulator setup.

3 Methods

3.1 Zygote Preparation

1. Superovulate female mice (3–4 weeks old for the best result or older than 8 weeks) by intraperitoneal injection of 5–7 IU of PMSG at around 1 pm, followed by 5–7 IU of hCG injection 48 h later; immediately place each female into the male's cage and mate overnight.
2. Check each female for a vaginal plug in the next day's morning; euthanize the females with plugs and collect zygote-cumulus complexes in M2 medium from the swollen part of the oviduct.
3. Prepare four 35 mm dishes for zygote wash and culture: (a) 1.5 ml of M2 + hyaluronidase; (b) 1.5 ml of M2 only; (c) two dishes with four drops (20–30 μl for each drop) of M2 overlaid with mineral oil and put away in a 37 °C incubator.
4. Transfer the zygote-cumulus complexes to the M2 + hyaluronidase dish to remove cumulus cells from zygotes. Wash zygotes in the M2 medium dish 5–10 times by mouth pipetting. Transfer them into one or more of the M2 droplets and culture at 37 °C.

3.2 Injection Plate and Wash Plate Setup

1. Prepare 5 μl of injection mix as indicated in Subheading 2.1, **item 15**, before injection.
2. Set up the injection and wash plates using a petri dish (Fig. 2). Use the lid for injection, and the bottom for wash. Caution to avoid RNase contamination for all steps.
3. Draw 1 ml of M2 into 1 ml syringe and attach a syringe filter (0.22 μm pore, 13 mm diameter).
4. Dispense drops of M2 (~30 μl for each drop) on the wash dish, overlay the dish with mineral oil, and put away in a 37 °C incubator (Fig. 2a). Reserve about 0.2 ml of M2 in syringe.
5. Remove the filter from the syringe, add 2 μl of 100× CB stock using a RNase-free filter P20 pipette tip, reattach the syringe filter, and mix CB into M2 by inverting syringe.
6. Dispense 8–10 drops of M2 + CB on the injection dish (Fig. 2b). The first few droplets may not contain enough CB.
7. Dispense at least three 6 μl drops of 10% PVP and one or more 5 μl drops of the Cas9/sgRNA/donor injection mix on the injection dish using RNase-free filter P20 pipette tips (Fig. 2b).

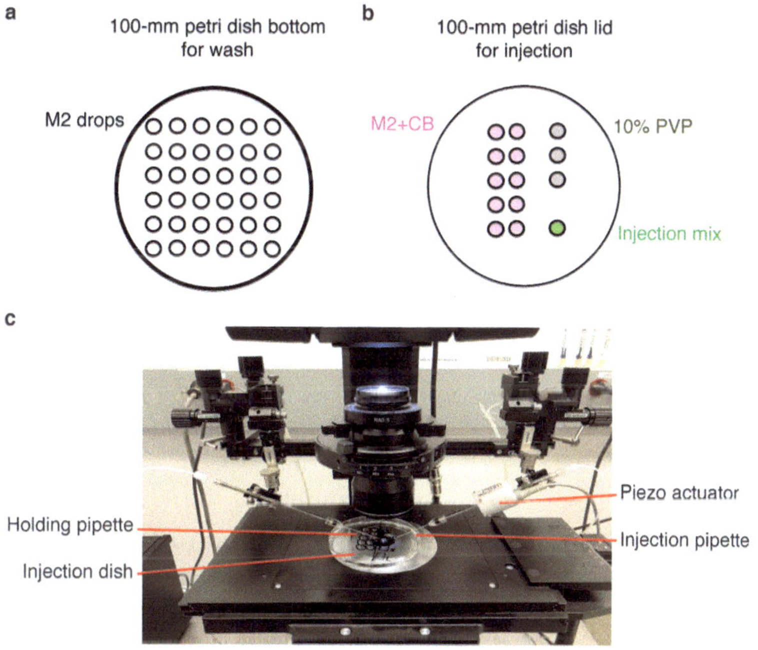

Fig. 2 The setup of injection and wash dishes. (**a**) Drops of filtered M2 medium (~30 μl for each drop) are dispensed on a petri dish bottom and overlaid with mineral oil. The dish is kept in a 37 °C incubator. (**b**) Drops of M2 + CB (~30 μl for each drop; pink circles), 10% PVP (~6 μl for each drop; gray circles), and the Cas9/sgRNA/donor injection mix (5 μl for each drop; green circle) are dispensed on a petri dish lid and overlaid with mineral oil. (**c**) An image of the injection dish on the microscope stage

8. Overlay the dish with mineral oil and place on the microscope stage (Fig. 2c).

3.3 Microinjection Setup

1. Use 4× objective. Transfer 50–100 zygotes to one M2 + CB droplet on the microscope stage and arrange them fairly in a straight line (*see* **Note 2**). Note that at least 5 min incubation time to soften zygotes is required before starting the injection.
2. While waiting, install the holding pipette and proceed to the next step. If reusing the holding pipette, wash it in a M2 + CB droplet several times and lift up to the air (*see* **Note 3**).
3. Backload Fluorinert into an injection pipette with a 1 ml syringe with 25 G × 5/8″ needle. Fluorinert should fill the end of the pipette with a length of 3–4 mm without air bubbles. Hold the pipette vertically with the tip end up to keep Fluorinert at the bottom.
4. Detach the injection pipette holder from the micromanipulator and hold vertically. Adjust the outflow knob until water reaches the opening of the holder by forming a dome. Gently lower the end of the injection pipette containing Fluorinert into the dome of water and tighten the injection pipette into place. Make sure that the interface between Fluorinert and water has no air bubbles or debris. Note that we use a water-type hydraulic microinjector. An air- or oil-type microinjector works fine, too.
5. Bring Fluorinert to the tip of the injection pipette by adjusting the knob slightly. Reattach the pipette holder to the micromanipulator and make sure that the injection pipette tip lays almost horizontally and in line with the holding pipette (*see* **Note 4**).
6. Carefully lower the injection pipette tip into a 10% PVP droplet. Aspirate in and out the PVP along the visible length of the injection tip several times and avoid losing Fluorinert.
7. Draw in the PVP and apply the piezo pulses at the highest intensity and speed (cleaning mode) for about 30 s continuously.
8. Bring the Fluorinert to the pipette tip and gently move the dish to position the pipette tip into the injection mix droplet.
9. Collect the injection mix by turning the knob a circle counterclockwise and wait ~15 s until the influx slows down. Stop the intake and produce a slow efflux by one-quarter clockwise turn of the knob. Note that this is based on a PrimeTech HDJ-M3 microinjector. The length and timing of the knob movement needs to be adjusted according to the specific microinjector used in individual laboratories.
10. Lift up the filled injection pipette to the air.

11. Move the droplet that contains zygotes to the center, adjust the focus, and gently lower the holding pipette to the bottom.
12. Gently lower the injection pipette into the same droplet. As soon as it enters the medium, check the outflow of the injection mix (the difference in the density between M2 and the injection mix allows you to see the outflow). If the outflow is too strong or not apparent, turn the knob immediately.

3.4 Microinjection Procedure (Fig. 3)

1. Switch to a 40× objective.
2. Hold a zygote from the end of the line (*see* **Notes 2** and **5**) and adjust the position until the perivitelline space is on the side of the injection pipette without a polar body in the path of the injection pipette to the cell body.
3. Test the piezo intensity of the injection pipette. The piezo pulse should be just powerful enough to pierce the zona pellucida (*see* **Note 6**).

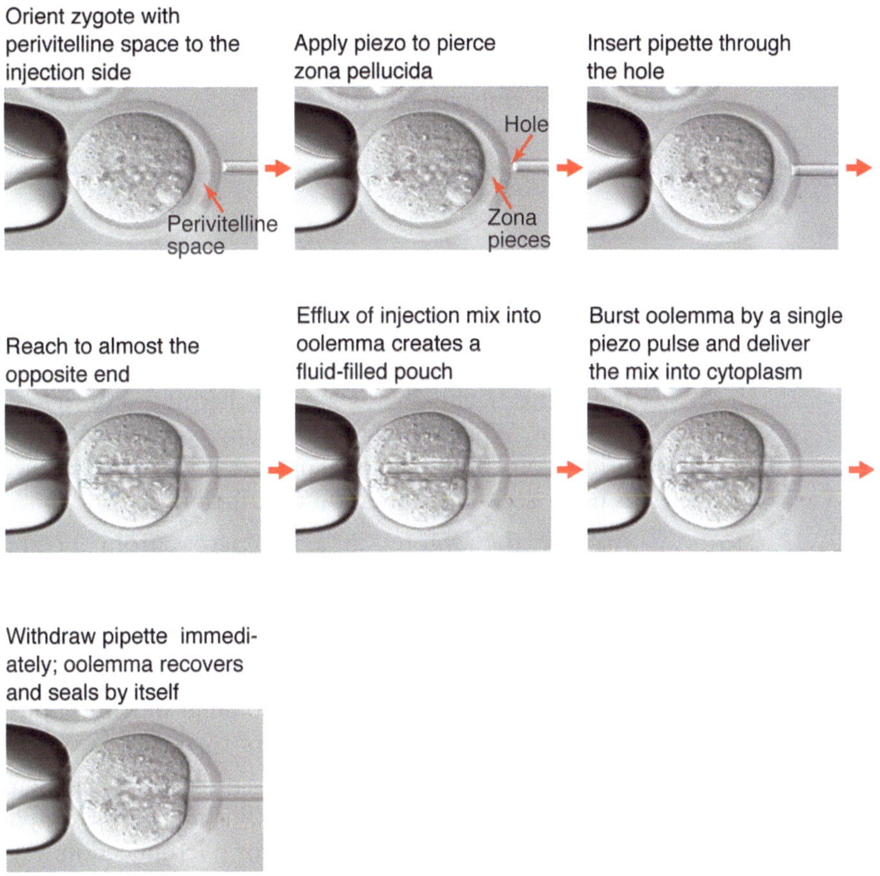

Fig. 3 Piezo-driven cytoplasmic zygote injection. This technique allows a large volume of the Cas9/sgRNA/donor injection mix to be delivered into the cytoplasm of the zygotes without damaging the embryos or causing the pipette tip clogging

4. Apply a few piezo pulses to pierce the zona pellucida and insert the pipette deeply into the body of the ovum following the midline, almost reaching the opposite end of the cell body. The displacement of the membrane around the pipette indicates the rate of injection mix outflow. The outflow should not be too strong, as this will damage the zygote (*see* **Note 7**). Break the membrane with a single piezo pulse, and at the same time the injection mix bursts into the cytoplasm of the zygote. Retract the pipette quickly before the membrane encapsulates the pipette tip.
5. Move the injected zygote to the side (*see* **Note 2**) and proceed to the next one until the current batch is finished. 40-60 zygotes are generally sufficient for one project (*see* **Note 8**).

3.5 Embryo Transfer

1. Transfer the injected zygotes to the wash dish and move along the M2 droplets at least ten times.
2. Washed zygotes are ready for oviduct transfer into pseudopregnant surrogate mothers for fetal development and birth. Or, culture them in KSOM medium at 37 °C in a 5% CO_2 incubator overnight until the 2-cell stage or 3 days until the blastocyst stage for embryo transfer.

4 Notes

1. Compared to pronuclear injection, this piezo-driven microinjection technique allows a larger amount of the CRISPR/Cas9 mix into the cytoplasm of the zygotes, which is expected to enhance the targeting efficiency. Despite a larger injection volume, this technique is neither harmful nor lethal to the embryos. While pipette tip clogging is common to techniques using a very fine injection pipette (e.g., pronuclear injection), it is never seen in the method described here.
2. Arrange the zygotes in a line across from 12 o'clock to 6 o'clock with a width of 2–3 zygotes in the M2 + CB droplet. Inject the zygotes starting at 6 o'clock progressing up the line toward 12 o'clock. The holding and injection pipettes act as a barrier separating the injected and uninjected zygotes. Uninjected zygotes are above the holding and injection needles; injected zygotes are pushed below by the injection pipette.
3. The holding pipette may be reused several times. Before reusing, wash the holding pipette by flushing with a droplet of M2 media. Do not use this droplet for microinjection.
4. When attaching the microinjection pipette holder to the micromanipulator, make sure to adjust the pipette so that it lies parallel to the surface of the dish, not quite touching the

bottom of the dish. The tips of the injection pipette and the holding pipette should mirror one another, crossing the diameter of the dish, as if hands at 9 and 3 o'clock.

5. Use the injection pipette to move zygotes toward the holding pipette or adjust the position. Avoid moving the holding pipette, which ensures it remains at the bottom of the dish in the correct focal plane.
6. When using PrimTech PMM-4G or PMM-150 Piezo Impact Drive, combined with Origio pipette (cat. no. PIZEO-8-15 sterile: 8 μm inner diameter and bent at a 15° angle), the proper range of piezo intensity is set between 5 and 9, and the speed is set at 3.
7. The outflow of the injection mix should remain steady enough to inject a series of 5–10 zygotes from one small move of the knob. The outflow rate will dissipate slightly over time and should be adjusted accordingly. One collection (Subheading 3, **step 15**) of the injection mix should be enough to inject ~100 zygotes. If multiple collections are needed, this is indicative of poor flow control or unnecessary high outflow. Be careful not to draw too much injection mix into the injection pipette. As the distance between the tip and the Fluorinert increases, the power of the piezo-driven vibration decreases. Keep in mind that all the element segments (i.e., water, Fluorinert, and injection mix) have to be intact and nicely separated. If any interruption occurs along the path from the pipette tip via the holder and the tube to the microinjector, such as air bubbles, oil droplets, excess media, will significantly impact the piezo-driven motion. If the Fluorinert segment is interrupted by water or injection mix, the piezo-driven tip movement becomes lateral, instead of longitudinal, which kills the embryos.
8. We have used the piezo-driven microinjection to generate the rodent mouse models of all types, including knockouts at up to four loci per microinjection, large deletions and inversions (~ 2.5 mb), knockins using ssDNA or dsDNA donors, and conditional alleles, with excellent efficiency [5]. Because of its superior embryo survival and birth rate, we normally use 40–60 zygotes for one project and produce more than enough gene-edited founder animals for the investigators. We have also compared this injection technique side-by-side with the pronuclear injection and found that this technique demonstrated a higher targeting efficiency for making knockouts or knockins using ssDNA donors. When a dsDNA donor plasmid is used for making knockins, both methods worked at a similar efficiency.

Acknowledgments

We thank current and former staff members, Yinhuai Chen, Huirong Xie, Alexandra Falcone, Susan Martin, Celvie Yuan, Evan Barr-Beare, Calista Falcone, Kendall Smith, and Kristen Martin for performing the services and generating the data for this article. This work was supported by Cincinnati Children's Research Foundation.

References

1. Wang H, Yang H, Shivalila CS, Dawlaty MM, Cheng AW, Zhang F, Jaenisch R (2013) One-step generation of mice carrying mutations in multiple genes by CRISPR/Cas-mediated genome engineering. Cell 153 (4):910–918. https://doi.org/10.1016/j.cell.2013.04.025
2. Yang H, Wang H, Shivalila CS, Cheng AW, Shi L, Jaenisch R (2013) One-step generation of mice carrying reporter and conditional alleles by CRISPR/Cas-mediated genome engineering. Cell 154(6):1370–1379. https://doi.org/10.1016/j.cell.2013.08.022
3. Li D, Qiu Z, Shao Y, Chen Y, Guan Y, Liu M, Li Y, Gao N, Wang L, Lu X, Zhao Y, Liu M (2013) Heritable gene targeting in the mouse and rat using a CRISPR-Cas system. Nat Biotechnol 31(8):681–683. https://doi.org/10.1038/nbt.2661
4. Li W, Teng F, Li T, Zhou Q (2013) Simultaneous generation and germline transmission of multiple gene mutations in rat using CRISPR-Cas systems. Nat Biotechnol 31(8):684–686. https://doi.org/10.1038/nbt.2652
5. Yuan CL, Hu YC (2017) A transgenic core facility's experience in genome editing revolution. Adv Exp Med Biol 1016:75–90. https://doi.org/10.1007/978-3-319-63904-8_4
6. Ittner LM, Gotz J (2007) Pronuclear injection for the production of transgenic mice. Nat Protoc 2(5):1206–1215. https://doi.org/10.1038/nprot.2007.145
7. Horii T, Arai Y, Yamazaki M, Morita S, Kimura M, Itoh M, Abe Y, Hatada I (2014) Validation of microinjection methods for generating knockout mice by CRISPR/Cas-mediated genome engineering. Sci Rep 4:4513. https://doi.org/10.1038/srep04513
8. Singh P, Schimenti JC, Bolcun-Filas E (2015) A mouse geneticist's practical guide to CRISPR applications. Genetics 199(1):1–15. https://doi.org/10.1534/genetics.114.169771
9. Hashimoto M, Takemoto T (2015) Electroporation enables the efficient mRNA delivery into the mouse zygotes and facilitates CRISPR/Cas9-based genome editing. Sci Rep 5:11315. https://doi.org/10.1038/srep11315
10. Kaneko T, Mashimo T (2015) Simple genome editing of rodent intact embryos by electroporation. PLoS One 10(11):e0142755. https://doi.org/10.1371/journal.pone.0142755
11. Qin W, Dion SL, Kutny PM, Zhang Y, Cheng AW, Jillette NL, Malhotra A, Geurts AM, Chen YG, Wang H (2015) Efficient CRISPR/Cas9-mediated genome editing in mice by zygote electroporation of nuclease. Genetics 200 (2):423–430. https://doi.org/10.1534/genetics.115.176594
12. Chen S, Lee B, Lee AY, Modzelewski AJ, He L (2016) Highly efficient mouse genome editing by CRISPR ribonucleoprotein electroporation of zygotes. J Biol Chem 291 (28):14457–14467. https://doi.org/10.1074/jbc.M116.733154
13. Yoshida N, Perry AC (2007) Piezo-actuated mouse intracytoplasmic sperm injection (ICSI). Nat Protoc 2(2):296–304. https://doi.org/10.1038/nprot.2007.7
14. Wakayama T, Perry AC, Zuccotti M, Johnson KR, Yanagimachi R (1998) Full-term development of mice from enucleated oocytes injected with cumulus cell nuclei. Nature 394 (6691):369–374. https://doi.org/10.1038/28615
15. Kimura Y, Yanagimachi R (1995) Intracytoplasmic sperm injection in the mouse. Biol Reprod 52(4):709–720
16. Kawase Y, Iwata T, Watanabe M, Kamada N, Ueda O, Suzuki H (2001) Application of the piezo-micromanipulator for injection of embryonic stem cells into mouse blastocysts. Contemp Top Lab Anim Sci 40(2):31–34
17. Nakao H, Harada T, Nakao K, Kiyonari H, Inoue K, Furuta Y, Aiba A (2016) A possible aid in targeted insertion of large DNA elements by CRISPR/Cas in mouse zygotes. Genesis 54 (2):65–77. https://doi.org/10.1002/dvg.22914

Chapter 10

Delivery of CRISPR-Cas9 into Mouse Zygotes by Electroporation

Wenning Qin and Haoyi Wang

Abstract

The CRISPR-Cas9 system in bacteria and archaea has recently been exploited for genome editing in various model organisms, including the mice. In this scheme, components of the CRISPR-Cas9 system are delivered into the mouse zygote and mutant mice carrying genetic modifications derived. Although microinjection has been the technology of choice, electroporation has also emerged and been proven to be effective delivering CRISPR-Cas9 reagents into the mouse zygote. Here, we describe the experimental protocol employing electroporation to deliver CRISPR-Cas9 reagents into mouse embryos and derive gene-edited mutant mice.

Key words CRISPR-Cas9, Electroporation, Genome editing, Mouse models

1 Introduction

Clustered Regularly Interspaced Palindromic Repeat (CRISPR) and CRISPR-associated endonuclease (Cas) are components of an acquired immune system operating in bacteria and archaea [1]. In recent years, it has been reformatted and repurposed into a powerful tool in gene editing [2–4] and various mutant animals, including the mice, have been created [5–9].

One of the most widely used CRISPR-Cas systems is that derived from *Streptococcus pyogenes* [2]. In strain SF370, serotype M1, the type II CRISPR-Cas locus consists of four structural genes (Cas9, cas1, cas2, and csn2), the tracrRNA gene, and the CRISPR sequence (it has seven short palindromic repeats separated by six spacer sequences) (Fig. 1a, b). The CRISPR sequence is transcribed, as directed by the AT-rich leader sequence located at its 5′ end, and the expressed CRISPR sequence further processed into individual crRNA sequences. In its natural setting, crRNA forms a duplex with the tracrRNA, which is then loaded into the Cas9 nuclease to form a ribonucleoprotein (RNP) complex. In a lab setting, however, crRNA and tracrRNA can be conveniently linked

Chengyu Liu and Yubin Du (eds.), *Microinjection: Methods and Protocols*, Methods in Molecular Biology, vol. 1874, https://doi.org/10.1007/978-1-4939-8831-0_10,

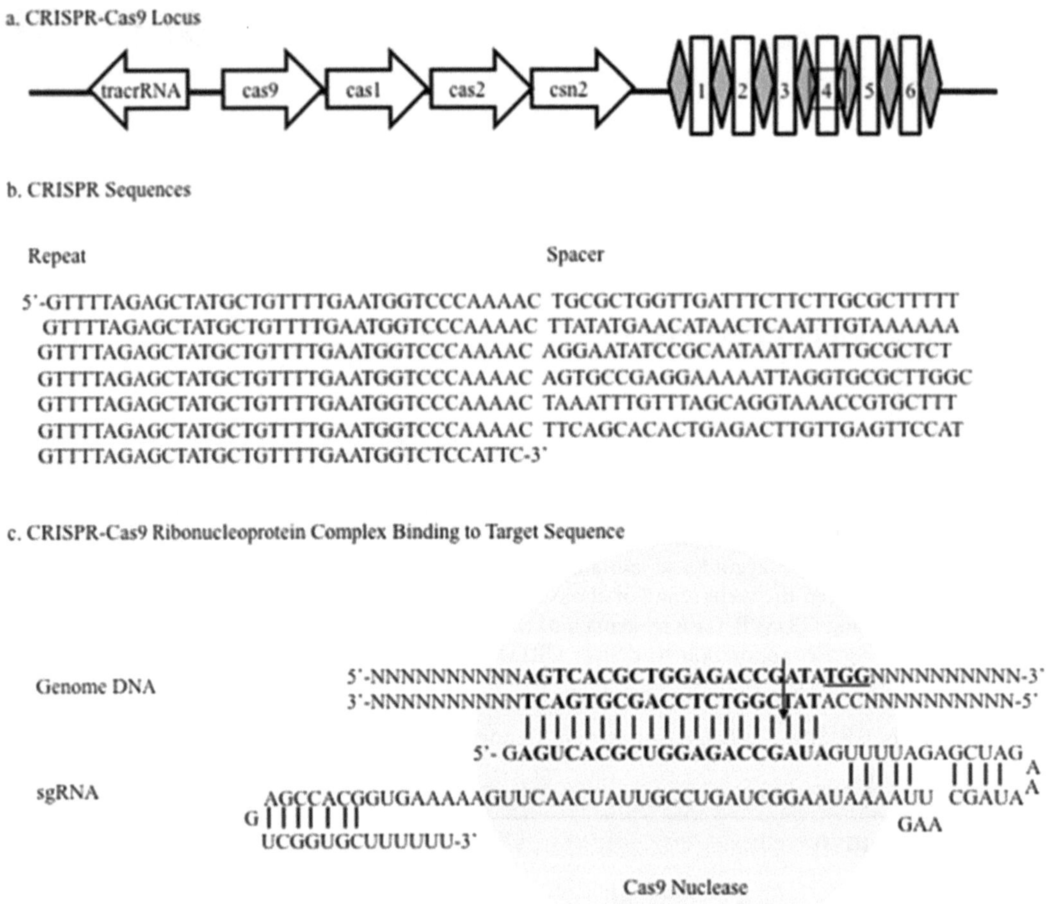

Fig. 1 The type II CRISPR-Cas9 system in *Streptococcus pyogenes*, strain SF370, serotype M1. (**a**) The CRISPR locus contains 4 structural genes (cas9, cas1, cas2, and csn2), six spacers separated by seven repeats, and a tracrRNA gene. The Cas9 gene encodes the effector nuclease Cas9. During the interference phase of the immune reaction, the CRISPR sequence is transcribed as directed by the 5′ AT-rich leader sequence and further processed into individual crRNA sequences. (**b**) The seven short palindromic repeats are interspaced by the six spacers (based on NC_002737, analyzed by CRISPRFinder program, http:/crispr.i2bc.paris-saclay.fr/Server/). (**c**) sgRNA is loaded into Cas9 nuclease and forms the ribonucleoprotein (RNP) complex, which then binds to the complementary strand of the target (protospacer) sequence. Between the two components of CRISPR-Cas9 RNP complex, the guide sequence (highlighted in bold and located at the 5′ end of the crRNA sequence) specifies target specificity by base pairing with the matching protospacer sequence (highlighted in bold among the double-stranded genomic DNA sequence), while the Cas9 nuclease makes the double-strand break. Another requirement for target specificity is the "NGG" PAM (protospacer adjacent motif) sequence that is contiguous from and 3′ to the protospacer sequence (highlighted in bold and underlined). Together, these 23 nucleotides define the site where the double-strand break should be made (indicated by arrow head)

and synthesized as a single-guide RNA molecule (sgRNA) [3]. The system, when transplanted into eukaryotic cells, has been found to work and genomes of human and mice edited [2, 4–8].

For the mouse genome to be edited, a DNA double-strand break must first be created, at a site chosen among the three billion nucleotides in the mouse genome. Between the two components of the CRISPR-Cas9 RNP complex, the guide RNA located at the 5′ end of the crRNA sequence specifies target specificity by base pairing with the complementary strand of the matching genomic sequence (protospacer), while the Cas9 nuclease makes the double-strand break (Fig. 1c). Another requirement is the 3 nt (nucleotide) PAM sequence (protospacer adjacent motif), "NGG," continuous from and 3′ to the protospacer sequence in the target genome. When a DNA double-strand break is created, mutations may be introduced at the break point as the cellular repair mechanisms of non-homologous end joining (NHEJ) or homology directed repair (HDR) are mobilized to repair the damage. To generate INDEL (insertion/deletion) mutations exploiting into the NHEJ mechanism, only sgRNA(s) and Cas9 nuclease need to be provided. However, if the goal is to generate precise mutations leveraging into HDR, a donor oligonucleotide (less than 200 nucleotides, for SNP incorporation or short tag insertion) or a donor plasmid (reporter insertion, etc.) carrying the intended mutation flanked by homology arms will be required, in addition to sgRNA(s) and Cas9 nuclease [10].

To genetically transform the host genome in a laboratory setting, historically, microinjection has been employed to deposit exogenous DNA into the pronucleus of mouse zygote to bring the exogenous DNA into direct contact with the host genome [11, 12]. The exogenous DNA is subsequently integrated into the host genome and through breeding, transmitted to the next generation. Although effective, microinjection is technically demanding, low in throughput, and requires significant upfront investment in equipment. To address these limitations, we and others explored into electroporation and showed that it is possible to deliver CRISPR-Cas9 reagents into the mouse zygote and derive mutant mice with high efficiency [13–17].

There are several practical considerations when delivering CRISPR-Cas9 reagents into mouse embryos using electroporation. The mouse zygote is coated with a layer of glycoprotein matrix, called zona pellucida, which may present a physical barrier for molecules to traverse [18]. Using FITC conjugated dextrans and lectins, it was determined that molecules greater than 110 kDa in molecular weight cannot penetrate through this glycoprotein layer [19]. Among the CRISPR-Cas9 reagents, sgRNA is about 100 nt in length and 33 kDa in weight. As such, it is expected to penetrate the intact zona. However, the Cas9 protein, at 165 kDa, is not expected to pass through the zona (with the addition of 3× FLAG tag and the two nuclear localization signals, it is 1423 aa). The same is expected of the Cas9 mRNA (4284 nt and 1395 kDa). Also, a donor plasmid of 10,000 nt in length has a molecular weight of

3000 kDa, which may make it even more challenging to pass through the zona. To make the zona more permeable, in the field of mouse embryology, a simple protocol was invented where mouse embryos are incubated briefly in acidic Tyrode's solution (pH 2.5) and zona "weakened" [20, 21]. In fact, zona "weakening" can be observed under a microscope and the degree of "weakening" controlled by strength of the acidic solution and exposure time. Various workers used this simple procedure in their protocol [13, 14, 16]. We will include this step in our protocol.

It is well accepted that "pores" of nanometer sizes form on plasma membrane when cells are subjected to electric pulses of sufficient voltage and duration [22]. Using digitized fluorescence microscopy, electroporation-mediated gene transfer was studied at the single-cell level in Chinese hamster ovary cells [23, 24]. It was discovered that propidium-iodide (PI), a fluorescent intercalating agent with a molecular weight of 668 Da, freely penetrates the plasma membrane and gets into the cells during and immediately after electroporation. Under the same electric field, the large, negatively charged DNA molecule was found to migrate toward the plasma membrane facing the cathode side of the electric field and accumulate by the plasma membrane by forming "localized spots." And it remained outside of the membrane 10 min after delivery of the pulses, when the "pores" should have been sealed. However, 30 min after delivery of the pulses, DNA fluorescence was found to distribute in the cytoplasm. This suggests that DNA was later translocated from the extracellular space to the cytoplasm by a yet unidentified mechanism. Based on these observations, among CRISPR-Cas9 reagents, sgRNA, which is relatively small, is expected to migrate into the cell through the pores by electrophoresis (Fig. 2b). However, the Cas9 protein (alone or in complex with sgRNA) and Cas9 mRNA may be too large to migrate directly through the pore (Fig. 2c). The fact that mutant mice are derived suggests that they are indeed translocated through plasma membrane. In this protocol, we describe delivery of the sgRNA/Cas9 RNP, with or without a donor oligonucleotide, into the mouse zygote for the purpose of deriving gene-edited mice.

2 Materials

2.1 Equipment

1. Spectrophotometer.
2. Centrifuge.
3. Benchtop incubator.
4. Micropipette puller.
5. Stereomicroscope.
6. ECM830 square wave electroporation system (UX-02894-18, Harvard Apparatus).

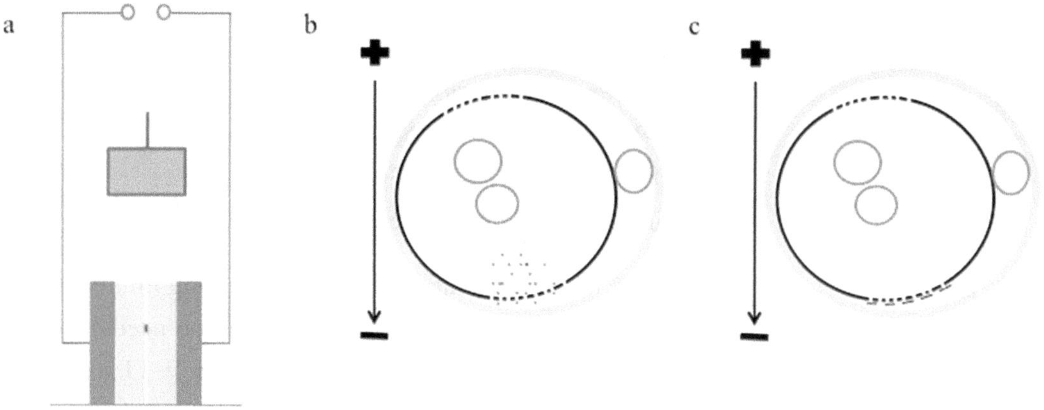

Fig. 2 Delivery of CRISPR-Cas9 reagents into mouse zygote by electroporation. (**a**) Mouse embryos (in phenol red containing Opti-MEM medium) are deposited into the chamber of an electroporation cuvette with 1 mm gap width and electric pulses delivered; (**b**) sgRNA, about 100 nt in length and 33 kDa in molecular weight, may be small enough and is expected to migrate through the zona and pores on the oolemma which may form during application of the electric pulses by electrophoresis; (**c**) In our protocol, the zona is weakened by treating the embryos for 10 s in acidic Tyrode's solution (pH 2.5). Cas9 protein, with 1423 aa and at 165 kDa, and in complex with sgRNA, may penetrate through the zona and reach the outer surface of the oolemma. From here, it is translocated to the cytoplasm by a yet uncharacterized mechanism

2.2 Supplies

1. RNase decontamination reagent.
2. RNase-free PCR tubes (0.2 mL).
3. RNase-free Microfuge Tubes (1.5 mL).
4. Microdissecting forceps.
5. Thin wall glass capillaries.
6. Cryogenic tube with round bottom, 1.8 mL.
7. Electroporation cuvette, 1 mm Gap (Part Number 45-0124, Harvard Apparatus).
8. 20 μL TipOne beveled sterile filter tip (Item No. 1120-1810, USA Scientific).
9. 35 × 10 mm Petri dish.

2.3 Chemicals and Reagents

1. TE buffer: 10 mM Tris–HCl, 0.1 mM EDTA, pH 7.5, DNase and RNase free.
2. Cas9 protein, 1000 ng/μL.
3. sgRNA, 1000 ng/μL.
4. Donor oligonucleotide: 1000 ng/μL.
5. Pregnant mare serum gonadotrophin (PMSG).
6. Human chorionic gonadotrophin (hCG).
7. Hyaluronidase (300 μg/mL).
8. M2 medium.

9. K-RVCL-50 medium (COOK Medical).
10. Mineral oil.
11. Acidic Tyrode's Solution.
12. Opti-MEM medium (P/N 31985, Gibco).
13. KSOMaa Evolve (ZEKS-050, Zenith Biotech).
14. Bovine serum albumin (used at 1 mg/mL with M2 medium).

2.4 Mice

1. Donor female mice (various strains, depending on needs).
2. Stud male mice (various strains, depending on needs).
3. Pseudopregnant female mice (various strains, 9–11 weeks).
4. Vasectomized male mice (various strains).

3 Methods

3.1 Collect Mouse Zygotes

Mouse embryos are collected and prepared as in a pronuclear injection experiment. Here we describe the protocol producing and collecting fertilized eggs from superovulated C57BL/6J female mice through natural mating (*see* **Note 1**). Please notice different strains of mice may respond differently to the regimen of PMSG and hCG and when working with a new strain, condition may need to be optimized [25].

1. Day −2, 10:00 am: inject female mice (21–35 days of age, or 12–14 g of body weight) 5 IU (IP, intraperitoneal) of PMSG.
2. Day 0, 9:00 am: inject female mice 5 IU (IP) of hCG.
3. Immediately after administration of hCG, mate each female mouse with one stud male mouse (8 weeks of age or older, that has been singly housed) (*see* **Note 2**).
4. Day E0.5 (embryonic day 0.5), 8:00 am: examine the female mice for presence of a copulation plug in the vaginal tract.
5. Euthanize the female mice displaying a copulation plug by cervical dislocation.
6. Excise oviducts and place into M2 medium and wash briefly.
7. Transfer the oviducts into M2 medium containing hyaluronidase (300 μg/mL).
8. Release the oocyte "clutch" (containing fertilized as well as unfertilized eggs) by puncturing the ampulla with a pair of microdissecting forceps and allow the clutch to incubate in the M2 medium containing hyaluronidase (300 μg/mL) (optimal time can be determined by until the cumulus cells become detached from the eggs but is generally less than 5 min).
9. Pick up the zygotes and transfer to a dish containing fresh M2 medium and examine the embryos for viability and fertilization

based on morphology, including presence of the second pronucleus and select those that are healthy and fertilized.

10. Pass the selected zygotes through two washes of fresh M2 medium and then place in microdrops of K-RVCL-50 medium that has been equilibrated under mineral oil for 24 h in a COOK MINC benchtop incubator until ready for electroporation (37 °C, 5% CO_2, 5% O_2).

3.2 Prepare CRISPR-Cas9 Mixture

For Indel mutations mediated by NHEJ, the microinjection mixture of Cas9 protein and sgRNA(s) is electroporated into mouse zygotes. For a knockin model mediated by HDR, a donor oligonucleotide is also required. Individual components of the CRISPR-Cas9 system, including Cas9 protein, sgRNA, and donor oligonucleotide when called for, are prepared at 2× of the final concentration and brought down to the final concentration when mixed with embryos in an equal volume. The CRISPR-Cas9 reagents are prepared fresh on the day of experiment (*see* **Note 3**).

1. Add CRISPR-Cas9 components, including sgRNA, Cas9 protein, and donor oligonucleotide when indicated, into an RNase-free PCR tube (0.2 mL), and bring to the final 10 μL with TE buffer. Both sgRNA and Cas9 protein will be brought to 500 ng/μL at this step.

   Add 5 μL of sgRNA (stock concentration 1000 ng/μL) (*see* **Note 4**).
   Add 5 μL of Cas9 protein (stock concentration 1000 ng/μL). (The total volume is 10 μL.)
   (For an HDR experiment, add oligonucleotide to a final concentration of 1000 ng/μL. Lyophilize the mixture and reconstitute to 10 μL total volume. *See* **Note 5**.)
2. Cap the tube immediately.
3. Incubate the mixture at room temperature for 15 min to allow loading of sgRNA into the Cas9 nuclease to form the RNP complex.
4. Place on ice.

3.3 Electroporate CRISPR-Cas9 Reagents into Mouse Zygote

1. Remove the embryos from K-RVCL-50 medium and wash in pre-warmed M2 medium.
2. Depending on comfort level of personnel handling the embryos, embryos can be processed in groups of 50–150 and placed in the acidic Tyrode's solution (pH 2.5) for 10 s (*see* **Note 6**), followed by washes through three 100 μL drops of pre-warmed M2 medium.
3. Immediately before electroporation, collect the embryos with a pipette and place them into a 10 μL drop of Opti-MEM

medium that has been pre-warmed to 37 °C and equilibrated with 5% CO_2 and 5% O_2 (*see* **Note 7**).

4. Add the 10 μL of embryos in Opti-MEM medium to the 10 μL of CRISPR-Cas9 reagents, containing Cas9 protein, sgRNA, and donor oligonucleotide (if relevant) reconstituted in TE buffer (pH 7.5). Pipet gently up and down a few times to disperse the embryos into uniform contact with the CRISPR-Cas9 reagents.
5. Pick up the 20 μL of embryos suspended in CRISPR-Cas9 reagents with a 20 μL pipet tip and release the content into the chamber of the electroporation cuvette (1 mm gap width) prechilled at 4 °C. Make sure the pipet tip is securely positioned between the two electrodes of the cuvette and deposit the mixture in a slow motion to avoid generating any air bubbles. Cap the cuvette and tap gently to collect all content to the bottom of the chamber, eliminating any bubbles that may have been formed. Use a soft paper towel to wipe the outside of the cuvette of any water droplet that may have adhered to the cuvette when it was prechilled on ice.
6. Position the cuvette between the two metal contacts in the safety stand and make sure it has good contact so current can flow from the pulse generator to the two electrodes of the cuvette.
7. Electroporate with the condition of 30 V, 1 ms pulse duration, and a string of two pulses separated by 100 ms pulse interval. Repeat up to eight times, separated by 3 s (*see* **Note 8**).
8. Immediately following delivery of the pulses, with the sterile plastic transfer pipette that is provided with the cuvette, pick up 100 μL of pre-warmed and pre equilibrated KSOMaa Evolve/BSA (1 mg/mL) medium and release into the cuvette.
9. Remove the embryos from the cuvette (total of 120 μL) and place into a 35 × 10 mm Petri dish.
10. Rinse the cuvette with additional 100 μL of KOSMaa Evolve/BSA (1 mg/mL) and add to the embryos in 35 × 10 mm Petri dish (*see* **Note 9**).

3.4 Transfer Embryos to Pseudopregnant Female Mice

1. Zygotes are removed from culture and placed into a 1.8 mL screw-cap cryogenic tube preloaded with 800 μL of pre-warmed M2 medium for transport to the surgical suite located within a barrier facility.
2. The zygotes are removed from the tube and placed into culture (K-RVCL-50 under oil, COOK MINC benchtop incubator, 37 °C, 5% CO_2, 5% O_2).
3. The zygotes are removed from culture and placed into pre-warmed M2 medium and transferred via the oviduct into day 0.5 pseudopregnant females (9–11 weeks) following

standard methodologies. We use only one oviduct and transfer up to 20 electroporated embryos for each pseudopregnant female mouse.

4 Notes

1. We have also used embryos produced from IVF with good results, when compared with embryos produced from natural mating. In our protocol, embryos produced from IVF are not treated with acidic Tyrode's solution, as they would have been previously treated with glutathione in preparation for fertilization [17, 22]. Notice that another group also reported the use of embryos generated from IVF and obtained good results in CRISPR-Cas9-mediated gene editing [26].
2. Ovulation occurs approximately 12 h after hCG administration, at which time the oocytes may be fertilized. When hCG is administered at 9:00 am on Day 0 and female mice placed with stud males, fertilization could occur at 9:00 pm on Day 0. As such, the embryos will be at Day E0.5 at 9:00 am when electroporated;
3. RNase is ubiquitous in a laboratory environment. Special care must be exercised not to introduce RNase into the samples during synthesis, purification, and electroporation. All counter tops, glassware, and plastic wares must be cleansed with solution formulated to eliminate RNase activities and tips and vials coming into contact with the samples certified to be RNase free.
4. It is always a good practice to examine the sgRNA sample and determine integrity and yield by separating it on an agarose gel by electrophoresis [10]. A good reading from a spectrophotometer does not necessarily indicate presence of the RNA or accurately measure the concentration of the RNA sample, if the sample is contaminated by nucleotides carried over from synthesis which contributes to absorbance at A260.
5. If stock concentrations of the CRISPR-Cas9 reagents are not sufficiently high, the RNP mixture may be in a volume greater than the required 10 μL. In this case, the mixture can be lyophilized and reconstituted with RNase-free water and brought to the final volume of 10 μL.
6. Due to possible lot to lot variation of acidic Tyrode's solution and shelve lives, optimal time length for bathing the embryos in acidic Tyrode's solution must be determined by observing thinning of the zona under a microscope. A good time length is, when compared with untreated embryos, zona is "thinned" and yet still observable. Overexposure will damage the embryo

and reduce embryo survival upon transfer to pseudopregnant female mice.

7. Although M2 medium has been used extensively to culture embryos, it was found to be incompatible with electroporation [27]. In our protocol, we suspended the embryos in Opti-MEM and combined it 1:1 in volume with the CRISPR-Cas reagents reconstituted in TE buffer, in preparation for electroporation [16]. This combination was shown to support embryo survival and compatible with our electroporation protocol.
8. Electroporation parameters for delivering the payload into mouse embryos vary slightly among laboratories. For delivery of dsRNA into zona weakened embryos, optimal parameters were found to be three series of four pulses of 1–2 ms duration and 20 V [27]. To deliver sgRNA/Cas9 RNP into zona weakened embryos, two pulses of 30 V for 3 ms, separated by a 100 ms interval, were found optimal [13]. To deliver sgRNA/Cas9 mRNA into untreated embryos, another group used a more elaborate scheme of three series of pulses [15]. The "poring pulse" was set to a voltage of 225 V, pulse duration of 2.5 ms, pulse interval of 50 ms, and consisting of four or more pulses. This is followed by two "transfer pulses," with polarity reversed between the two pulses, of a voltage of 20 V, pulse duration of 50 ms, pulse interval of 50 ms, and consisting of five pulses. We extensively tested the parameters and found the condition of a voltage of 30 V, pulse duration of 1 ms, and two pulses separated by pulse interval of 100 ms to be compatible with embryo survival, yet efficient in delivering sgRNA/Cas9 mRNA/donor oligonucleotide into zona weakened mouse embryos [16]. In a follow-up study, we improved targeting efficiency by using a string of up to eight pulses for delivery of the sgRNA/Cas9 protein RNP [17].
9. If it should be found that not all embryos are recovered from the electroporation cuvette, they may have adhered to transfer tips or the sides of the cuvette chamber. Be sure to use the plastic transfer pipette that is provided with the cuvette to transfer the embryos. Also, the embryos can be recovered by an additional round of rinsing of the cuvette chamber.
10. If no targeting or low targeting is observed, consider choosing another guide or carry more than one guide in parallel if possible to maximize the chance of successful delivery of the model at the end of the process. Also before launching a project at full scale, it may be a good idea to screen the guides in a cell line, in mouse embryonic stem cells, or in mouse embryos. If zygote is chosen as the setting in which to screen the guides,

the embryos can be cultured in vitro for 4 days and blastocysts (containing a couple hundred cells) harvested and analyzed.

11. If convenient, Cas9 mRNA can be used in place of Cas9 protein. In this case, it is used at a final concentration of 600 ng/μL [16]. However, one must bear in mind that these two choices may not work out the same. By electroporating mCherry mRNA into mouse zygotes generated from IVF at various time points (E0.25, E0.5, or E0.75), it was shown that mCherry fluorescence only increased after E0.8 (19 h after fertilization), suggesting that translation was only fully activated after E0.8 [26]. The G1 phase of the first cell cycle in the mouse embryo may last 9 h, S phase 6 h, and G2 very short (1 h). Considering that most zygotes enter the "S" phase of the cell cycle well before E0.8, the mouse genome may be duplicating or have already been duplicated around E0.8 when Cas9 mRNA will have been translated and Cas9 protein becomes available. As such, when mRNA is provided, it may be more likely that mice carrying more than two different mutant alleles may be produced [26]. If the intent is to minimize the appearance of multiple mutant alleles in each founder mouse, it may be more desirable to deliver Cas9 protein instead.

Acknowledgment

We would like to thank the Genetic Engineering Technologies and the Reproductive sciences groups of the Jackson Laboratory for their partnership and contribution exploring into the CRISPR-Cas9 technology. Research reported in this publication was partially supported by the National Cancer Institute under award number P30CA034196. The content is solely the responsibility of the authors and does not necessarily represent the official views of the NIH. H.W. is supported by the Strategic Priority Research Program of the Chinese Academy of Sciences (No. XDA16010205), "National Natural Science Foundation of China" (31471215), and the National Key Research and Development Program of China (No. 2016YFA0101402).

References

1. Horvath P, Barrangou R (2010) CRISPR/Cas, the immune system of bacteria and archaea. Science 327(5962):167–170
2. Cong L, Ran FA, Cox D, Lin S, Barretto R, Habib N et al (2013) Multiplex genome engineering using CRISPR/Cas systems. Science 339(6121):819–823 http://www.ncbi.nlm.nih.gov/pubmed/23287718%5Cnhttp://www.pubmedcentral.nih.gov/articlerender.fcgi?artid=PMC3795411
3. Jinek M, Chylinski K, Fonfara I, Hauer M, Doudna JA, Charpentier E (2012) A programmable dual-RNA – guided. Science 337:816–822
4. Mali P, Yang L, Esvelt KM, Aach J, Guell M, DiCarlo JE et al (2013) RNA-guided human

genome engineering via Cas9. Science 339 (6121):823–826

5. Hai T, Teng F, Guo R, Li W, Zhou Q (2014) One-step generation of knockout pigs by zygote injection of CRISPR/Cas system. Cell Res 24(3):372
6. Hwang WY, Fu Y, Reyon D, Maeder ML, Tsai SQ, Sander JD et al (2013) Efficient genome editing in zebrafish using a CRISPR-Cas system. Nat Biotechnol 31(3):227
7. Niu Y, Shen B, Cui Y, Chen Y, Wang J, Wang L et al (2014) Generation of gene-modified cynomolgus monkey via Cas9/RNA-mediated gene targeting in one-cell embryos. Cell 156 (4):836–843
8. Wang H, Yang H, Shivalila CS, Dawlaty MM, Cheng AW, Zhang F et al (2013) One-step generation of mice carrying mutations in multiple genes by CRISPR/cas-mediated genome engineering. Cell 153(4):910–918
9. Yang H, Wang H, Shivalila CS, Cheng AW, Shi L, Jaenisch R (2013) One-step generation of mice carrying reporter and conditional alleles by CRISPR/Cas-mediated genome engineering. Cell 154(6):1370–1379
10. Qin W, Kutny PM, Maser RS, Dion SL, Lamont JD, Zhang Y et al (2016) Generating mouse models using CRISPR-Cas9-mediated genome editing. Curr Protoc Mouse Biol 6:39–66
11. Gordon JW, Scangos GA, Plotkin DJ, Barbosa JA, Ruddle FH (1980) Genetic transformation of mouse embryos by microinjection of purified DNA. Proc Natl Acad Sci 77 (12):7380–7384
12. Palmiter RD, Brinster RL, Hammer RE, Trumbauer ME, Rosenfeld MG, Birnberg NC et al (1982) Dramatic growth of mice that develop from eggs microinjected with metallothionein–growth hormone fusion genes. Nature 300(5893):611
13. Chen S, Lee B, Lee AY-F, Modzelewski AJ, He L (2016) Highly efficient mouse genome editing by CRISPR ribonucleoprotein electroporation of zygotes. J Biol Chem 291 (28):14457–14467
14. Hashimoto M, Takemoto T (2015) Electroporation enables the efficient mRNA delivery into the mouse zygotes and facilitates CRISPR/Cas9-based genome editing. Sci Rep 5:11315
15. Kaneko T, Mashimo T (2015) Simple genome editing of rodent intact embryos by electroporation. PLoS One 10(11):e0142755
16. Qin W, Dion SL, Kutny PM, Zhang Y, Cheng AW, Jillette NL et al (2015) Efficient CRISPR/Cas9-mediated genome editing in mice by zygote electroporation of nuclease. Genetics 200(2):423–430
17. Wang W, Kutny PM, Byers SL, Longstaff CJ, DaCosta MJ, Pang C et al (2016) Delivery of Cas9 protein into mouse zygotes through a series of electroporation dramatically increases the efficiency of model creation. J Genet Genomics 43(5):319–327
18. Wassarman PM (1988) Zona pellucida glycoproteins. Annu Rev Biochem 57(1):415–442
19. Legge M (1995) Oocyte and zygote zona pellucida permeability to macromolecules. J Exp Zool Part A Ecol Genet Physiol 271 (2):145–150
20. Nagy A, Gertsenstein M, Vintersten K, Behringer R (2006) Removal of zona pellucida. Cold Spring Harb Protoc 2006(3):pdb-prot4421
21. Nicolson GL, Yanagimachi R, Yanagimachi H (1975) Ultrastructural localization of lectin-binding sites on the zonae pellucidae and plasma membranes of mammalian eggs. J Cell Biol 66(2):263–274
22. Escoffre J-M, Portet T, Wasungu L, Teissié J, Dean D, Rols M-P (2009) What is (still not) known of the mechanism by which electroporation mediates gene transfer and expression in cells and tissues. Mol Biotechnol 41 (3):286–295
23. Paganin-Gioanni A, Bellard E, Escoffre JM, Rols MP, Teissie J, Golzio M (2011) Direct visualization at the single-cell level of siRNA electrotransfer into cancer cells. Proc Natl Acad Sci 108(26):10443–10447
24. Golzio M, Teissié J, Rols M-P (2002) Direct visualization at the single-cell level of electrically mediated gene delivery. Proc Natl Acad Sci 99(3):1292–1297
25. Byers SL, Payson SJ, Taft RA (2006) Performance of ten inbred mouse strains following assisted reproductive technologies (ARTs). Theriogenology 65(9):1716–1726
26. Hashimoto M, Yamashita Y, Takemoto T (2016) Electroporation of Cas9 protein/sgRNA into early pronuclear zygotes generates non-mosaic mutants in the mouse. Dev Biol 418(1):1–9
27. Grabarek JB, Plusa B, Glover DM, Zernicka-Goetz M (2002) Efficient delivery of dsRNA into zona-enclosed mouse oocytes and preimplantation embryos by electroporation. Genesis 32(4):269–276

Chapter 11

Generation of Conditional Knockout Mice by Sequential Insertion of Two loxP Sites *In Cis* Using CRISPR/Cas9 and Single-Stranded DNA Oligonucleotides

Ye Liu, Yubin Du, Wen Xie, Fan Zhang, Douglas Forrest, and Chengyu Liu

Abstract

Conditional knockout (cKO) mice are extremely valuable for biomedical research because they enable detailed analyses of gene functions in a tissue- or temporally-specific fashion. The conventional method for generating cKO mice is time consuming and labor intensive, which involves making a large gene-targeting construct, transfecting and screening many embryonic stem (ES) cell clones, injecting positive ES clones into blastocysts to produce chimeric mice, and breeding the chimeras to transmit the targeted gene through the germline. Recently developed CRISPR technology has revolutionized the way genetically engineered organisms are created. Knockout and knockin mice can now be made by directly injecting zygotes with Cas9, sgRNA, and donor DNA. In theory, cKO mice can be generated by simultaneously inserting two loxP sites using two sgRNAs and two oligonucleotides as donors, but in practice the probability of obtaining cKO mice in one step is still very low, partly because the efficiency of oligo-mediated knockin is much lower than non-homologous end joining (NHEJ) and partly because co-cutting juxtaposed sites in one allele at the same time often leads to the deletion of the entire fragment between the two cutting sites. Therefore, many laboratories prefer to insert the two loxP sites sequentially, i.e., generating mice with one loxP first and then use embryos collected from these mice to insert the second loxP site. In this chapter, we describe our procedures and timeline using this sequential method to make a Six6 cKO mouse line as a demonstration of its feasibility.

Key words Microinjection, CRISPR, Conditional knockout, loxP, Embryo transfer, Sequential method, Six6

1 Introduction

The knockout mouse technology has played a fundamentally important role in biomedical research, and Drs. Martin Evans [1], Oliver Smithies [2–4], and Mario Capecchi [5, 6] were recognized with a Nobel Prize in 2007 for their ground breaking work in developing this technology. The so-called conditional knockout (cKO) approach has further increased the versatility of these methods by enabling gene knockout in a tissue-specific or temporally-

Chengyu Liu and Yubin Du (eds.), *Microinjection: Methods and Protocols*, Methods in Molecular Biology, vol. 1874, https://doi.org/10.1007/978-1-4939-8831-0_11, © Springer Science+Business Media, LLC, part of Springer Nature 2019

specific manner [7]. Traditionally, cKO mice are generated by electroporating a large gene-targeting construct into germline-competent embryonic stem (ES) cells. After screening a few hundred ES cell clones, the clones with the correctly targeted gene are microinjected into blastocyst-stage embryos for producing chimeric mice. The chimeras are then bred with wild-type mice to pass the targeted gene through the germline. Next, germline transmitted heterozygous mice are crossed with a mouse line expressing a recombinase (often Flp but sometimes Cre) for removing the selection marker gene (often neomycin-resistance gene) which was required for selecting ES cell clones earlier. Only after that, the floxed mice can be crossed with Cre-driver lines for conducting tissue-specific knockout studies.

Recently developed CRISPR technology is a revolutionary method that has produced a profound impact on biomedical research [8–10]. It has dramatically speeded up the process for creating knockout and knockin mice [11], and has also been successfully used for cKO mouse generation [12]. However, in practice, it still remains challenging to insert simultaneously both loxP sites into the same allele. As depicted in Fig. 1, several possible outcomes exist when two juxtaposed sites are targeted simultaneously. At this time, the efficiency for oligonucleotide-mediated knockin is still much lower than NHEJ, so the probability of obtaining mice with both loxP insertions is still low. Most mice with one correct loxP often have indels (deletions or insertions) at

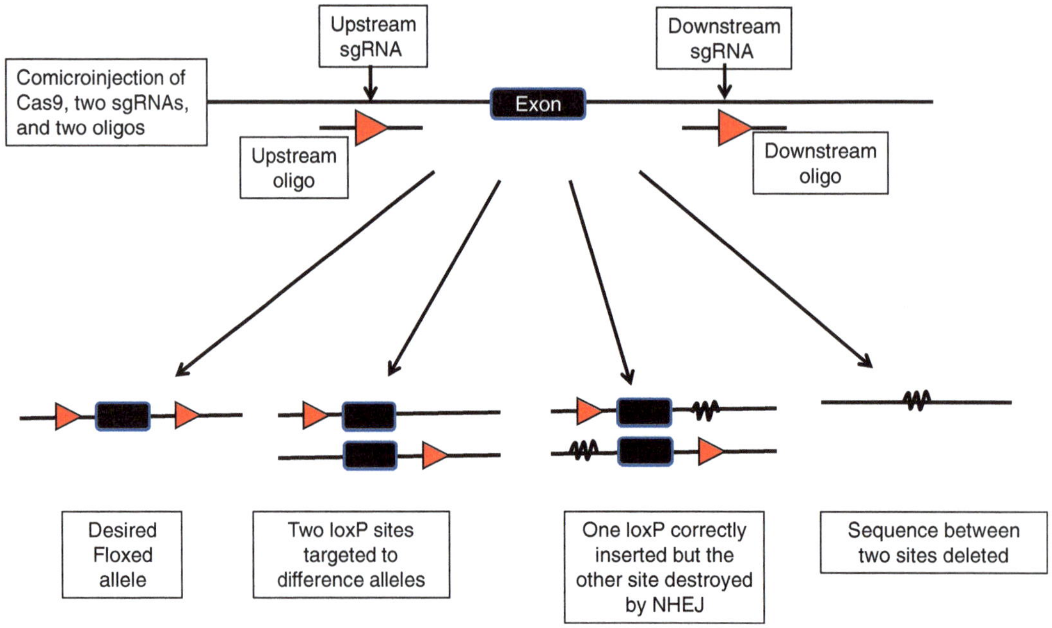

Fig. 1 Possible outcomes of co-injecting CRISPR reagents for simultaneously inserting both loxP sites

the other sgRNA cutting site, which is counterproductive because this renders the mice useless for subsequent use to insert the second loxP site. In the cases when both sites are successfully knocked in, the two loxPs are inserted into different alleles in about half of founder mice. Another outcome is that the entire sequence between the two cutting sites is deleted, which occurs at a surprisingly high efficiency. If each Cas9/sgRNA complex makes the cut and the cut is immediately repaired by NHEJ or HDR as an independent event, one would expect that big deletions only happen somewhat rarely when the two CRISPR sgRNAs cut at the exact same moment. However, in reality such big deletions occur frequently, impling that Cas9/sgRNA complexes that sit in juxtaposed positions may not cut independent of each other, or that they do not repair double-strand break immediately after cutting.

Given the challenges of simultaneously inserting both loxP sites using two sgRNAs and two oligos, various CRISPR-mediated strategies have been developed for generating tissue-specific knockout mice, including tissue-specific expression of Cas9 through Cre-activated gene expression [13] or tissue-specific promoters [14], inserting a Cre-regulated artificial small intron into the gene [15]; electroporating the two loxP sites into zygotes and 2-cell embryos on 2 consecutive days [16], and using a long single-stranded DNA that contains both loxP sites as a donor [17, 18]. However, these methods are still considered at the exploratory stage, and many transgenic facilities are still using the sequential method to insert the two loxP sites in two steps, i.e., generating mice with one loxP first and then use these carrier mice as embryo donors for microinjection to insert the second loxP site. Even though, the sequential approach requires one or two (if breeding is required for expanding the colony with one loxP) generations of breeding, it is advantageous over the traditional ES cell-mediated method for the following reasons: (1) loxP sites can be inserted using single-stranded oligonucleotides and consequently eliminate the need for making a large gene-targeting construct; (2) CRISPR reagents and donor DNAs can be directly microinjected into fertilized eggs and therefore eliminate the need for using ES cells which not only saves time and labor but also removes much of the uncertainty associated with germline competency of the final ES clones; (3) CRISPR method does not require generating chimeric mice nor crossing with Flp (sometimes Cre) mice for removing the selection marker gene, and therefore save two generations of mouse breeding.

In this chapter, we describe the generation of mice carrying a floxed allele of the Six6 gene as an example of our procedures and timeline (Fig. 2) and as a demonstration of the feasibility of using the CRISPR/Cas9 technology to sequentially insert two loxP sites.

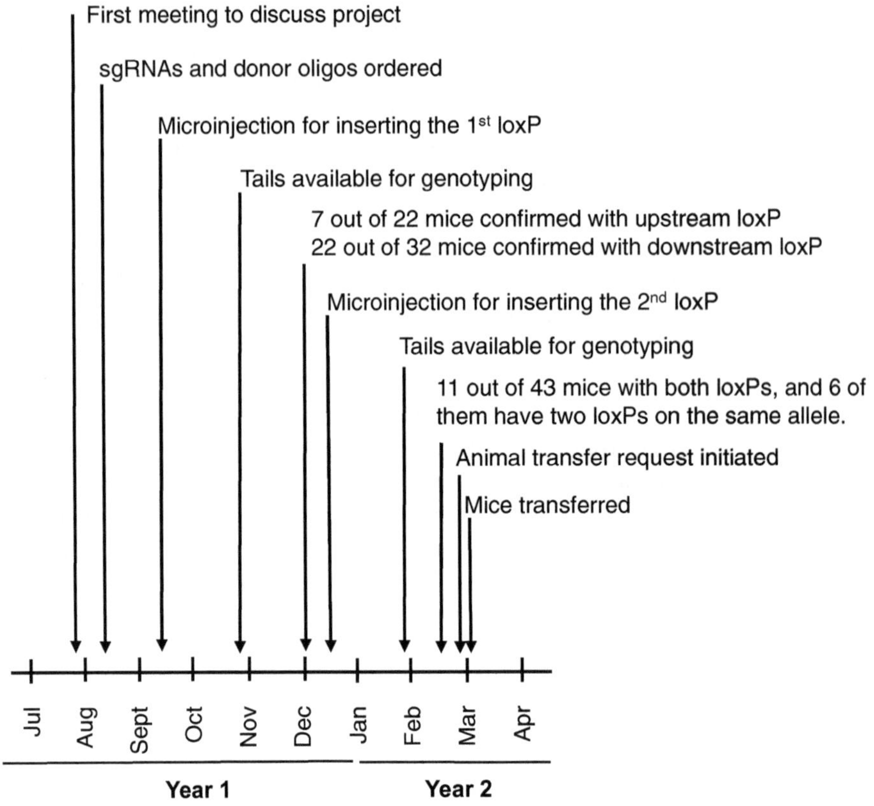

Fig. 2 Timeline for the generation of the Six6 cKO mice by sequentially inserting the two loxP sites

2 Materials

2.1 Mice

Male and female B6D2F1 mice (The Jackson Laboratory) as embryo donors; CD-1 (Charles River) vasectomized males and CD-1 females for producing foster mothers (*see* **Note 1** for choosing mouse strains).

2.2 Equipment and Tools for Microinjection and Mouse Surgery

1. Zeiss Axiovert S100 inverted microscope with 5×, 10×, 20×, and 40× objectives.
2. Two Eppendorf TransferMan NK-2 micromanipulators.
3. Eppendorf FemtoJet4i.
4. Eppendorf CellTram Vario.
5. Air table.
6. Sutter P-97 horizontal pipette puller.
7. Stereo microscopes for embryo collection and embryo transfer procedures.
8. CO_2 incubators.

9. Autoclave machine for sterilizing surgical instruments.
10. Hot beads sterilizer for sterilizing instruments between animals when more than one animal is used during a session.
11. Mouth pipette for transferring embryos.
12. Hair clipper.
13. Wound clip applier.
14. Suturing needle holder.
15. Eppendorf VacuTip holding pipette.
16. 5-0 Vicryl suture.
17. 26–29 G disposable hypodermic needles.
18. T/Pump heating water blanket.
19. Povidone-Iodine swabsticks.
20. Micro dissecting scissors.
21. Iris forceps.
22. Micro dissecting tweezers #5.
23. Dieffenbach micro clamp.
24. Surgical instruments sterilization tray.
25. Microloader pipette tips.
26. Alcohol swab.
27. Sterile surgical gloves.
28. Glass capillary tubing for pulling microinjection needles: Borosil, 1.0 mm OD × 0.75 mm ID with omega dot fiber for rapid filling.

2.3 Media and Reagents for Embryo Culture and Animal Surgery

1. M2 culture medium.
2. M16 culture medium.
3. Ketamine/Xylazine anesthetics: mixing 0.2 mL Ketamine stock solution (100 mg/mL) and 0.1 mL Xylazine stock solution (20 mg/mL) with 1.7 mL 0.9% sodium chloride solution before ip injection (*see* **Note 2**).
4. Pregnant mare's serum gonadotropin (PMSG).
5. Human chorionic gonadotropin (hCG).
6. Hyaluronidase: 10× concentrated solution is prepared by dissolving 100 mg in 20 mL M2 medium. Filter through a 0.22 μm filter, aliquot into microcentrifuge tubes, and store at −20 °C in the dark.
7. Embryo culture-tested mineral oil.
8. Bupivacaine: 2.5 mg/mL solution.
9. Meloxicam: 5 mg/mL solution.

2.4 Reagents for CRISPR/Cas9-Mediated Gene-Editing

1. Cas9 mRNA.
2. sgRNAs:

   Six-Up (ACCTTAACTGCACGGCATTG) for inserting the upstream loxP.

   Six-In (GCATAGGCCGACTAGCTTGC) for inserting the loxP in intron 1.

3. Donor single-stranded DNA oligos for using with corresponding sgRNAs to insert the upstream and downstream loxP sequences.

   Six-Up: CTCTGCAGGAGGGGTCAGTGACCCGTGGCT CTGACCAATCTACCTTGCCACCTTAACTGCACGG CATAACTTCGTATAATGTATGCTATACGAAGTTAT GAATTCATTGTGGGATGCAGCCTGCACTTTATCT TGCTGGCACTTCTGATCTAAAA.

   Six-In: GTTTGGGCCTGAGTTGCAAGCTCCAAAAGCTG GCGAATGAAAAGCATAGGCCGACTAGCGGATCCA TAACTTCGTATAGCATACATTATACGAAGTTATTTG CTGGGTGGAGACAGGATAGTTGTCATGAGGAAA-CACGTCTTTCTGCCTCTTAGAATC.

4. Microinjection buffer: 10 mM Tris–HCl (pH 7.5), 0.1 mM EDTA, 100 mM NaCl.

2.5 Genotyping Equipment and Reagents

1. PCR thermal cycler.
2. Electrophoresis power supply.
3. Horizontal DNA gel electrophoresis chamber.
4. Gel imaging system.
5. Agarose.
6. 1× TBE electrophoresis buffer: 89 mM Tris-borate, 2 mM EDTA, pH 8.3.
7. Ethidium bromide (EtBr) stock solution: 10 mg/mL.
8. Alkaline lysis buffer: 50 mM NaOH.
9. Tail neutralization buffer: 1 M Tris–HCl, pH 7.4.
10. Proteinase K.
11. Taq DNA polymerase: usually contained in a master mix containing Taq, reaction buffer, $MgCl_2$, dNTPs, and loading dyes.
12. 10× standard Taq reaction buffer with 15 mM $MgCl_2$.
13. 100 mM dNTPs (if not included in PCR master mix).
14. EcoRI, BamHI, and 10× restriction reaction buffers.
15. 1 Kb DNA ladder.
16. Gel loading dye (if not included in PCR master mix).

17. PCR DNA purification kit.
18. Genomic DNA purification kit.
19. Phenol:Chloroform:Isoamyl Alcohol 25:24:1 saturated with 10 mM Tris, pH 8.0, 1 mM EDTA.
20. Sodium acetate solution (3 M), pH 5.2.
21. Isopropanol (optional).
22. Ethanol.
23. 5 M Betaine.
24. Phusion high fidelity DNA polymerase.
25. PCR primers for genotyping the upstream loxP site (Six-Up):
    Sixup-F100 (common forward), 5′-TTGGCCAGAAGGCAG AGTGC-3′;
    Sixup-R26 (common reverse), 5′-AGATCAGAAGTGCCA GCAAG-3′;
    Sixup-EIR (allele-specific reverse), 5′-CAGGCTGCATCCC$ ACAATGAAT-3′.
26. PCR primers for genotyping the downstream intronic loxP site (Six-In):
    Sixin-F56 (common forward), 5′-CTAAGAGGCAGAAA GACGTG-3′;
    Sixin-R63 (common reverse), 5′-CTGAGGCTAGAGGAT GCTTG-3′;
    Sixin-BIR (allele-specific reverse), 5′-CATAGGCCGACTA GCGGATC-3′.
    For flox allele identification, Sixin-BIR is paired with Sixup-lxpF (forward): 5′-TTAACTGCACGGCATAACTTCG-3′.

3 Methods

3.1 Designing and Generating CRISPR Reagents

1. Selection of loxP insertion sites: A general strategy for creating cKO mouse lines is to flox an exon around the 5′ end/translational start codon of the gene with two loxP sites, but floxing the first exon is usually avoided because inserting a loxP site upstream of the first exon risks disrupting the promoter function and therefore affects gene expression before Cre-mediated deletion occurs (*see* **Note 3**). However, as shown in Fig. 3, the mouse Six6 gene contains only two exons and the majority of the coding sequence is contained in exon 1, therefore making it necessary to flox the first exon. In order to minimize the chance of disrupting promoter function, the upstream regions across difference vertebrate classes were aligned for avoiding inserting loxP into highly conserved regions.
2. After selecting the insertion regions, sequences are submitted to MIT's CRISPR design website (http://crispr.mit.edu) for

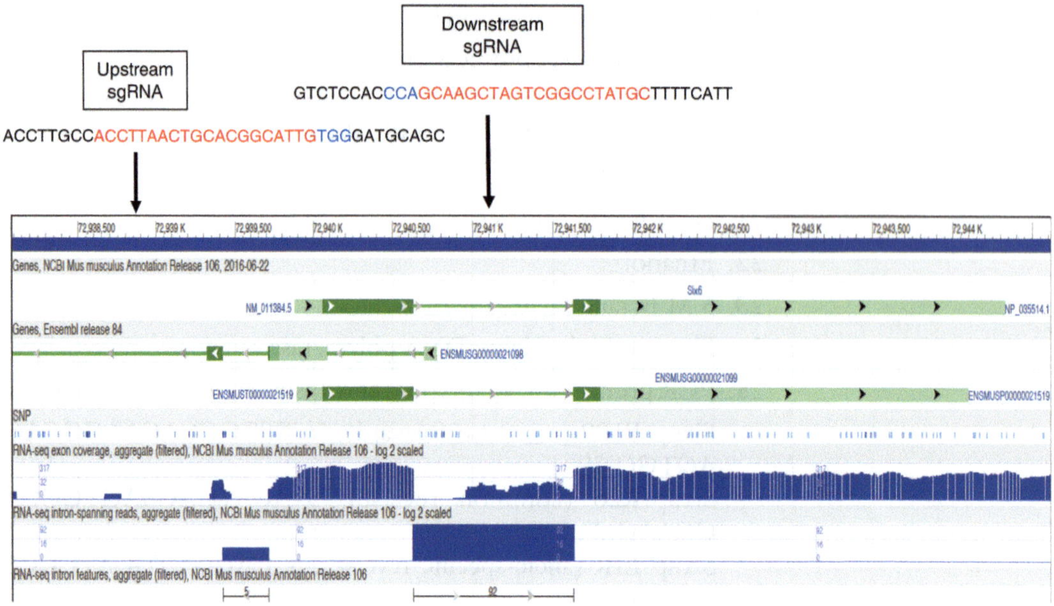

Fig. 3 Positions of the upstream and downstream CRISPR target sites for inserting the two loxP sites flanking the Exon 1 of the mouse Six6 gene

searching and analyzing possible sgRNAs. Ideal sgRNAs should have a score of >60 (higher score indicating fewer highly homologous sequences throughout the genome and consequently lowering the chance of off-target activity), a GC content of 50–70%, and should be devoid of homopolymers (a stretch of the same nucleotide).

3. Purchase in vitro transcribed (IVT) sgRNAs from ThermoFisher, and further purify the sgRNAs by pheno/chloroform extraction and ethanol precipitation.
4. Design donor oligos: single-stranded DNA oligos are used as donors for inserting the loxP sequences. We prefer to insert the loxP at the exact CRISPR cutting site (between 3 and 4 bp upstream of NGG (PAM) for SpCas9). We usually insert a restriction site adjacent to the loxP to facilitate genotyping. In this particular project, an EcoRI site is included in the Six-Up oligo and a BamHI site is included in the Six-In oligo. On each side of the insert (loxP + restriction site), we include a 50–60 bp homologous arm (*see* **Note 4**).

3.2 Microinjection of CRISPR Reagents for Inserting the First loxP

In theory, only the CRISPR reagents (Cas9 mRNA, sgRNA, and the corresponding oligos) targeting one loxP site need to be microinjected for obtaining mice with the first loxP insertion. However, in practice, it is strongly advisable to perform in parallel two separate microinjection experiments, one for each loxP site, as outlined in Fig. 4. Occasionally, some sgRNAs do not work efficiently or

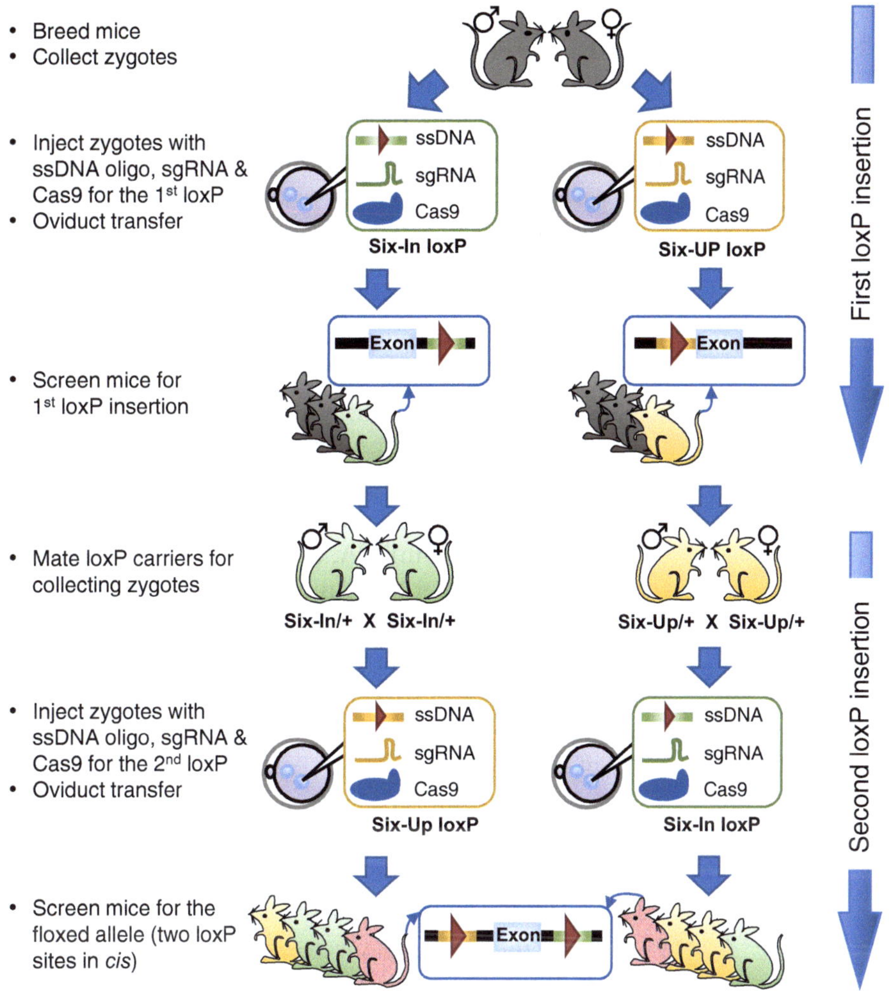

Fig. 4 Schematic outline of experimental procedures for sequentially inserting the two loxP sites in parallel

may be toxic to embryos, but there is no easy and reliable method to pretest the sgRNA (*see* **Note 5**) before microinjection. Injecting the two sgRNAs in parallel enables the identification of bad sgRNAs without causing a severe delay to the project, because as long as one of the two sgRNA works well, the project can proceed while the bad sgRNA is replaced in a repeat attempt for that loxP insertion. If both sgRNAs work efficiently at the first attempt, the mice with either loxP can be used for reciprocal injection of the second loxP site. Our microinjection and embryo transfer procedures are outlined below:

1. Three days prior to the scheduled microinjection, inject (i.p.) ten B6D2F1 female mice with 5 units of PMS. About 46 h (44–48 h) later, inject the same female mice with 5 units of hCG each. Then each female mouse is paired with an individually caged stud male.

2. In the next morning, check the females for vaginal plugs. Euthanize plugged females with CO_2 and quickly dissect out both oviducts from each female and place all dissected oviducts in a 35 mm culture dish containing M2 medium.
3. Using a pair of forceps and a 29-G needle tear open the ampulla of each oviduct to allow the cumulus mass to extrude spontaneously into the media.
4. Add 1/10 volume of concentrated (10×) hyaluronidase solution and incubate for 2–5 min at room temperature so that individual eggs are released from the masses.
5. Use a mouth operated embryo transfer pipette to pick up the eggs and wash them through three drops of M2 medium. The eggs are now ready for microinjection.
6. Pull fresh microinjection needles using the P-97 pipette puller and load each needle with 2–3 μL of CRISPR reagents (100 ng/μL Cas9 mRNA, 50 ng/μL of sgRNA, and 100 ng/μL of oligos dissolved in microinjection buffer) using an Eppendorf microloader tip.
7. Place a drop of M2 in the injection chamber and cover it under mineral oil. Load 30–40 eggs into the drop using an embryo transfer pipet. Place the injection chamber on the microscope stage.
8. Use the CellTram vario micromanipulator and holding pipette to grasp an egg. Insert the injection needle into its cytoplasm and quickly deliver the CRISPR reagents by pushing the foot pedal connected to the FemtoJet 4i.
9. After all the eggs are injected, they are washed through 2–3 M16 drops and then cultured in M16 medium overnight to allow them to develop into 2-cell stage embryos (*see* **Note 6** for alternative embryo transfer strategies).
10. On the day of microinjection, mate CD-1 female mice with vasectomized males.
11. Check vaginal plugs in the next morning, and use plugged CD-1 females as surrogate mothers for embryo transfer.
12. Anesthetize the recipient mothers by ip injection of diluted Ketamine/Xylazine solution at 10 mL/Kg of body weight. Clip hair off from a generous area on back and disinfect the clipped area by alternating applications of Betadine and 70% alcohol.
13. Cut open the skin (~10 mm) and muscle body wall (5–10 mm), and use a pair of iris forceps carefully take out the ovary, oviduct, and a portion of the uterine horn.
14. Under a dissecting microscope, carefully implant about ten injected embryos into one of the oviducts, and repeat the procedure for the oviduct on the other side.

15. Close the abdominal wall with a cruciate suture utilizing absorbable 5-0 Vicryl suture. Drop one or two drops of 0.25% Bupivacaine solution on the muscle at the surgical site using a 26-G needle.
16. Close the skin incision with stainless steel surgical wound clips. Inject the mouse with diluted (in 0.9% sodium chloride) Meloxicam (2–5 mg/kg of body weight), which serves as a long-lasting analgesic for alleviating pain.
17. Place the mouse in a pre-warmed mouse cage placed on a circulating water blanket. The mouse will normally awake 30–60 min after surgery.
18. Wound clips are removed 10–14 days post-surgery, and pups are usually born 19–20 days post implantation. At weaning (18–20 days old), young mice are marked with ear-tags and their tail tips are cut for genotyping.

3.3 Genotyping for the First loxP Site Integration

1. Design a primer pair targeting outside of 5′ and 3′ homology arms of the donor oligo (*see* Fig. 5). These primers are designated as "common flanking primers," as they can amplify both wild-type and loxP-integrated alleles (*see* **Note** 7).

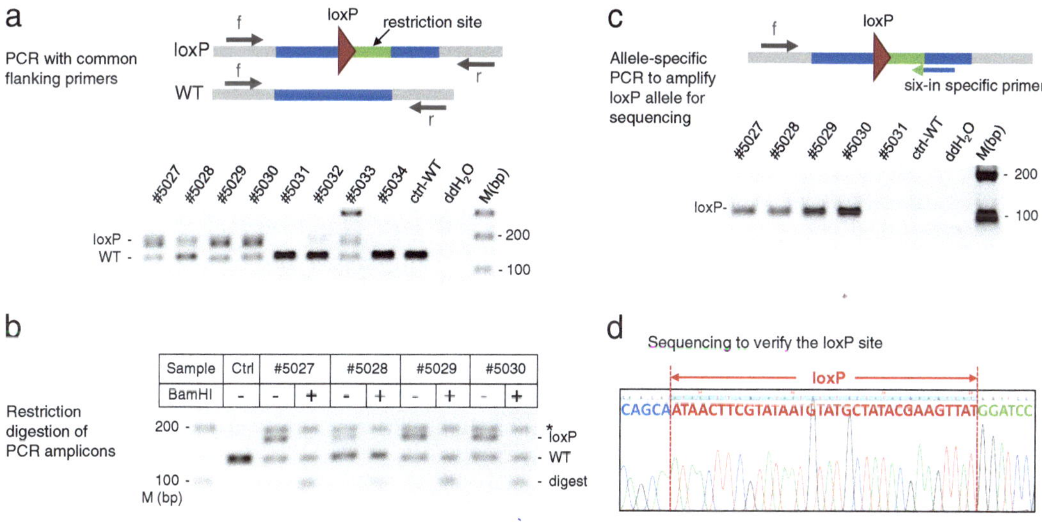

Fig. 5 Identification of mice with insertion of the first loxP site. (**a**) Upper panel: schematic presentation of the PCR strategy for identifying the integrated loxP site at the target region. Gray bar: genomic DNA; blue bar: regions complementary to oligo's homology arms; green bar: restriction site; maroon triangle: loxP site; arrows: common forward (f) and reverse primers (r). Lower panel: gel analysis of PCR products of the Six-In locus to separate the 179 bp loxP amplicon from the 138 bp wild-type (WT) amplicon. M (bp), DNA size marker. (**b**) Restriction digestion (BamHI) of Six-In PCR products to confirm the integration of donor oligos. Ctrl: wild-type; digest: loxP digested fragments of 93 bp and 86 bp; a star (*) points to heteroduplexes formed between wild-type and loxP amplicons. (**c**) Six-In loxP allele-specific PCR (loxP, 111 bp) using common forward primer F56 and loxP-specific reverse primer Sixin-BIR. Upper panel: schematic indication of the position and direction of PCR primers relative to the inserted loxP. Lower panel: gel analysis of the PCR products. (**d**) DNA sequencing chromatogram for verifying perfect loxP sequence (maroon)

2. Place tail tips into 8-strip PCR tubes. Extract genomic DNA using a modified HotSHOT method [19].
3. Add 100 μL of alkaline lysis buffer to each sample tube and incubate at 95 °C for 20 min. After cooling at 4 °C for 5 min, adjust pH by adding 20 μL of tail neutralization buffer and thoroughly mix. The supernatant can be directly used for PCR as DNA template (*see* **Note 8**).
4. Set up a 20 μL PCR reaction for each sample by mixing 10 μL of GoTaq green master mix (2×), 1 μL each of the common primers (10 μM), 1 μL of tail DNA prep, and 7 μL of ddH_2O.
5. Use standard parameters for PCR: initial denaturation at 94 °C for 2 min, followed by 35 cycles of 94 °C for 30 s, 58 °C for 30 s, and 72 °C for 30 s, and a final extension at 72 °C for 10 min (*see* **Note 9**).
6. Make a 2.5–3% agarose gel with 1× TBE buffer and 0.5 μg/mL ethidium bromide (*see* **Note 10**). Load 6 μL PCR products per well alongside a DNA size marker. Perform electrophoresis until the wild-type (lower band) *vs.* loxP (upper band) amplicons of expected sizes are well separated. Visualize DNA by UV transilluminator and document gel pictures using a standard gel imaging system (*see* Fig. 5a lower panel).
7. To confirm the larger PCR amplicon indeed has loxP site integration, set up restriction digests for upper band-positive samples: mix 8 μL of PCR product, 1.2 μL of 10× Reaction Buffer, 0.5 μL of restriction enzyme (EcoRI for Six-Up; BamHI-HF for Six-In), and 10.3 μL of ddH_2O in a 20 μL reaction.
8. After incubating at 37 °C for 1–2 h, load each digested (20 μL) and undigested (~6 μL) sample pairwise onto a 3% agarose gel and then conduct electrophoresis. A sample carrying the loxP integrated allele displays a band shift after the restriction digest: the loxP amplicon, previously appearing as the upper band, disappears (or becomes weaker), while its digestion products, predicted as smaller than wild-type, now appear on the gel (*see* Fig. 5b and **Note 11**).
9. Design a loxP allele-specific reverse PCR primer that spans the junction of genomic sequence and the inserted restriction enzyme site, with the 3′ terminal 4–5 nt of the primer matching the restriction enzyme site. This primer thus only recognizes the loxP-integrated but not the wild-type allele. Pair this allele-specific primer with a common forward primer and set up PCR reactions. Amplicons should only appear in loxP-positive but not in wild-type or samples with indels (*see* Fig. 5c and **Note 12**).

Table 1
Summary of targeting experiments for the first loxP site

Target site	No. of embryos injected/implanted	Mice weaned	No. of loxP carrier mice	Targeting efficiency[a]
Six-Up	140/75	22	7	7/22 (32%)
Six-In	144/99	32	22	22/32 (69%)

[a]Targeting efficiency is defined as total numbers of mice that have correctly targeted loxP over total numbers of founder mice

10. Purify the loxP-specific PCR amplicon by column-based (e.g., Qiaquick PCR purification kit) or magnetic bead-based (e.g., Agencourt AMPure XP beads) method, following the manufacturer's protocol. Sequence purified DNA using the common forward primer to confirm the loxP site (*see* Fig. 5d).
11. Analyze sequencing results and select the verified loxP-positive mice to use for the second loxP injection. As shown in Table 1 and Fig. 5, of the 140 Six-Up CRISPR injected zygotes, 75 developed to the 2-cell stage and were implanted, which result in a total of 22 live pups. Upon genotyping, eight mice were identified as potential loxP carriers, with three being heterozygous and five hemi- or homozygous for the loxP allele (*see* **Note 13**). DNA sequencing analyses showed that all but one mouse have the correct loxP site. Of 144 Six-In CRISPR injected embryos, 99 developed into 2-cell stage and were implanted, 32 pups were born, 22 were identified as loxP carriers including 16 heterozygotes and six hemi- or homozygotes. LoxP sites in all 22 mice were confirmed by sequencing.

3.4 CRISPR-Mediated Integration of the Second loxP Site

1. Male mice with correct loxP sites were separated into individual cages when they reach 7 weeks of age. We normally set up 3–7 stud males for mating with females to collect zygotes for the second injection (*see* **Note 14**).
2. Female loxP carrier mice were superovulated (*see* Subheading 3.2) and mated with males carrying the same loxP allele (*see* **steps 1–5** of Subheading 3.2). If there are insufficient female loxP carriers, wild-type females can also be used. In the latter case, the chance of obtaining mice with both loxPs integrated on the same chromosome is lower but it is usually sufficient for obtaining the desired cKO mice.
3. Fertilized eggs are collected and microinjected as described in **steps 6–9** of Subheading 3.2, but with reciprocal CRISPR reagents, i.e., Six-Up mice injected with Six-In CRISPR and Six-In mice injected with Six-Up reagents.

4. Two-cell stage embryos are implanted as described in **steps 10–18** of Subheading 3.2, and offspring are ear-tagged and genotyped as described in Subheading 3.5 (below).

3.5 Genotyping Analysis for the Presence of Both loxP Sites

1. Take a tail biopsy from each mouse at weaning. Cut 1 mm of tail tip for crude genomic DNA extraction, which is usually adequate for PCR genotyping. Save the remaining tail biopsies at −20 °C as a source of high-quality genomic DNA for long-range PCR that will be performed later.
2. Set up genotyping PCR for the first loxP site using the common primer pair as described in Subheading 3.3. Save only samples that contain the first loxP insertion for further analysis.
3. Follow the same streamlined procedures as in Subheading 3.3 to screen for correct integration of the second loxP site: perform PCR using a common primer pair for the target site, select samples positive for the second loxP amplicon of predicted size, perform the second loxP allele-specific PCR on selected samples, then purify PCR amplicons for sequencing. Mice containing both loxP sites are potential carriers for the floxed allele (two loxP sites in *cis* configuration) and are subject to next-step analysis (*see* **Note 15**).
4. Retrieve tail biopsies frozen in **step 1** of Subheading 3.5 from potential floxed allele carriers. Digest each sample in 200 μL of digestion buffer with proteinase K added fresh to 0.5 mg/mL final concentration. Incubate at 55 °C with shaking until clumps of tissue disappear (3 h to overnight). Extract DNA with the QiaAmp DNA Mini Kit to obtain high-quality genomic DNA that is suitable for long-range genomic PCR (*see* **Note 16**).
5. To identify alleles with two loxP sites integrated in *cis*, pair the Six-Up loxP-specific forward primer with the Six-In loxP-specific reverse primer to amplify regions spanning the two loxP sites (*see* Fig. 6a and **Note 17**): mix 10 μL of GoTaq green master mix (2×), 5 μL of Betaine (5 M), 0.2 μL of Phusion DNA polymerase, 1 μL of each primer (10 μM), 1 μL of high-quality genomic DNA (20 ng/μL), and 1.8 μL of ddH_2O for a 20 μL reaction.
6. Perform PCR with an initial denaturation at 95 °C for 1 min, followed by 30 cycles of 95 °C for 25 s, 58 °C for 25 s, and 72 °C for 2 min and 40 s, and a final extension at 72 °C for 10 min.
7. A parallel PCR (amplicon size >500 bp) can be performed to amplify an unrelated genomic locus, serving as a control for the quality and quantity of input genomic DNA (*see* Fig. 6b lower panel).

8. PCR products are analyzed by agarose gel (1%) electrophoresis. A positive band of expected size from the allele-specific PCR indicates the presence of the floxed allele. A disproportionately faint band (taking into consideration of the intensity of control PCR bands) implies mosaicism (Fig. 6b).
9. Analyze PCR results and confirm by DNA sequencing: as summarized in Table 2, seven out of the 32 weaned Six-In mice injected with Six-Up CRIPSR have both loxP sites integrated. Among them, three have loxP sites in *trans* (unwanted) and four have loxP sites in *cis* (selected for breeding), including a potentially mosaic mouse. Of the 11 weaned Six-Up mice injected with Six-In CRISPR, four have both loxP sites, and two of these have loxP sites in *cis* (selected for breeding). Genotyping analysis of F1 offspring from the six flox-positive carrier mice indicates that three are heterozygous

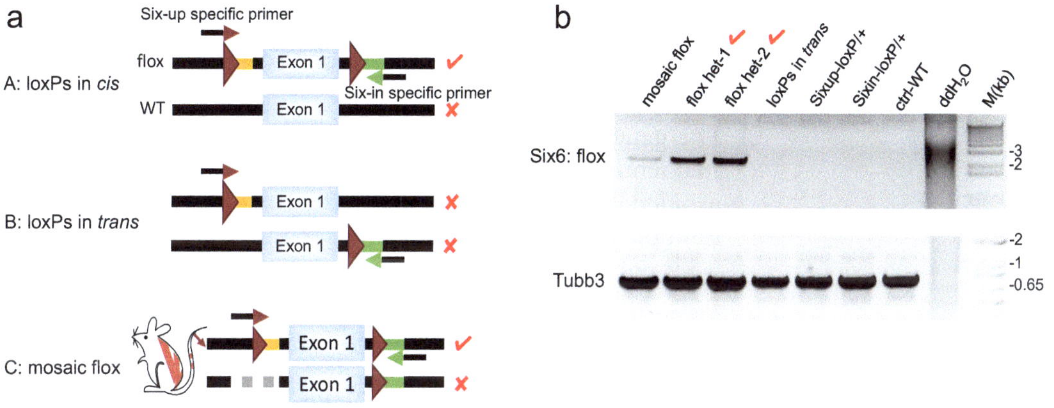

Fig. 6 Identification of the floxed (*cis*-loxP sites) allele. (**a**) Schematic diagram of allele-specific PCR strategy to amplify DNA strand with two loxP sites integrated in *cis* configuration. Mouse C shows a founder mouse carrying a mosaic flox allele that does not have germline transmission. Arrow with maroon arrowhead: a forward primer with 3′ terminus complementary to the upstream loxP site; arrow with green arrowhead: a reverse primer with 3′ terminus complementary to the downstream loxP-linked restriction site. Red check mark: successful PCR amplification resulted from annealing of both primers to the DNA molecule. Red cross mark: no amplification as no primer or only one primer anneals to DNA. (**b**) Upper panel: representative allele-specific PCR analysis for the Six6 flox allele, with an upstream loxP-specific forward primer (Sixup-lxpF) and a downstream loxP site-specific reverse primer (Sixin-BIR) for amplification of the 2.4 Kb loxP-floxed region. Lower panel: PCR amplification of the Tubb3 gene, which serves as a control for PCR reaction

Table 2
Summary of targeting experiments for the second loxP site

Target site	No. of embryos injected/implanted	Mice weaned	Mice w/two loxPs	Mice w/*cis* loxPs	Germline transmission (het/hm)
Six-Up	105/74	32	7	4	3 (1/2)
Six-In	51/36	11	4	2	2 (2/0)

(het) and two are homozygous (hm) for the floxed allele of the Six6 gene, while the mosaic mouse has not transmitted the floxed allele to its offspring.

4 Notes

1. It is strongly recommended to investigate potential strain background for known phenotypical traits or deleterious genes before initiating a long-term project. For example, for studies of visual function or retinal differentiation, a number of retinal degeneration genes (rd) are known. Some of these rd genes are common in certain mouse strains, such as the rd1 mutation ($Pde6b^{rd1}$) which is present in many commonly used mouse lines, including the FVB and C3H inbred strains [20]. The rd8 mutation ($Crb1^{rd8}$) is present in all substrains of C57BL/6N, an NIH subline of C57BL/6 mice [21].
2. Ketamine is a controlled substance in the United States. It must be approved, then stored in a locked drug box and its usage clearly recorded.
3. For generating conditional knockout mouse lines, the most frequently floxed exon is Exon 2. However, Exon 2 should be avoided if the number of nucleotides contained in this exon is a multiple of three, which only leads to the deletion of the amino acids encoded by this exon instead of a frameshift null mutation. Under this circumstance, an alternative exon further downstream can be floxed. Multiple exons can be floxed if one is concerned with leaving too many amino acids in an open reading frame prior to the floxed exon, which can produce a truncated protein. When using conventional ES-cell-mediated gene-targeting method, the distance between the two loxP sites is often quite small because of the size limit of the targeting DNA constructs. However, for the CRISPR-mediated method, the two loxP sites are inserted independently and therefore multiple exons or even the entire gene can be floxed. This advantage comes in handy when the gene contains multiple splicing isoforms, but it is not recommended to flox a very large region without specific reasons because: 1) concerns exist that the efficiency of Cre-mediated deletions may decrease when the two loxP sites are too far apart; and 2) it is now generally believed that >80% of the noncoding genome plays some sort of functional role and therefore it is not advisable to remove unnecessarily large pieces of DNA from the chromosomes.
4. When using single-stranded oligos as donors for knocking in DNA sequences, one common question is whether to use the strand that is complementary or noncomplementary to the

sgRNA. We sometimes prefer one strand over the other depending on whether the sgRNA and oligos can form DNA-RNA duplexes and the location of the mismatches relative to the position of CRISPR cutting site, but for the purpose of inserting loxP using the strategy we have described, we find that either strand works efficiently.

5. We normally do not pretest sgRNAs before microinjecting them into zygotes, but some laboratories test guide sgRNA using *in vitro* (incubating Cas9 protein, template DNA, and sgRNA in test tubes) or *ex vivo* (transfecting into culture cells) methods. We do not believe these testing strategies can reliably predict sgRNA efficiency in a zygote after microinjection because CRISPR efficiency is also dependent on the delivery method. Culturing injected zygotes into blastocysts and genotyping them by PCR can be a reliable method for confirming the efficiency of guide RNAs, but establishing a reliable PCR protocol for genotyping blastocysts requires a significant amount of work and therefore this culturing procedure is used only in extremely important or time-sensitive projects.
6. We prefer to culture the injected zygotes overnight and transfer 2-cell embryos the next morning because: 1) a typical microinjection day is already very busy, so culturing them overnight can alleviate some of the workload to the next day; 2) a significant portion of sgRNAs (5–10% under our standard microinjection conditions) are lethal to early embryos by blocking injected embryos from developing into 2-cell stage embryos. Culturing overnight allows earlier detection of such toxic sgRNAs and redesigning them in a timely manner.
7. The loxP-integrated amplicon is around 40 bp larger than the wild-type amplicon. Given this relatively small size difference, both amplicons should be designed to be less than 300 bp to allow good separation by electrophoresis on an agarose gel. It is also advisable to test the primer sets on wild-type DNA to optimize PCR conditions in advance.
8. In addition to the HotSHOT method, other DNA extraction methods should also work. However, the HotSHOT method is fast and cost-effective, and works well for short-amplicon PCR analysis. Caution should be exercised because DNA prepared in this way is in a crude and unstable condition. For long-term storage, freeze an aliquot at -20 °C.
9. Adjust the annealing temperature if necessary to achieve optimal PCR results. The upper (loxP) and lower (wild-type) band should display a relatively consistent ratio of intensity across all samples. A sample with an unusually weak loxP band may indicate mosaicism, that is only some of the cells in the mouse carry the loxP allele, and thus the germline of the mouse may or

may not carry the desired mutation. Avoid using such mouse for the second targeting if possible (*see* Fig. 5a mouse #5032).

10. Gels with 2.5–3.0% agarose can be used to separate short DNA fragments (<500 bp) efficiently.
11. Often part of the loxP and WT amplicons form heteroduplex DNA (hDNA), which migrates as a band close to the loxP band on the gel, and is resistant to restriction digest (*see* bands marked with * in Fig. 5b). As a result, the digested sample will have a weaker rather than a complete absence of the "upper band." The appearance of a smaller-than-WT digested band(s), therefore, is more reliable than loss of the upper band, for indicating successful restriction digestion.
12. Mixed amplicons can be problematic for sequencing. Allele-specific PCR selectively amplifies the loxP allele, eliminating the contamination of wild-type amplicon. An alternative approach is to subclone mixed PCR amplicons into a plasmid vector (e.g., TOPO-cloning kit) and subsequently sequence 6–10 individual clones for ensuring representation of all alleles. The latter method can be time-consuming if large numbers of mice are to be screened.
13. Some of the injected mice only display the loxP but not wild-type amplicon. This could either mean both alleles have loxP (homozygote), or one allele has loxP integration while the other allele has a big deletion such that the common PCR primers now lack a complementary genomic binding site and therefore cannot amplify that allele (hemizygote). For differentiating between these two possibilities, one can design PCR primers further away from the CRISPR cutting site which enables the detection of bigger deletions.
14. We prefer to set up at least three breeding pairs of mice in order to obtain enough zygotes for each microinjection session. On average, only one half to two thirds of injected zygotes can reach the 2-cell stage of development, and after implantation only about half of the embryos can lead to birth of pups at full term. When a surrogate mother is implanted with too few embryos, the litter size is small and pups often cannot survive long. In the case that only a small number of positive mice are available, it is necessary to breed them with wild-type mice to expand the colony before the second injection can take place. An alternative strategy is to use sperm collected from one loxP-positive male for in vitro fertilization (IVF) to generate a large number of zygotes. With recent improvements in IVF efficiency (*see* Chapter 14 of this book), this method will likely be increasingly used in coming years. Both loxP-positive and wild-type females can be used as egg donors for IVF, although the eggs from wild-type females are less efficient than those

from loxP-carrier females for generating the final floxed mice. Recently developed HyperOva-stimulated ovulation method (*see* Chapter 13 of this volume) makes it possible for using only a few egg donors for each IVF session, and therefore greatly increases the chance of using only loxP-positive females.

15. The PCR protocol described here is suitable for identifying floxed mice when the two loxPs are a few kilobases apart. When the length of the DNA between the two loxPs is too big to be amplified by long-range PCR, it may be necessary to breed candidate mice with wild-type mice and then genotype their offspring. If the loxP sites are *in cis* (floxed), some offspring (approximately 50% or lower if mosaic) will carry both the upstream and downstream loxPs. However, if the loxP sites are *in trans*, they will be segregated into different offspring.
16. Alternatively, genomic DNA can be extracted using standard phenol-chloroform extraction method after proteinase K digestion.
17. The upstream loxP-specific forward primer (Sixup-lxpF) has 9 nt at the 3′ terminus complementary to the loxP sequence, while the downstream loxP-specific reverse primer has 5 nt at the 3′ terminus complementary to BamHI restriction enzyme site. PCR parameters may be adjusted based on actual reactions. Betaine is a PCR enhancing agent, Phusion DNA polymerase offers high processivity. In our experience, a mixture of Phusion and Taq polymerases usually gives more robust results for long-range PCR than Phusion alone. An alternative method is to set up long-range genomic PCR using common primers that amplify the entire "floxed" region. Then subclone the PCR products into a plasmid vector (e.g., TOPO-clone kit) and sequence individual clones.

Acknowledgements

This work was supported in part by the Intramural Research Programs at the National Institute of Diabetes and Digestive and Kidney Diseases (Y.L. and D.F.), and the National Heart, Lung, and Blood Institute (Y.D., W.X., F.Z., and C.L.) at the National Institutes of Health.

References

1. Evans MJ, Kaufman MH (1981) Establishment in culture of pluripotential cells from mouse embryos. Nature 292:154–156
2. Kucherlapati RS, Eves EM, Song KY, Morse BS, Smithies O (1984) Homologous recombination between plasmids in mammalian cells can be enhanced by treatment of input DNA. Proc Natl Acad Sci U S A 81:3153–3157
3. Smithies O, Gregg RG, Boggs SS, Koralewski MA, Kucherlapati RS (1985) Insertion of DNA sequences into the human chromosomal beta-globin locus by homologous recombination. Nature 317:230–234

4. Doetschman T, Gregg RG, Maeda N, Hooper ML, Melton DW, Thompson S, Smithies O (1987) Targetted correction of a mutant HPRT gene in mouse embryonic stem cells. Nature 330:576–578. https://doi.org/10.1038/330576a0
5. Mansour SL, Thomas KR, Capecchi MR (1988) Disruption of the proto-oncogene int-2 in mouse embryo-derived stem cells: a general strategy for targeting mutations to non-selectable genes. Nature 336:348–352. https://doi.org/10.1038/336348a0
6. Thomas KR, Capecchi MR (1987) Site-directed mutagenesis by gene targeting in mouse embryo-derived stem cells. Cell 51:503–512
7. Gu H, Zou YR, Rajewsky K (1993) Independent control of immunoglobulin switch recombination at individual switch regions evidenced through Cre-loxP-mediated gene targeting. Cell 73:1155–1164
8. Cong L, Ran FA, Cox D, Lin S, Barretto R, Habib N, Hsu PD, Wu X, Jiang W, Marraffini LA, Zhang F (2013) Multiplex genome engineering using CRISPR/Cas systems. Science 339:819–823. https://doi.org/10.1126/science.1231143
9. Jinek M, East A, Cheng A, Lin S, Ma E, Doudna J (2013) RNA-programmed genome editing in human cells. elife 2:e00471. https://doi.org/10.7554/eLife.00471
10. Mali P, Yang L, Esvelt KM, Aach J, Guell M, DiCarlo JE, Norville JE, Church GM (2013) RNA-guided human genome engineering via Cas9. Science 339:823–826. https://doi.org/10.1126/science.1232033
11. Wang H, Yang H, Shivalila CS, Dawlaty MM, Cheng AW, Zhang F, Jaenisch R (2013) One-step generation of mice carrying mutations in multiple genes by CRISPR/Cas-mediated genome engineering. Cell 153:910–918. https://doi.org/10.1016/j.cell.2013.04.025
12. Yang H, Wang H, Shivalila CS, Cheng AW, Shi L, Jaenisch R (2013) One-step generation of mice carrying reporter and conditional alleles by CRISPR/Cas-mediated genome engineering. Cell 154:1370–1379. https://doi.org/10.1016/j.cell.2013.08.022
13. Chen J, Du Y, He X, Huang X, Shi YS (2017) A convenient Cas9-based conditional knockout strategy for simultaneously targeting multiple genes in mouse. Sci Rep 7:517. https://doi.org/10.1038/s41598-017-00654-2
14. Mandasari M, Sawangarun W, Katsube K, Kayamori K, Yamaguchi A, Sakamoto K (2016) A facile one-step strategy for the generation of conditional knockout mice to explore the role of Notch1 in oroesophageal tumorigenesis. Biochem Biophys Res Commun 469:761–767. https://doi.org/10.1016/j.bbrc.2015.12.006
15. Guzzardo PM, Rashkova C, Dos Santos RL, Tehrani R, Collin P, Burckstummer T (2017) A small cassette enables conditional gene inactivation by CRISPR/Cas9. Sci Rep 7:16770. https://doi.org/10.1038/s41598-017-16931-z
16. Horii T, Morita S, Kimura M, Terawaki N, Shibutani M, Hatada I (2017) Efficient generation of conditional knockout mice via sequential introduction of lox sites. Sci Rep 7:7891. https://doi.org/10.1038/s41598-017-08496-8
17. Miura H, Quadros RM, Gurumurthy CB, Ohtsuka M (2018) Easi-CRISPR for creating knock-in and conditional knockout mouse models using long ssDNA donors. Nat Protoc 13:195–215. https://doi.org/10.1038/nprot.2017.153
18. Quadros RM, Miura H, Harms DW, Akatsuka H, Sato T, Aida T, Redder R, Richardson GP, Inagaki Y, Sakai D, Buckley SM, Seshacharyulu P, Batra SK, Behlke MA, Zeiner SA, Jacobi AM, Izu Y, Thoreson WB, Urness LD, Mansour SL, Ohtsuka M, Gurumurthy CB (2017) Easi-CRISPR: a robust method for one-step generation of mice carrying conditional and insertion alleles using long ssDNA donors and CRISPR ribonucleoproteins. Genome Biol 18:92. https://doi.org/10.1186/s13059-017-1220-4
19. Truett GE, Heeger P, Mynatt RL, Truett AA, Walker JA, Warman ML (2000) Preparation of PCR-quality mouse genomic DNA with hot sodium hydroxide and tris (HotSHOT). BioTechniques 29(52):54
20. Chang B, Hurd R, Wang J, Nishina P (2013) Survey of common eye diseases in laboratory mouse strains. Invest Ophthalmol Vis Sci 54:4974–4981. https://doi.org/10.1167/iovs.13-12289
21. Mattapallil MJ, Wawrousek EF, Chan CC, Zhao H, Roychoudhury J, Ferguson TA, Caspi RR (2012) The Rd8 mutation of the Crb1 gene is present in vendor lines of C57BL/6N mice and embryonic stem cells, and confounds ocular induced mutant phenotypes. Invest Ophthalmol Vis Sci 53:2921–2927. https://doi.org/10.1167/iovs.12-9662

Chapter 12

Improvement of Mouse Cloning from Any Type of Cell by Nuclear Injection

Sayaka Wakayama, Satoshi Kishigami, and Teruhiko Wakayama

Abstract

Somatic cell nuclear transfer (SCNT) technology has become a useful tool for animal cloning, gene manipulation, and genomic reprograming research. The original SCNT was performed using cell fusion between the donor cell and oocyte. This method remains very popular, but we have recently developed an alternative method that relies on nuclear injection rather than cell fusion. The advantages of nuclear injection include a shortened experimental procedure and reduced contamination of donor cytoplasm in the oocyte. In particular, only this method allows us to perform SCNT using dead cells or naked nuclei such as those from cadavers or body wastes. This chapter describes a basic protocol for the production of cloned mice by the nuclear injection method using a piezo-actuated micromanipulator as well as our recent advances in SCNT using noninvasively collected donor cells such as urine-derived somatic cells. This technique will greatly help not only SCNT but also other forms of micromanipulation, including sperm microinjection into oocytes and embryonic stem cell injection into blastocysts.

Key words Nuclear transfer, Cloning, Piezo, Injection

1 Introduction

Since 1997, when cloning of Dolly the sheep was first reported [1], many mammalian species have been successfully cloned using somatic cell nuclear transfer (SCNT). However, due to the low success rate of cloning, this technique has not yet been applied to industries or agricultural practices, and efforts have been made in many laboratories to develop a simpler method with higher efficiency. Previously, we developed a new nuclear transfer method in which the nuclei were directly injected into oocytes rather than via cell fusion. In this method, we used a piezo impact drive unit (hereafter termed "piezo unit") for nuclear injection [2]. It can aid in bypassing several steps of the original cell fusion procedure. In general, SCNT involves a number of steps: cutting the zona pellucida, inserting a donor cell into the perivitelline space, transferring cells and enucleated oocytes into an electrofusion machine,

Chengyu Liu and Yubin Du (eds.), *Microinjection: Methods and Protocols*, Methods in Molecular Biology, vol. 1874, https://doi.org/10.1007/978-1-4939-8831-0_12, © Springer Science+Business Media, LLC, part of Springer Nature 2019

applying electrofusion, and confirming cell fusion. The utilization of a piezo unit markedly simplifies these steps. Initially, other laboratories found it difficult to reproduce this technique without extensive training. However, many laboratories can now continuously produce not only cloned mice but also other species such as pigs [3] and sheep [4]. Once the piezo unit is set up properly, it will greatly assist not only in animal cloning but also in other forms of micromanipulation such as intracytoplasmic sperm injection into oocytes [5, 6] or embryonic stem (ES) cell injection into blastocysts [7]. Furthermore, the piezo unit simplifies pipette preparation as it allows the use of blunt-tipped pipettes without any additional modifications. In addition, the nuclear injection method can be used for SCNT, regardless of the condition of donor cells. For example, healthy cloned mice could be generated even from frozen mouse cadavers stored at −20 °C for more than 16 years [8]. In that study, no live cells were present and the cell membrane was broken; therefore, naked nuclei were directly injected into oocytes, which would not have been possible using the cell fusion method.

On the other hand, the success rate of cloned mice was still low even when nuclear injection method was used [9]. Interestingly, we found that the efficiency of mouse cloning could be enhanced by up to fivefold by the addition of the histone deacetylase inhibitor (HDACi) trichostatin A (TSA) into the oocyte activation medium [10–13] despite the toxicity of TSA to the embryos [14, 15]. On the other hand, when cytochalasin B (a cytokinesis inhibiter essential for SCNT to prevent DNA loss) was replaced with latrunculin A (Lat A), the procedure was simplified and the efficiency of mouse cloning was increased without epigenetic modifications [16, 17]. These results suggest that cloning efficiency could be enhanced either by correcting epigenetic abnormalities or technical improvements and that it could soon be possible to produce cloned mice with a high success rate. Using this improved method, we succeeded in serial-cloning over 25 generations or generated more than 500 cloned mice from one donor mouse [18], which we were unable to achieve previously [19].

Here, we describe the nuclear injection method and our improved approaches for the production of cloned mice from several types of somatic cells.

2 Materials

2.1 Equipment

1. Inverted microscope with Hoffman or Nomarski optics.
2. Micromanipulator set.
3. Microforge.
4. Pipette puller.

5. Glass pipettes.
6. Warm plate.
7. Piezo impact drive system.
8. Humidified incubator set at 37 °C, 5% CO_2 in air.
9. Tissue culture hood.
10. Centrifuge.

2.2 Mouse Strains

1. Nucleus donor: The most popular hybrid mouse strains such as B6D2F1 (C57BL/6 × DBA/2) or B6C3F1 (C57BL/6 × C_3H/He) can be used as donors. Inbred strains such as C57BL/6 or C_3H/He can be used, but they have a lower embryonic development rate than hybrid strains [20, 21].
2. Oocytes: B6D2F1 mice (approximately 2–3 months old) are preferred sources of oocytes. B6D2F1 oocytes are very translucent, and the metaphase spindles are easy to find. Other strains such as ICR (CD-1) can also be used as recipient oocytes, but the enucleation step can be difficult and the in vitro developmental potential of cloned embryos is lower than that of F1 oocytes [22].
3. Recipients, foster mother, and vasectomized male mice: The ICR (CD-1) strain is used to generate pseudopregnant surrogate mothers, lactating foster mothers, and vasectomized males.

2.3 Donor Cell Types

Probably, most cell types can be used as the nucleus donor; however, in this protocol, we used the most popular cell types, cumulus cells, and tail tip fibroblasts.

2.4 Chemical and Stock Solution

1. Strontium chloride ($SrCl_2$) stock solution: 100 mM (20× stock solution). Dissolve $SrCl_2·6H_2O$ in distilled water (DW) to make up a total volume of 100 mM and store in aliquots at room temperature. The final concentration will be 5 mM.
2. EGTA stock solution: 40 mM (20× stock solution). Dissolve EGTA in DW to make up a total volume of 40 mM and store in aliquots at 4 °C. The final concentration will be 2 mM.
3. Cytochalasin B (CB) stock solution: 500 μg/mL (100× CB stock solution). Add 2 mL DMSO to a vial with 1 mg CB. Aliquot into small tubes (10–20 μL) and store at −30 °C. The final concentration will be 5 μg/mL.
4. Lat A stock solution: 500 μM (100× Lat A stock solution). Add 475 μL DMSO to a vial with 100 μg Lat A. Aliquot into small tubes (5–10 μL) and store at −30 °C. The final concentration will be 5 μM [16]. This Lat A stock solution is optional and is not used in this protocol.

5. TSA stock solution: 5 μM (100× TSA stock solution). Add 3.307 mL DMSO to a vial with 1 mg TSA to make a 1 mM stock solution (first stock solution). Aliquot into small tubes (5–10 μL and/or 200 μL) and store at −30 °C (1 mM TSA stock solution). Then, take 2 μL of the first stock solution and dilute with 398 μL of DMSO. Aliquot into small tubes (5–10 μL) and store at −30 °C. The final concentration will be 50 nM. This TSA stock solution is optional and is not used in this protocol.
6. L + T (Lat A and TSA) stock solution: 5 μM TSA and 500 μM Lat A (100× L + T stock solution). Take 3 μL of the first stock solution of TSA and dilute it with 597 μL of DMSO (5 μM). Add 475 μL of this solution to a vial with 100 μg Lat A (500 μM) to make a final stock solution containing both TSA and Lat A. Aliquot into small tubes (5–10 μL) and store at −30 °C. The final concentration will be 50 nM of TSA and 5 μM of Lat A.
7. Hyaluronidase: 10% (100× stock solution). Dissolve 0.1 mg of hyaluronidase in 1 mL of HEPES-buffered Chatot, Ziomek, and Bavister (H-CZB) medium. Aliquot into small tubes (5–10 μL) and store at −30 °C. The final concentration will be 0.1% concentration.
8. Phenylmethanesulfonyl fluoride (PMSF) stock solution: 50 mM (100× stock solution). Dissolve PMSF in ethanol and store in aliquots at 4 °C.
9. Pregnant mare serum gonadotropin (PMSG) and human chorionic gonadotropin (hCG): 50 IU/mL. Dissolve PMSG and hCG individually in normal saline and store in aliquots at −30 °C.
10. Anesthetic stock solution: 2.5% Avertine. Dissolve 5 g of 2,2,2,-tribromoethyl alcohol in 5 mL of 2-methyl-2-butanol in a 50 °C water bath. Then, add 1.25 mL of this mixture to 50 mL of normal saline and store at 4 °C. Recently we are trying to use following agents: 0.75/4/5 mg/kg medetomidine/midazolam/butorphanol mixed agents. Dissolve 1.875 mL of medetomidine (1 mg/mL), 2 mL of midazolam (5 mg/mL), and 2.5 mL of butorphanol (5 mg/mL) in 18.625 mL of normal saline. The final volume will be 25 mL. Use 0.1 mL/10 g of body weight. 0.75 mg/kg atipamezole (5 mg/mL) in normal saline can be used to wake up the mice [23].

2.5 Culture Media

1. Oocyte and embryo culture media: CZB medium [24] is used for oocyte/embryo culture. H-CZB medium is used for gamete handling and SCNT under air. H-CZB is not suitable for

use in a CO_2 incubator. Ca^{2+}-free CZB medium is used for oocyte activation.

2. Fibroblast cell culture media: tail tip or embryonic fibroblast (EF) cells are cultured in EF medium, which is Dulbecco's Modified Eagle's Medium supplemented with 10% fetal bovine serum and 1% penicillin–streptomycin solution.
3. Nuclear isolation medium (NIM): NIM is used for frozen donor cells and is composed of 123.0 mM KCl, 2.6 mM NaCl, 7.8 mM NaH_2PO_4, 1.4 mM KH_2PO_4, 3 mM EDTA disodium salt, and 0.5 mM PMSF (use 100× stock solution). The pH is adjusted to 7.2 by the addition of a small quantity of 1 M KOH [25].
4. Enucleation media: Add 2 μL of CB stock solution to 198 μL of H-CZB medium. The final concentration of CB will be 5 μg/mL.
5. Reconstructed oocyte activation and reprograming media: Add 10 μL of $SrCl_2$ stock solution (final concentration, 5 mM) and 2 μL of L + T stock solution (final concentration, 5 μM Lat A and 50 nM TSA) to 188 μL of Ca^{2+}-free CZB medium. If Ca^{2+}-free medium is not available, add 10 μL of EGTA stock solution (final concentration, 2 mM) to 178 μL of CZB medium [26] before adding the $SrCl_2$ stock solution and L + T stock solution.

3 Methods

The outline of the procedure is shown in Figs. 1 and 2.

3.1 Preparation of Holding, Enucleation, and Injection Micropipettes

1. Pull the glass pipette by pipette puller.
2. Cut the tip of pipette at the appropriate diameter (Fig. 3a, b) (*see* **Note 1**).
3. Bend all pipettes close to the tip (approximately 300 μm away) at 10–25° (this will depend on the setup of the micromanipulator) using a microforge (Fig. 3c).
4. Insert a small volume of mercury or PMM Operation Liquid into the back of the pipette using a 1 mL syringe and a 30 G needle (approximately 1–2 mm long column) (Fig. 3d), which will enhance the power of the piezo unit (*see* **Note 2**).
5. Store pipettes in a 10 cm dish at room temperature for many days. However, when POL is used, it will evaporate within a day; therefore, POL should be injected into the pipette immediately before use.

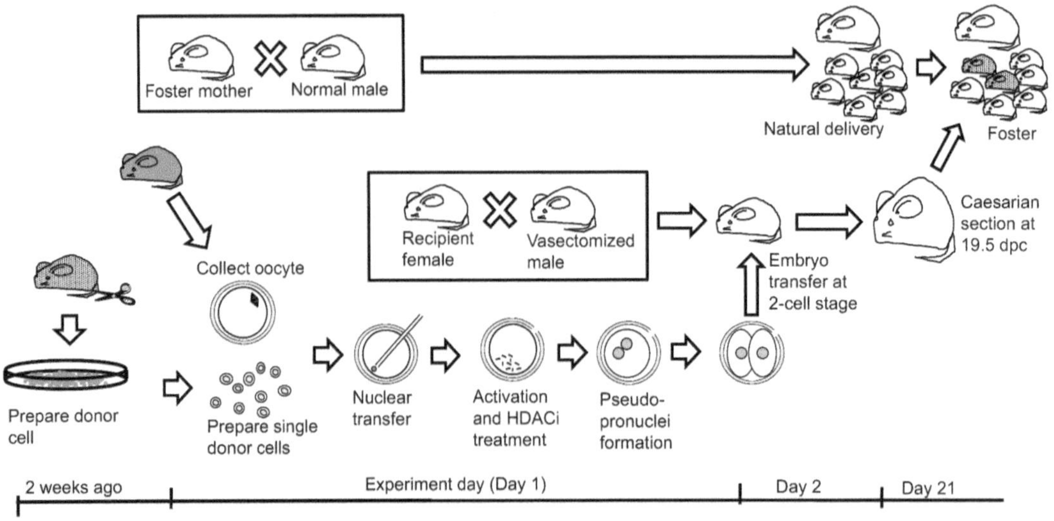

Fig. 1 Overall outline of the SCNT procedure

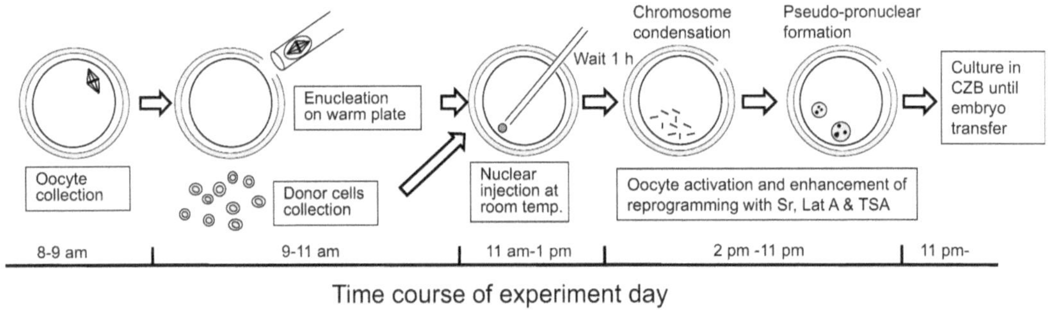

Fig. 2 Outline of the procedures on the day of the experiment

3.2 Preparation of Media Dishes

1. Oocyte culture dish during manipulation: place 20 (e.g., 4 × 5) or more droplets of CZB medium (approximately 10 μL each) on a 6 cm cell culture dish and cover this with sterile mineral oil (Fig. 4a) (*see* **Note 3**). This dish is used from oocyte collection to artificial activation; thus, it must be prepared at the beginning of the protocol and warmed in a 37 °C incubator with 5% CO_2.
2. Manipulation chamber: the manipulation dish includes three types of media: H-CZB, enucleation medium, and 10% polyvinylpyrrolidone (360 kDa). Each droplet should be 10–15 μL in volume, should be placed on the top of a lid of a 6 cm dish (the bottom of the dish cannot be used since the high edges will prevent the appropriate pipette setup) as shown in Fig. 4b, and should be covered with mineral oil. Drawing lines on the outside of the dish will allow the user to easily distinguish the different types of media.

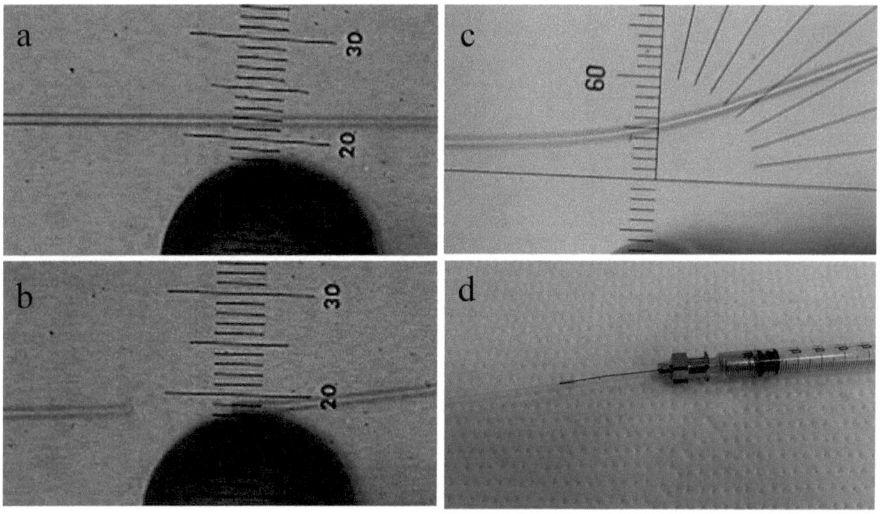

Fig. 3 Generation of a manipulation pipette. (**a**) Set up the pipette to the microfuge. (**b**) The pipette after cutting. (**c**) A pipette bent at 20–25° with the bend located approximately 300 μm from the tip. (**d**) Injection of 1–2 mm mercury into the pipette from the end using a 1 mL syringe with a 26 G needle

3. Oocyte activation and reprograming dish: place 12 (e.g., 4 × 3) or more droplets of oocyte activation and reprograming medium droplets and eight (e.g., 4 × 2) droplets of CZB medium for washing embryos on a 6 cm dish (Fig. 4c). Each droplet should be 10–15 μL in volume. Following this, cover this dish with mineral oil and draw lines to separate the droplet types on the outside of the dish.
4. Culture dish: place 12 or more droplets of CZB medium (Fig. 4d) on a cell culture dish. The top eight droplets are used for embryo washing and the last droplets is used for long-term culture of the embryos until they are transferred to recipient mice. Each droplet should be 10–15 μL in volume. Following this, cover this dish with mineral oil.

3.3 Oocyte Production

1. Injecting PMSG: female mice (8–10 weeks) are superovulated by an injection of 5 IU of PMSG into the abdominal cavity 3 days before the experiment. Typically, this is performed between 5 and 6 pm.
2. Injecting hCG: those mice are then injected with hCG (5 IU) 48 h later (1 day before the experiment). Typically, this is performed between 5 and 6 pm, as same as PMSG injection.
3. Collection of oocytes: Collect oocyte cumulus cell complexes from the oviducal ampullae at 14–15 h after hCG injection (usually, oocytes are collected between 8 and 9 am, Fig. 2).

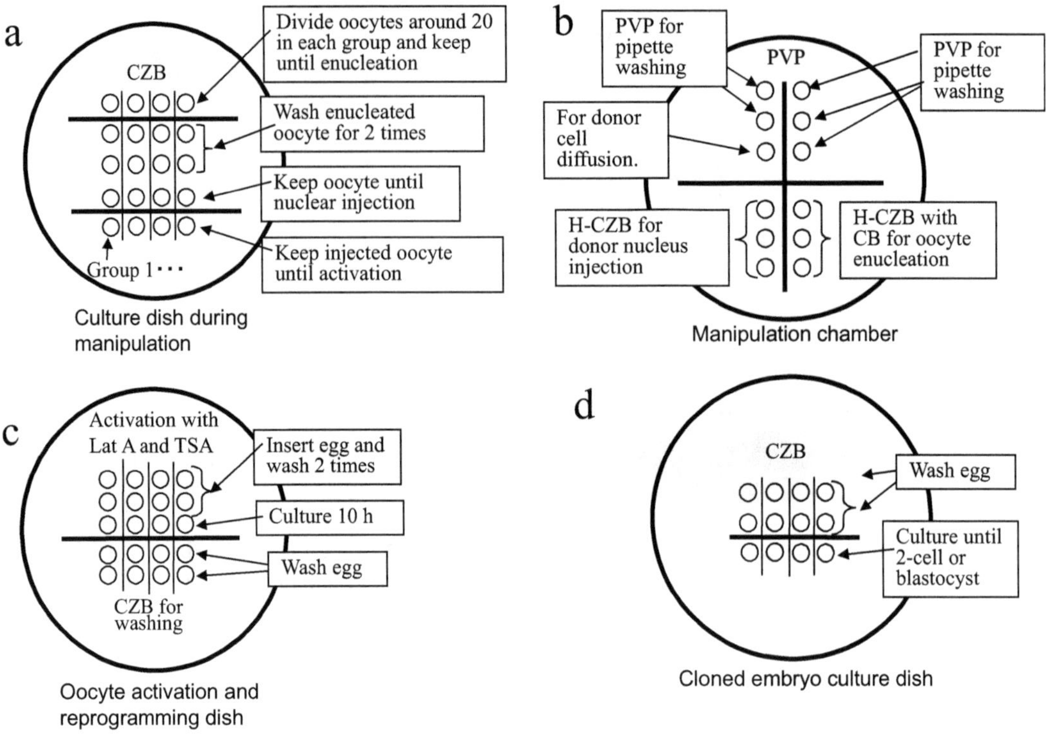

Fig. 4 Culture, manipulation, and activation dish. (**a**) A 6 cm culture dish for collected oocytes, enucleated oocytes, and reconstructed oocytes before activation. (**b**) The lid of a 6 cm dish is used as a micromanipulation chamber for SCNT. (**c**) Reconstructed oocytes are exposed to TSA- and Lat A-containing activation medium for 9–10 h. Wash embryos twice in CZB medium after treatment. (**d**) Move the cloned embryos to the long-term culture dish until transfer into recipient females

4. Hyaluronidase treatment: transfer oocyte-cumulus complexes into a 50 μL droplet of H-CZB containing 0.1% hyaluronidase (use 100× stock solution) for 5 min.
5. Select good oocytes (*see* **Note 4**): wash oocytes from hyaluronidase medium using H-CZB twice, and place them onto an oocyte culture dish prepared as mentioned above (Fig. 4a). The number of oocytes in a drop will depend on the level of skill and the type of experiment (*see* **Note 5**).

3.4 Donor Cell Preparation

1. Cumulus cells: Cumulus cells are the easiest to prepare as nucleus donors because they can be collected and used immediately after oocyte selection without washing. In addition, there is no need to remove hyaluronidase from the medium [2].
2. Sertoli cells: Sertoli cells are testicular sustentacular cells and the male counterpart of cumulus cells. Sertoli cells collected from neonatal mouse testes (immature Sertoli cells) are suitable nucleus donors for SCNT and usually give better results than

cumulus cells [21, 27]. In contrast, Sertoli cells from adult testes are too large to be injected; therefore, the age of the donors will affect the success rate. We recommend the use of newborn males younger than postnatal day 6.

3. Tail tip fibroblasts: tail tip cells, probably fibroblast have to prepare before 10 days of the day of experiment, and not so easy to perform nuclear transfer due to the quite hard cell membrane, these cells can be easily collect either male or female, young or adult, wild or mutant mouse without special treatment [28].
4. ES cells: ES cells are the popular cell type for NT experiments because they have been demonstrated to yield the best success rate in the production of full-term offspring. However, ES cells are pluripotent cells and not differentiated somatic cells; therefore, they are not appropriate for genomic reprograming experiments. In addition, each ES cell line, even if from the same genetic background, will react to this procedure to a different degree and will yield different results for embryonic development [29]. The number of passages of ES cell lines will also affect the success rate.
5. Blood cell: probably lymphocyte can be collected from only a few microliter of peripheral blood from the tail vein and easy to use as same as cumulus cells [30].
6. Urine-derived cells: urine cell can be collected causing injury to the animal. For example, most of the mice discharge their urine immediately after being caught, and 10–100 cells are included in the urine [31].
7. Cells from cadavers: healthy mice can be generated from frozen dead and long-term preserved cadavers. This technique can be applied to generate unique mutant mice which are dead before generating their offspring and are preserved under freezer [8].

3.5 Step-by-Step Protocol for Preparing Cumulus Cells

1. Pick 1–3 μL of condensed cell suspension from hyaluronidase medium by mouth pipette.
2. Place the cells in a PVP medium droplet inside the micromanipulation chamber (Fig. 4b: For donor cell diffusion).
3. Immediately after introducing the cells to PVP medium, mix the cells with PVP medium gently but completely using sharp forceps (*see* **Note 6**). Take care not to scratch the bottom of the chamber.

3.6 Step-by-Step Protocol for Preparing Tail Tip Fibroblasts

1. Cut a part of the tail (at least 2 cm long) two weeks before the planned nuclear transfer experiment.
2. Wash the tails carefully in 70% ethanol.

3. Remove the skin in a sterile tissue culture hood and cut the tail into as many small pieces as possible (1–2 mm) in a 6 cm plastic dish.
4. Culture the fragments in 10 mL DMEM in a 5% CO_2, 37 °C incubator until they are used. There is no need to passage the cells if used within 2 weeks.
5. The day of experiment, remove the culture medium from the dish or flask and wash them in Dulbecco's phosphate-buffered saline [D-PBS(−)].
6. Remove PBS and incubate in trypsin-EDTA for 20–30 min (*see* **Note 7**) in a 37 °C incubator with 5% CO_2.
7. Add EF medium and triturate the cells to produce a single-cell suspension. Spin down the cells in a centrifuge at 300 × *g* for 5 min.
8. Wash the cells at least three times by centrifugation with EF medium. Trypsin is very toxic at the time of nuclear injection; therefore, the donor cells must be washed thoroughly.
9. Prepare a very concentrated cell suspension in EF medium. The final volume should be less than 10 μL.
10. Mix the cells into PVP medium in the same manner as that described for cumulus cells. (Replacing the blue part with the cumulus cell method).

3.7 Setting Up the Micromanipulator

1. Attach the holding pipette to a side of the micromanipulator.
2. On the opposite side, attach the enucleation pipette to the pipette holder of the piezo unit and fix the piezo unit onto the micromanipulator. The top of the pipette holder must be screwed on tightly (Fig. 5c: arrow). If this is not screwed tightly, the cutting of the zona pellucida of the oocyte by the piezo pulses will not be smooth.
3. Expel any air and oil along with a few drops of mercury or POL from the enucleation pipette for washing in PVP medium.
4. Wash both the inside and outside of the pipette with PVP medium until no oil remains (*see* **Note 8**).

3.8 Enucleation of Oocytes

1. Place one group of oocytes in a droplet of enucleation medium into the micromanipulation chamber. Incubate for 5 min before starting the enucleation because CB in the enucleation medium can make the oolemma more flexible and reduce the risk of oocyte lysis.
2. Find oocytes spindle inside ooplasm (*see* **Note 9**) and rotate the oocyte so that the spindle is placed between 8 and 10 o'clock positions and then attach the oocyte firmly to the holding pipette (Fig. 6a).

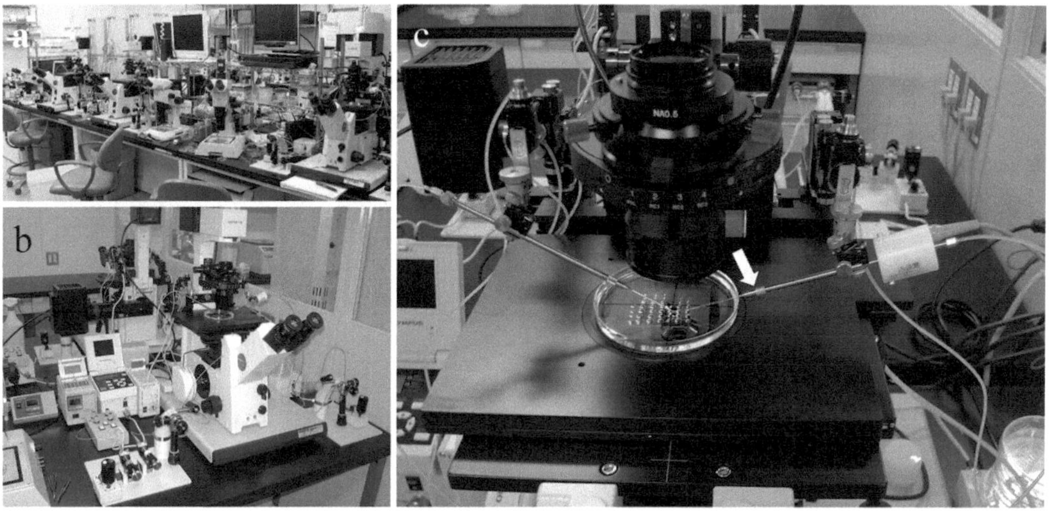

Fig. 5 Setup of the pipette for the pipette holder of the micromanipulator. (**a**, **b**) Micromanipulator set in our lab. (**c**) The pipette was adjusted onto the micromanipulator with an attached piezo drive. The arrow indicates the top of the pipette holder, which must be screwed on tightly

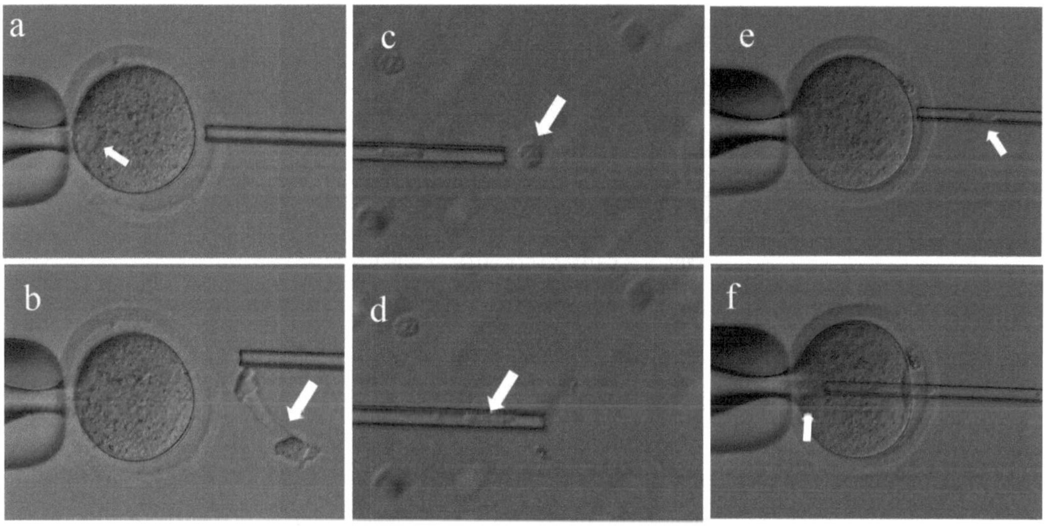

Fig. 6 Enucleation, donor nucleus collection, and injection. (**a**) Rotate the MII oocyte to find the spindle and then hold it between the 8 and 10 o'clock positions (*see* arrow). (**b**) Insert the enucleation pipette into the oocyte, suck MII spindle inside the pipette, and pull the pipette away from the oocyte gently. (**c**, **d**) Gently aspirate the donor cell nucleus in and out of the injection pipette until the nucleus is largely devoid of visible cytoplasmic material. Five to ten nuclei could be collected simultaneously. (**e**, **f**) The enucleated oocyte is held by the holding pipette, and the zona pellucida is cut using few piezo pulses. Insert the injection pipette into the oocyte and break the oocyte membrane using a single weak piezo pulse. Release the donor nucleus into the oocyte (see arrow). Carefully pull the injection pipette away from the oocyte

3. Using a few piezo pulses, cut through the zona pellucida; the addition of a slight negative pressure inside the pipette will increase the power of the pulses. To avoid damaging the

oocyte, ensure there is a large space between the zona pellucida and the oolemma; this space should be approximately as thick as the zona pellucida itself (*see* **Note 10**).

4. Insert the enucleation pipette into the oocyte without using piezo pulses to avoid breaking the oolemma.
5. Remove the MII spindle by aspiration with a minimal volume of cytoplasm. Aspirating with a large volume of cytoplasm may hamper embryonic development.
6. Pinch off the oocyte membrane and the MII spindle slowly, drawing the needle out carefully until the oolemma seals (Fig. 6b) (*see* **Note 11**).
7. When all oocytes in the group have been enucleated, wash them twice in CZB medium (Fig. 4a) to completely remove CB.
8. Return them to the incubator in CZB medium for at least 30 min before starting donor cell injection (Fig. 6a). If CB is not completely washed out, many oocytes will lyse after injection.
9. Intense concentration will be required in the next step (injection of somatic or ES cells); therefore, once all the enucleation of oocytes is complete, short rest is advised before starting the injection.

3.9 Donor Nucleus Injection

1. Transfer a group (10–20) of enucleated oocytes into H-CZB medium. The number of oocytes per droplet should depend on the skill level of an individual, such that each group of injections is completed within 15 min.
2. Remove the donor nuclei from the cells by gently aspirating them in and out of the injection pipette until each nucleus is clearly separated from any visible cytoplasmic material (Fig. 6c, d). We usually take up 5–10 nuclei simultaneously into each injection pipette.
3. Move the injection pipette to the H-CZB droplet containing the enucleated oocytes.
4. Stabilize an enucleated oocyte using a holding pipette and cut the zona pellucida with a few piezo pulses (power level 2–3 and speed 5–6) (Fig. 6e).
5. Reduce the power level of the piezo unit (power level 1–2 and speed 1) because the oolemma is weaker than the zona pellucida and the survival rate of oocytes after injection will be better with this reduced power.
6. Push one nucleus near the tip of the pipette while simultaneously advancing the pipette until it almost reaches the opposite side of the oocyte cortex.

7. Apply only one reduced piezo pulse to puncture the oolemma at the pipette tip. The puncture will be indicated by a rapid relaxation of the oocyte membrane (Fig. 6f) (*see* **Note 12**).
8. Immediately release the donor nucleus into the ooplasm with a minimal amount of PVP medium (Fig. 6f). Gently withdraw the injection pipette from the oocyte (*see* **Note 13**).
9. When all the nuclei have been injected into the oocytes, wash the injection pipette in a PVP dish by expelling some mercury or POL and applying power from the piezo unit. This washing step is important to prevent the pipette from getting sticky (*see* **Note 8**).
10. When a group of oocytes has been used for NT, keep the oocytes in this drop for 10–15 min before transferring them into CZB medium (Fig. 4a). During this period, the injection damage of oocyte will be fixed completely (*see* **Note 14**).
11. Culture the surviving oocytes for at least 30 min in a 5% CO_2 incubator before activation.
12. When the donor cell or nucleus is too large to inject into oocytes, such as a nucleus from a cadaver, many injected oocytes lyse immediately after injection. In this case, nuclear injection should be performed in H-CZB with CB medium and the hole of the membrane should be removed when withdrawing the injection pipette, similar to that in case of MII spindle enucleation.

3.10 Oocyte Activation and Embryo Culture

1. The medium and culture dish (Fig. 4c) should be prepared at least 30 min before use, and the dish must be placed in a CO_2 incubator for equilibration.
2. Transfer the oocytes into the medium, wash them twice, and then culture them for 10 h in a 5% CO_2 incubator at 37 °C.
3. After 9 h of activation, all embryos must be washed twice in CZB medium droplets (*see* **Note 15**). This should be followed by examination of the rate of oocyte activation. If NT and activation are done properly, the oocytes should each possess two or three pseudopronuclei.
4. Move the cloned embryos to a new CZB medium dish for long-term culture until the 2-cell or blastocyst stage is reached (Fig. 4d) (*see* **Note 16**).

3.11 Embryo Transfer

1. Foster mothers will be generated by mating the estrous ICR female mice with normal males on the same day or 1–2 days before the experiment (Fig. 1). These mice would have delivered offspring before the full-term development of cloned pups. Then, lactating mothers will be used to care for the cloned pups delivered by cesarean section at E19.5. Foster

mothers are necessary for cloned pups because the cloned mice litters will be too small to stimulate natural delivery in surrogate mothers.

2. Pseudopregnant mothers will be prepared by mating estrous ICR female mice with vasectomized males on the same day as that of the experiment (Fig. 1).
3. Cloned embryos can be transferred at different stages. Embryos at the 2-cell (24 h after NT) or 4- to 8-cell (48 h after NT) stage can be transferred into the oviducts of the recipients at 0.5 days post copulation (dpc). On the other hand, morulae/blastocysts (72 h after NT) or blastocysts (96 h after NT) can be transferred into the uterus of pseudopregnant mice at 2.5 dpc.

3.12 Cesarean Section

1. Euthanize the surrogate mother at 18.5 or 19.5 dpc by CO_2 exposure (usually at 19.5 dpc for cloned fetuses but at 18.5 dpc for fertilized fetuses).
2. Quickly open the abdomen with a pair of sharp scissors.
3. Remove the uterus and dissect out the cloned pups with their placentas.
4. Carefully wipe away the amniotic fluid from the skin, mouth, and nostrils.
5. Stimulate the pups to breathe by rubbing their backs or pinching them gently with blunt forceps (*see* **Note 17**).
6. Place the pups warm on a 37 °C hotplate (for how long?).
7. Remove the foster mother from her cage.
8. Take some soiled bedding from the cage and nestle the cloned pups in the bedding material so that they take on the odor of the bedding.
9. Mix in the cloned pups with some of the pups from the foster mother's litter.

4 Notes

1. For the holding pipette, the outside diameter (OD) should be smaller than that of the oocyte [e.g., OD, 70–80 μm; inner diameter (ID), 10 μm]. ID of the enucleation pipette should be 8–10 μm. ID of the injection pipette depends on the nucleus donor cell type: 5–6 μm for blood-derived cells, 6–7 μm for cumulus cells, 7–8 μm for fibroblasts or ES cells, and 8–9 μm for naked nucleus injection. Modified pipettes can also be ordered from several companies to save time.

2. Mercury will significantly enhance the power of the piezo unit when introduced inside the micropipette. However, this operation must be done in a working fume hood with appropriate protective gloves because mercury is volatile and toxic if absorbed by breathing or through the skin. Instead, we recommend using POL, which is known to enhance the power of the piezo unit without toxicity.
3. Certain batches of mineral oil have been found to be toxic. Therefore, it is advisable to test all the new batches of oil by culturing embryos in drops of CZB medium overlaid with the new oil. Always set up a control group of embryos cultured under mineral oil of known quality.
4. Some practice necessary to select good oocytes. Those oocytes possess large cytoplasm compare to bad one.
5. We recommend that all the oocytes in a droplet should be finished manipulation within 15 min. If you are a beginner, place approximately five oocytes per drop, and if you have enough experience, place approximately 20 oocytes per drop.
6. When donor cells are transferred to the PVP droplet, mix them with PVP medium using sharp tweezers for at least 30 s. If the donor cells are not mixed thoroughly with PVP medium, they will aggregate and it will be difficult to isolate single cells. ES cells are particularly sensitive and fragile in PVP medium; therefore, it is better to make a new ES cell suspension every 30 min.
7. Fibroblasts have one of the toughest cell membranes, which are very difficult to break. If you fail to break the cell membrane before injection into oocytes, the reconstructed oocytes will not form pseudo-pronuclei after activation. Fibroblasts treated with trypsin for 20–30 min will have cell membranes that are more capable of breaking, thereby increasing the rate of pseudo-pronuclear formation compared with that by usual trypsin treatment.
8. During micromanipulation, the pipette could become sticky. The condition of the pipette is very important; it will affect not only oocyte survival rate but also embryo development after NT. To avoid this problem, wash the pipette in PVP medium completely; PVP will cover both the inside and outside of the pipette to keep the surface slick. Without this step, the pipette soils rapidly and needs to be changed often.
9. The key point in the enucleation step is finding the metaphase II (MII) spindle of the oocyte. It can be identified using Nomarski differential interference or Hoffman modulation contrast optics. Although you can use Hoechst DNA dyes to stain the nuclei, they are harmful to the oocyte and hamper embryonic development [32]. Sometimes, if you cannot detect

the MII spindle at the enucleation step, then the oocytes should be cultured in a 37 °C incubator for approximately 30 min. It has been shown that when oocytes are exposed to room temperature, the spindle microtubules de-polymerize and become difficult to visualize. However, when the oocytes are warmed, the spindles become visible again. Oocyte transparency is also dependent on the mouse strain; oocytes from B6D2F1 are more visible than those from other strains.

10. When you use the piezo unit for the first time, it should be tested whether you can cut the zona pellucida smoothly. If you cannot do so, check the connection between the pipette and the pipette holder. The top of the pipette holder must be screwed on tightly. Expel all oil inside the pipette because this might have reduced the power transmission of the piezo unit. There should be a slight negative pressure inside the pipette to enhance the power of the piezo unit. In contrast, if the power is too strong even at the lowest setting, add a few drops of oil inside the pipette to reduce the power.
11. The MII spindle will appear like a small sphere under Nomarski or Hoffman optics. Moreover, the spindle is harder than the cytoplasm; therefore, the change in consistency will be felt through the micromanipulator.
12. No piezo pulses should be applied until the pipette is near the opposite side. If the piezo pulse is applied with the tip of the pipette in the middle of the oocyte, the oocyte will die after injection.
13. Sometimes, it becomes difficult to release the donor nucleus from the pipette because the pipette may be too dirty; hence, the pipette must be washed frequently using PVP medium. This can be done by expelling some mercury or POL and applying power from the piezo unit. Otherwise, the pipette should be changed.
14. There are many factors that can induce oocyte lysis after injection, such as the pipette having a very large ID, the room temperature being excessively high, or the pipette insertion being too shallow. A large pipette causes the power of the piezo unit to greatly increase, which can increase the rate of oocyte lysis. The process of NT should be performed at room temperature (25–26 °C). Do not use a warm plate for the microinjection process. The injection pipette must be inserted deeply into the oocyte before applying the piezo pulse. The rate of lysis will improve with training. If you are a novice, all the oocytes will lyse immediately after injection. Typically, after 1 month of training, approximately 50% survival may be observed, and after 1 year of practice, approximately 80% survival will be observed. Reconstructed oocytes should be placed

in H-CZB medium for at least 10 min. If they are transferred into CZB medium immediately after injection, nearly 15% of them will undergo lysis from the damage of injection.

15. It is impossible to completely avoid cloned embryo death during activation. During strontium treatment, up to 10% of the oocytes will die and the medium will become dirty. However, this is normal and the surviving oocytes are usually undamaged. Therefore, check the activation medium with fresh intact oocytes as necessary.
16. When the activation is complete, check the rate of pseudo-pronuclear formation in the cloned embryos. If you cannot confirm this formation, it means that the procedure has failed. There are several reasons why oocytes do not form pseudo-pronuclei. Usually, this is because of failure to break the donor cell membrane or failure to activate oocytes. The injection pipette must be smaller than the donor cell. If the donor cell has a tough cell membrane (e.g., tail tip fibroblasts), apply a piezo pulse to break the donor cell membrane at the time of cell pickup or treat the cells with trypsin for a longer period.
17. To date, all cloned mice have been born with abnormal and hypertrophic placentas and they often die just after birth from respiratory failure. At present, there is no way to avoid this lethal phenotype.

References

1. Wilmut I, Schnieke AE, McWhir J, Kind AJ, Campbell KH (1997) Viable offspring derived from fetal and adult mammalian cells. Nature 385:810–813
2. Wakayama T, Perry AC, Zuccotti M, Johnson KR, Yanagimachi R (1998) Full-term development of mice from enucleated oocytes injected with cumulus cell nuclei. Nature 394:369–374
3. Onishi A, Iwamoto M, Akita T, Mikawa S, Takeda K, Awata T et al (2000) Pig cloning by microinjection of fetal fibroblast nuclei. Science 289:1188–1190
4. Loi P, Ptak G, Barboni B, Fulka J Jr, Cappai P, Clinton M (2001) Genetic rescue of an endangered mammal by cross-species nuclear transfer using post-mortem somatic cells. Nat Biotechnol 19:962–964
5. Kimura Y, Yanagimachi R (1995) Intracytoplasmic sperm injection in the mouse. Biol Reprod 52:709–720
6. Wakayama T, Yanagimachi R (1998) Development of normal mice from oocytes injected with freeze-dried spermatozoa. Nat Biotechnol 16:639–641
7. Kawase Y, Iwata T, Watanabe M, Kamada N, Ueda O, Suzuki H (2001) Application of the piezo-micromanipulator for injection of embryonic stem cells into mouse blastocysts. Contemp Top Lab Anim Sci 40:31–34
8. Wakayama S, Ohta H, Hikichi T, Mizutani E, Iwaki T, Kanagawa O et al (2008) Production of healthy cloned mice from bodies frozen at −20 degrees C for 16 years. Proc Natl Acad Sci U S A 105:17318–17322
9. Thuan NV, Kishigami S, Wakayama T (2010) How to improve the success rate of mouse cloning technology. J Reprod Dev 56:20–30
10. Kishigami S, Bui HT, Wakayama S, Tokunaga K, Van Thuan N, Hikichi T et al (2007) Successful mouse cloning of an outbred strain by trichostatin A treatment after somatic nuclear transfer. J Reprod Dev 53:165–170
11. Kishigami S, Mizutani E, Ohta H, Hikichi T, Thuan NV, Wakayama S et al (2006) Significant improvement of mouse cloning technique by treatment with trichostatin A after somatic nuclear transfer. Biochem Biophys Res Commun 340:183–189

12. Van Thuan N, Bui HT, Kim JH, Hikichi T, Wakayama S, Kishigami S et al (2009) The histone deacetylase inhibitor scriptaid enhances nascent mRNA production and rescues full-term development in cloned inbred mice. Reproduction 138:309–317
13. Ono T, Li C, Mizutani E, Terashita Y, Yamagata K, Wakayama T (2010) Inhibition of class IIb histone deacetylase significantly improves cloning efficiency in mice. Biol Reprod 83:929–937
14. Svensson K, Mattsson R, James TC, Wentzel P, Pilartz M, MacLaughlin J et al (1998) The paternal allele of the H19 gene is progressively silenced during early mouse development: the acetylation status of histones may be involved in the generation of variegated expression patterns. Development 125:61–69
15. Kishigami S, Ohta H, Mizutani E, Wakayama S, Bui HT, Thuan NV et al (2006) Harmful or not: trichostatin A treatment of embryos generated by ICSI or ROSI. Cent Eur J Biol 1:376–385
16. Terashita Y, Wakayama S, Yamagata K, Li C, Sato E, Wakayama T (2012) Latrunculin a can improve the birth rate of cloned mice and simplify the nuclear transfer protocol by gently inhibiting actin polymerization. Biol Reprod 86:180
17. Terashita Y, Yamagata K, Tokoro M, Itoi F, Wakayama S, Li C et al (2013) Latrunculin a treatment prevents abnormal chromosome segregation for successful development of cloned embryos. PLoS One 8:e78380
18. Wakayama S, Kohda T, Obokata H, Tokoro M, Li C, Terashita Y et al (2013) Successful serial recloning in the mouse over multiple generations. Cell Stem Cell 12:293–297
19. Wakayama T, Shinkai Y, Tamashiro KL, Niida H, Blanchard DC, Blanchard RJ et al (2000) Cloning of mice to six generations. Nature 407:318–319
20. Wakayama T, Yanagimachi R (2001) Mouse cloning with nucleus donor cells of different age and type. Mol Reprod Dev 58:376–383
21. Inoue K, Ogonuki N, Mochida K, Yamamoto Y, Takano K, Kohda T et al (2003) Effects of donor cell type and genotype on the efficiency of mouse somatic cell cloning. Biol Reprod 69:1394–1400
22. Tanabe Y, Kuwayama H, Wakayama S, Nagatomo H, Ooga M, Kamimura S et al (2017) Production of cloned mice using oocytes derived from ICR outbred strain. Reproduction 154(6):859–866
23. Kawai S, Takagi Y, Kaneko S, Kurosawa T (2011) Effect of three types of mixed anesthetic agents alternate to ketamine in mice. Exp Anim 60:481–487
24. Chatot CL, Ziomek CA, Bavister BD, Lewis JL, Torres I (1989) An improved culture medium supports development of random-bred 1-cell mouse embryos in vitro. J Reprod Fertil 86:679–688
25. Kuretake S, Kimura Y, Hoshi K, Yanagimachi R (1996) Fertilization and development of mouse oocytes injected with isolated sperm heads. Biol Reprod 55:789–795
26. Kishigami S, Wakayama T (2007) Efficient strontium-induced activation of mouse oocytes in standard culture media by chelating calcium. J Reprod Dev 53:1207–1215
27. Ogura A, Inoue K, Ogonuki N, Noguchi A, Takano K, Nagano R et al (2000) Production of male cloned mice from fresh, cultured, and cryopreserved immature Sertoli cells. Biol Reprod 62:1579–1584
28. Wakayama T, Yanagimachi R (1999) Cloning of male mice from adult tail-tip cells. Nat Genet 22:127–128
29. Wakayama T, Rodriguez I, Perry AC, Yanagimachi R, Mombaerts P (1999) Mice cloned from embryonic stem cells. Proc Natl Acad Sci U S A 96:14984–14989
30. Kamimura S, Inoue K, Ogonuki N, Hirose M, Oikawa M, Yo M et al (2013) Mouse cloning using a drop of peripheral blood. Biol Reprod 89:24
31. Mizutani E, Torikai K, Wakayama S, Nagatomo H, Ohinata Y, Kishigami S et al (2016) Generation of cloned mice and nuclear transfer embryonic stem cell lines from urine-derived cells. Sci Rep 6:23808
32. Terashita Y, Li C, Yamagata K, Sato E, Wakayama T (2011) Effect of fluorescent mercury light irradiation on in vitro and in vivo development of mouse oocytes after parthenogenetic activation or sperm microinjection. J Reprod Dev 57(5):564–571

Chapter 13

The CARD Method for Simple Vitrification of Mouse Oocytes: Advantages and Applications

Naomi Nakagata, Jorge Sztein, and Toru Takeo

Abstract

A light modification on the mouse embryos CARD simple vitrification protocol makes it possible the oocyte freezing. This rapid and handy method for cryopreservation of mouse oocytes has many advantages and multiple applications like banking the female gamete for future use, controlling the IVF timing, and facilitating the transgenic production. In this chapter we described the protocol for cryopreserving oocytes by simple vitrification method and the in-vitro fertilization carried out using fresh or frozen sperm. With this method, nearly all of the cryopreserved oocytes were recovered, resulting in more than 90% morphologically normal. From those oocytes used for in vitro fertilization, 89% of the oocytes were fertilized.

Key words Oocytes, Vitrification, Cryopreservation, IVF, Genetic resources

1 Introduction

Nowadays, the number of genetically engineered mouse (GEM) models used worldwide in biomedical research use has grown exponentially [1–5]. To protect and share those precious animal models numerous mouse genetic resources banks have been established in different countries [6, 7]. The primary goal of a mouse germplasm repository is the preservation and supply of mouse strains; originally it was done by embryo freezing and in recent years those banks included sperm and ovaries in their frozen stocks, knowing that they are dealing with a haplotype version of the mouse line. Today, it is also possible to bank oocytes with high success. Since the first successful reports of mouse oocyte cryopreservation [8, 9], many papers describing variations of the method for freezing mouse oocytes were published [10–14]. In general, the C57BL/6 is the inbred mouse strain of choice for a transgenic or knockout study. Moreover, the current common genetic background for GEM's is mainly the C57BL/6N strain or hybrids between C57BL/6 and other strains. Therefore, having frozen oocytes of the target strain will allow researchers to have the

Chengyu Liu and Yubin Du (eds.), *Microinjection: Methods and Protocols*, Methods in Molecular Biology, vol. 1874, https://doi.org/10.1007/978-1-4939-8831-0_13,

opportunity of expanding the C57BL/6 mouse colony via IVF and ET techniques for use in their studies.

To efficiently perform an IVF for production or colony expansion it requires have access to a considerable number of oocytes, therefore, a large number of females from where collect a sufficient number of embryos. To synchronize these activities and get enough number of oocytes females must be superovulated. In the case that the IVF is performed during the working week, the superovulation with PMSG and hCG of the female mice is normally scheduled on Mondays and Wednesdays, which allowed us to perform the IVF on Thursdays and then to freeze or transfer the produced 2-cell embryos to recipients on Fridays. However, if we have access to a stock of frozen oocytes, we will be able to carry out an IVF at any time.

This chapter described a simple vitrification method [15] for unfertilized C57BL/6 mouse oocytes that is used in the CARD service (Center for Animal Resources and Development: CARD) (http://card.medic.kumamoto-u.ac.jp/card/english/sigen/manual/cryooocyte.html).

With this method we obtain excellent survival rates after freezing and warming and high fertilization rates after in-vitro fertilization (IVF) [16]. Although not explained in this chapter, other application for the oocytes cryopreserved by this technique is the production of transgenic mice by microinjection of the frozen thawed oocytes [17] and ICSI [18].

The technique for sperm freezing has slowly takeover on the more expensive embryo cryopreservation for the maintenance of GEM strains [7, 17, 19–21]. One of the reasons is that the number of strains carrying a single allele is larger than the mutations, inbred, congenic, or other strains with multiple alleles of interest; the other are the technical advantages like less time consuming and more economical. Nevertheless, the need for IVF of fresh oocyte production by superovulation could be a problematic step. It is then when the oocyte cryopreservation could play an important role to our advantage. The use of oocytes immediately after warming will be a solution for many labs doing IVF.

In our institute, CARD, we have established a transport system for epididymal sperm at refrigerated temperatures [22]. Therefore having frozen oocytes ready for an IVF immediately after receiving GEM strains from other centers, makes it very simple and cheaper than transporting live animals.

2 Materials

2.1 Collection of Oocytes

1. Animals: Superovulated mature female mouse (8–20 weeks old) with pregnant mare serum gonadotropin (PMSG) in saline (7.5 IU) or CARD Hyperova (Kyudo Co. Ltd., https://www.cosmobio.com/contents/hyperova.html, (0.2 mL).

Table 1
Composition of mHTF

Reagent name	mg/100 mL
NaCl	593.8
KCl	35.0
$MgSO_4 \cdot 7H_2O$	4.9
KH_2PO_4	5.4
$CaCl_2$	57.0
$NaHCO_3$	210.0
Glucose	50.0
Na-lactate[a]	0.34 mL
Na-pyruvate	3.7
Penicillin G	7.5
Streptomycin	5.0
BSA	400
0.5% phenol red	0.04 mL

[a]Assay; 70%

2. Human chorionic gonadotropin (hCG) in saline (7.5 IU).
3. Plastic dishes.
4. Modified HTF containing high calcium (mHTF) (Table 1).
5. Liquid paraffin.
6. Hyaluronidase 1%.
7. mHTF [23] containing 20% fetal calf serum (FCS).
8. Incubator (37 °C, 5% CO_2 in air).

2.2 Simple Vitrification

1. PB1 (Table 2) [24].
2. 1 M DMSO (Table 3).
3. DAP213 (Table 4).
4. Plastic dish.
5. Filter unit (Millex-GV 0.22 μm).
6. Gel loading tip.
7. Transfer pipettes.
8. Conical bottom cryo tubes.
9. Micropipette.
10. Vial canes.
11. Block Cooler.
12. Liquid nitrogen.

Table 2
Composition of PB1

Reagent name	mg/100 mL
NaCl	800.0
KCl	20.0
$CaCl_2$	12.0
KH_2PO_4	20.0
$MgCl_2{\cdot}6H_2O$	10.0
Na_2HPO_4	115.0
Na-pyruvate	3.6
Glucose	100.0
Penicillin	7.5
Streptomycin	5.0
BSA	300.0

Table 3
Composition of 1 M DMSO

Reagent name	mL[a]
DMSO	1.56
PB1	18.44

[a]Final volume: 20 mL

2.3 Recovering Vitrified Oocytes

1. mHTF.
2. Liquid Parafin.
3. 0.25 M sucrose (Table 5).
4. Incubator (37 °C, 5% CO_2 in air).

2.4 Collection of Spermatozoa

1. Plastic dishes.
2. FERTIUP® (PM) (Kyudo Co., Ltd., Japan).
3. Liquid paraffin.
4. Incubator (37 °C, 5% CO_2 in air).

2.5 Insemination

1. CARD MEDIUM (Kyudo Co., Ltd., Japan).
2. mHTF.
3. Liquid paraffin.
4. Pipette tips.

Table 4
Method of preparing DAP213

1. Solution A and Solution B are first prepared and are each fully dissolved	
2. Equal volumes of A and B are then mixed to form DAP213	
Solution A	
Reagent name	**mL**[a]
PB1	2.3088
DMSO	3.1252
Propylene glycol (PG)	4.556
Caution: The solution may become cloudy when DMSO is added	
Solution B	
Reagent	**mg**[b]
Acetamide (AA)	1181.4

[a]Final volume: 10 mL
[b]mg/10 mL in PB1

Table 5
Composition of 0.25 M sucrose

Reagent name	mg[a]
Sucrose	1711.5

[a]mg/20 mL in PB1

5. KSOM/AA media (Table 6) [25].
6. Incubator (37 °C, 5% CO_2 in air).

2.6 Embryo Transfer

1. Female mice (10–16 weeks old) on Day 1 of pseudopregnancy (the day on which a vaginal plug is observed).
2. Micro spring scissors (5 mm blade).
3. Pair of watchmaker's #5 forceps.
4. Serrefine clamp.
5. Wound clip and clip applicator.
6. Plastic dish.
7. Glass capillary for embryo transfer.

3 Methods

3.1 Preparation of Dishes and Oocytes Collection

1. Before collecting oocytes, prepare dishes containing 200 and 100 μL drops of modified HTF containing high calcium (mHTF) (Table 1) covered with liquid paraffin. Keep them in an incubator (37 °C, 5% CO_2 in air) for at least 30 min.

Table 6
Composition of KSOM/AA

Reagent name	mg/100 mL	Vendor	Catalog number
NaCl	555	Sigma	S 5886
KCl	18.5	Sigma	P 5405
KH_2PO_4	4.75	Sigma	P 5655
$MgSO_4{\cdot}7H_2O$	4.95	Sigma	M 2773
$CaCl_2{\cdot}2H_2O$	25	Sigma	C 7902
$NaHCO_3$	210	Sigma	S 5761
Glucose	3.6	Sigma	G 6152
Na-Pyruvate	2.2	Sigma	P 4562
DL-Lactic acid, sodium salt	0.174 mL	Sigma	L 1375
10 mM EDTA	100 μL	Sigma	E 6635
Streptomycin	5	Sigma	S 9137
Penicillin	6.3	Sigma	P 7794
0.5% phenol red	0.1 mL	Sigma	P 0290
L-Glutamine	14.6	Sigma	G 8540
MEM Essential Amino Acids	1 mL	GIBCO	11130-051
MEM Nonessential AA	0.5 mL	Sigma	M 7145
BSA	100	Sigma	A 4378

2. Euthanize the superovulated mature female mouse (8–20 weeks old) approximately 15–17 h after administering hCG.
3. Dissect the mouse to expose the abdominal cavity.
4. Move the digestive tract from inside the abdomen and expose the uterus, oviducts, and ovaries.
5. Remove the uterus, oviducts and ovaries, and place them on sterile filter paper.
6. Remove the oviducts (ampullae) only, avoiding as much fat, blood, and fluids as possible.
7. Immerse the removed oviducts in liquid paraffin contained within a dish (Fig. 1a).
8. Use forceps to hold the oviduct against the base of the dish, and then use a dissecting needle to tear open the ampulla of the oviduct and release the cumulus-oocytes-complexes (COCs). Drag them into the drop of mHTF (200 μL) (Fig. 1a) (*see* **Note 1**).

9. Add 20 μL of 1% hyaluronidase to the drop of mHTF containing the COCs. Keep the dish in an incubator (37 °C, 5% CO_2 in air) for 1 min (Fig. 1b).
10. Promptly collect and transfer cumulus free oocytes into a 100 μL drop of mHTF, and wash them in turn in the drops in the washing dish (Fig. 1b) (*see* **Note 2**).
11. Place the cumulus-free oocytes in the washing dish into a drop (100 μL) of mHTF containing 20% fetal calf serum (FCS). Incubate them in an incubator for 10–20 min to prevent zona hardening.

3.2 Simple Vitrification [15, 16]

3.2.1 Preparation of a Block Cooler and Cryotubes

1. A day before use, place a block cooler in a freezer at −20 °C.
2. About 10 min before commencing the vitrification procedure, take the block cooler out of the freezer.
3. Stand some cryotubes in the block cooler.

   About 40 oocytes/tube are easy to handle; in other words, when you want to vitrify 120 oocytes, you need to stand three cryotubes in the block cooler.
4. Just before starting the procedure, check that the temperature inside the tubes is at 0 °C.

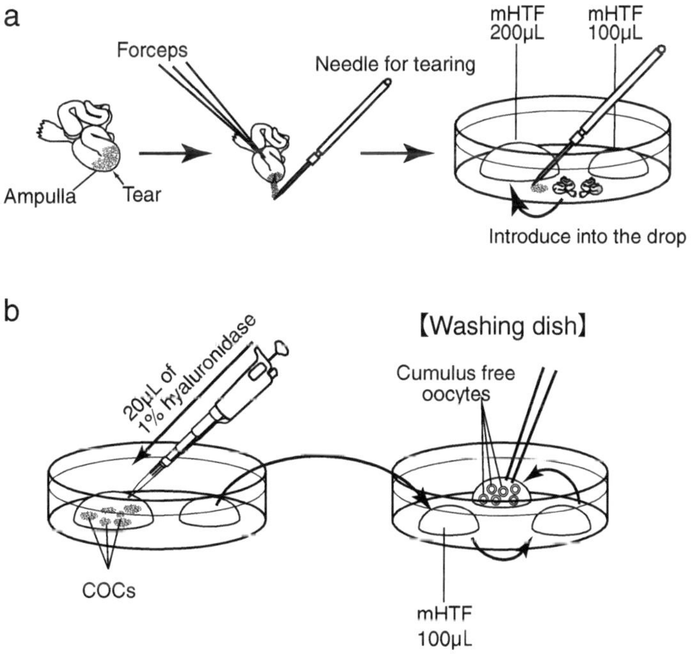

Fig. 1 Oocyte collection. (**a**) Localization of the ampulla and how to tear the COC's out. Oviducts into the oil pushing the COC's into the media drop. (**b**) Adding hyaluronidase to the COCs drop and washing the oocytes. Transferring the oocytes to a drop of mHTF with FCS

3.2.2 Vitrification

1. Filter 1 M DMSO and put four drops of it (100 μL/drop) into a dish. One drop is to wash the oocytes taken from the collection medium, while the others are to hold the washed oocytes.
2. Place a group of oocytes in the drop of mHTF containing 20% FCS into one of the four drops to rinse them. Divide the rinsed oocytes equally between the other drops. These aliquots will eventually be transferred to a storage vial. For example, if one were to collect 120 oocytes and vitrify them in 40-oocyte aliquots, the oocytes would first be placed together in the rinse drop and then divided equally among the three drops (Fig. 2a).
3. Using a 20 μL pipette and a gel-loading tip, transfer the oocytes contained within 5 μL of 1 M DMSO solution into a cryotube. Once transferred, put the cryotube into the block cooler at 0 °C and wait for 5 min (Fig. 2b) (*see* **Notes 3** and **4**).
4. Add 45 μL of cryoprotective solution (DAP213) at 0 °C into the cryotube and equilibrate for 5 min in the 0 °C block cooler (Fig. 2c) (*see* **Note 5**).
5. Quickly set the cryotubes on a cane and plunge the samples directly into the liquid nitrogen (Fig. 2c).

3.3 Recovering Vitrified Oocytes

Preparation for Warming

1. Put three drops (100 μL/drop) of mHTF into a dish and cover them with liquid paraffin. Place the dish in an incubator (37 °C, 5% CO_2 in air) for at least 30 min.
2. Warm 0.25 M sucrose in an incubator (37 °C, 5% CO_2 in air) before use.
3. Remove the required sample from the liquid nitrogen and open the cryotube cap. Discard any liquid nitrogen in the tube and allow it to stand at room temperature for 30 s.
4. Add 0.9 mL of 0.25 M sucrose (preheated to 37 °C) to the cryotube and warm the sample quickly via pipetting. When pipetting, take care not to generate large amounts of bubbles and to not physically damage oocytes by pipetting too quickly. Once warmed, transfer the contents of the cryotube into a culture dish (*see* **Notes 6** and 7).
5. Place 0.4–0.5 mL of 0.25 M sucrose into the cryotube, and transfer the contents into the culture dish. This further dilutes the cryoprotectant and ensures that all of the oocytes have been transferred.
6. Aspirate the oocytes from the liquid and carefully transfer them into a drop of mHTF (washing dish), then keep them in an incubator (37 °C, 5% CO_2 in air).
7. After 10 min, wash the oocytes with two cycles of fresh mHTF in the washing dish.

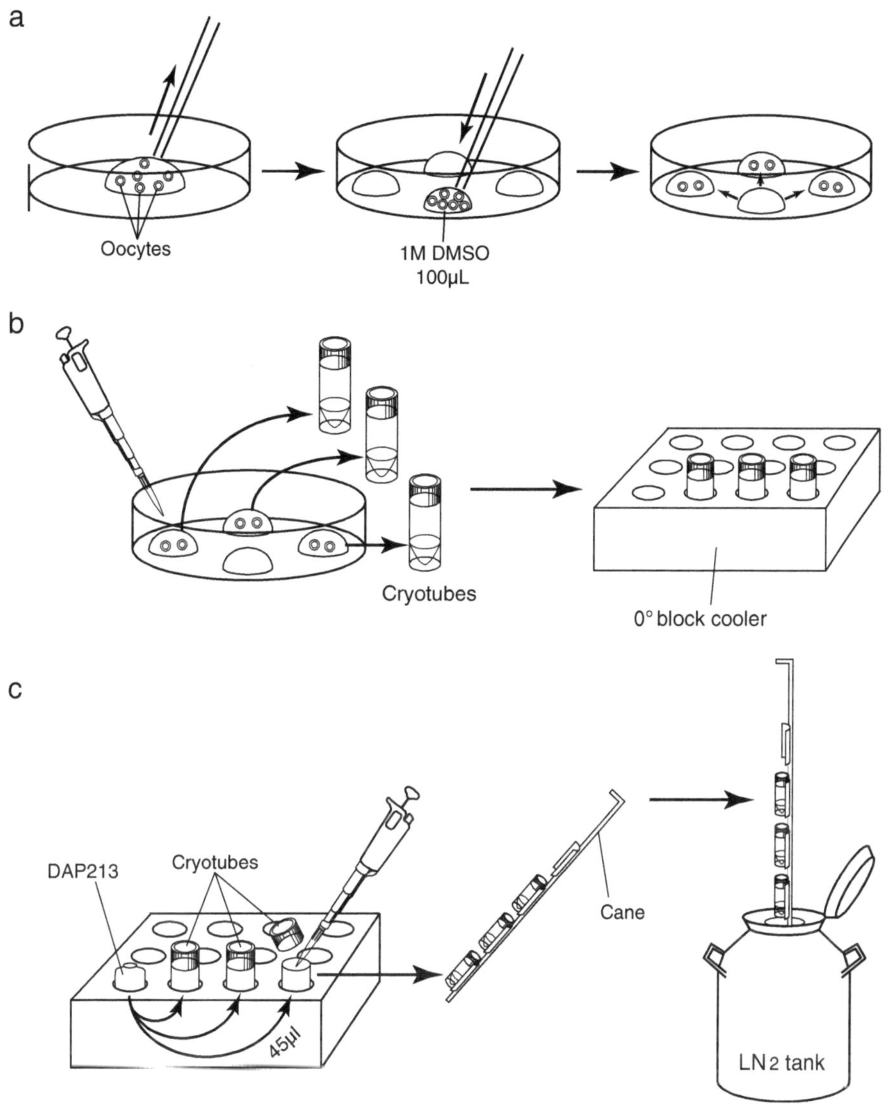

Fig. 2 Oocyte vitrification. (**a**) Transferring oocytes to 1 M DMSO and dividing in groups. (**b**) Transferring oocytes into cryotubes and chill to 0 °C. (**c**) Adding 45 μL DAP213 into each cryotube and transferring into the LN_2 tank

3.4 Collection of Spermatozoa

Preparation of Dishes and Sperm Collection

1. Before collecting spermatozoa, prepare dishes containing a 100 μL drop of FERTIUP® (PM) covered with liquid paraffin. Keep them in an incubator (37 °C, 5% CO_2 in air) for 30 min (sperm dish).
2. Euthanize a mature male mouse (3–6 months old) and remove their cauda epididymides, avoiding as much fat, blood, and tissue fluid as possible.

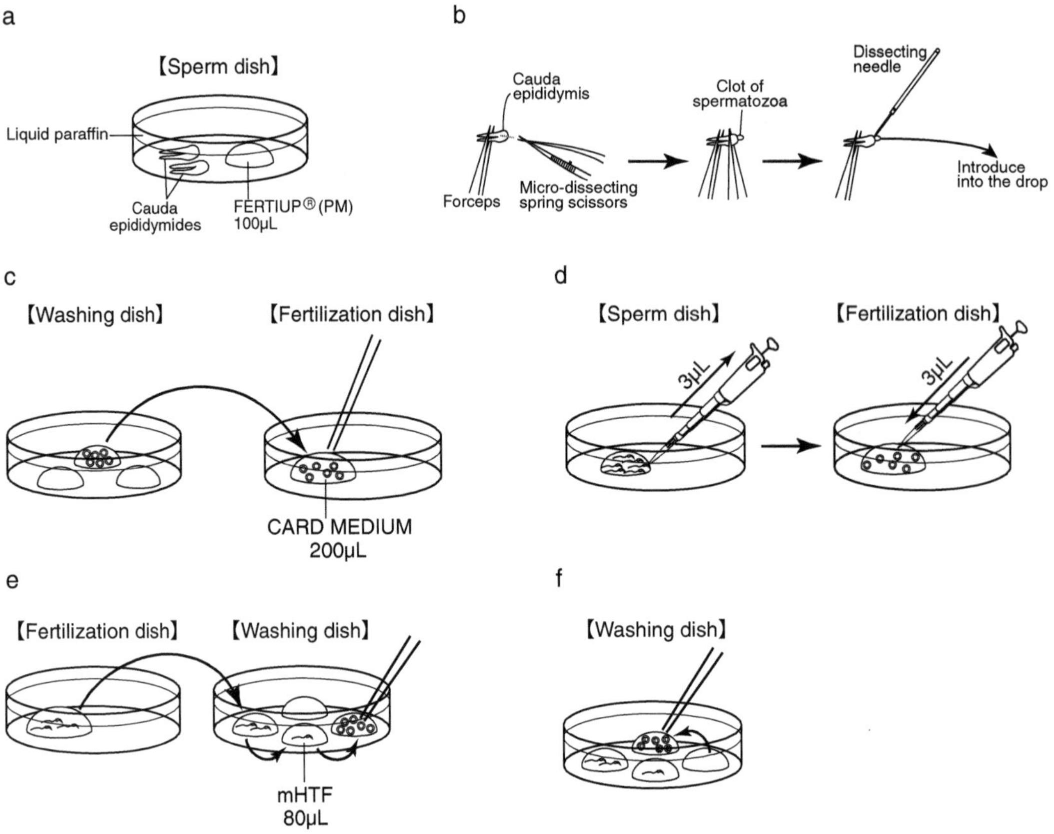

Fig. 3 Collecting sperm and steps for an IVF. (**a**) Caudas into the liquid paraffin. (**b**) Sequence of sperm collection. (**c**) Transfer the oocytes from the washing dish into the fertilization drop. (**d**) Adding sperm. (**e**) Transferring oocytes from fertilization to washing dish and washing oocytes sequence. (**f**) Transferring 2-cell embryos to the fourth drop in the washing dish

3. Place the tissue on sterile filter paper to blot away any blood and fluid.
4. Place removed cauda epididymides in a sperm dish containing liquid paraffin (Fig. 3a).
5. Cut the duct of each cauda epididymis using a pair of micro-dissecting scissors, and then use a dissecting needle to gently press the surface of the cauda epididymis and release the sperm. Use a dissecting needle to introduce the clots of spermatozoa released from the cauda epididymides into the drop of FERTIUP® (PM) (Fig. 3b).
6. Allow the sperm to capacitate by placing the suspension in an incubator for 60 min (37 °C, 5% CO_2 in air).

3.5 Insemination

3.5.1 Preparation of Dishes

1. Fertilization dish.
   Put one drop (200 μL/drop) of CARD MEDIUM into a dish and cover it with liquid paraffin 10 min before collecting oocytes. Place the dish in an incubator (fertilization dish) (*see* **Note 8**).
2. Washing dish.
   Put four drops (80 μL/drop) of mHTF into a dish and cover them with liquid paraffin. Place the dish in an incubator for at least 30 min (washing dish).

3.5.2 Procedure

1. Place the morphologically normal oocytes in the drop of mHTF (washing dish) into the drop of CARD MEDIUM in a fertilization dish and keep it in an incubator (37 °C, 5% CO_2 in air) for 30–60 min before insemination (Fig. 3c).
2. Use a pipette tip to add appropriate amounts (usually about 3 μL) of the sperm suspension to the drop of mHTF containing the morphologically normal oocytes until the final sperm cell concentration is 500–1000 cells/μL (Fig. 3d).
3. Place the fertilization dish in an incubator (37 °C, 5% CO_2 in air).
4. Three hours after insemination, wash the oocytes gently three times in fresh mHTF (80 μL) in a washing dish, removing the sperm attached to the surface of zona pellucida and avoiding the transfer of CARD MEDIUM (Fig. 3e).
5. Six hours after insemination, observe the oocytes in the third drop of mHTF and remove any parthenogenetic oocytes which have only one pronucleus (*see* **Note 9**).
6. After overnight culture of the oocytes, transfer the obtained 2-cell stage embryos only to the fourth drop of mHTF in a washing dish (Fig. 3f). These embryos can be transferred to recipient females, or cultured to the blastocyst stage in KSOM/AA.

3.6 Embryo Transfer

In CARD, the 2-cell embryos transfer is through the wall of the oviduct of pseudopregnant recipients [26] instead of through the infundibulum. This procedure is much easier and simpler to conduct than the conventional embryo transfer procedure, and is therefore suitable for inexperienced users.

3.6.1 Preparation of Mice

1. Anesthetize a female mouse. (Anesthesia should be according to your IACUC approved working protocol.)
2. Pull out the ovary, oviduct, and part of the uterine horn as per the conventional procedure (*see* **Note 10**).
3. Clip a serrafine clamp onto the fat pad which is attached to the ovarian bursa.

3.6.2 Preparation of Embryos and Glass Capillary

1. Make a 200 μL drop of KSOM/AA in a dish (without liquid paraffin), and introduce 20 embryos into the drop. It is also recommended the use M2 media for such procedure because it contain Hepes buffer for maintaining pH outside the incubator.
2. Aspirate air and medium in alternate intervals of 2–3 mm into a glass capillary in preparation for embryo transfer. Draw ten embryos into the glass capillary (*see* **Note 11**).

3.6.3 Embryo Transfer

1. Using a pair of watchmaker's #5 forceps and micro-spring scissors, make a small incision in the wall of the oviduct between the infundibulum and ampulla.
2. Insert the tip of the capillary containing the embryos into the slit, then push the capillary further into the slit toward the ampulla.
3. Use the forceps to hold the portion of the oviduct into which the capillary was inserted.
4. Expel the embryos and 2–3 of the air bubbles into the ampulla. If performed successfully, you should be able to see air bubbles through the wall of the ampulla (*see* **Note 12**).
5. Withdraw the capillary gently from the slit.
6. Gently push the ovary, oviduct, and uterine horn back into the abdomen and close the wound and treat the anesthetized female following your institutional approved protocol.
7. Repeat the process to transfer the remaining ten embryos into the other oviduct as described.
8. Keep the mouse warm on a 37 °C warming plate until the mouse recovers from the effects of the anesthesia.

4 Notes

1. Be sure to carry out all operations, from sacrificing the female and removing her oviducts to introducing the COCs into a drop of mHTF, in the shortest time possible (within 30 s). Moreover, when carrying out this process alone, do not sacrifice multiple mice at once; instead, sacrifice one mouse and swiftly remove its oviducts before moving on to the next mouse.
2. If some cumulus cells become attached to the zona pellucida of the oocytes, they can be removed by manipulating the glass capillary.
3. It is possible to keep the cryotubes in the block cooler at 0 °C for longer than 5 min (<20 min).

4. If the oocytes are pushed together in the center of the drop, it is easy to suck them all up in 5 μL of the 1 M DMSO solution.
5. Do not fasten the caps too tightly after adding the DAP213, or they will be too difficult to remove quickly when samples are recovered from the liquid nitrogen.
6. The tip must not touch the bottom of the cryotube. If it does so, the 0.25 M sucrose in the tip will freeze and you will not be able to inject the 0.25 M sucrose into the cryotube.
7. On the first press, the full volume of 0.9 mL of 0.25 M sucrose will be expelled. On subsequent presses, around ten amounts of 0.25 M sucrose which you have sucked up will be expelled. Pipetting smaller volumes in this way prevents bubbles from generating in the 0.25 M sucrose.
8. There are three different methods of preparing CARD MEDIUM, depending on whether IVF will be carried out using fresh, frozen-thawed, or cold-temperature transported spermatozoa. Please refer to the CARD MEDIUM instruction manual (http://www.kyudo.co.jp/Fertiup/methods.htm).
9. At this stage it is important that you identify and remove any parthenogenetic oocytes. Please note that if you do not remove the parthenogenetic oocytes at this stage, the next day they develop to the 2-cell stage, at which point it will be impossible to distinguish the fertilized oocytes from the parthenogenetic oocytes. The fertilized oocyte has both a male and female pronucleus. On the other hand, the parthenogenetic oocyte has only one pronucleus and the unfertilized oocyte does not have any pronuclei.
10. It is not the intention of this chapter to describe a full embryo transfer surgery, but only the modifications applied on it.
11. When the glass capillary is first inserted into the drop, some liquid paraffin will remain on the outer surface of the drop. The embryos should be drawn into the glass capillary from the opposite side of the drop to avoid sucking up any liquid paraffin. Evidence suggests that liquid paraffin which passes into the oviduct may have adverse effects on the development of the embryos into offspring.
12. If you cannot expel the embryos and the air bubbles into the oviduct, pull the capillary back out just a little from the slit and expel them again.

References

1. Jaenisch R (1988) Transgenic animals. Science 240:1468–1473
2. Bedell MA, Largaespada DA, Jenkins NA, Copeland NG (1997) Mouse models of human disease. Part II: Recent progress and future directions. Genes Dev 11:11–43
3. Simpson EM, Linder CC, Sargent EE, Davisson MT, Mobraaten LE et al (1997) Genetic

variation among 129 substrains and its importance for targeted mutagenesis in mice. Nat Genet 16:19–27

4. Brown SDM, Nolan PM (1998) Mouse mutagenesis – systematic studies of mammalian gene function. Hum Mol Genet 7(10):1627–1633
5. Hrabe de Angelis M, Balling R (1998) Large scale ENU screens in the mouse: genetics meets genomics. Mutat Res 400:25–32
6. Davisson M (2006) FIMRe: Federation of International Mouse Resources: global networking of resource centers. Mamm Genome 17:363–364
7. Donahue LR, Hrabe de Angelis M, Hagn M, Franklin C, Kent Lloyd KC, Magnuson T, McKerlie C, Nakagata N, Obata Y, Read S, Wurst W, Hörlein A, Davisson T (2012) Centralized mouse repositories. Mamm Genome 23(0):559–571
8. Parkening TA, Chang MC (1977) Effects of cooling rates and maturity of the animal on the recovery and fertilization of frozen-thawed rodent eggs. Biol Reprod 17:527–531
9. Whittingham DG (1977) Fertilization in vitro and development to term of unfertilized mouse oocytes previously stored at −196 degrees C. J Reprod Fertil 49:89–94
10. Nakagata N (1989) High survival rate of unfertilized mouse oocytes after vitrification. J Reprod Fertil 87(2):479–483
11. Nakagata N (1990) Cryopreservation of unfertilized mouse oocytes from inbred strains by ultrarapid freezing. Jikken Dobutsu 39 (2):303–305
12. Nakagata N (1993) Production of normal young following transfer of mouse embryos obtained by *in vitro* fertilization between cryopreserved gametes. J Reprod Fertil 99:77–80
13. Shaw JM, Nakagata N (2002) Cryopreservation of transgenic mouse lines. Methods Mol Biol 180:207–228
14. Kohaya N, Fujiwara K, Ito J, Kashiwazaki N (2011) High developmental rates of mouse oocytes cryopreserved by an optimized vitrification protocol: the effects of cryoprotectants, calcium and cumulus cells. J Reprod 57 (6):675–680
15. Nakao K, Nakagata N, Katsuki M (1997) Simple and efficient procedure for cryopreservation of mouse embryos by simple vitrification. Exp Anim 46:231–234
16. Nakagata N, Takeo T, Fukumoto K, Kondo T, Haruguchi Y, Takeshita Y, Nakamuta Y, Matsunaga H, Tsuchiyama S, Ishizuka Y, Araki K (2013) Applications of cryopreserved unfertilized mouse oocytes for in vitro fertilization. Cryobiology 67(2):188–192
17. Nakagata N (1996) Use of cryopreservation techniques of embryos and spermatozoa for production of transgenic (Tg) mice and for maintenance of Tg mouse lines. Lab Anim Sci 46(2):236–238
18. Sakamoto W, Kaneko T, Nakagata N (2005) Use of frozen-thawed oocytes for efficient production of normal offspring from cryopreserved mouse spermatozoa showing low fertility. Comp Med 55(2):136–139
19. Nakagata N (2000) Cryopreservation of mouse spermatozoa. Mamm Genome 11:572–576
20. Nakagata N (2011) Cryopreservation of mouse spermatozoa and in vitro fertilization. Methods Mol Biol 693:57–73
21. Takeo T, Nakagata N (2011) Reduced glutathione enhances fertility of frozen/thawed C57BL/6 mouse sperm after exposure to methyl-beta-cyclodextrin. Biol Reprod 85 (5):1066–1072
22. Takeo T, Tsutsumi A, Omaru T, Fukumoto K, Haruguchi Y, Kondo T, Nakamuta Y, Takeshita Y, Matsunaga H, Tsuchiyama S, Sakoh K, Nakao S, Yoshimoto H, Shimizu N, Nakagata N (2012) Establishment of a transport system for mouse epididymal sperm at refrigerated temperatures. Cryobiology 65 (3):163–168
23. Quinn P, Kerin JF, Warnes GM (1985) Improved pregnancy rate in human *in vitro* fertilization with the use of a medium based on the composition of human tubal fluid. Fertil Steril 44:493–498
24. Whittingham DG (1974) Embryo banks in the future of developmental genetics. Genetics 78 (1):395–402
25. Ho Y, Wigglesworth K, Eppig JJ, Schultz RM (1995) Preimplantation development of mouse embryos in KSOM: augmentation by amino acids and analysis of gene expression. Mol Reprod Dev 41:232–238
26. Nakagata N (1992) Embryo transfer through the wall of the Fallopian tube in mice. Exp Anim 41(3):387–388

Chapter 14

The CARD Method for Mouse Sperm Cryopreservation and In Vitro Fertilization Using Frozen-Thawed Sperm

Toru Takeo, Jorge Sztein, and Naomi Nakagata

Abstract

In the last few years sperm cryopreservation was rapidly established as the technique to efficiently manage production, preservation, and transportation of genetically engineered mice. However, occasionally, the reduced fertility of the frozen-thawed mouse sperm can make it difficult to revitalize the mouse by in vitro fertilization (IVF). In particular, the frozen-thawed sperm of C57BL/6 mice, widely used as the background of choice for genetically engineered strains, show very low fertility after freezing and thawing. To overcome this problem, we have developed a new protocol for sperm cryopreservation and IVF with frozen-thawed C57BL/6 sperm as well as other mouse strains. This protocol has the following three modifications: (1) addition of L-glutamine to the sperm cryoprotectant, (2) addition of methyl-β-cyclodextrin to the sperm preincubation medium, and (3) addition of reduced glutathione to the fertilization medium. These modifications greatly enhanced the fertility of frozen-thawed C57BL/6 sperm, resulting in a stable fertilization rate >80% in IVF. Our results indicate that this robust protocol for sperm cryopreservation may improve the archiving and distributing system for genetically engineered mice.

Key words Mouse, Sperm, Cryopreservation, In vitro fertilization, Repositories

1 Introduction

Successful mouse sperm cryopreservation was first carried out in 1990 by three independent groups of Japanese scientists [1–3]. They described the cryoprotectant agents (CPAs) consisting of raffinose pentahydrate combined with skim milk and/or permeable organic compounds (e.g., dimethyl sulfoxide, glycerol) that enabled in vitro fertilization (IVF) or artificial insemination using frozen-thawed mouse sperm. Subsequently, our group optimized and improved the method in a simple and practical way [4–11]. In the year 2000, we developed a sperm cryopreservation protocol known as the Nakagata or CARD method [12].

Numerous genetically engineered mouse strains have been cryopreserved and archived as sperm in genetic resource banks [13, 14], and the Nakagata method is now widely used around

Chengyu Liu and Yubin Du (eds.), *Microinjection: Methods and Protocols*, Methods in Molecular Biology, vol. 1874, https://doi.org/10.1007/978-1-4939-8831-0_14, © Springer Science+Business Media, LLC, part of Springer Nature 2019

the world. The technique is simple, fast, and inexpensive, allowing sperm collected from a large number of males to be frozen within a short period of time. However, a drawback of this method was the instability of frozen-thawed sperm and the poor fertility (0–20%) of frozen-thawed sperm derived from C57BL/6 mice in conventional IVF [12, 15, 16]. The low fertility of frozen-thawed C57BL/6 sperm is a particularly serious problem because most genetically engineered mouse strains have a C57BL/6 background [17, 18]. This problem is a major barrier to the efficient use of archived sperm in genetic resource banks [19].

To overcome this problem, we modified the original protocol to improve the methods for sperm cryopreservation, sperm preincubation, and IVF [20–23]. We found that L-glutamine enhanced the protective effects of 18% raffinose pentahydrate combined with 3% skim milk (R18S3) [21]. L-Glutamine prevents membrane damage and motility loss after freezing and thawing in the sperm of several mammalian species [24–29]. In addition, methyl-β-cyclodextrin (MBCD) greatly enhances the fertility of frozen-thawed C57BL/6 sperm [20]. MBCD promotes plasma membrane cholesterol efflux, which induces sperm capacitation [30–32]. We also demonstrated that reduced glutathione (GSH) improves fertilization with frozen-thawed C57BL/6 sperm by breaking disulfide bonds in the zona pellucida, thereby loosening the network of glycoprotein filaments [22]. In addition, GSH protects against oxidative stress, which may have contributed to the increased fertility of the frozen-thawed sperm [33]. By combining these modifications, we were able to achieve a high and stable fertilization rate with IVF using frozen-thawed C57BL/6 sperm.

At the Center for Animal Resources and Development (CARD) at Kumamoto University, a major mouse genetic resources bank in Japan, we produce many embryos by IVF using frozen-thawed sperm of genetically engineered mice [14, 34]. We have achieved a stable fertilization rate exceeding 80% using frozen-thawed sperm from genetically engineered mice with the C57BL/6 background.

Sperm cryopreservation may facilitate the sharing of genetically engineered mouse strains. Researchers can locate genetically engineered mouse strains archived in genetic resource banks around the world by using the International Mouse Strain Resource database [35]; however, the transport of live animals is carefully regulated. Mice sometimes escape or die during transit and can spread infections. In addition, animal activist groups have made it more difficult to ship live animals [36]. Shipping cryopreserved sperm is one solution to this problem. However, the quality of the cryopreserved sperm is critical. Sharing information to improve mouse reproductive technology can help establish a global infrastructure for mouse genome science. To that end we have made available these optimized mouse reproductive technologies on our website: http://card.medic.kumamoto-u.ac.jp/card/english/sigen/manual/onlinemanual.html.

In this chapter, we introduce our improved protocols for mouse sperm cryopreservation and IVF using frozen-thawed sperm. Our technological improvements have greatly reduced the amount of wasted material, time, and effort and increased the efficiency of archiving genetically engineered mouse strains and recovering the mice by IVF.

2 Materials

2.1 Materials for Sperm Cryopreservation

2.1.1 Animals

Male mice (3–6 months old).

2.1.2 Equipment and Supplies

1. Sperm CPA (R18S3) containing L-glutamine (Table 1) [21].
2. Modified human tubal fluid (mHTF) (Table 2) [37, 38].
3. Paraffin liquid.
4. Filter paper.
5. Plastic culture dish, 35 mm × 10 mm.
6. Micropipettes.
7. Pipette tips.
8. Hot plate (set at 37 °C).
9. Pair of watchmaker's #5 forceps.
10. Fine scissors.
11. Micro-spring scissors (5 mm blade).
12. Polyethylene terephthalate glycol (PETG) cotton-plugged sperm straws, 0.25 mL.
13. Straw connector (Fig. 1a).
14. Impulse sealer.
15. Freezing canister (Fig. 1b).
16. Triangular cassette.
17. Liquid nitrogen tank.

2.2 Materials and Supplies for Sperm Thawing and Preincubation

1. Sperm preincubation medium: TYH with MBCD (Table 3) [20, 39].
2. Plastic culture dish, 35 mm × 10 mm.
3. Micropipettes.
4. Pipette tips.

Table 1
Composition of sperm cryoprotectant

Reagent	Amount per 10 mL solution (mg)	Amount per 20 mL solution (mg)
Raffinose pentahydrate	1800	3600
Skim milk	300	600
L-Glutamine	146	292

Table 2
Composition of modified Human Tubal Fluid (mHTF)

Reagents	Amount per 100 mL solution	Amount per 200 mL solution
NaCl, mg	593.8	1187.6
KCl, mg	35.0	70.0
$CaCl_2$, mg	57.0	114.0
Glucose, mg	50.0	100.0
$MgSO_4 \cdot 7H_2O$, mg	4.9	9.8
KH_2PO_4, mg	5.4	10.8
Sodium pyruvate, mg	3.5	7.0
60% sodium lactate, mL	0.34	0.68
$NaHCO_3$, mg	210.0	420.0
Penicillin G, mg	7.5	15.0
Streptomycin, mg	5.0	10.0
Bovine serum albumin, mg	400.0	800.0
0.5% phenol red, mL	0.04	0.08

5. Paraffin liquid.
6. Water bath.
7. Floating container (Fig. 1c).
8. Straw connector (Fig. 1a).
9. Humidified incubator (37 °C, 5% CO_2).

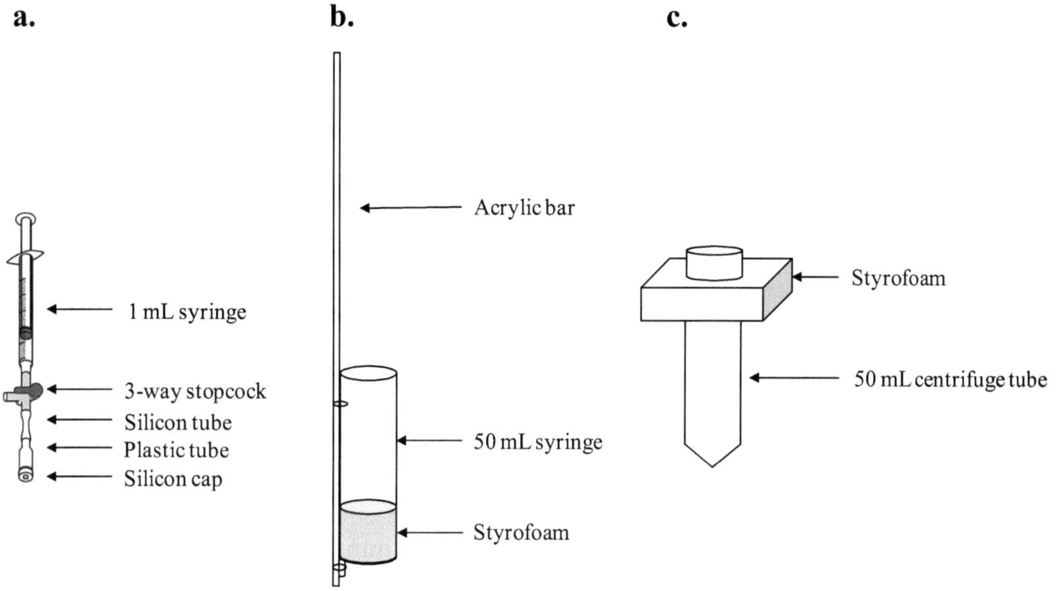

Fig. 1 Preparation of devices for sperm freezing and thawing. (**a**) To prepare the straw connector, connect the following devices in this order: 1.0 mL syringe, three-way stopcock, silicone tube (3 mm internal diameter, 5 mm external diameter), plastic tube, and silicone tip for bulb assembly. (**b**) To prepare the freezing canister: (1) insert a 30 mm thick piece of styrofoam tightly into the bottom of a 50 mL syringe, (2) heat-seal the outlet of the syringe tip, and (3) fix the syringe to an acrylic bar (5 mm × 5 mm × 500 mm). (**c**) To prepare the floating container, make a hole in a flat piece styrofoam and insert a 50 mL centrifuge tube

Table 3
Composition of TYH with methyl-β-cyclodextrin

Reagents	Amount per 100 mL (mg)	Amount per 200 mL (mg)
NaCl	697.6	1395.2
KCl	35.6	71.2
$CaCl_2 \cdot 2H_2O$	25.1	50.2
Glucose	100.0	200.0
$MgSO_4 \cdot 7H_2O$	29.3	58.6
KH_2PO_4	16.2	32.4
Sodium pyruvate	5.5	11.0
$NaHCO_3$	210.6	421.2
Penicillin G	7.5	15.0
Streptomycin	5.0	10.0
Polyvinyl alcohol	100.0	200.0
Methyl-β-cyclodextrin	98.3	196.6

2.3 Materials for In Vitro Fertilization Using Frozen-Thawed Sperm

2.3.1 Animals

Female mice (4 or 8–20 weeks old).

2.3.2 Equipment and Supplies

1. Pregnant mare serum gonadotropin (PMSG) in saline (37.5 IU/mL) or CARD Hyperova (Kyudo Co. Ltd., https://www.cosmobio.com/contents/hyperova.html, [40]).
2. Human chorionic gonadotropin (hCG) in saline (37.5 IU/mL).
3. Fertilization medium (mHTF with reduced GSH) [22].
4. Washing and culture medium (mHTF) [37].
5. Paraffin liquid.
6. Plastic dishes, 35 mm × 10 mm.
7. Pipette tips.
8. Micropipettes.
9. Humidified incubator (37 °C, 5% CO_2).
10. Filter paper.
11. Pair of watchmaker's #5 forceps.
12. Fine scissors.
13. Dissecting needle.
14. Pipette tip for insemination.

3 Methods

3.1 Methods for Sperm Cryopreservation

3.1.1 Preparation of Sperm CPA

1. Dissolve 3.6 g raffinose pentahydrate, 0.6 g skim milk, and 0.292 g L-glutamine into 20 mL distilled water at 60 °C in a 50 mL centrifuge tube (Table 1).
2. Incubate the tube in a water bath at 60 °C for 90 min, vortexing every 30 min.
3. Transfer 1.0 mL aliquots of the solution into 1.5 mL microcentrifuge tubes.
4. Centrifuge the solution at 10,000 × *g* for 60 min.
5. Carefully collect supernatant using a micropipette (*see* **Note 1**).
6. Filter the supernatant with a 0.22 μm disposable filter and use as sperm CPA.
7. Transfer the CPA into 1.0 mL ampules, fill the space above the liquid with nitrogen gas, and store at room temperature (*see* **Note 2**).

3.1.2 Sperm Cryopreservation

1. Place a 60 μL drop of CPA in a 35 mm culture dish and cover with paraffin liquid.
2. Place another 60 μL aliquot of CPA on top of the first drop (total 120 μL CPA/male) to make a tall drop (*see* **Note 3**).
3. Euthanize a 3- to 6-month-old male mouse by cervical dislocation and aseptically remove the two cauda epididymides (Fig. 2a).
4. Place the cauda epididymides on sterile filter paper. Completely remove adipose tissue and blood under a microscope using a pair of watchmaker's #5 forceps and fine scissors.
5. Transfer the cauda epididymides into the CPA drop (Fig. 2b). Make five or six deep cuts in the epididymides using a micro-spring scissors under the microscope (Fig. 2c).
6. Place the dish on a hot plate at 37 °C for 3 min. Gently swirl the dish every minute to uniformly disperse the sperm in the CPA (Fig. 2d).
7. During equilibration of the sperm in the CPA, attach the straw connector (Fig. 1a) to the end of the straw. Carefully aspirate 100 μL mHTF and 15 mm air into the straw (Fig. 3a). Ten

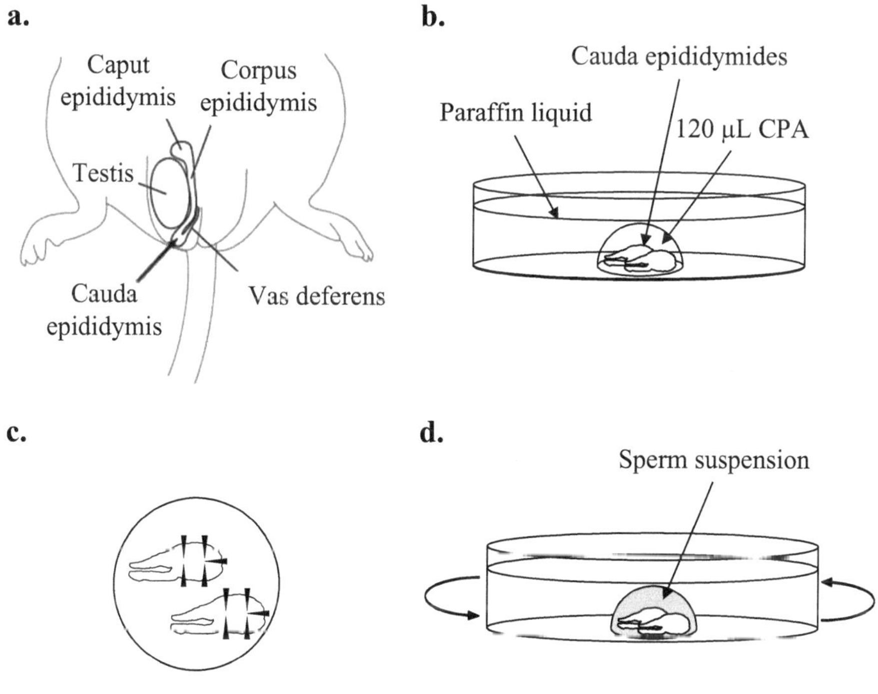

Fig. 2 Preparation of the sperm suspension. (**a**) Collection of cauda epididymides. (**b**) Introduction of cauda epididymides in 120 μL CPA drop covered with paraffin oil. (**c**) Making five or six deep cuts in the epididymides using a micro-spring scissors under the microscope. (**d**) Equilibrium of the sperm on a hot plate at 37 °C for 3 min (swirl the dish every minute)

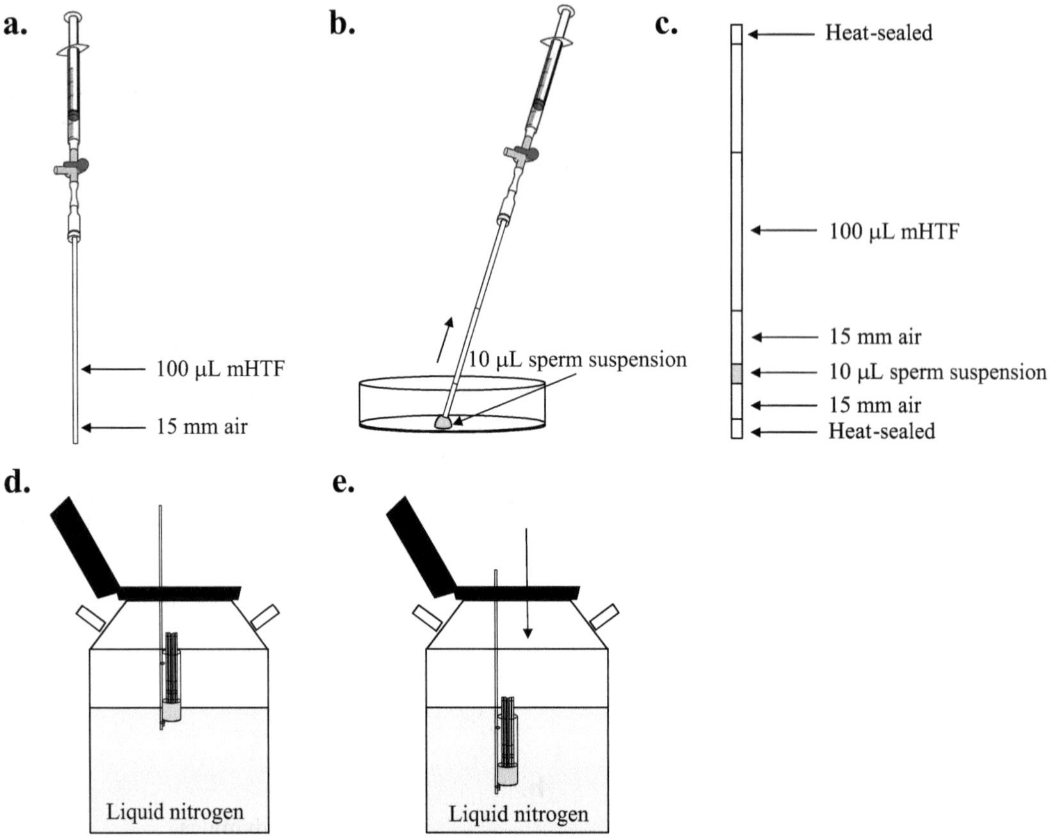

Fig. 3 Sperm cryopreservation. (**a**) Attachment of sperm freezing straw with the straw connector and loading 100 μL mHTF and 15 mm air into the straw. (**b**) Collection of a 10 μL drop of sperm and 15 mm air into the straw. (**c**) Sealing both ends of the straw using the impulse sealer. (**d**) Floating the canister on the surface of the liquid nitrogen in a liquid nitrogen tank for 10 min. (**e**) Immersion of the freezing canister into the liquid nitrogen

straws can be prepared from the two epididymides of a single male.

8. Using a micropipette, place ten 10 μL aliquots of sperm in a plastic culture.
9. Aspirate one 10 μL drop of sperm and 15 mm air into each of the ten straws dish (Fig. 3b). Carefully seal both ends of the straw using the impulse sealer (Fig. 3c) (*see* **Note 4**).
10. Place the sealed straws into the freezing canister (Fig. 1b) and float the canister on the surface of the liquid nitrogen in a liquid nitrogen tank for 10 min (Fig. 3d).
11. After 10 min, completely immerse the freezing canister into the liquid nitrogen (Fig. 3e).

12. Transfer the straws to a triangular cassette that was precooled in liquid nitrogen and then immerse the cassette in the tank for long-term storage (*see* **Note 5**).

3.2 Methods for Sperm Thawing and Preincubation

1. Place a 90 μL drop of TYH with MBCD in a 35 mm culture dish and cover it with paraffin liquid (sperm preincubation drop). Equilibrate the medium in an incubator (37 °C, 5% CO_2) for 30 min before use.
2. Remove a frozen straw of sperm from the liquid nitrogen tank and hold it in the air for 5 s.
3. Place the frozen straw in the floating container (Fig. 1c) and incubate in a 37 °C water bath for 10 min to thaw (Fig. 4a).
4. Remove the straw from the water bath and dry with a paper towel.
5. Cut the straw between the mHTF medium and the seal. Then insert the straw into the straw connector and turn the stopcock to release internal pressure (Fig. 4b). Finally, cut the sealed end of straw nearest from sperm.
6. Push the plunger of the straw connector and transfer only the sperm suspension (10 μL) to the 90 μL drop of TYH with MBCD in the culture dish (Fig. 4c). Place the dish in a humidified incubator (37 °C, 5% CO_2) for 30 min.

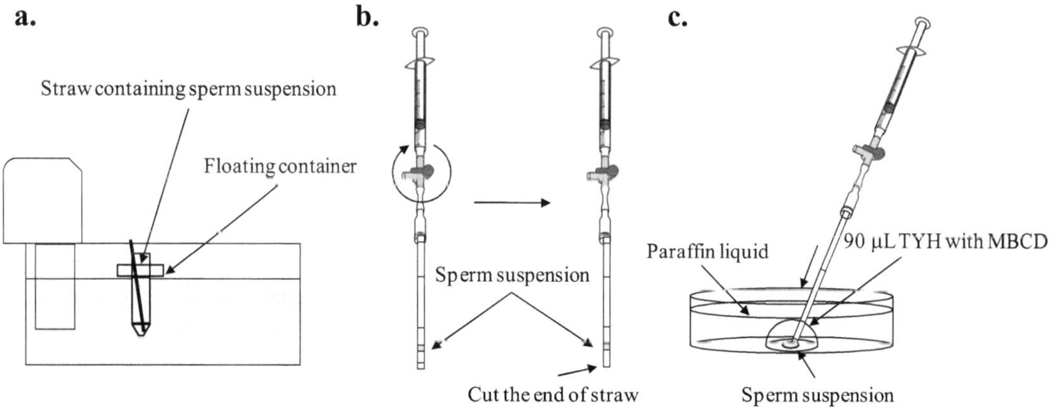

Fig. 4 Sperm thawing and preincubation. (**a**) Incubation of the frozen straw in the floating container for 10 min at 37 °C. (**b**) Attachment of the straw with the straw connector and releasing internal pressure of the straw by turning the stopcock. (**c**) Introduction of the sperm suspension (10 μL) to the 90 μL drop of TYH with MBCD in the culture dish

3.3 Methods for In Vitro Fertilization Using Frozen-Thawed Sperm

3.3.1 Preparation of Hormones and Superovulation

1. Dissolve 3000 IU PMSG in 80 mL saline, and 3000 IU hCG in 80 mL saline. Store the solutions of PMSG and hCG at 4 °C or −20 °C (*see* **Note 6**).
2. Three days before IVF, inject 7.5 IU PMSG (0.2 mL) or CARD HyperOva (0.2 mL) into the abdominal cavity of mature female mice between 4 and 6 PM.
3. One day before IVF, inject 7.5 IU hCG (0.2 mL) into the abdominal cavity of the female mice at between 4 and 6 PM.

3.3.2 In Vitro Fertilization

1. The morning before IVF, dissolve 30.7 mg reduced GSH in 1.0 mL mHTF (100 mM GSH) in a 1.5 mL microcentrifuge tube and gently mix the solution.
2. Dilute the medium 100-fold with mHTF so that the final concentration of GSH is 1.0 mM (e.g., mix 10 μL mHTF with 100 mM GSH and 990 μL mHTF).
3. Place a 90 μL drop of HTF with 1.0 mM GSH in a plastic culture dish and cover with paraffin liquid (fertilization drop). Equilibrate the medium in a humidified incubator (37 °C, 5% CO_2) for 10 min.
4. Place four drops of mHTF in a plastic culture dish and cover with paraffin liquid (washing drops). Incubate this washing dish in a humidified incubator with (37 °C, 5% CO_2) for at least 30 min.
5. Fifteen to seventeen hours after the hCG injection, euthanize the superovulated mice and quickly remove oviducts without adipose tissue.
6. Place the oviducts on sterile filter paper. Remove any blood or fluid from the tissue under a microscope.
7. Immerse the oviducts in the paraffin liquid contained in the culture dish with the fertilization drop (mHTF with 1.0 mM GSH).
8. Under a microscope, hold each oviduct with forceps and gently tear the swollen ampulla of the oviduct with a dissecting needle; after opening the swollen ampulla, wait for the cumulus-oocyte complexes (COCs) to be released into the paraffin liquid (Fig. 5a).
9. Drag the COCs into the drop of fertilization medium using the dissecting needle (PMSG-hCG: 4–6 COCs/drop or CARD HyperOva-hCG: one COCs/drop) (*see* **Note 7**). Place the dish in a humidified incubator (37 °C, 5% CO_2) and incubate for 30–60 min before insemination (*see* **Note 8**).
10. After sperm preincubation in the incubator, aspirate 10 μL motile sperm from the edge of the drop (TYH with MBCD using a pipette tip for insemination (Fig. 5b)) (*see* **Note 9**).

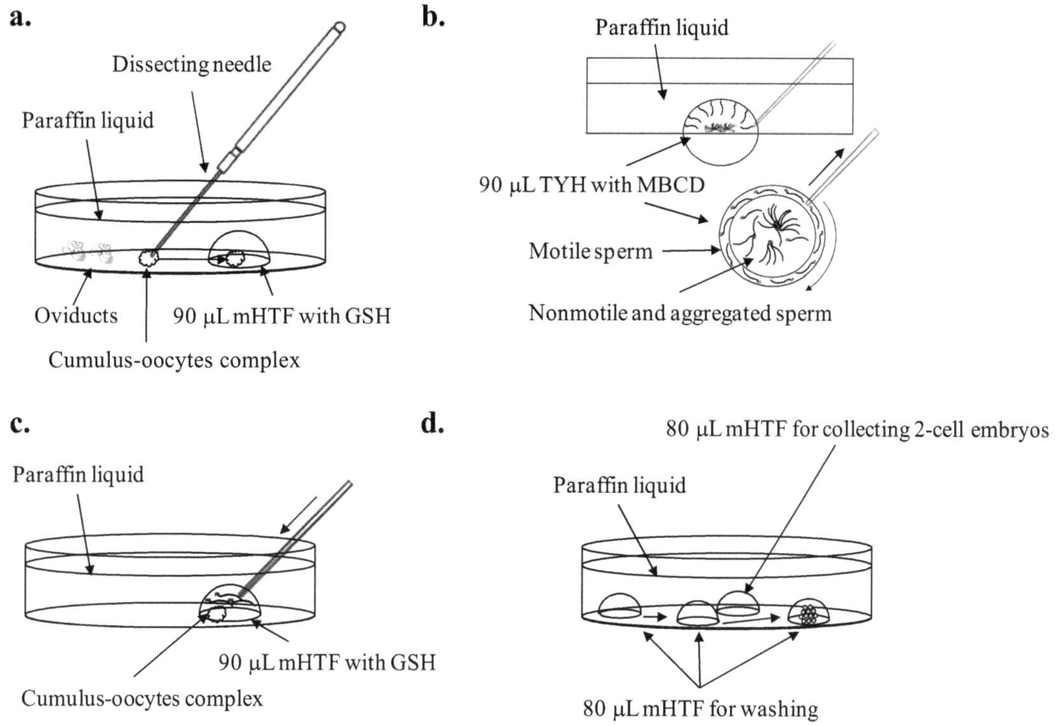

Fig. 5 In vitro fertilization using frozen-thawed sperm. (**a**) Introduction of cumulus-oocytes complexes into a 90 μL drop of mHTF with 1.0 mM GSH. (**b**) Collection of a 10 μL aliquot of sperm suspension from the edge of the drop (TYH with MBCD) using a pipette tip for insemination. (**c**) Addition of the sperm suspension to the fertilization drop (mHTF with 1.0 mM GSH) containing COCs. (**d**) After incubating for 3 h, washing the oocytes three times in the 80 μL drops of fresh mHTF in the washing dish

11. Add the sperm suspension to the fertilization drop (mHTF with 1.0 mM GSH) containing COCs (Fig. 5c). Place the culture dish in a humidified incubator (37 °C, 5% CO_2).
12. After incubating for 3 h, wash the oocytes three times in the 80 μL drops of fresh mHTF in the washing dish (Fig. 5d) (*see* **Note 10**).
13. Six hours after insemination, observe oocytes in the third drop of mHTF and remove any parthenogenetic oocytes that have only one pronucleus.
14. After overnight culture of the oocytes, transfer only the 2-cell embryos to the fourth drop of mHTF in the washing dish. These embryos are ready for embryo cryopreservation or embryo transfer into pseudopregnant females.

4 Notes

1. CPA preparation: if the supernatant is not clear, centrifuge until the solution becomes clear.
2. Once prepared use it within 3 months.
3. The tall drop, cylindrical, results in a more uniform sperm suspension and reduces the amount of residual sperm remaining in the dish after collecting the sperm.
4. Confirm that both ends of the straw are well sealed to prevent liquid nitrogen from entering during storage, which could result in the straw exploding during warming.
5. To assess the motility of the collected sperm, transfer a 1.0 μL aliquot of sperm suspension from the CPA drop into a 100 μL drop of mHTF covered with paraffin liquid in a plastic culture dish. Incubate the sperm in a CO_2 incubator for 10 min before observing sperm motility under the microscope.
6. Use the hormones within 3 weeks of its preparation.
7. When using CARD HyperOva-hCG use only one ampulla per fertilization drop.
8. Perform all operations—from sacrificing the female mice to introducing the COCs into the fertilization medium—in the shortest time possible, preferably within 30 s. If carrying out this process alone, sacrifice one mouse at a time and swiftly remove its oviducts before moving on to the next mouse.
9. Carefully collect motile sperm from the periphery of the drop. It is possible to collect four 10 μL aliquots of the sperm suspension per drop.
10. If many spermatozoa are attached to the zona pellucida of the oocytes, they can be removed by gently pipetting up and down (using a 20 μL micropipette and a tip) 20–30 times in the same fertilization drop before washing.

Acknowledgments

The authors thank our CARD research group members Kiyoko Fukumoto, Yukie Haruguchi, Tomoko Kondo, Yuko Nakamuta, Yumi Takeshita, Hiroko Matsunaga, Manami Nishimura, Tomoko Umeno, Eri Ishida, Fumi Takahashi, Mari Iwamoto, Eri Kohagura Tomoko Matsumoto and Shuuji Tsuchiyama, my students: Aki Tsutsumi, Taichi Omaru, Kazuhito Sakoh, Yuta Ishizuka, Satohiro Nakao, Hidetaka Yoshimoto, Yuki Sakai, and Yuka Horikoshi, Ayumi Mukunoki for their technical assistance and helpful discussions.

References

1. Yokoyama M, Akiba H, Katsuki M, Nomura T (1990) Production of normal young following transfer of mouse embryos obtained by in vitro fertilization using cryopreserved spermatozoa. Jikken Dobutsu 39:125–128
2. Tada N, Sato M, Yamanoi J, Mizorogi T, Kasai K, Ogawa S (1990) Cryopreservation of mouse spermatozoa in the presence of raffinose and glycerol. J Reprod Fertil 89:511–516
3. Okuyama M, Isogai S, Saga M, Hamada H, Ogawa S (1990) In vitro fertilization (IVF) and artificial insemination (AI) by cryopreserved spermatozoa in mouse. J Fertil Implant 7:116–119
4. Takeshima T, Nakagata N, Ogawa S (1991) Cryopreservation of mouse spermatozoa. Jikken Dobutsu 40:493–497
5. Nakagata N, Takeshima T (1992) High fertilizing ability of mouse spermatozoa diluted slowly after cryoprservation. Theriogenology 37:1283–1291
6. Nakagata N (1992) Production of normal young following insemination of frozen-thawed mouse spermatozoa into fallopian tubes of pseudopregnant females. Jikken Dobutsu 41:519–522
7. Nakagata N, Matsumoto K, Anzai M, Takahashi A, Takahashi Y, Matsuzaki Y, Miyata K (1992) Cryopreservation of spermatozoa of a transgenic mouse. Jikken Dobutsu 41:537–540
8. Nakagata N, Takeshima T (1993) Cryopreservation of mouse spermatozoa from inbred and F1 hybrid strains. Jikken Dobutsu 42:317–320
9. Nakagata N (1993) Production of normal young following transfer of mouse embryos obtained by in vitro fertilization between cryopreserved gametes. J Reprod Fertil 99:77–80
10. Nakagata N, Ueda S, Yamanouchi K, Okamoto M, Matsuda Y, Tsuchiya K, Nishimura M, Oda S, Koyasu K, Azuma S, Toyoda Y (1995) Cryopreservation of wild mouse spermatozoa. Theriogenology 43:635–643
11. Nakagata N, Okamoto M, Ueda O, Suzuki H (1997) Positive effect of partial zona-pellucida dissection on the in vitro fertilizing capacity of cryopreserved C57BL/6J transgenic mouse spermatozoa of low motility. Biol Reprod 57:1050–1055
12. Nakagata N (2000) Cryopreservation of mouse spermatozoa. Mamm Genome 11:572–576
13. Davisson M (2006) FIMRe: Federation of International Mouse Resources: global networking of resource centers. Mamm Genome 17:363–364
14. Donahue L, Angelis M, Hagn M, Franklin C, Lloyd K, Magnuson T, McKerlie C, Nakagata N, Obata Y, Read S, Wurst W, Hörlein A et al (2012) Centralized mouse repositories. Mamm Genome 23(0):559–571
15. Nishizono H, Shioda M, Takeo T, Irie T, Nakagata N (2004) Decrease of fertilizing ability of mouse spermatozoa after freezing and thawing is related to cellular injury. Biol Reprod 71:973–978
16. Songsasen N, Leibo SP (1997) Cryopreservation of mouse spermatozoa. II. Relationship between survival after cryopreservation and osmotic tolerance of spermatozoa from three strains of mice. Cryobiology 35:255–269
17. Sztein JM, Farley JS, Mobraaten LE (2000) In vitro fertilization with cryopreserved inbred mouse sperm. Biol Reprod 63:1774–1780
18. Ostermeier GC, Wiles MV, Farley JS, Taft RA (2008) Conserving, distributing and managing genetically modified mouse lines by sperm cryopreservation. PLoS One 3:e2792
19. Fray MD (2009) Biological methods for archiving and maintaining mutant laboratory mice. Part I: Conserving mutant strains. Methods Mol Biol 561:301–319
20. Takeo T, Hoshii T, Kondo Y, Toyodome H, Arima H, Yamamura K, Irie T, Nakagata N (2008) Methyl-beta-cyclodextrin improves fertilizing ability of C57BL/6 mouse sperm after freezing and thawing by facilitating cholesterol efflux from the cells. Biol Reprod 78:546–551
21. Takeo T, Nakagata N (2010) Combination medium of cryoprotective agents containing L-glutamine and methyl-{beta}-cyclodextrin in a preincubation medium yields a high fertilization rate for cryopreserved C57BL/6J mouse sperm. Lab Anim 44(2):132–137
22. Takeo T, Nakagata N (2011) Reduced glutathione enhances fertility of frozen/thawed C57BL/6 mouse sperm after exposure to methyl-beta-cyclodextrin. Biol Reprod 85 (5):1066–1072
23. Nakagata N, Takeo T, Fukumoto K et al (2014) Rescue in vitro fertilization method for legacy stock of frozen mouse sperm. J Reprod Dev 60:168–171
24. Renard P, Grizard G, Griveau JF, Sion B, Boucher D, Le Lannou D (1996) Improvement of motility and fertilization potential of postthaw human sperm using glutamine. Cryobiology 33:311–319

25. Trimeche A, Yvon JM, Vidament M, Palmer E, Magistrini M (1999) Effects of glutamine, proline, histidine and betaine on post-thaw motility of stallion spermatozoa. Theriogenology 52:181–191
26. Kundu CN, Das K, Majumder GC (2001) Effect of amino acids on goat cauda epididymal sperm cryopreservation using a chemically defined model system. Cryobiology 42:21–27
27. Khlifaoui M, Battut I, Bruyas JF, Chatagnon G, Trimeche A, Tainturier D (2005) Effects of glutamine on post-thaw motility of stallion spermatozoa: an approach of the mechanism of action at spermatozoa level. Theriogenology 63:138–149
28. Al Ahmad MZ, Chatagnon G, Amirat-Briand-L, Moussa M, Tainturier D, Anton M, Fieni F (2008) Use of glutamine and low density lipoproteins isolated from egg yolk to improve buck semen freezing. Reprod Domest Anim 43:429–436
29. Amirat-Briand L, Bencharif D, Vera-Munoz O, Bel Hadj Ali H, Destrumelle S, Desherces S, Schmidt E, Anton M, Tainturier D (2009) Effect of glutamine on post-thaw motility of bull spermatozoa after association with LDL (low density lipoproteins) extender: preliminary results. Theriogenology 71:1209–1214
30. Ohtani Y, Irie T, Uekama K, Fukunaga K, Pitha J (1989) Differential effects of alpha-, beta- and gamma-cyclodextrins on human erythrocytes. Eur J Biochem 186:17–22
31. Choi YH, Toyoda Y (1998) Cyclodextrin removes cholesterol from mouse sperm and induces capacitation in a protein-free medium. Biol Reprod 59:1328–1333
32. Visconti PE, Galantino-Homer H, Ning X, Moore GD, Valenzuela JP, Jorgez CJ, Alvarez JG, Kopf GS (1999) Cholesterol efflux-mediated signal transduction in mammalian sperm: beta-cyclodextrins initiate transmembrane signaling leading to an increase in protein tyrosine phosphorylation and capacitation. J Biol Chem 274:3235–3242
33. Bath ML (2010) Inhibition of in vitro fertilizing capacity of cryopreserved mouse sperm by factors released by damaged sperm, and stimulation by glutathione. PLoS One 5(2):e9387
34. Nakagata N, Yamamura K (2009) Current activities of CARD as an international core center for mouse resources. Exp Anim 58:343–350
35. Eppig JT, Strivens M (1999) Finding a mouse: the International Mouse Strain Resource (IMSR). Trends Genet 15:81–82
36. Wadman M (2012) Lab-animal flights squeezed. Nature 489:344–345
37. Quinn P, Kerin JF, Warnes GM (1985) Improved pregnancy rate in human in vitro fertilization with the use of a medium based on the composition of human tubal fluid. Fertil Steril 44:493–498
38. Nakagata N (2011) Cryopreservation of mouse spermatozoa and in vitro fertilization. Methods Mol Biol 693:57–73
39. Toyoda Y, Yokoyama M, Hoshi T (1971) Studies on the fertilization of mouse eggs in vitro: I. In vitro fertilization of eggs by fresh epididymal sperm. Jpn J Anim Reprod 16:147–151
40. Takeo T, Nakagata N (2015) Superovulation using the combined administration of inhibin antiserum and equine chorionic gonadotropin increases the number of ovulated oocytes in C57BL/6 female mice. PLoS One 10: e0128330

Chapter 15

Isolation and Analysis of a Genome-Edited Single-Hepatocyte from a Cas9 Transgenic Mouse Line

Takayuki Sakurai, Akiko Kamiyoshi, Masato Ohtsuka, Channabasavaiah B. Gurumurthy, Masahiro Sato, and Takayuki Shindo

Abstract

The primary cells isolated from the freshly dissected organ are thought to be different from those cultured for a long time in vitro. For instance, hepatocytes isolated in situ from the liver, display the ability to produce albumin, cultured for about a week often tend to cease production of albumin, including loss of proliferation capability. Thus, it is difficult to perform genome editing (i.e., production of genome-edited hepatocytes by in vitro gene delivery) in such cultured cells. Furthermore, hepatic cell lines available so far do not produce albumin and they would also have lost several characteristics of native liver cells. This poses a serious disadvantage when researchers want to study gene expression profiles under specific experimental settings, for example before and after genome editing. However, this demerit can be overcome if genome-editing is performed in situ in liver and single hepatocytes (both genome-edited and wild-type) can be isolated for analysis immediately following transient gene editing. Previously, we demonstrated successful isolation of genome-edited single hepatocytes, using mice expressing systemic Cas9 transgene (called "sCAT" mouse) and by tail-vein-mediated hydrodynamics-based gene delivery of gRNA targeted to *Albumin* gene (Sakurai et al., Sci Rep 6:20011, 2016). Here, we describe the detailed protocols for collection and analysis of single genome-edited hepatocytes, which will be useful for many types of hepatocyte functional studies.

Key words CRISPR/Cas9, Transgenic mouse, Hepatocyte, In vivo genome editing, gRNA, Single-cell isolation, T7 endonuclease I assay, Whole genome amplification, Surveyor assay, Hydrodynamics-mediated gene delivery

1 Introduction

Targeted genome-editing technology is of great interest for biomedical research, including its potential for clinical applications to treat or improve symptoms of some genetic diseases. In particular, CRISPR/Cas9-based genome editing enables rapid genetic manipulation of any genomic locus without the need for cumbersome gene-targeting via homologous recombination. CRISPR/Cas9 system comprises a Cas9 endonuclease and guide RNA (gRNA), which is composed of 20 nucleotides complementary to the

Chengyu Liu and Yubin Du (eds.), *Microinjection: Methods and Protocols*, Methods in Molecular Biology, vol. 1874, https://doi.org/10.1007/978-1-4939-8831-0_15,

genomic site(s) of interest, and an NGG protospacer-adjacent motif (PAM) that directs Cas9 to induce double-stranded breaks (DSBs) in target DNA [1]. These Cas9-mediated DSBs are repaired through either non-homologous end joining (NHEJ) or homologous recombination (HR). Although NHEJ results in an insertion and deletion (indel) mutation and may cause frameshift in the wild-type reading frame, HR-mediated repair may lead to the incorporation of a homologous fragment when it is available near the DSB sites [2, 3].

Presently, several laboratories have created transgenic (Tg) mice expressing Cas9 under a tissue-specific or ubiquitous promoter system [4–11]. With Tg mice, one can induce mutations in the desired tissue/cells in vivo using relatively simplified methods such as tail vein-mediated delivery of gRNA expression vectors, direct injection of gRNA-carrying viral vectors via needle or micropipette, and in vivo electroporation followed by injection of gRNA into tissues. For instance, Platt et al. [4] created conditional expressing Cas9 line through the knockin of Cas9 expression unit into *Rosa26* locus at chromosome 6. With this line, these authors demonstrated the generation of tumors in multiple tissues upon systemic administration of viral DNA-containing *Kras*, *Trp53 (p53)*, and *Stk11* gRNAs via tail vein. Thus, the expression of the endogenous tumor suppressor *Trp53* and *Stk11* genes is blocked following in vivo transfection of cells with gRNAs for tumorigenesis-related genes. Genome-edited cells consequently exhibit continuous cell proliferation owing to the loss of anti-cancer activity, thereby progressing toward tumorigenesis. Since the study conducted by Platt et al. [4], several groups have reported the successful in vivo induction of mutations in target tissues/cells using Cas9 Tg animals through the introduction of a solution containing gRNAs alone [5–7].

The most important concern with the use of Cas9 Tg animals is if these mice are phenotypically normal and fertile; in other words, whether Cas9 overexpression affects the normal cellular functions in tissues. Platt et al. [4] showed that the mice produced were quite normal and displayed no abnormalities. Independent of this report, we created a mouse line expressing multiple copies *s*ystemic *Ca*s9 *t*ransgene (hereafter referred to as "sCAT") and demonstrated that these Cas9 mice are normal and fertile (Fig. 1, Fig. S4, Table S1, and Table S2 in Sakurai et al. [6]).

Isolated hepatocytes offer a very useful approach for studying liver function under an in vitro setting. We successfully demonstrated isolation of single fluorescent genome-edited hepatocytes from sCAT mouse that was administered, via hydrodynamics-based gene delivery approach, with a solution containing gRNA alone and an enhanced green fluorescent protein (EGFP) expression vector (used as a reporter showing successful gene delivery)

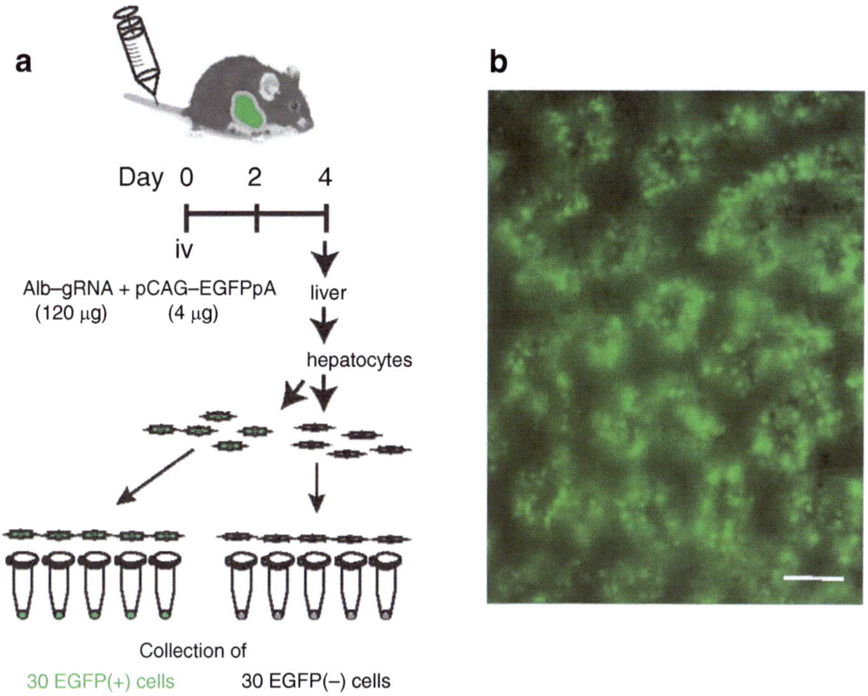

Fig. 1 Schematic of the experimental methods for single Cas9/+ hepatocyte isolation after transfection with *Alb*-gRNA alone by hydrodynamic-based gene delivery. (**a**) Hydrodynamic-based gene delivery of Cas9/+ Tg female mouse. Four days later, the mouse was subjected to perfusion using a solution containing collagenase I under anesthesia, as described in Subheading 3.3. The mouse was sacrificed and the liver dissected. The dissected liver was dissociated into single cells by pipetting, followed by mincing using scalpels under anaerobic condition. Isolation of fluorescent and non-fluorescent hepatocytes was performed using the standard microinjection system attached to Olympus IX70 inverted fluorescence microscope. Each single cell was transferred to a tube for molecular biological analysis. (**b**) Fluorescence observation of liver dissected 4 days after gene delivery showing several fluorescent foci comprising EGFP(+) hepatocytes were observed. Bar = 200 μm

[6]. This can be a very valuable method for studying gene function in hepatocytes before and after genome editing.

Here, we describe a detailed protocol for the isolation of genome-edited hepatocytes from the transiently genome-edited sCAT mouse and subsequent molecular biological analysis as to whether the isolated fluorescent cells contain indel mutations at the target locus.

2 Materials

2.1 Surgical Instruments

All surgical instruments were sterilized by immersing in 70% ethanol solution.

1. No. 5 INOX Forceps.
2. Aorta-Klemme (40 mm length).

3. Scissors.
4. Suture wound clips.
5. Micro scissors.
6. Disposable scalpel.

2.2 Equipment

1. Olympus SZ11 dissecting microscope.
2. Olympus SZX12 stereomicroscope.
3. Olympus KL1600 LED Stereo microscope lighting.
4. Narishige-Olympus MMO-202ND microinjection system.
5. Microinjection pipette: This is the same apparatus used for DNA microinjection into zygotes in a typical transgenic mouse. The glass capillary (ID.0.6 mm/OD 1 mm) was pulled by Sutter Instruments P-97/IVF micropipette puller to make a pointed end. The preparation of microinjection needles was performed as described by Nagy et al. [12] with some modifications in the needle-pulling settings: heat, 790; pull, 35; velocity, 50; pressure, 500; time, 100.
6. Micropipette for embryo handling used to transfer embryos from one dish to another. We used handmade finely drawn pipettes from microdispenser capillaries (ID.0.93 mm/OD 1.37 mm) by pulling them in opposite directions after flaming at the center.
7. Micropipette for the isolation of single cells: This is the same as the micropipettes used for embryo handling, except the inner diameter was narrower than the embryo handling micropipette.
8. Thin-walled polymerase chain reaction (PCR) tube (0.2 mL) with dome cap.
9. 60 mm tissue culture dish.
10. 100 mm tissue culture dish.
11. 2.5 mL luer lock type syringe.
12. 27-Gauge needle.
13. ABI GeneAmp System 9700 PCR thermal cycler.
14. ABI Genetic Analyzer 3130 DNA sequencer.
15. Genetyx-Mac (ver.13) sequence analysis software.
16. Olympus IX70 inverted fluorescence microscope with U-MWIBA2 filter set.
17. Peristaltic perfusion pump.
18. 26-Gauge needle.
19. Polyethylene tube (ID.0.5 mm/OD 0.8 mm).

20. Handmade 26-gauge extension tube: Feeding tube used a perfusion tubing (length 50 cm, ID. 2.1 mm) connected to the 26-gauge needle.
21. Terufusion three-way stopcock.
22. Cell strainer (40 μm pore size).
23. Water bath.
24. 50 mL conical tube.

2.3 Reagents

1. 70% ethanol.
2. Equine chorionic gonadotropin (eCG).
3. Human chorionic gonadotropin (hCG).
4. Dulbecco's phosphate-buffered saline (DPBS) without Ca^{2+} and Mg^{2+}.
5. Fetal bovine serum (FBS).
6. Collagenase I.
7. Trypsin inhibitor.
8. Ethylene glycol tetraacetic acid (EGTA).
9. Hank's balanced salt solution (HBSS) without Ca^{2+} and Mg^{2+}.
10. Percoll.
11. Three types of mixed anesthetic agents (medetomidine, midazolam, and butorphanol) [13].
12. Lysis buffer (125 μg/mL proteinase K, 100 mM Tris–HCl [pH 8.3], 100 mM potassium chloride [KCl], 0.02% gelatin, 0.45% Tween-20, and 60 μg/mL yeast tRNA) for the isolation of genomic DNA.
13. Takara Tks Gflex DNA polymerase.
14. Ambion MEGAshortscript T7 RNA polymerase.
15. pCAG-NFCas9 plasmid carrying NFCas9 gene (wherein SV40-derived nuclear location signal [NLS] and FLAG sequences or FLAG sequence alone were placed immediately after ATG codon of humanized [h] Cas9 sequence) under the ubiquitous promoter CAG (cytomegalovirus enhancer + chicken β-actin-based promoter) [6].
16. *pAlb*-gRNA plasmid carrying T7 promoter and gRNA targeted to *Albumin* (*Alb*) gene for preparing *Alb*-gRNA [6].
17. pCAG-EGFPpA carrying EGFP cDNA under CAG promoter system and used as a reference gene to monitor the co-introduction of *Alb*-gRNA.
18. Agarose.
19. QIAGEN QIAquick gel extraction spin columns.
20. Mirus TransIT-QR hydrodynamic delivery solution.

21. QIAGEN REPLI-g phage Φ29 DNA polymerase.
22. TAE buffer.
23. Ethidium bromide.
24. T7 endonuclease I (T7EI).
25. Perfusion buffer (HBSS + 0.5 mM EGTA) warmed for 30 min in the water bath at 37 °C before perfusion.
26. Collagenase I solution (HBSS + collagenase I [0.65 mg/mL] + trypsin inhibitor [0.5 mg/mL] + 5.4 mM calcium chloride [$CaCl_2$]) warmed for 30 min in the water bath at 37 °C before perfusion.
27. Percoll solution (×2) (13.5 mL of Percoll + 1.5 mL of 10× HBSS) cooled at 4 °C before use.
28. ABI BigDye Terminator v.3.1 Cycle Sequencing.

2.4 Mice

We purchased ICR females (aged 8–15 weeks) and males (aged 10–20 weeks), C57BL/6J mice (hereafter referred to as B6; aged 10–20 weeks), and BDF1 females (hybrid between C57BL/6N and DBA/2N; aged 8–15 weeks). Cas9 Tg mice (sCAT) were in-house produced (*see* Subheading 3.1).

3 Methods

3.1 Production of sCAT by General Pronuclear Injection-Based Method

The sCAT mouse was produced via standard pronuclear microinjection according to the method described by Nagy et al. [12], and is genotyped using the primer set CAG1620-S (5′-GCTCTAGAGCCTCTGCTAACCATGT-3′)/CAS9215-A (5′-CGCCGTGCTGTTCTTTTGAGC-3′), which corresponds to the coding region of hCas9 gene (*see* **Note 1**).

1. Excise the CAG-NFCas9 expression cassette, by digesting pCAG-NFCas9 plasmid with *Not* I, which cuts the two *Not* I sites located at the 5′ end of the cytomegalovirus enhancer and the 3′ end of the poly(A) sites of pCAG-NFCas9.
2. Size-fractionate, in 0.8% agarose gels, the *Not* I-digested samples, and isolate the fragment containing CAG-NFCas9. The isolated fragment is purified by QIAGEN QIAquick gel extraction spin column.
3. Prior to microinjection, the purified DNA fragment is dissolved in purified water at a concentration of 2 ng/μL.
4. Collect zygotes using a standard superovulation and in vitro fertilization (IVF) protocol by fertilizing BDF1-derived oocytes with B6-derived spermatozoa. Zygotes were subjected to pronuclear microinjection of the DNA fragment containing CAG-NFCas9 (2 ng/μL) by using Narishige-Olympus microinjection system.

5. Microinjected zygotes were transferred into the oviducts of pseudopregnant ICR surrogate females (on Day 0.7 of gestation), according to the method described by Nagy et al. [12]. Day 0 of pregnancy was designated as the day when plug was recognized. Pseudopregnant ICR surrogate females were prepared by mating ICR females with vasectomized ICR males 1 day before egg transfer. The method for making vasectomized ICR males was previously described by Nagy et al. [12].
6. Founder (F0) Cas9 Tg mice and their progenies were identified with PCR using the primer set CAG1620-S/CAS9215-A, corresponding to the coding region of hCas9 gene.
7. After obtaining F0 Tg mice, the offspring was first crossed with B6 mice to obtain F1 pups. F1 pups that were identified as heterozygous Tg mice (defined hereafter as Cas9/+ mice) were mated with B6 mice to obtain G2 pups. Female Cas9/+ mice at 2–3 generations (aged 10–15 weeks) were usually used for the gene delivery to hepatocytes (*see* **Note 2**).

3.2 Hydrodynamic gRNA Delivery into Cas9/+ Tg Mice

Hydrodynamic-based transfection can be performed using Mirus TransIT-QR hydrodynamic delivery solution, according to the method of Liu et al. [14].

1. The schematic procedure for hydrodynamic gRNA delivery is shown in Fig. 1a.
2. Prepare *Alb*-gRNA from an *EcoR* I-linearized *pAlb*-gRNA plasmid using Ambion MEGAshortscript T7 polymerase.
3. Dissolve *Alb*-gRNA (120 μg) and pCAG-EGFPpA (4 μg) in Mirus TransIT-QR HD solution equivalent to one-tenth of the mouse weight (for example, for a mouse weighing 30 g).
4. Anesthetize Cas9/+ females (Experimental group) or non-Tg females (Control group) [13], and administer the guide RNA solution using a 2.5 mL luer lock type syringe fitted with a 27-gauge needle at a constant speed via the tail vein, and the procedure completed within 10 s (*see* **Note 3**). If this transfection is successful, the liver exhibits green fluorescence over its surface, as shown in Fig. 1b.

3.3 Enzymatic Dissociation of Hepatocytes

This method is based on the two-step collagenase perfusion technique, which involves the sequential perfusion of liver with perfusion buffer and collagenase I solution, followed by centrifugation with 45% Percoll solution [15].

1. The handmade 26-gauge extension tube (Fig. 2a) was connected to the peristaltic perfusion pump via perfusion tubing (Fig. 2b).

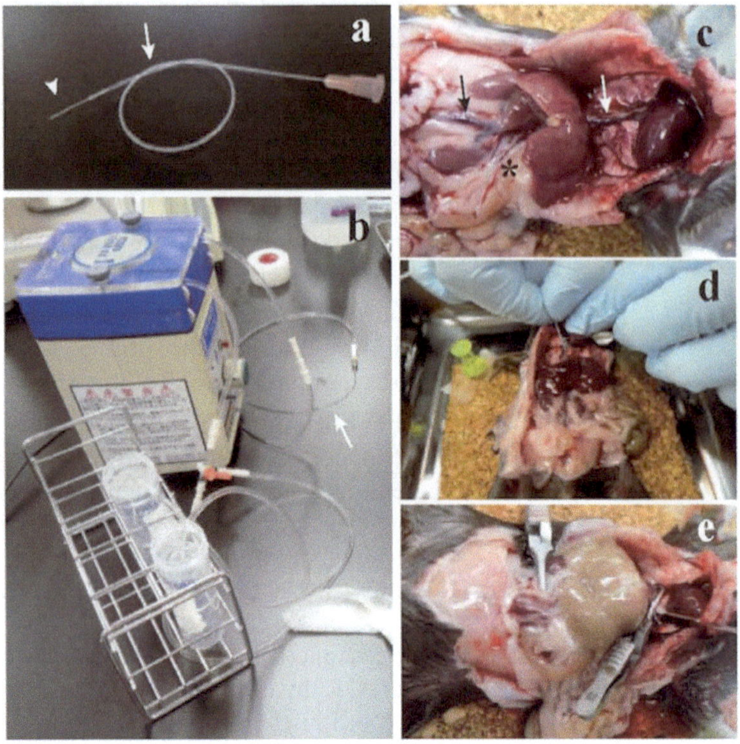

Fig. 2 Enzymatic treatment of sCAT mouse liver. (**a**) A handmade 26-gauge extension tube. The 26-gauge needle alone (shown by arrowhead) was connected to a polyethylene tube (ID.0.5 mm/OD 0.8 mm) (shown by arrow). (**b**) Liver perfusion system. The perfusate tube was connected to a peristaltic perfusion pump and handmade 26-gauge extension tube (shown by arrow). (**c**) The position of the inferior vena cava and hepatic portal vein. Abdominal inferior vena cava (black arrow), heart inferior vena cava of heart (white arrow), and hepatic portal vein (asterisk) are shown. (**d**) Image showing the insertion of the handmade 26-gauge extension tube into the heart inferior vena cava of the heart. (**e**) A picture showing post-perfusion. Note a pale color of the liver

2. Prepare the mice after 4 days after hydrodynamic gRNA delivery for hepatocyte isolation.
3. After laparotomy surgical procedure of the anesthetized mouse, the abdominal cavity is exposed. Expose the hepatic portal vein by carefully moving the viscera to the right and outside of the abdominal cavity (Fig. 2c).
4. After thoracotomy, insert the 26-gauge extension tube into the heart inferior vena cava (Fig. 2d). Then, both the blood vessel and 26-gauge extension tube were tied using Aorta-Klemme. Initiate the infusion in situ at a low flow rate (~10 mL/min) with ~50 mL of pre-warmed (37 °C) perfusion buffer.
5. Make a cut at the abdominal inferior vena cava to allow efflux for the release of blood. Bleeding was stopped from the abdominal inferior vena cava with Aorta-Klemme. The portal

vein was cut to remove blood. All lobes of the liver became pale in color (Fig. 2e).

6. Switch the perfusion solution with ~50 mL collagenase I solution without any interruption in the flow (~10 mL/min). During this process, the liver showed quick swelling.
7. When liver begins to look mushy, the liver is dissected and placed in a 100 mm tissue culture dish containing ~8 mL of cooled HBSS (*see* **Note 4**).
8. Loosen the liver using two scalpels to allow cells to disperse.
9. Filter the cell dispersion through a cell strainer (with 40 μm pore size) into a 50 mL conical tube in order to remove connective tissues and undigested tissue fragments.
10. Recover the filtrate by centrifugation at 50 × *g* for 2 min. Dissolve the pellet in 10–15 mL cooled HBSS and centrifuge the solution. Repeat this step three times (*see* **Note 5**).
11. Next, add equal volume of cooled Percoll solution to the cell suspension, and mix the solution gently. Centrifuge the solution at 50 × *g* for 10 min. Discard the supernatant (that typically contains dead cells).
12. Resuspend the cell precipitate in 30 mL of ice-cold DPBS + 10% FBS, and centrifuge at 50 × *g* for 10 min.
13. Resuspend the cell pellet in 10–15 mL ice-cold DPBS + 10% FBS and then keep the tube on ice (*see* **Note 6**).

3.4 Separation of Fluorescent and Non-Fluorescent Hepatocytes

Collect single hepatocytes using a standard microinjection apparatus attached to an Olympus IX70 inverted fluorescence microscope with a U-MWIBA2 filter set.

1. Place a drop (~200 μL) of the hepatocyte suspension (isolated in Subheading 3.3) onto the lid of 60 mm tissue culture dish (Fig. 3a). Cells were diluted with ice-cooled DPBS + 10% FBS to obtain a cell density of ~4000 cells per drop. Another fresh drop (~100 μL) of DPBS + 10% FBS is placed for isolating single cells (Fig. 3a). In the drops, several green fluorescent cells and non-fluorescent cells can be observed under the microscope (Fig. 3b).
2. While looking through the UV illumination using Olympus SZX12 stereomicroscope, pick several cells containing green fluorescent cells by a sucking pipette attached to the standard microinjection system and transfer them to the fresh drop (~100 μL) of DPBS + 10% FBS on 60 mm tissue culture dish (Fig. 3c; *see* **Note 7**).
3. Using Olympus SZX12 stereomicroscope without the UV illumination, the solution (0.1–0.3 μL) containing a single-fluorescent hepatocyte and transfer it to the inner wall near

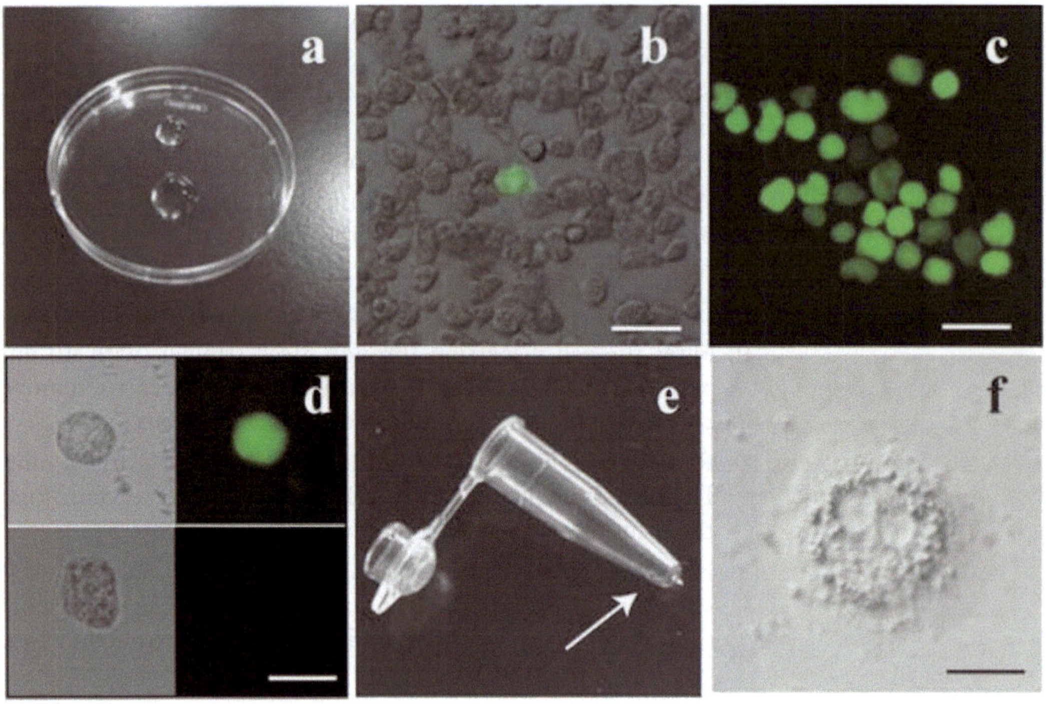

Fig. 3 Isolation of fluorescent single hepatocytes. (**a**) Drops of hepatocyte suspension on a lid of a 60 mm tissue culture dish for collection of single hepatocytes. (**b**) A microscopic image of the drop in (**a**) showing a single-fluorescent hepatocyte among the non-fluorescent cells. Bar = 40 μm. (**c**) The single EGFP(+) hepatocyte selected from the cells in (**b**). Bar = 40 μm. (**d**) Image of the isolated single hepatocyte from the microscopic filed in (**b**). Upper panels indicate a single EGFP(+) hepatocyte. Lower panels indicate a single EGFP(−) hepatocyte. Bar = 20 μm. (**e**) A PCR tube (0.2 mL) carrying the transferred single hepatocyte. Arrow indicates the drop (~0.3 μL) containing the single hepatocyte. (**f**) Hepatocytes 2 days after the cultivation of the single hepatocytes. Bar = 10 μm

the bottom of 0.2 mL PCR tube with a micropipette (upper panels of Fig. 3d and e; *see* **Note 8**). Store the PCR tube on ice. Collect non-fluorescent hepatocytes to use as control (lower panels of Fig. 3d). Repeat these steps (of single hepatocyte isolation) several times to collect over 30 fluorescent cells (*see* **Notes 9** and **10**).

3.5 Isolation of Genomic DNA from Single Hepatocytes

Crude DNA derived from a single hepatocyte can be prepared for use as a PCR template following the method described in Sakurai et al. [6] with some modifications.

1. Add 2 μL lysis buffer into the PCR tubes containing the single hepatocytes (Subheading 3.4, **step 3**). Incubate each PCR tube at 56 °C for 10 min and then at 95 °C for 10 min in ABI GeneAmp System 9700 PCR thermal cycler to inactivate any proteolytic activity.
2. Store the resulting crude DNA solution at −20 °C until use.

3.6 Genotyping of Single Hepatocytes

In this protocol, we describe two rounds of PCR genotyping strategy to amplify an 855 bp region of the murine *Alb* gene [6], to detect CRISPR/Cas9-meditaed targeted mutation.

1. For the first round of PCR, prepare a 23 μL reaction mixture, in a PCR tube, containing 1× Tks Gflex buffer, 0.4 μM each primer [Alb-7S (5′-CTAACGCCTGCTGGTGGTCACAGT-3′) and Alb-7A (5′-TGTCTAGCCCCTGCTTCTCT-3′)], and 1.25 U of Takara Tks Gflex DNA polymerase and 2 μL of crude DNA (prepared in Subheading **3.5**) (*see* **Note 11**).
2. Perform PCRs using the following conditions: denaturation at 94 °C for 1 min, followed by 35 cycles of 98 °C for 5 s, 68 °C for 20 s, and 68 °C for 5 min using ABI GeneAmp System 9700 PCR thermal cycler.
3. For second rounds of PCR, 2 μL PCR product from first and second PCR rounds is added to 23 μL reaction mixture, as described above. Perform PCR reactions using the same conditions described above.
4. Check the PCR products (2.5 μL) using a 1% agarose gel/TAE buffer system, stained with ethidium bromide, and photographed under UV illumination.

3.7 Indel Mutational Analysis of Single Hepatocytes

CRISPR/Cas9-induced *indel* mutations are detected using T7EI assay and direct sequencing of the PCR products.

1. Place 10 μL 1× T7EI reaction buffer containing 200–400 ng PCR product in 0.2 mL PCR tube. Incubate the tube at 95 °C for 5 min for denaturation using ABI GeneAmp System 9700 PCR thermal cycler and then incubate for 0.5–1 h at room temperature (24 °C) for re-annealing. Add 1 μL of T7EI (2–2.5 U/μL) to the denatured/re-annealed sample, and incubate the PCR tube at 37 °C for 1–2 h.
2. Analyze the nuclease-treated solutions using a 1% agarose gel/-TAE buffer system, stained with ethidium bromide, and photographed under UV illumination (Fig. 4; *see* **Note 12**).
3. For direct sequencing of PCR products, 10 ng PCR product was amplified using ABI BigDye Terminator v.3.1 Cycle Sequencing with Alb-7S or Alb-7S primer and sequenced with ABI Genetic Analyzer 3130 DNA sequencer. For cloning sequencing of PCR products, it may be preferable to use TA cloning method.
4. Perform sequence analysis using Genetyx-Mac (ver.13) sequence analysis software.

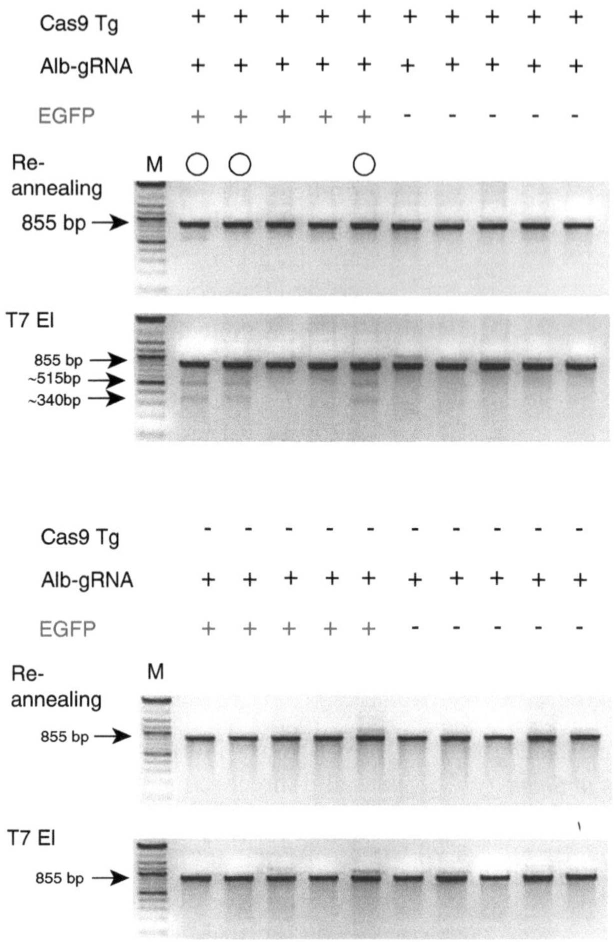

Fig. 4 Indel mutation of the *Albumin* (*Alb*) gene in Cas9/+ hepatocytes after hydrodynamic-based gene delivery of *Alb*-gRNA. Upper panels: T7EI assay gel images of indel mutation in single *Cas9* Tg hepatocytes co-transfected with *Alb*-gRNA and pCAG-EGFPpA. Lower panels: T7EI assay gel images of indel mutation in single wild-type hepatocytes co-transfected with *Alb*-gRNA and pCAG-EGFPpA. Circles on each lane in upper panels indicate the samples showing successful genome editing at the target *Alb* gene

3.8 Whole Genome Amplification (WGA; as an Optional Approach)

The whole genome can be amplified for WGA analysis. WGA method shown here is based on the method described by Sakurai et al. [16], which uses QIAGEN REPLI-g phage Φ29 DNA polymerase with single blastocyst-derived crude DNA.

1. Perform WGA according to the manufacturer's protocol. Briefly, mix 2 μL crude DNA solution with 2.4 μL Buffer D2. Incubate the mixture on ice for 10 min and the reaction stopped with the addition of 2.4 μL stop solution.
2. Add 27.2 μL REPLI-g master solution to the mixture and incubate at 30 °C for 16 h and at 65 °C for 3 min.

3. Store the amplified DNA solution at −20 °C until use (*see* **Note 13**).
4. Check the PCR amplification by analyzing 1–2 μL of product on 1% agarose gel/TAE buffer system, stained with ethidium bromide, and photographed under UV illumination.

4 Notes

1. All procedures involving laboratory animals were performed according to the institutional and national guidelines and legislation.
2. With this system, we obtained a total of four Tg founder mice [6]. All of these mice exhibited the expression of hCas9 mRNA in their major organs; however, the expression level of hCas9 mRNA varied among mice. We selected one line showing the highest expression of hCas9 mRNA and named it as "sCAT (systemic Cas9 expressed Tg mouse)." This line displayed transgenes in exon 9 of GTP-binding protein 10 (*Gtpbp10*) gene in chromosome 5.
3. The success of the experiment depended on (1) the use of mice aged 5–6–week old and weighing 18–25 g and (2) the completion of the tail vein-mediated injection of Mirus TransIT-QR HD solution within 10 s at a constant speed.
4. The success of the perfusion process may be assessed by the following two criteria: (1) pale color of the liver and (2) quick swelling in all lobes of the liver.
5. The supernatant contains non-parenchymal cells, while the precipitate is usually enriched with hepatocytes.
6. It is desirable to finish the isolation of single hepatocytes within 1 h from the preparation of hepatocyte suspensions.
7. UV exposure should be minimized.
8. Single-isolation procedure is completed within 15 min per run. As isolated hepatocytes are easily susceptible to cell damage, the isolation process is completed within 1 h.
9. In our previous study [6], single hepatocytes were collected using the standard microinjection system attached to an inverted fluorescence microscope. However, it is possible to use a mouth-controlled suction glass pipette that is commonly used as an egg-handling equipment.
10. With our experience, it is possible to culture the freshly isolated hepatocytes for a week in a relatively good condition (Fig. 3d).
11. PCR was performed using entire crude DNA from a single hepatocyte to evaluate the presence/absence of indel

mutations in the target *Alb* gene. Multiplex PCR using a single hepatocyte-derived DNA requires increased amount of DNA for WGA (Subheading 3.8).

12. The hydrodynamic-based gRNA delivery method yielded EGFP(+) hepatocytes at 0.1% (1/1000 cells). T7EI assays using genomic DNA from single hepatocytes revealed that only EGFP-positive Tg hepatocytes harbored indel mutations in *Alb* gene at an average rate of 50% (15/30) (*see* Fig. 2 in Sakurai et al. [6]), indicating that the percentage of mutated hepatocytes was about 0.05% (1/2000 cells). No indel mutations were observed in EGFP-negative Tg, EGFP-positive non-Tg, and EGFP-negative non-Tg hepatocytes. Altogether these findings show that in vivo gRNA administration results in the induction of indel mutations in the endogenous target gene of Cas9 mice.
13. As WGA is always associated with the amplification of very small amount (even at femotogram level) of DNA, it requires control reaction using a solution containing no template.

Acknowledgments

This study was supported by JSPS KAKENHI Grants (Nos. 26293085, 26670153 and 16K15233) from the Japanese Ministry of Education, Culture, Sports, Science, and Technology.

References

1. Jinek M, Chylinski K, Fonfara I et al (2012) A programmable dual-RNA-guided DNA endonuclease in adaptive bacterial immunity. Science 337:816–821. https://doi.org/10.1126/science.1225829
2. Cong L, Ran FA, Cox D et al (2013) Multiplex genome engineering using CRISPR/Cas systems. Science 339:819–823. https://doi.org/10.1126/science.1231143
3. Mali P, Yang L, Esvelt KM et al (2013) RNA-guided human genome engineering via Cas9. Science 339:823–826. https://doi.org/10.1126/science.1232033
4. Platt RJ, Chen S, Zhou Y et al (2014) CRISPR-Cas9 knockin mice for genome editing and cancer modeling. Cell 159:440–455. https://doi.org/10.1016/j.cell.2014.09.014
5. Dow LE, Fisher J, O'Rourke KP et al (2015) Inducible in vivo genome editing with CRISPR-Cas9. Nat Biotechnol 33:390–394. https://doi.org/10.1038/nbt.3155
6. Sakurai T, Kamiyoshi A, Kawate H et al (2016) A non-inheritable maternal Cas9-based multiple-gene editing system in mice. Sci Rep 6:20011. https://doi.org/10.1038/srep20011
7. Carroll KJ, Makarewich CA, McAnally J et al (2016) A mouse model for adult cardiac-specific gene deletion with CRISPR/Cas9. Proc Natl Acad Sci U S A 113:338–343. https://doi.org/10.1073/pnas.1523918113
8. Chu VT, Graf R, Wirtz T et al (2016) Efficient CRISPR-mediated mutagenesis in primary immune cells using CrispRGold and a C57BL/6 Cas9 transgenic mouse line. Proc Natl Acad Sci U S A 113:12514–12519. https://doi.org/10.1073/pnas.1613884113
9. Nakao H, Harada T, Nakao K et al (2016) A possible aid in targeted insertion of large DNA elements by CRISPR/Cas in mouse zygotes. Genesis 54:65–77. https://doi.org/10.1002/dvg.22914
10. Zhang L, Zhou J, Han J et al (2016) Generation of an oocyte-specific Cas9 transgenic mouse for genome editing. PLoS One 11:

e0154364. https://doi.org/10.1371/journal.pone.0154364
11. Cebrian-Serrano A, Zha S, Hanssen L et al (2017) Maternal supply of Cas9 to zygotes facilitates the efficient generation of site-specific mutant mouse models. PLoS One 12: e0169887. https://doi.org/10.1371/journal.pone.0169887
12. Nagy A, Marina Gertsenstein M, Kristina Vintersten K et al (eds) (2003) Manipulating the mouse embryo: a laboratory manual, 3rd edn. Cold Spring Harbor Laboratory Press, New York
13. Kawai S, Takagi Y, Kaneko S et al (2011) Effect of three types of mixed anesthetic agents alternate to ketamine in mice. Exp Anim 60:481–487
14. Liu F, Song Y, Liu D (1999) Hydrodynamics-based transfection in animals by systemic administration of plasmid DNA. Gene Ther 6:1258–1266. https://doi.org/10.1038/sj.gt.3300947
15. Kamiyoshi A, Sakurai T, Ichikawa-Shindo Y et al (2006) Endogenous alphaCGRP protects against concanavalin A-induced hepatitis in mice. Biochem Biophys Res Commun 343:152–158. https://doi.org/10.1016/j.bbrc.2006.02.132
16. Sakurai T, Watanabe S, Kamiyoshi A et al (2014) A single blastocyst assay optimized for detecting CRISPR/Cas9 system-induced indel mutations in mice. BMC Biotechnol 14:69. https://doi.org/10.1186/1472-6750-14-69

Chapter 16

Microinjection and Oviduct Transfer Procedures for Rat Model Generation with CRISPR-Cas9 Technology

Tuija M. Alcantar and Xin Y. Rairdan

Abstract

Since the first knockout rat model was generated with zinc-finger nucleases (ZFNs) by Geurt's group in 2009, the demand for making targeted rat models has increased tremendously. The advent of the clustered regularly interspaced short palindromic repeats-CRISPR associated protein 9 (CRISPR-Cas9) system provides researchers with a more efficient method for producing modified animals, which has since then been developed and applied in rat. Since we established a rat model production system at our facility in 2014, we have consistently generated rat models. Due to differences in physiology and embryology between mouse and rat, species-specific protocols for superovulation conditions, microinjection, and embryo transfer (among others) are required. There are over 100 rat strains, and Sprague Dawley is one of the commonly used outbred strains in biomedical research. In this chapter, we describe in detail a range of topics including donor and recipient preparation, microinjection setup, CRISPR reagent preparation, and oviduct transfer procedures for making rat models in the Sprague Dawley background.

Key words Sprague Dawley rat, Superovulation, Pronuclear and cytoplasmic microinjection, CRISPR-Cas9, Embryo transfer

1 Introduction

The rat, *Rattus norvegicus*, has been used in the laboratory as an animal model for human disease for over 150 years [1]. Accordingly, a PubMed data search revealed over one million publications that used the rat [2]. Its size, cognitive, and physiological characteristics make the rat a great model for cardiovascular disease [3], neurological disorders [4], and metabolic disorders [5, 6]. Rat models are superior to mouse models for testing the pharmacodynamics, pharmacokinetics, and toxicity, partially because many of their detoxifying enzymes are very similar to those in humans [7]. However, generation of rat models was precluded by a lack of robust rat ES cells that could stably contribute to germline chimeras. The first germline-competent rat ES cell lines were reported in 2008 [8, 9], which is almost two decades after the first mouse model was

Chengyu Liu and Yubin Du (eds.), *Microinjection: Methods and Protocols*, Methods in Molecular Biology, vol. 1874,
https://doi.org/10.1007/978-1-4939-8831-0_16,

generated with ES cells [10–12]. With the breakthrough of nuclease technology in 2009, the first targeted knockout rat with a germline-transmitted mutation was generated with ZFNs [13]. The next generation genome editing tool, CRISPR-Cas technology, makes generating rat models much more efficient and cost effective [14]. Naturally occurring CRISPR-Cas systems have evolved in bacteria and archaea as an RNA-based adaptive immune system against viral and plasmid invasion [15, 16]. These defense systems rely on small RNAs for sequence-specific detection and destruction of foreign nucleic acids. An adapted CRISPR-Cas9 system, composed of the Cas9 endonuclease of Streptococcus pyogenes and a synthetic guide RNA (sgRNA) of choice, is generally used to produce genetically modified organisms. The sgRNA binds to complementary sequences in the targeted genome and guides the Cas9 endonuclease there to generate double-stranded breaks [17]. Although the mouse remains the preferred mammalian genetic model, the rat was once the most widely used organism in medical research, and the gap may be closing due to successful gene editing with CRISPR-Cas9 technology. One drawback of the technology is the production of mutant animals exhibiting mosaicism, which can be mitigated by combining CRISPR/Cas9 technology with spermatogonial stem cell advances to generate targeted germline modifications in rats [18]. However, spermatogonial stem cell isolation is not a well-established method in most transgenic facilities, posing new challenges if implementation is desired. Alternatively, successful facility implementation of rat embryo modification is more feasible given its similarity to the mouse and the availability of protocols. In this chapter, we describe details about preparing donor and recipient rats, superovulation regime, microinjection setup, CRISPR reagent handling, pronuclear and cytoplasmic microinjection procedures, and embryo transfer surgery. The protocols described here have been solely developed and validated in the Sprague Dawley strain. Therefore, modifications might be necessary when using other strains, especially inbred stains such as Fisher 344 and Brown Norway. We hope this chapter will assist you in establishing a rat model production system.

2 Materials

2.1 Rats

1. Ten Sprague Dawley stud males, 2–6 months old, individually caged.
2. Ten Sprague Dawley donor females, 4–5 weeks old, group-housed according to regulations for size of caging.
3. Six vasectomized Sprague Dawley stud males, 2–6 months old, individually caged. We perform vasectomy procedures in house.
4. Six Sprague Dawley recipient females, 8–12 weeks old, group-housed according to regulations for size of caging.

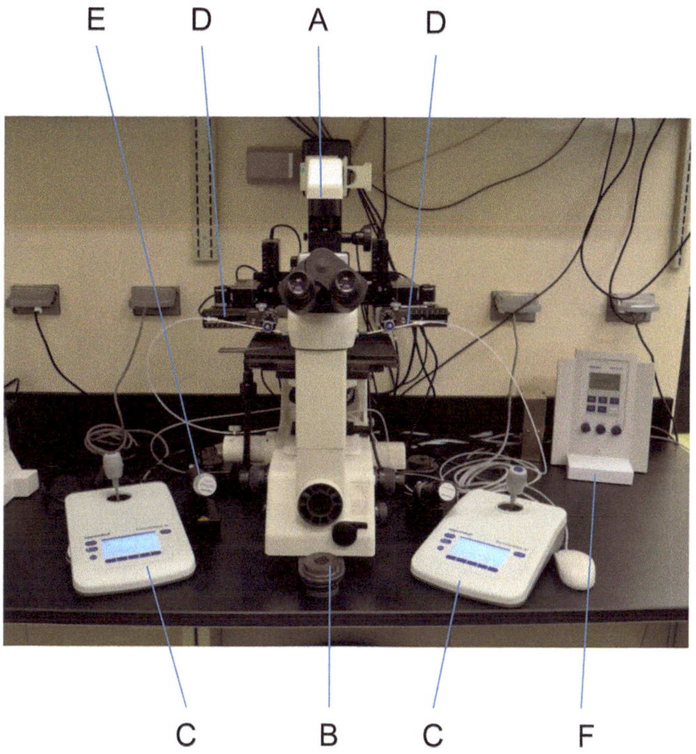

Fig. 1 Microinjection setup. Our system includes: (**A**) Nikon TE300 inverted microscope with 4×, 10×, 20×, and 40× objectives set on the top of (**B**) an anti-vibration pad. (**C**) Two Eppendorf TransferMan 4r micromanipulators enable controlled movement of the microinjection needle and holding pipette attached to (**D**) capillary holders. (**E**) Eppendorf Cell-Tram Air and (**F**) FemtoJet provide suction/control of the holding pipette and regulation of air pressure delivered to the injection needle, respectively

2.2 Microinjection Setup

As shown in Fig. 1, the microinjection system consists of the following components:

1. An inverted microscope with 4×, 10×, 20×, and 40× objectives for pronuclear and cytoplasmic microinjection.
2. Two Eppendorf TransferMan 4r micromanipulators or alternatively, Narishige four-axis hanging joystick oil hydraulic micromanipulators.
3. Eppendorf CellTram Air for holding pipettes.
4. FemtoJet for injection.

2.3 Anesthesia System

As shown in Fig. 2, the anesthesia system consists of the following components:

1. Mobile non-rebreathing anesthesia machine.
2. Isoflurane Vaporizer.

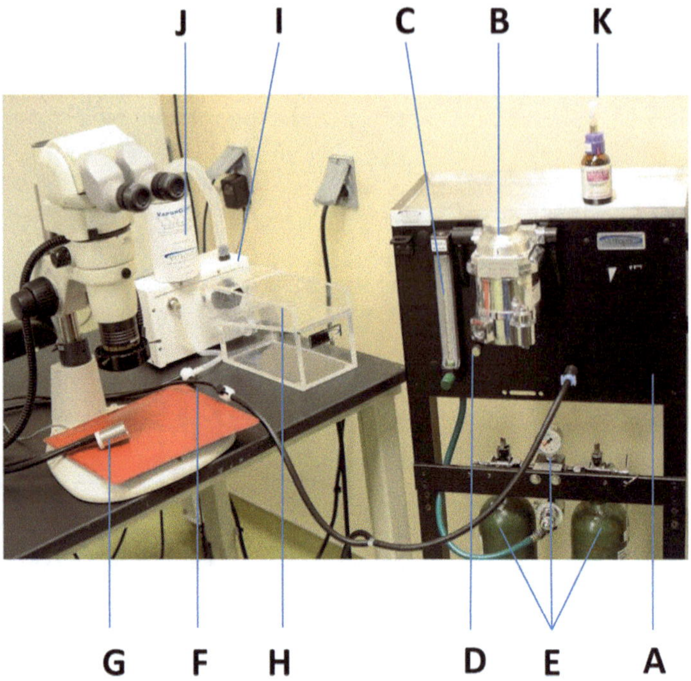

Fig. 2 Anesthesia delivery system. Our mobile non-rebreathing anesthesia system consists of the following components: (**A**) Mobile non-rebreathing anesthesia machine, Model V-10. (**B**) Isoflurane vaporizer for delivering specified isoflurane concentrations. (**C**) Oxygen regulator and flow meter for delivering specified oxygen concentrations. (**D**) Oxygen flush valve. (**E**) Two cylinders of medical grade oxygen with a cylinder yoke/regulator and an oxygen flow meter. (**F**) Non-rebreathing anesthetic circuit with 2-way-valves to open and close extraction points. (**G**) 14 mm rat nose cone. (**H**) 2-L vented induction chamber. (**I**) Scavenging cube. (**J**) VaporGuard activated charcoal adsorption filter. (**K**) Isoflurane bottle with an anti-spill bottle-top adapter

3. Cylinder yoke/regulator and oxygen flow meter.
4. 2 L vented induction chamber.
5. 14 mm rat nosecone.
6. Activated charcoal absorption filter.
7. Scavenging cube.
8. Medical grade oxygen tank.
9. Isoflurane.

2.4 Other Equipment

1. Stereo Microscope: Used for embryo collection and embryo transfer procedures. These microscopes are equipped with both top and bottom light sources.
2. Inverted Microscope: Used for prescreening zygotes with clear pronuclei for microinjection (400× magnification power).

3. Programmable Microinjector: An injector that has an independent pressure supply via built-in compressor is used for pronuclear and cytoplasmic microinjections.
4. CO_2 Incubators: Used for embryo culture. They are set at 37 °C, 6% CO_2, and humidified with a tray containing distilled water.
5. Biosafety cabinet: Class II biological safety cabinet is used to handle reagents for microinjection.
6. Isotemp Oven: Used for baking slides, cover slips and capillary tubes for microinjection procedures.
7. Pipette Puller: A horizontal pipette puller is used for pulling microinjection needles.
8. A table-top autoclave for sterilizing surgical instruments.
9. A bead sterilizer is used to sterilize instruments during embryo transfer surgery as needed.

2.5 Tools and Supplies

1. Gloves for handling reagents and animals.
2. 1.0 ml insulin syringe.
3. 26 G1/2 needle.
4. 15 cm, 6 cm, and 35 mm culture dishes for oviduct collection and embryo culture.
5. RNase-and DNase-free 1.7 ml microcentrifuge tube.
6. 20, 200, and 1000 μl RNase and DNase-free, pre-sterilized filter pipette tips.
7. 2, 20, 200, and 1000 μl pipettes.
8. Polypropylene centrifuge tube filter: 0.22 μm, RNase, DNase-free, Sterile.
9. Depression slide.
10. Microslides for assembling microinjection chamber.
11. Micro cover glass: 22 × 22 mm.
12. Plexiglass blocks: 3 × 3 × 10 mm.
13. Embryo handling pipette: Pulled over an alcohol burner from 100 μl calibrated pipets ranging between 150 and 200 μm in diameter.
14. Holding pipette for microinjection.
15. Microinjection needle (as shown in Fig. 3): Pulled from a 1.0 mm OD × 0.75 mm ID glass capillary tubing with filament using a horizontal pipette puller.
16. Embryo transfer pipette: Assembled by connecting a flat aspirator mouthpiece, 30 cm of latex tubing, a 25 mm disposable syringe filter, another 30 cm of latex tubing, a micropipette holder, and a stripper tip.

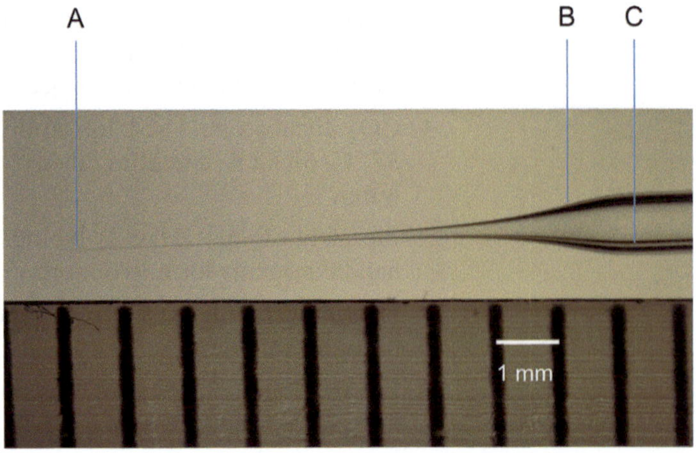

Fig. 3 Rat pronuclear/cytoplasmic injection needle. (**A**) Long and thin tip. (**B**) Shoulder. (**C**) Filament provides capillary action for quick filling of the injection needle, while minimizing air bubble formation

17. Balance scales.
18. Mini fur trimmer.
19. Straight fine scissors.
20. Straight micro dissecting forceps.
21. Straight fine sharp micro dissecting forceps.
22. 45°-angled sharp forceps.
23. Micro dissecting tweezers.
24. Angled Dietrich bulldog clamp.
25. Cotton-tipped applicators.
26. 5-0 braided silk suture.
27. 9 mm autoclip applier.
28. Autoclip wound clips.
29. Slides warmer.
30. Low-foaming, phosphate-free, powdered detergent.
31. Self-seal sterilization pouch.

2.6 Media and Chemicals

1. 0.9% Sodium Chloride (NaCl): Pharmaceutical grade for hormone preparation.
2. Pregnant mare serum gonadotropin: Each vial contains 2000 IU of lyophilized PMSG. To make 100 IU/ml stock solution, dissolve PMSG in 20 ml 0.9% NaCl in a 50 ml centrifuge tube. Aliquot solution by adding 2.7 ml into 5 ml polypropylene round-bottom tubes and store at −20 °C in the dark for up to 6 months. Each aliquot can treat up to six rats (*see* **Note 1**).

3. Human chorionic gonadotropin: Each vial contains 1000 IU of lyophilized hCG. Use 10 ml of 0.9% NaCl to make 100 IU/ml solution. Aliquot solution by adding 2.7 ml into 5 ml polypropylene round-bottom tubes and store at −20 °C in the dark for up to 6 months. Each aliquot contains enough for up to six rats.
4. Luteinizing hormone: Each vial contains 5 mg of lyophilized LHRHa. Use 50 ml of 0.9% NaCl to make 100 μg/ml working concentration. Aliquot solution by adding 2.7 ml into 5 ml polypropylene round-bottom tubes and store at −20 °C in the dark for up to 6 months. Each aliquot contains enough for up to six rats.
5. Ovine hyaluronidase: Each vial contains 100 KU of lyophilized ovine hyaluronidase. To make 3000 U/ml stock solution, dissolve one vial with 33 ml of M2 culture medium. Make aliquots by adding 1 ml into microcentrifuge tubes and store at −20 °C in the dark for up to 6 months.
6. Bovine Serum Albumin.
7. M2 culture medium: Add 4 mg/ml BSA before use and store at 4 °C for 2 weeks.
8. KSOMaa without BSA: Add 4 mg/ml BSA before use and store at 4 °C for 2 weeks (KSOM).
9. mR1ECM culture medium.
10. Cytochalasin B (CB): Each vial contains 5 mg of CB. Inject 5 ml of DMSO into the vial with a hypodermic needle to make 1 mg/ml stock. Make aliquots of 5 μl in 1.5 ml microcentrifuge tubes. Store at −20 °C in the dark.
11. Mineral oil (*see* **Note 2**).
12. Isopropyl Alcohol (IPA), 70% (v/v).
13. 100× Tris EDTA, pH 8.0.
14. Embryo-tested water.
15. Cas9 mRNA: Add 45 μl of mRNA injection buffer into one vial with 15 μl of Cas9 mRNA (1 μg/μl) to make 250 ng/μl stock. Mix well by pipetting up and down. Aliquot 4.5 μl into 1.5 ml RNase, DNase-free microcentrifuge tubes and freeze at −80 °C.
16. Donor oligo: Can be purchased from commercial vendors.
17. sgRNA: Can be ordered from many commercial vendors.
18. RNase decontamination agent.
19. Betadine solution.
20. Sterile ocular lubricant, petroleum ophthalmic ointment.
21. Meloxicam SR analgesic.

3 Methods

3.1 Recipient Preparation

1. Four days prior to mating, inject six recipient rats with 0.4 ml of 100 μg/ml LHRHa (40 μg) at 8:00 am in order to synchronize estrus (*see* **Note 3**).
2. One day prior to microinjection, set up one recipient rat with one vasectomized stud male at 8:00 am. Vasectomized stud males should be rotated approximately every 6 months for optimal performance.
3. Check plugs early in the morning (7:00–8:00 am) the following day (the day of microinjection) to ensure plugged females are detected before the plugs fall out. Separate the plugged females from the non-plugged females. In general, one can expect to find 3–4 plugged females out of a total of 6 mated females (50–67% plug rate).

3.2 Zygote Collection

1. Three days prior to the scheduled microinjection, inject 0.4 ml of 100 IU/ml PMSG (40 IU) into each donor rat by IP injection at 11:00 am (*see* **Note 4**).
2. Two days (46–48 h) later, inject 0.4 ml of 100 IU/ml hCG (40 IU) into each donor rat by IP injection.
3. Immediately after hCG injection, set up each donor female with one singly-housed stud male for mating. Each stud male should only be mated once a week to ensure high mating and fertilization rates (*see* **Note 5**).
4. Check plugs at 8:00 am the following day and separate plugged females from non-plugged females.
5. Prepare separate culture dishes for holding uninjected and injected embryos, each with several small drops (approximately 30 μl) of KSOM covered with mineral oil, and place them inside a 37 °C, 6% CO_2 incubator.
6. Euthanize plugged females using CO_2 per guidelines set by the American Veterinary Medical Association. Adjust CO_2 flow rate according to recommended setting for rats and the specific cage type. To ensure death prior to oviduct collection, cervical dislocation is performed on all donor rats.
7. Disinfect the abdominal skin with 70% IPA and dissect out both oviducts from each donor female with a pair of sterile scissors and forceps. Place the harvested oviducts into a collection dish and add enough ovine hyaluronidase to cover the tissue (*see* **Note 6**).
8. After grasping an oviduct with forceps, use the point of a needle to pierce the ampulla, which is the swollen portion of the oviduct that contains the cumulus-oocyte mass, or clutch. The clutch should ooze out of the ampulla but if not, drag it

out using the needle. Repeat this until all oviducts have been harvested; usually 16–20 are harvested at a time.

9. Focus the view at the bottom of the dish where the zygotes lie and begin collecting all the denuded zygotes. Depending on the thickness of the cumulus cell layer, some zygotes may require longer incubation in hyaluronidase and/or pipetting to complete denudation.
10. Transfer the collected zygotes into an M2 drop.
11. Subsequently, pass them through at least three more M2 drops to wash off the remaining cumulus cells and hyaluronidase.
12. Transfer the zygotes into the "uninjected" culture dish with preincubated (at least 30 min) KSOM drops covered with mineral oil, and leave the dish in the incubator until injection.

3.3 CRISPR Reagents Preparation

1. All supplies should be RNase free when working with RNA. Use a biosafety cabinet to make mRNA injection buffer, the final injection mixture, and aliquots of all reagents. Clean the biosafety cabinet regularly with 70% IPA and RNase decontamination agent (*see* **Note 7**).
2. To prevent freeze/thaw cycles, make 10× or 20× aliquots for Cas9 mRNA, Cas9 protein, donor oligo, and sgRNA and store them at −80 °C. Take one aliquot out for each injection day.
3. The day of microinjection, make fresh 1× mRNA injection buffer by adding 5 μl of 100× Tris-EDTA into 495 μl of embryo tested water in an RNase, DNase-free microcentrifuge tube. Pipette ten times before use. Discard leftover mRNA injection buffer.
4. Remove one aliquot of each component from the −80 °C freezer and make the final microinjection mixture: 25 ng/μl of Cas9 mRNA, 13 ng/μl of sgRNA, and 100 ng/μl of donor oligo (as needed) in an RNase, DNase-free microcentrifuge tube. Preparing 40 μl of injection mixture is sufficient for injecting 300–500 oocytes.
5. Transfer the diluted injection mixture into a polypropylene centrifuge tube filter and centrifuge it for 5 min at 200 × *g* to get rid of debris that might block the injection needles. Do not use the polypropylene centrifuge tube filter if you are using a plasmid donor that is over 10 kb (*see* **Note 8**).
6. Keep diluted reagents on ice during microinjection.

3.4 Injection Needle Preparation

1. All injection and embryo handling needles must be baked in the oven at 232 °C for 2 h prior to injection day. This ensures that all possible RNases have been eliminated and will not decrease injection success by breaking down the Cas9 mRNA and sgRNA.

2. Before starting, all surfaces and equipment used for RNA injection (including the micropipette puller) need to be wiped down with RNase decontamination agent.
3. Pull 10–15 injection needles with glass capillaries on a horizontal pipette puller. Our settings for pulling thin, longer tapered injection needles (better for rat injections) are as follows: Pressure = 300, heat = 240, pull = 160, velocity = 110, and time = 160. Our puller has a 3 mm trough filament (FT330B) and every time the filament is changed, a new program setting has to be configured (*see* **Note 9**).
4. Place pulled needles into DNAse and RNAse free 1.7 ml microcentrifuge tubes, 4–6 needles per tube. Do not touch the tips as they are very fine and will break even from the slightest touch.
5. Mix the diluted CRISPR reagents in the tube by tapping the tube. Using a mouth pipette with a clean baked needle attached, first draw up reagent mixture into the needle. Subsequently, fill each injection needle by depositing a small column of reagent mixture directly inside each injection needle. Finally, wait approximately 2 min and visually confirm that the CRISPR reagent mixture has drawn up to the tip of each injection needle. Only fill as many needles as you can use within an hour of injection at a time and make sure to place the CRISPR reagent mixture back on ice when not in use.

3.5 Microinjection Chamber Preparation

1. Make the microinjection medium by adding 1 ml of M2 into an RNase, DNase-free microcentrifuge tube containing 5 μl of 1 mg/ml CB solution for a final concentration of 5 μg/ml. Mix well (*see* **Note 10**).
2. There are several methods for setting up a microinjection chamber. We use a depression slide injection chamber (approximately a 30 μl droplet of injection medium in the center of a depression slide, covered with mineral oil) and a custom designed hanging drop chamber (*see* **Note 11**). As shown in Fig. 4, this injection chamber is assembled by first applying a thin layer of vacuum grease on two opposite sides of two custom-made plexiglass blocks (*see* **Note 12**). Next, attach the two blocks opposite of each other but parallel to the long axis of a siliconized slide and fill the area between the blocks with a strip of injection medium. Then place a siliconized coverslip (ends on one side only) on top and press slightly to attach the ends to the greased blocks so that the media merges and forms a chamber between the slide and the coverslip. Finally, apply a thin coat of mineral oil with a needle and syringe to the two exposed sides of the media column to seal the chamber and prevent evaporation.

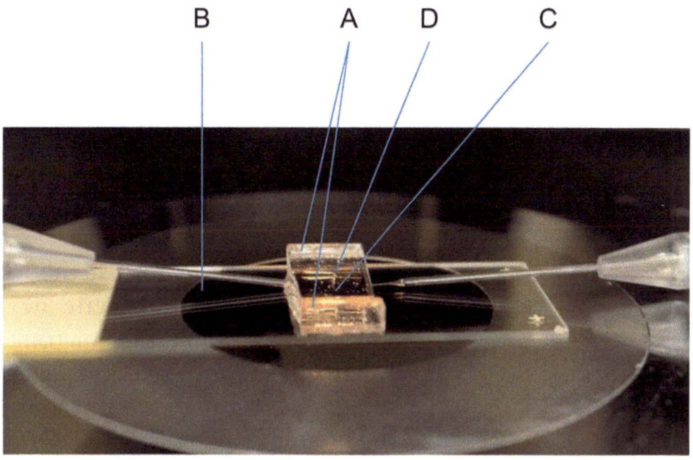

Fig. 4 Custom microinjection chamber assembly. Apply a thin layer of vacuum grease on two opposite sides of (**A**) two custom-made Plexiglass blocks. Attach the two blocks opposite each other while remaining parallel to the long axis of (**B**) a sigmacote-coated slide. Fill the area between the blocks with a strip of (**C**) injection medium. Place the sigmacote-coated ends of (**D**) a coverslip (two opposing ends of one side only) over the greased blocks and apply light pressure until the medium merges and forms a chamber between the slide and the coverslip. Finally, apply a thin mineral oil coating to the two exposed sides of the medium column with a needle and syringe to seal the chamber and prevent evaporation

3.6 Pronuclear/ Cytoplasmic Microinjection

1. Sort through the KSOM dish containing pronuclear stage embryos using the inverted culture microscope and pick up as many zygotes with two visible pronuclei as you can inject in a half an hour (generally 25–50).
2. While using the 4× objective, load the embryos into the injection chamber (or the droplet) on the injection station, arranging them in a vertical column starting from the bottom of the chamber.
3. Allow the embryos to equilibrate in the injection medium containing CB for 10 min before starting injections.
4. After 10 min, move the stage so that your holding pipette and injection needle are on top of the first zygote in your column, and change the objective to 40×.
5. With tips of both the holding pipette and injection needle in the center of the field of view and lined up, bring them into sharp focus and then gently brush the very tip of the injection needle against the tip of the holding pipette to break a small opening in the injection needle. With a programmable microinjector, the best results are achieved when the opening of the injection needle is small enough that you see a small droplet forming inside the zygote when injecting under the constant

pressure of 2–3 PSI and applying a manual injection setting as needed (10–20 PSI). Constant pressure also prevents capillary effect backflow of injection medium into the injection needle.

6. The desired reagent flow rate is indicated by a slow, controlled swelling of the pronucleus or a small deposit inside the cytoplasm. Injecting excess reagent into the zygote will result in lysis.

7. Aspirate a single zygote at the top of the embryo column by moving the holding pipette next to it and applying gentle suction by turning the micromanipulator counterclockwise. Rotate the zygote until the male pronucleus (larger one of the two) is located in the center and in line with the axis of the holding pipette (Fig. 5). Once the desired positioning has been achieved, tighten the hold of the zygote by applying additional suction.

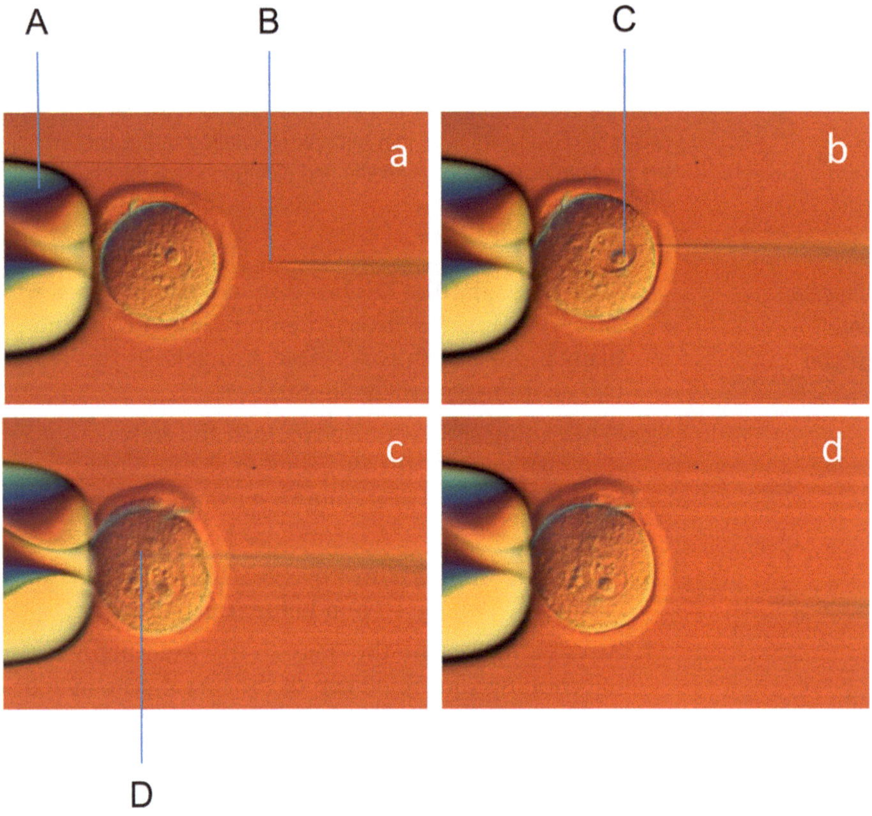

Fig. 5 Pronuclear and cytoplasmic microinjection of 1-cell embryos. (**a**) 1-cell embryo positioned with suction from the (**A**) holding pipette for pronuclear microinjection. (**B**) *Injection needle loaded with the microinjection mixture aligned with the holding needle on the opposite side of the embryo.* (**b**) The expanded (**C**) pronucleus after the tip of the injection needle was inserted and the contents injected. (**c**) *Cytoplasmic injection targeting the (**D**) region in an area away from the two pronucei;* notice the swelling of the injected area. (**d**) *Quick and smooth removal of the needle* once swelling of the pronucleus/cytoplasm has occurred

8. Bring the pronucleus and the tip of the injection needle into sharp focus. Carefully guide the needle into either the pronucleus (avoid touching the sticky nucleoli) or the cytoplasm (avoid piercing the nuclei), depending on the desired injection method (*see* **Note 13**).
9. Once successfully entered, carefully observe the swelling of the pronucleus/cytoplasm. With the constant flow setting, manual flow adjustment may be unnecessary if already sufficient. However, if no flow is detected, manually adjust flow rate as necessary. Depending on the size of the injection needle opening, the injection time and/or use of manual injection pressure will have to be adjusted. A successful injection is indicated by visible swelling of the pronucleus/cytoplasm. Failure to penetrate the membrane surrounding the pronucleus/cytoplasm will result in a small droplet forming on the outside of the membrane. If this occurs try pushing the needle a little further toward the holding pipette to break through the membrane until you finally see the swelling inside the pronucleus/cytoplasm.
10. Once you have observed the swelling of the pronucleus/cytoplasm, pull the needle out in a quick, smooth motion.
11. Release suction on the holding pipette by turning the micromanipulator knob clockwise, and push the injected zygote above and away from the holding pipette and the uninjected zygotes, respectively.
12. Move on to injecting the next zygote in the column and down the column until all zygotes have been injected. Needle replacement is required if any of the following occur: the needle opening becomes too large, and needle begins gushing out injection solution, the tip becomes covered with sticky debris, or the needle becomes clogged and no swelling of the pronucleus/cytoplasm is observed.
13. Once the end of the zygotes column is reached, remove all injected zygotes from the slide and place them in the "injected" KSOM dish in the incubator.

3.7 Prepare Surgery Station for Oviduct Transfer

1. Turn on a slide warmer set at 37 °C, and place clean cages with 9% fat diet, water, and enrichment on top to warm.
2. Plug in the heating pad placed on the top of the transfer microscope stage.
3. Place sterile towels on the rat surgery preparation area, stereomicroscope stage, surgical tool holding area, and the induction chamber. Lay out autoclaved surgical tools on the sterile towels and put cotton swabs into Betadine and 70% IPA jars to soak.

3.8 Prepare Isoflurane Machine

1. Fill vaporizer with isoflurane by carefully pouring into the filling port until the meniscus has reached just below the full mark.
2. Weigh the carbon filter canister to make sure no more than the indicated maximum weight has been added to the original measurement. Replace if needed and record the starting date and weight in the canister log.
3. Make sure the induction chamber is attached to the evacuation system and turn on the power switch of the evacuation system unit.
4. Open up both oxygen cylinder valves all the way by turning the handles counterclockwise and check the level of the pressure gauge to make sure enough oxygen is left to complete the surgery. If the pressure gauge reads less than 200 PSI, replace the oxygen cylinders.
5. Turn the oxygen flow meter knob counterclockwise until the center of the indicator ball is set to two.
6. Turn the vaporizer knob until it reaches the maximum concentration of 5% by pressing and holding the lever down while turning the big vaporizer knob in a counterclockwise direction.
7. Open up the valves in the tubing by twisting until the valves are parallel to the length of the tubing to release gas into the induction chamber and the nose cone.

3.9 Oviduct Transfer

1. On the day of injection, bring the plugged recipient rats to the surgery suite (*see* **Note 14**) and ensure that there are no mice in the room (*see* **Note 15**). In addition, make sure that the recipients in the surgery suite are being held in cages with adequate air supply as rats, especially when group housed, can quickly deplete the oxygen reserve inside a microisolator cage that has been removed from its air supply.
2. Place a recipient rat (plugged the day of injection) into the induction chamber and wait until it is fully asleep.
3. Weigh recipient rat.
4. Shave fur off the back of the rat between flank and ribcage.
5. Apply ophthalmic ointment to both eyes of the rat.
6. Lay the rat stomach down, head facing away from you, and nose inserted into the nose piece located on top of the stereoscope stage. Verify absence of a toe-pinch reflex. If recipient rat is at the correct depth of anesthesia, the isoflurane should be turned down to 2–3%. Isoflurane level should be carefully monitored throughout the surgery and adjusted accordingly to maintain sedation level without exposing the recipient to unnecessarily high levels of isoflurane.

7. Using soaked cotton swabs, clean shaved skin first with betadine followed by 70% IPA.
8. Make a small longitudinal incision in the midline at the level of the last ribs and insert the tips of fine scissors between the skin and body wall in the left side and separate the two by repeatedly opening and closing the scissors. Pick up the skin on the left side with a pair of forceps in your left hand, and with another pair of forceps in your right hand, reach deep into the left side and grab a hold of the body wall (left-handed technicians will follow the opposite instructions). Holding on to the body wall with the forceps in the non-dominant hand, grab a pair of scissors with your dominant hand. In order to prevent excessive bleeding, first make a small cut into the body wall and then enlarge the incision by inserting the tips of the scissors into the hole and opening them.
9. Use blunt forceps to pick up ovarian fat pad and gently pull out the ovary and upper part of the uterus.
10. Clip a bulldog clamp onto the fat pad and lay it down over the middle of the back on top of sterile cotton gauze.
11. Check that the recipient is at the correct stage of pseudopregnancy by verifying that a swollen ampulla can be visualized in the ovarian end of the oviduct.
12. Reorient the ovary and oviducts by twisting the bulldog clamp toward yourself until the infundibular opening inside the bursa, between the ovary and the coiled oviduct, is in the middle of your microscope field of vision and can be easily accessed with a transfer pipette from the ovarian side.
13. Load transfer pipette with M2 medium and 5–8 small bubbles followed by 10–15 embryos and two final small bubbles at the tip of the pipette.
14. With a pair of fine forceps in the non-dominant hand, gently grab a hold of the oviduct half way between the ampulla and the infundibulum. With the dominant hand, hold the transfer pipette filled with embryos and a 30 G needle and use the needle to make a hole in the oviduct in front of the forceps. Keeping an eye on the pierced hole, insert the tip of the transfer pipette through the pierced hole making sure that the tip is not pushing against the wall of the oviduct (which may risk embryos flowing back out) and expel the embryos. Observe the marker bubbles inside the oviduct (Fig. 6) to ensure that the embryos have ended up preferably inside the ampulla or close by (*see* **Note 16**).
15. Remove the clip and gently push ovary and fat pad back into the body cavity.

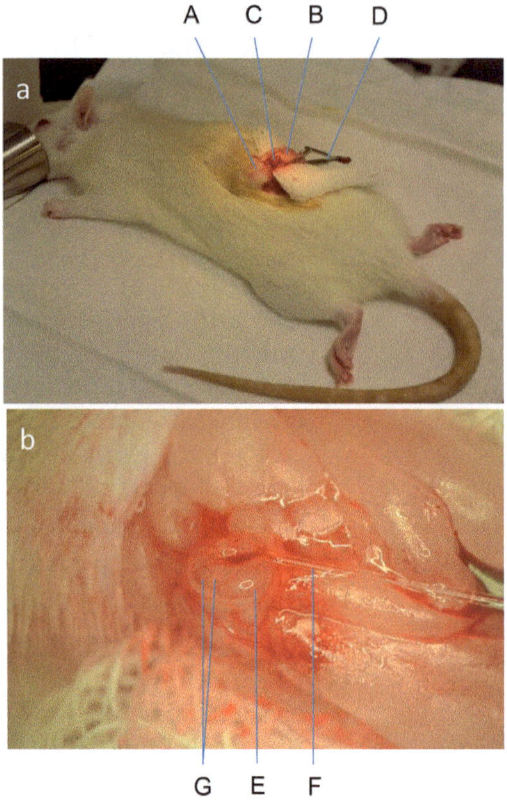

Fig. 6 Oviduct transfer. (**a**) Rat preparation for oviduct transfer. With a pair of dissection scissors, make a small longitudinal incision along the midline at the level of the last rib pair and another incision through the (**A**) body wall directly over the (**B**) ovarian fat pad. Use blunt forceps to gently exteriorize the ovarian fat pad, (**C**) ovary, and upper part of the uterus. Clip (**D**) a bulldog clamp onto the fat pad. (**b**) Using forceps, gently grab the oviduct half way between the (**E**) ampulla and the infundibulum, make a hole in the oviduct with a 30 G needle, insert the tip of the (**F**) transfer pipette through the pierced hole, and expel the (**G**) marker air bubbles and injected embryos into the ampulla

16. Close the hole by applying one suture to prevent the ovary from escaping the body cavity.
17. Check the toe-pinch reflex again to ensure that the rat is still fully anesthetized. If the reflex is detected, isoflurane needs to be adjusted accordingly.
18. Repeat the procedure on the right side (*see* **Note 17**).
19. Close the incision with a 9 mm wound clip.
20. Inject the rat with an analgesic (Meloxicam SR, 4 mg/kg) subcutaneously.
21. Place the rat in a clean cage on the warming stage until ambulatory. Recipient rats are housed one per cage due to their large litter sizes.

3.10 Clean Up Surgery Station

1. Turn off the isoflurane by pressing and holding the lever down while turning the big vaporizer knob clockwise until it reads 0.
2. Turn the oxygen flow meter knob clockwise until the center of the indicator ball has come all the way down to 0.
3. Close both oxygen cylinder valves by turning handles clockwise and check the level of the pressure gauge to make sure it reads 0.
4. Press the oxygen release button to release all the leftover oxygen from the system.
5. Turn off the evacuation system and turn off valves by twisting until they are perpendicular to the tubes.
6. Wipe down the induction chamber.
7. All work surfaces should be wiped down with 70% IPA and tools soaked overnight in low-foaming, phosphate-free, powdered detergent solution. Tools will be cleaned the next day and autoclaved for future use.

4 Notes

1. PMSG is temperature sensitive and care should be taken to prepare the working solution on ice to prevent potential hormone degradation. When diluting PMSG, the vial should be washed at least three times with saline to get all the PMSG out of the vial. Finally, the working solution should be mixed gently by pipetting before making aliquots for storage at −20 °C.
2. Mineral oil should be batch tested before using for embryo culture. In an incubator (37 °C, 6% CO_2), culture 20 zygotes or 2-cell stage embryos in a 90 μl microdrop of KSOM covered with mineral oil until the blastocyst stage. Only mineral oil exceeding a 90% blastocyst formation rate passes our quality threshold for use in embryo culture.
3. Reliable production of pseudopregnant recipient rats can be challenging. It is best to use females weighing 200–300 g as embryo transfer surgeries with heavier females complicate the surgery due to increased fat stores and potential for bleeding. We inject each recipient rat with LHRHa to synchronize estrus, which greatly reduces the number of females needed compared with using non-synchronized females. Detecting vaginal plugs in rats is also more difficult than in mice. First, plugs fall out much quicker than in mice and therefore, performing plug checks as early as possible in the morning is recommended. Additionally, some facilities house pseudopregnant recipients on wire bottom cages, enabling detection of plugs that fall out earlier in the morning. However, we have not adopted the

aforementioned housing practice as our plug detection rates suffice. Secondly, the plugs are often located fairly deep in the vaginal canal and cannot be easily detected without using a probe. To do this, the rat will often need to be stabilized inside a tube or rat retreat as rats will not hold still. We typically detect about 50–65% plug rates and close to 100% pregnancy rates.

4. In order to obtain the highest number of fertilized, injectable oocytes, we compared superovulation by injection of 20 IU of PMSG followed by 30 IU of HCG (P20/H30) with injection of 40 IU of PMSG followed by 40 IU of HCG (P40/H40). In our hands, P40/H40 gave an average of 33 oocytes per donor female while P20/H30 only resulted in an average of 19 oocytes per donor. Therefore, we have routinely been using P40/H40 for rat superovulation (P20/H30 reported by Filipiak and Saunders [19]).
5. We have observed a lower than usual embryo yield when mating virgin males as such, stud males are "primed" with non-usable females (offspring from previous injections or matings) by mating them once during a 1–2 week window before being used for microinjection matings. After these "priming matings," embryo yields are generally normal when these males are mated again.
6. Compared to mouse, rat zygotes are often surrounded by a much thicker layer of cumulus cells and their removal can be challenging. We have found that using ovine hyaluronidase instead of bovine hyaluronidase, which is often used to denude mouse zygotes, seems to be more effective for rat zygotes. However, hyaluronidase treatment alone is inadequate to remove all the cumulus cells from the oocytes. Therefore, we use a transfer pipette that is 90–100 μm in diameter to strip more of the remaining cumulus cells. It is often not possible to remove all the cumulus cells without damaging the zygote but injection can be performed even with a few remaining cumulus cells attached.
7. It is important to use RNase, DNase-free reagents and supplies when working with RNA to prevent degradation and achieve the highest mutation rate. As such, we handle all reagents for microinjection in a biosafety cabinet, use RNase, DNase-free 100× TE, and prepare mRNA injection buffer with embryo-tested water. To prevent freeze-thaw cycles, all reagents are aliquoted and stored in RNase-free, DNase-free, and low protein and nuclei acid binding microcentrifuge tubes. Fresh mRNA microinjection buffer is always prepared the day of microinjection.
8. To remove small particles in the injection mixture that might block the fine injection needles, we filter the injection mixture

with a 0.22 μm polypropylene centrifuge tube filter for 5 min at 200 × *g* before microinjection. The polypropylene centrifuge tube filters we use come with cellulose acetate membranes that have a very low binding affinity for protein and DNA. Polypropylene centrifuge tube filters cannot be used for DNA sizes over 10 kb.

9. Microinjection of rat zygotes is more challenging than mouse zygotes because of the flexibility of the oolamellar and pronuclear membranes in the rat oocyte, which make it much harder to penetrate with a conventional pronuclear microinjection needle. Therefore, rat microinjection needles have to be fabricated with a very fine tip and a longer taper than conventional mouse injection needles. The thinner injection needles will also clog easier so it is important to filter all reagents used for injection [19].

10. Due to the flexibility of the rat oocyte, we have observed a higher survival rate of microinjected rat oocytes relative to mouse. However, addition of 5 μg/ml of CB into M2 injection medium can further increase zygote survival without impairing development. CB is a toxic microfilament-inhibiting agent that relaxes the cytoskeleton and makes zygotes more flexible thus enhancing their survival during microinjection [20]. We simply add embryos into our microinjection chamber filled with 5 μg/ml of CB in M2 and let them equilibrate for 10 min before beginning injection. As a result of the relaxed cytoskeleton, embryos may exhibit blebbing which is not detrimental and will disappear once the embryos have been returned to KSOM culture and are incubated.

11. Some technicians in our lab prefer using hanging drop chambers over depression slides for injection. The hanging drop chamber allows for injection on a larger flat surface and a minimal injection needle angle relative to the microscope stage.

12. It was not an easy task to find a vendor to manufacture the custom plexiglass blocks (3 × 3 × 10 mm) which we use for assembling the hanging drop microinjection chambers. Meisner Gallery Inc. is specialized in acrylic casting (631-777-1711) and sells each block for $2.00.

13. Whether performing pronuclear or cytoplasmic injections, the major impediment with rat injections is the difficulty in penetrating the membranes due to their extreme elasticity. As previously described, the best survival for both types of injections is achieved by using the modified thinner injection needle, which enables successful piercing after a single insertion. By the afternoon, rat pronuclei are obscure and fairly large and irregular in shape complicating visualization during injection. It is often necessary to insert the injection needle quite far into

the cytoplasm and/or the pronucleus before the elastic membranes are pierced and injection of CRISPR reagents can be visualized inside the cytoplasm/pronucleus.

14. When recipients are not available for same-day transfer of injected zygotes, embryos may be cultured overnight in KSOM and transferred at 2-cell stage the following morning. If further culture is required, 2-cell stage embryos must be transferred into mR1ECM medium to avoid the 2- to 4-cell stage block of embryos [21].
15. Managing multiple species in a facility can be challenging, especially when space is limited. Co-housing mice and rats has been studied by many researchers and the results are controversial [22]. Since rats are natural predators of mice, the *Guide for the Care and Use of Laboratory Animals* recommends housing rats and mice separately to reduce the potential environmental stress to mice. However, most facilities have only one surgery suite and therefore, embryo transfer surgeries should be planned ahead of time to avoid having both species in the surgery room at the same time.
16. During rat embryo transfers, it is crucial to minimize bleeding as it can jeopardize the transfer procedure and the survival of the rat. To avoid bleeding caused by tearing of the bursa during embryo transfer procedures, we bypass entry of the bursa and make a tiny hole for the transfer pipette directly through the oviduct instead. We also use a more flexible silicone needle, a stripper tip, which aids in the transfer through the coiled oviduct as it is the perfect size to prevent backflow of the embryos during the transfer procedure.
17. The rigidity of the isoflurane anesthesia machine and large size of rats make transfer surgeries in rats more challenging than in mice. While a transfer procedure can be completed on both sides with the rat's head facing forward, we prefer reorienting the rat's head to face the technician for the transfer procedure on the right side of the rat (left side for left-handed technicians). We use long tubing for isoflurane delivery to the nose cone so that we are able to move the nose cone to the bottom of the stereomicroscope stage when reorienting the rat for transfer surgery on the right side.

Acknowledgments

We thank the Technology Development Group, Genotyping Analysis Lab, Colony Resource Group, Animal Resource Group, and Transgenic Technology Co-Op group at Genentech for their collaboration and support for producing rat models. We would also like to thank Dr. Juan M. Reyes for his suggestions and assistance with editing this chapter, and Natasha O'Neil for photography.

References

1. Lindsey JR (1979) Historical foundations in the laboratory rat. The laboratory rat. Academic Press, New York, NY
2. Iannaccone PM, Jacob HJ (2009) Rats! Dis Model Mech 2(5–6):206–210. https://doi.org/10.1242/dmm.002733
3. Bader M, Bohnemeier H, Zollmann FS, Lockley-Jones OE, Ganten D (2000) Transgenic animals in cardiovascular disease research. Exp Physiol 85(6):713–731
4. Hamilton SM, Green JR, Veeraragavan S, Yuva L, McCoy A, Wu Y, Warren J, Little L, Ji D, Cui X, Weinstein E, Paylor R (2014) Fmr1 and Nlgn3 knockout rats: novel tools for investigating autism spectrum disorders. Behav Neurosci 128(2):103–109. https://doi.org/10.1037/a0035988
5. Scarda A, Franzin C, Milan G, Sanna M, Dal Pra C, Pagano C, Boldrin L, Piccoli M, Trevellin E, Granzotto M, Gamba P, Federspil G, De Coppi P, Vettor R (2010) Increased adipogenic conversion of muscle satellite cells in obese Zucker rats. Int J Obes 34 (8):1319–1327. https://doi.org/10.1038/ijo.2010.47
6. Tarry-Adkins JL, Fernandez-Twinn DS, Chen JH, Hargreaves IP, Neergheen V, Aiken CE, Ozanne SE (2016) Poor maternal nutrition and accelerated postnatal growth induces an accelerated aging phenotype and oxidative stress in skeletal muscle of male rats. Dis Model Mech 9(10):1221–1229. https://doi.org/10.1242/dmm.026591
7. Naidu JR, Ismail R, Sasidharan S (2014) Acute oral toxicity and brine shrimp lethality of methanol extract of Mentha Spicata L (Lamiaceae). Trop J Pharm Res 13(1):101–107. https://doi.org/10.4314/tjpr.v13i1.15
8. Buehr M, Meek S, Blair K, Yang J, Ure J, Silva J, McLay R, Hall J, Ying QL, Smith A (2008) Capture of authentic embryonic stem cells from rat blastocysts. Cell 135 (7):1287–1298. https://doi.org/10.1016/j.cell.2008.12.007
9. Li P, Tong C, Mehrian-Shai R, Jia L, Wu N, Yan Y, Maxson RE, Schulze EN, Song H, Hsieh CL, Pera MF, Ying QL (2008) Germline competent embryonic stem cells derived from rat blastocysts. Cell 135(7):1299–1310. https://doi.org/10.1016/j.cell.2008.12.006
10. Koller BH, Hagemann LJ, Doetschman T, Hagaman JR, Huang S, Williams PJ, First NL, Maeda N, Smithies O (1989) Germ-line transmission of a planned alteration made in a hypoxanthine phosphoribosyltransferase gene by homologous recombination in embryonic stem cells. Proc Natl Acad Sci U S A 86 (22):8927–8931
11. Thompson S, Clarke AR, Pow AM, Hooper ML, Melton DW (1989) Germ line transmission and expression of a corrected HPRT gene produced by gene targeting in embryonic stem cells. Cell 56(2):313–321
12. Zijlstra M, Li E, Sajjadi F, Subramani S, Jaenisch R (1989) Germ-line transmission of a disrupted beta 2-microglobulin gene produced by homologous recombination in embryonic stem cells. Nature 342(6248):435–438. https://doi.org/10.1038/342435a0
13. Geurts AM, Cost GJ, Freyvert Y, Zeitler B, Miller JC, Choi VM, Jenkins SS, Wood A, Cui X, Meng X, Vincent A, Lam S, Michalkiewicz M, Schilling R, Foeckler J, Kalloway S, Weiler H, Menoret S, Anegon I, Davis GD, Zhang L, Rebar EJ, Gregory PD, Urnov FD, Jacob HJ, Buelow R (2009) Knockout rats via embryo microinjection of zinc-finger nucleases. Science 325(5939):433. https://doi.org/10.1126/science.1172447
14. Li D, Qiu Z, Shao Y, Chen Y, Guan Y, Liu M, Li Y, Gao N, Wang L, Lu X, Zhao Y, Liu M (2013) Heritable gene targeting in the mouse and rat using a CRISPR-Cas system. Nat Biotechnol 31(8):681–683. https://doi.org/10.1038/nbt.2661
15. Gudbergsdottir S, Deng L, Chen Z, Jensen JV, Jensen LR, She Q, Garrett RA (2011) Dynamic properties of the Sulfolobus CRISPR/Cas and CRISPR/Cmr systems when challenged with vector-borne viral and plasmid genes and protospacers. Mol Microbiol 79(1):35–49. https://doi.org/10.1111/j.1365-2958.2010.07452.x
16. Manica A, Zebec Z, Teichmann D, Schleper C (2011) In vivo activity of CRISPR-mediated virus defence in a hyperthermophilic archaeon. Mol Microbiol 80(2):481–491. https://doi.org/10.1111/j.1365-2958.2011.07586.x
17. Jinek M, Chylinski K, Fonfara I, Hauer M, Doudna JA, Charpentier E (2012) A programmable dual-RNA-guided DNA endonuclease in adaptive bacterial immunity. Science 337 (6096):816–821. https://doi.org/10.1126/science.1225829
18. Chapman KM, Medrano GA, Jaichander P, Chaudhary J, Waits AE, Nobrega MA, Hotaling JM, Ober C, Hamra FK (2015) Targeted germline modifications in rats using CRISPR/Cas9 and spermatogonial stem cells. Cell Rep 10(11):1828–1835. https://doi.org/10.1016/j.celrep.2015.02.040

19. Filipiak WE, Saunders TL (2006) Advances in transgenic rat production. Transgenic Res 15 (6):673–686. https://doi.org/10.1007/s11248-006-9002-x

20. Hu LL, Shen XH, Zheng Z, Wang ZD, Liu ZH, Jin LH, Lei L (2012) Cytochalasin B treatment of mouse oocytes during intracytoplasmic sperm injection (ICSI) increases embryo survival without impairment of development. Zygote 20(4):361–369. https://doi.org/10.1017/S0967199411000438

21. Zhou Y, Galat V, Garton R, Taborn G, Niwa K, Iannaccone P (2003) Two-phase chemically defined culture system for preimplantation rat embryos. Genesis 36(3):129–133. https://doi.org/10.1002/gene.10203

22. Greene TM, Redding CL, Birkett MA (2014) Effects of rat visual, olfactory, or combined stimuli during cohousing on stress-related physiology and behavior in C57BL/6NCrl mice. J Am Assoc Lab Anim Sci 53(6): 647–652

Chapter 17

Molecular Aspects of Zinc Finger Nucleases (ZFNs)-Mediated Gene Editing in Rat Embryos

Xiaoxia Cui

Abstract

This chapter contains a collection of protocols involved in using ZFNs to create rat models with various types of genome editing, including simple knockout, point mutation, large deletions, floxing, and insertions. The protocols cover ZFN and donor design criteria, in vitro transcription of ZFNs, validation of ZFNs activity in cultured cells, RNA stability test, microinjection sample preparation, genotyping and in vitro confirmation of floxed alleles, and Southern blot analysis, most of which are not limited to using ZFNs. Instead they apply to model creation in general. When appropriate, a comparison between ZFNs and CRISPR is provided. Standard pronuclear microinjection per se is not discussed.

Key words Zinc finger nucleases, Pronuclear microinjection, Single-cell embryos, Knockout, Knockin, Conditional knockout, Nonhomologous end joining, Homology directed repair

1 Introduction

The first knockout mice were created in 1987 by targeting embryonic stem (ES) cells in culture and injecting targeted ES cells into recipient blastocysts for germline contribution [1–3]. The rat ES cells were successfully cultured and demonstrated germline transmission competency only after more than two decades later in 2008 [4, 5], largely contributing to the fact that the mouse is the most used species for research models. However, before the rat ES cells had ample time to be widely used as for the mouse ES cells, development of programmable nucleases changed the genome engineering landscape.

The ES cell technology relies on low-frequency random breaks in the target site to mediate homologous recombination. Even with antibiotic selection, some genomic loci are more difficult than others to target. Additionally, targeted ES cells do not always go germline when injected into recipient blastocysts, rendering unpredictable model completion timeline. On the contrary, programmable nucleases cleave genomic DNA sequence specifically with

Chengyu Liu and Yubin Du (eds.), *Microinjection: Methods and Protocols*, Methods in Molecular Biology, vol. 1874, https://doi.org/10.1007/978-1-4939-8831-0_17, © Springer Science+Business Media, LLC, part of Springer Nature 2019

significantly higher efficiency and lead to non-homologous end joining-mediated, undefined in/dels, or stimulate precise homology-mediated repair. The meganuclease I-SceI was first shown in 1994 to introduce targeted double-strand breaks in mouse cells [6] and is still widely used in DNA repair studies [7]. Later, a meganuclease was engineered to target a novel sequence, demonstrating promising programmability of meganucleases [8]. However, they are in general difficult to engineer and retarget, primarily because the DNA binding domain that determines specificity overlaps with nuclease activity motif, and necessary mutations to change binding sequence also have a high chance to inactivate nuclease activity [9].

ZFNs are fusion proteins between zinc finger protein and the nuclease domain of FokI type II restriction enzyme. Zinc finger proteins are composed of zinc finger motifs, each about 30 amino acids long forming an alpha helical structure. The amino acids on one side of the helix directly interact with three nucleotides. Multiple fingers can be assembled together to form a fusion protein for binding a longer sequence with higher specificity and affinity [10]. However, home-brew ZFNs, even only containing three fingers (to recognize 9 bp), are difficult to build [11]. The low success rate for functional fusion proteins was improved by using two-finger modules, which have been pre-validated to bind six basepairs, minimizing the number of fusion junctions. Even though screening multiple pairs of ZFNs is necessary to obtain an active pair, ZFNs were a revolutionary technology, proving the power of programmable nucleases for genome editing [12, 13]. Improvements by mutating residues critical to dimerization in the FokI domain increased specificity [14, 15] and activity [15]. The commercialization of custom ZFNs, pricy but with validated activity, allowed the easy creation of modified cell lines [16], zebra fish [17], and notably, the first ever targeted knockout rats [18]. Soon after many knockouts were made in many species, including mouse [19], rabbits [20], pigs [21], catfish [22], to name a few. Proof of concept on creating knockin [23] as well as conditional knockout [24] in the rat were also both first obtained using ZFNs. Importantly, by manipulating single-cell embryos, modified alleles detected in founders almost always transmit through the germline.

Transcription activator-like effector nucleases (TALENs) were first published in 2010 [25] and adapted to mammalian system in 2011 [26]. Like ZFNs, TALENs are fusions between the FokI nuclease domain and the TAL effector DNA binding domains, composed of highly conserved repeats each directly interacting with a nucleotide. The design of a pair of TALENs is much simpler than that of a pair of ZFNs. TALENs have been used for rat model creation, as successfully as with ZFNs [27]. The once unclear IP landscape partially impeded the technology to quickly reach its full potential. Merely a few years later, CRISPR technology almost at once took over as the dominant technology for gene editing.

CRISPR technology has exploded in its usage like no other before [28, 29]. Without involving protein engineering as all other nuclease technologies demand, targeting a new sequence is as easy as changing 20 bases in the guide RNA to base pair to the target site next to a short consensus sequence (such as a NGG as a PAM for s. pyrogen Cas9). Practically every lab has the potential to master the design, production, and validation of CRISPR reagents at low cost. The cleavage efficiencies are generally good, and different delivery formats are available, including as expression plasmid DNA, as in vitro transcripts, as ribonucleoprotein complexes, or by viral means. Combined with the use of single-stranded oligo donors, an incredible number of cell lines and animal models have been created in the past couple years. Knockout, small modifications, and large deletions can all be done with high efficiency. More recently, several papers were published on using electroporation to deliver CRISPR targeting reagents into single-cell embryos [30–33]. Likely ZFN mRNAs can be done so as well. However, no one so far has been able to electroporate donor plasmid into the embryos.

Even though CRISPR reagents are much cheaper and easier to produce and often highly efficient at introducing KOs, large deletions, and small KIs, active ZFN pairs in our hands mediate kb-sized insertions more effectively, such as site-specific Cre lines [24] and reporters into Rosa26 site (unpublished). ZFNs, as well as TALENs, should be treated as valuable alternative technologies. Here, I summarized protocols involved in using ZFNs to create rat models and their genetic validation.

2 Materials

2.1 ZFN In Vitro Transcription

1. ZFNs.
2. *Xba*I or *Xho*I restriction enzyme, TE saturated phenol.
3. In vitro transcription kit, RNase inhibitor, and Poly (A) polymerase tailing kit.
4. 5 M Ammonium Sulfate.
5. Agarose gel.

2.2 Validation of ZFNs

1. Rat C6 cells.
2. Complete growth medium: Ham's F12K medium, 2.5% of fetal bovine serum, 15% of horse serum.
3. Nucleofector II 96-well shuttle system and SF solution.
4. Extraction buffer.

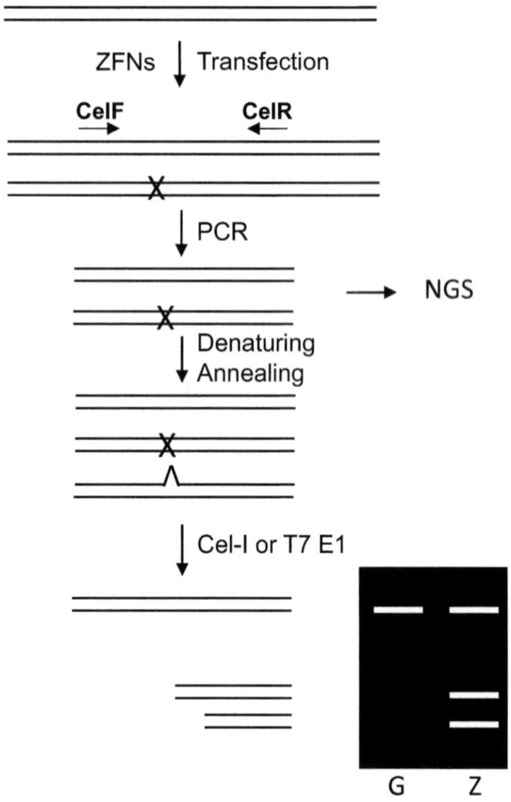

Fig. 1 ZFN activity validation. Paired ZFN in vitro transcripts are transfected to rat C6 cells. Active ZFNs modify a portion of the target site in the transfected pool. Using primers flanking the target site, CelF and CelR, both wild-type and modified sequences are amplified. The PCR mixture is heat-denatured and slowly cooled to allow wild-type and modified molecules to form heteroduplexes that are cleaved by a single-strand-specific endonuclease, such as Cel-I or T7 E1. The digested products are resolved on a gel. G: GFP transfection control. Z: ZFN transfected sample. Relative ratio of intensities of cleaved bands to parental band in lane Z is indicative of ZFN activity. Alternatively, the PCR mixture can be subjected to next generation sequencing (NGS) for an accurate measure of NHEJ rate

5. Primers 300–400 bp apart flanking the target site: CelF, CelR (*see* Fig. 1).
6. Mutation detection kit.
7. 10% TBE PAGE gel.
8. Image J software (https://imagej.nih.gov/ij/download.html).

2.3 Prepare Microinjection Samples

1. Endotoxin-Free Plasmid Maxiprep kit.
2. Microinjection buffer: 1 mM Tris–HCl, pH 7.4, 0.25 mM EDTA, filtered by using 0.2 μm syringe filters.

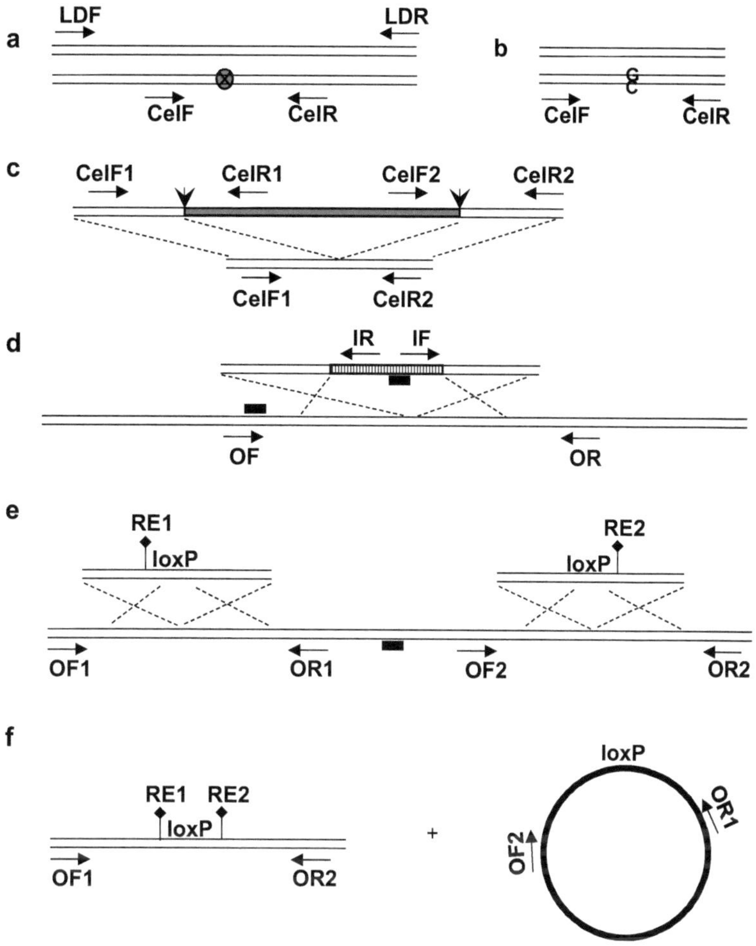

Fig. 2 Genotyping design for different types of modifications in the rat. Simple in/dels (**a**) and small homology directed modifications such as point mutations (**b**) can be identified by PCR amplifying genomic DNA with CelF and CelR primers as in Fig. 1, followed by sequencing of the PCR products. Large deletions are detected by using primers LDF and LDR, each about 1 kb from the cleavage site. Deletions mediated by two pairs of ZFNs can be identified by using primers CelF1 and CelR2, flanking the two cut sites (arrow heads) (**c**). (**d**) Insertion confirmation requires junction PCRs, using one primer outside of the homology arm (OF, OR) and the other within the insert (IR, IF). Both junctions should be detectable. For Southern blotting, probes should be designed both outside of homology arms and within the insert (black rectangles). **(e)** LoxP sites should be detected using at least one primer outside of homology arms and PCR products to be digested by the restriction enzyme with recognition site inserted along the loxP site. The restriction sites are also handy if Southern is planned. The genomic DNA can be digested by the two restriction enzymes and use a probe that is between the loxP site (black rectangle). **(f)** Presence of two loxP sites on the same chromosome can be proven by in vitro Cre excision, resulting in deletion between loxP site and circularization of the deleted sequence. The DNA circle can be detected by using primers OR1 and OF2

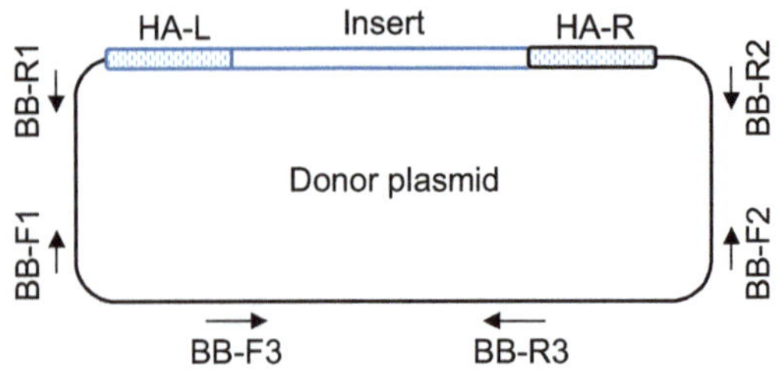

Fig. 3 Detection of donor random integration in knockin founder animals. Three sets of primers each 500 bp to 1 Kb apart are designed to spread along the donor backbone (BB-F1 & BB-R1, BB-F2 & BB-R2, BBF3 & BBR3). Given random integration has to occur after the donor plasmid linearizes, and terminal recession, 3 pairs should detect all possibilities. Genomic DNA negative for all 3 PCRs are likely without random integration

2.4 Genotyping by PCR

1. High Fidelity Taq Polymerase.
2. Primers: CelF, CelR, LDF, LDR, CelF1, CelR2, OF, IR, IF, OR, OF1, OR1, OF2, OR2 (*see* Fig. 2) and BB-F1, BB-R1, BB-F2, BB-R2, BB-F3, BB-R3 (*see* Fig. 3).
3. Cre recombinase (NEB).

2.5 Southern Blots

1. Mammalian genomic DNA kit.
2. Restriction enzymes (*see* **Note 1**), TE-saturated phenol:chloroform:isoamyl alcohol.
3. 0.8% agarose gel in 1×TBE.
4. UV crosslinker.
5. PCR DIG Probe Synthesis Kit, nylon membrane, DIG Easy Hyb Granules, DIG Detection Kit (*see* **Note 2**).
6. Paper towel and filter paper, transfer buffer reservoir.
7. 20× SSC as transfer buffer.
8. Washing solution I: 2× SSC + 0.5% SSC; Washing solution II: 0.5× SSC + 0.1% SDS.
9. Imaging System.

3 Methods

3.1 Considerations for ZFN and Donor Design

Even though Sigma does the design, the more you know about the target gene and the more specific requests you provide, the more likely they will design the ZFNs that best fit your need.

3.1.1 For Generation of Knockouts

1. Target the first two thirds of the coding sequence but avoid the very beginning of the coding sequence (*see* **Note 3**).
2. Target sequence coding for the critical functional domains.

3. If targeting the second to last exon, frameshift may not cause premature stop codon to trigger non-sense-mediated decay (*see* **Note 4**).
4. Target the common exons if alternative splicing isoforms are present.
5. If targeting close to intron and exon junction, it is preferred to target exons with the number of base pairs that is not the multiple of three so that exon skipping leads to frameshift.
6. Large deletions can be achieved using two pairs of ZFNs (*see* Fig. 2c). Usually, additional base pairs are deleted or inserted between the two cut sites.
7. Compared to using CRISPR, it is more challenging to use two pairs of ZFNs to create large deletions over 10 kb, partially due to the difficulty to obtain equally active ZFNs for both sites, and partially, the competition from modifications at each individual target site.
8. For genes lack common exons among all splicing isoforms, in addition to using two pairs of ZFNs to delete more than one exon, insertion of a polyA signal early in the coding sequence to terminate transcription is another option.

3.1.2 For Generation of Knockins

1. Choose cleavage sites as close to the site of modification as possible.
2. If after insertion, the ZFN target site is intact, silent mutations should be designed in the donor so that the donor plasmid will not be cleaved, and the targeted allele will not be recut and introduce undesired in/dels.
3. For introducing point mutations, it helps with genotyping if a restriction site can be engineered by incorporating silent mutations.
4. Avoid high GC content, high repeat sequences in homology arms when possible, even if it means to use shorter homology. Homology arms of 800 bp each are most frequently used (*see* **Note 5**).
5. When using a 2A peptide to create a cistronic fusion, take into consideration the ORF upstream will end up with a C-terminal addition of 20 amino acids or so 2A sequence, which can inactivate certain proteins where a free C-terminus is critical to function.
6. IRES can lead to unpredictable expression level of the downstream ORF and may also affect the upstream ORF [24]. Insert IRES-ORF immediately after stop codon to minimize the potential impact on 3′ UTR.

7. If a cDNA is inserted, expect lower than endogenous level of expression, even with artificial or endogenous introns.

3.1.3 For Floxing Alleles

1. Target functionally critical exons that are shared among different splicing isoforms.
2. Avoid floxing coding sequences with base pairs that are the multiple of three, if possible, so that frameshift is introduced upon Cre-mediated excision.
3. When possible insert at least 250 bp from intron exon junctions to avoid interfering with splicing.
4. Avoid inserting loxP sites into 5′ UTR or 3′ UTR if possible, unless the UTRs are known to tolerate insertions.
5. Format of donor: two separate plasmid or oligo donors, each delivering a loxP site, can be used (*see* **Note 6**). Alternatively, a single-plasmid donor with both loxP sites can be used.

3.2 ZFN mRNA Prep and Donor Plasmid Prep

1. ZFN plasmid is linearized at the X*ba*I or X*ho*I site, both located at the 3′ end of the FokI ORF (*see* **Note 7**), phenol extracted, precipitated, and resuspended in TE or H_2O at 0.18 μg/μl.
2. 5′ capped message RNA was prepared using an in vitro transcription kit containing T7 RNA polymerase according to the manufacturer's instructions, such as Cellscript's MessageMax T7 capped transcription kit by adding in the following order: 5.5 μl of template from **step 1**, 2 μl of T7 transcription buffer, 8 μl NTP mix, 2 μl 100 mM DTT, 0.5 μl RNAGuard, and 2 μl T7 RNA Polymerase and incubating at 37 °C for 30 min to 1 h.
3. Synthesize polyA tail by using a Poly(A) polymerase tailing kit, such as from Lucigen, by adding the following to the in vitro transcription reaction: 55.5 μl of H_2O, 10 μl of buffer, 2.5 μl of ScriptGuard RNase Inhibitor, and 2 μl of Poly(A) Polymerase and incubating at 37 °C for at least 1 h.
4. Add an equal volume of cold 5 M NH_4OAc, mix and incubate at ice for 15 min, and spin at top speed (>12,000 × *g*) at 4 °C for 15 min. Wash the pellet with 70% ethanol twice, dry and then dissolve in injection buffer (*see* **Note 8**).
5. Measure mRNA concentration.

3.3 Endotoxin-Free Donor Plasmid Purification

1. Follow the instructions for an Endotoxin-Free Plasmid Maxiprep kit, and include optional wash steps, if there are any.
2. Plasmid or oligo donors have to be tested for lack of RNase activity.

3.4 Stability Assay (See Note 9)

1. Incubate ZFN mRNA at 1 μg/μl concentration at 37 °C for 3 h.

2. If donors are to be used, incubate both ZFN mRNA alone and ZFN + donor in parallel.
3. Run on the gel alongside equal amount of mRNA without incubation.
4. Flash gel or thin agarose gel run fast. Observe under UV right away to assess the integrity of the RNA.

3.5 Activity Validation in Cultured Cells

1. Maintain a culture of the rat C6 cells and use between passage 3 to passage 25.
2. Harvest cells at 70–80% confluence by trypsinization and neutralize with growth medium.
3. Count the cells and pellet 200,000 cells per reaction and always include a GFP control.
4. Wash twice with at least equal volume of PBS as growth medium.
5. While waiting for centrifugation, set the Nucleofector on program FF-127.
6. Resuspend cells in 20 μl per reaction of SF solution and aliquot 20 μl of cells to each tube.
7. Combine the ZFN mRNA pair at 1:1 mass ratio.
8. Add 1 μg mRNA mixture onto each cell aliquot.
9. Immediately load onto the preset nucleofector and nucleofect (*see* **Note 10**).
10. Transfer cells to 96-well culture plate with pre-warmed medium.
11. Harvest the cells 20 h after transfection to analyze cleavage.

3.6 Mutation Detection Assay

1. Remove culture medium from above transfected cells, wash the cells with PBS, and incubate in 15 μl of trypsin per well for 5 min at 37 °C.
2. Transfer the cell suspension into 100 μl of extraction buffer and incubate accordingly, such as Lucigen's QuickExtract at 68 °C for 10 min, followed by at 98 °C for 3 min.
3. Assemble a PCR reaction using a high fidelity Taq polymerase, such as ThermoFisher's AccuPrime Taq polymerase high fidelity by combining the following for PCR: 1 μl of template in **step 2**, 2.5 μl of Buffer II, 2.5 μl of 10 μM each primer CelF and CelR, 0.25 μl of AccuPrime Taq polymerase high fidelity, and16.25 μl of H_2O.
4. PCR amplify using the following program or optimized program for a specific sequence: 95 °C, 5 min, 35 cycles of 95 °C, 30 s, 60 °C, 30 s, and 68 °C, 45 s, and then 68 °C, 5 min.

5. Incubate 10 μl of the above PCR reaction under the following program: 95 °C, 10 min, 95–85 °C, at −2 °C/s, 85–25 °C at −0.1 °C/s. If a different polymerase is used, conditions should be adjusted according to the manufacturer's instructions.
6. Add 1 μl of each nuclease S (Cel-I) and enhancer to the above reaction and incubate at 42 °C for 20 min. The mixture is resolved on a 10% polyacrylamide TBE gel. Relative density of parental and cut bands were used to estimate efficiency using Image J software (Fig. 1) (*see* **Note 11**).

3.7 Sample Preparation for Microinjection

1. To prepare microinjection samples, add each ZFN mRNA to injection buffer to a final concentration of 5–10 ng/μl (*see* **Note 12**) in 50–100 μl volume. If donor is needed, donor plasmid is added at 1–2 ng/μl final concentration (*see* **Note 13**), or single-strand oligo donor at 1–100 ng/μl (*see* **Note 14**).
2. Spin the sample at top speed (>12,000 × *g*) for 5 min at RT or 4 °C and take the top 2/3 volume to a new tube for injection (*see* **Note 15**).

3.8 Founder Identification

3.8.1 Methods for Genomic DNA Preparation

1. Add 100 μl of QuickExtract to each tube containing a toe clip and incubate at 55 °C for 30 min, followed by 68 °C for 10 min and 98 °C for 3 min (*see* **Note 16**).
2. Tail clips or ear notches are both convenient samples to collect. For Southern blot analysis. Tissues were first incubated in lysis buffer (100 mM Tris–HCl, pH 8.8; 50 mM EDTA, 0.5% SDS; 200 mM NaCl; 300 μg/ml proteinase K) at 55 °C for 2–5 h with occasional inversions. Supernatant was collected and precipitated. The washed and dried pellet was then dissolved in TE buffer (10 mM Tris–HCl, pH 8.0, 1 mM EDTA).
3. Harvest and weigh out up to 25 mg of tissue for each column according to instructions of a genomic DNA miniprep kit. Use multiple columns for each sample if necessary to avoid overloading columns. Follow the kit instructions closely.

3.8.2 Knockout

ZFNs generate both small in/dels and larger deletions to 1 kb range, depending on the target locus [19]. To identify small in/dels, the target region is amplified as in **steps 3** and **4** under Subheading 3.4. Sequence the PCR samples for detailed NHEJ modifications. If the founders are highly mosaic, TA cloning followed by sequencing is necessary.

To detect large deletions, an example PCR is shown below. If a different Taq polymerase is used, adjust conditions as necessary.

1. Combine 1 μl of tissue lysis, 2.5 μl of Buffer II, 2.5 μl 10 μM of each primer LDF and LDR, 0.25 μl of AccuPrime Taq Polymerase High Fidelity and 16.25 μl H_2O.
2. Amplify with the following program: 95 °C, 5′, 35 cycles of 95 °C, 30 s, 62 °C, 30 s, and 68 °C, 2 min, 68 °C, 10 min.

3. Resolve the samples on 1% agarose gel and band isolate products smaller than parental target bands for sequencing.

3.8.3 Knockin

As shown in Fig. 2d, junction PCR primer pairs OF and IR and IF and OR are designed to only amplify the site-specific targeted integration, not the donor. Make sure OF and OR primers do not anneal to the donor (*see* **Note 17**).

3.8.4 Floxed Alleles

Animals are first tested to identify those with two loxP sites by using primers OF and OR pairs to amplify each target site. If restriction sites are used, digest the PCR products with respective restriction enzyme (*see* **Note 18**). If plasmid donors are used, one primer has to be outside of the homology arms.

Those positive for both loxP insertions can carry loxP sites in separate alleles. To confirm the presence of floxed alleles, breeding to wild type to obtain offspring with two loxP sites is the ultimate test. However, if the floxed region is small, PCR across the floxed region and confirm by either sequencing or restriction digest (*see* **Note 18**). With longer floxed regions, a Southern blot on genomic DNA digested at the inserted restriction sites next to loxP sites or detection of in vitro Cre-mediated recombination can both demonstrate whether loxP-containing alleles are in fact floxed.

In vitro Cre-mediated recombination:

1. Combine 2 μl of 10× Cre buffer, and 1 unit of recombinant Cre recombinase with 0.1 μg or 1 μg of genomic DNA in two separate reactions in final volume of 20 μl. Incubate at 37 °C for 30 min.
2. Combine the following for PCR: 1 μl of the above Cre reaction, 2.5 μl of each 10 μM OF1 and OR2 primers (for detection of deletion between loxP sites, Fig. 2e) or OR1 and OF2 primers (for detection of excised circle, Fig. 2f).
3. Amplify with the following program: 95 °C, 5′, 35 cycles of 95 °C, 30 s, 62 °C, 30 s, and 68 °C, 2 min, 68 °C, 10 min. OR1 and OF2 primers will only amplify when sequence between loxP sites circularizes.
4. Sequence the PCR products to confirm.

At low DNA concentration, single-molecule Cre-mediated recombination is preferred over two-molecule reactions. Both deletion and excision PCRs should be positive for Cre reactions with 0.1 μg or 1 μg of genomic DNA.

3.8.5 Detection of Donor Random Integration Using PCR of Donor Backbone

Even when donor plasmid is delivered as supercoiled DNA, random integration of the plasmid still occurs. A simple method to detect random integration is to PCR amplify genomic DNA using at least three sets of primers on the donor backbone, each about 500 bp apart (Fig. 3). Southern blot is a complementary method to backbone PCRs.

3.9 Southern Blots

Southern is still the gold standard for proving both correct targeted integration and lack of random integration, especially with larger insertions.

3.9.1 Digestion with Restriction Enzymes

1. Restriction sites should be chosen so that different band patterns will result in wild type and targeted animals (*see* **Note 1**).
2. Digest 10 μg of genomic DNA in 100 μl of final volume using up to 9 μl of enzymes combined. For most enzymes, 2 h at 37 °C should be sufficient.
3. Phenol extract the digest and precipitate with 10 μl of 3 M NaOAc and 250 μl of 100% ethanol. Centrifuge at top speed for 15 min at 4 °C. Discard the supernatant.
4. Wash the pellet with 70% ethanol.
5. Resuspend the air dried pellet in 21.5 μl of TE and remeasure concentration (see **Note 19**).

3.9.2 Electrophoresis and Transfer

1. Pour a 0.8% 1×TBE agarose gel.
2. Load 2 μl of DIG DNA Marker (lambda/H*ind*III).
3. Load all samples along with plasmid positive controls if necessary and run at 5 V/cm.
4. Incubate the gel in 100 ml of 0.25 N HCl for 15 min with slow rotating motion at room temperature.
5. Discard HCl solution and replace with 100 ml of denaturing solution for 15 min. Discard the solution and replace with 100 ml of denaturing solution for another 15 min.
6. Discard the denaturing solution and replace with 100 ml of neutralization solution for 15 min and repeat once.
7. Equilibrate the gel in 100 ml of 20× SSC for 5 min.
8. Assemble the transfer as shown in Fig. 4.
9. Transfer 2 h to overnight.

3.9.3 Probe Design

1. An internal probe from the insertion sequence and a target probe outside of the homology arms are needed (Fig. 2d). For floxing alleles, design a probe in the floxed region (between the two inserted restriction sites, Fig. 2e).
2. Scan respective region for 200–300 bp stretch with 45–60% GC content. Try multiple probes if necessary.
3. Calculate hybridization temperature (T_{hyb}) using the following formula: $T_m = 49.82 + 0.41 \times (GC\%) - 600/\text{number of bp in probe}$. For example, T_m of a 300 bp probe with 60% GC is $T_m = 49.82 + 0.41 \times 60–600/300$ $T_{hyb} = T_m - 25$ °C.

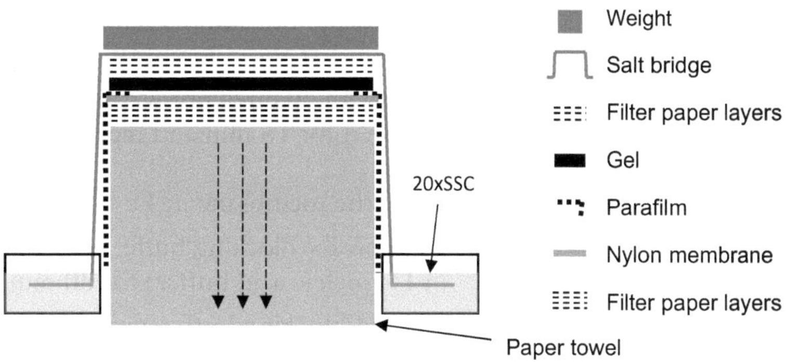

Fig. 4 The setup for downward transfer of DNA from agarose gel to nylon membrane for Southern blot analysis. From bottom to top: a generous stack of paper towel (gray rectangle) is placed on benchtop and covered with at least three layers of dry filter papers (black dashed lines). An additional layer of filter paper is pre-wetted with 20xSSC and is then placed on top, covered by pre-wetted nylon membrane (gray thin line), which in turn is covered by processed agarose gel (black rectangle). Another 3 layers of filter paper, salt bridge, and a light weight, such as a gel tray. The bottom portions of the salt bridge should be immerged in transfer buffer (gray-filled rectangles) in the two reservoirs (black lined rectangles) for transfer buffer (gray-filled rectangles)

3.9.4 Probe Synthesis

1. PCR amplify the probe region from genomic DNA.
2. Use 1 μl of 1:10 dilution of the PCR reaction as template in labeling reaction: add 5 μl of vial 3, 5 μl of vial 2, 5 μl of primer mix (10 μM), 0.75 μl of vial 1 and add 33.25 μl of H_2O.
3. Cycle with the following conditions: 95 °C, 5′; 30 cycles of 95 °C, 30″, 60 °C, 30″ and 72 °C, 40″; 72 °C, 10′ and then keep at 4 °C.
4. Run products on the gel next to unlabeled PCR products. A shift to larger size indicates successful labeling.

3.9.5 Hybridization

1. After transfer, equilibrate the membrane in 2× SSC for 5 min.
2. Add 64 ml water to a bottle of hybridization granule and heat up to help dissolve.
3. Place the membrane on a piece of filter paper to air dry for 5 min.
4. Crosslink the membrane on setting 1200 J/cm^2.
5. Prehybridize in 10 ml of solution from **step 2** at T_{hyb} for 30 min to overnight.
6. Heat probe to 98 °C for 5 min and snap cool on ice.
7. Discard prehybridization solution and add 10–20 μl of probe to 10 ml of preheated hybridization solution (*see* **Note 20**). Hybridize at least 16 h.

8. Discard hybridization solution. Wash with 20 ml of washing solution I at room temperature for 15 min twice in the oven.
9. Wash with 20 ml of pre-warmed washing solution II at 68 °C in the oven for 15 min and repeat.

3.9.6 Detection

1. Rinse the membrane in 1× Meleic acid buffer.
2. Block in 1× blocking buffer (1:10 dilution from the 10× stock in 1× meleic acid buffer) for 30 min to 3 h.
3. Discard blocking buffer and add 20 ml of blocking buffer.
4. Centrifuge the DIG antibody at top speed for 5 min and add 2 μl to the fresh blocking buffer.
5. Wash twice in 1× washing buffer for 15 min at room temperature *f.* Equilibrate the membrane in 1× detection buffer for 5 min.
6. Place the membrane between two layers of protection films and add 1–2 ml of CDP-Star onto the membrane.
7. Incubate at room temperature for 5 min.
8. Capture fluorescent signal. On ChemiDoc, use chemiluminescence setting, no filter, and automatic capture every 5 min, up to an hour.

4 Notes

1. We found that genomic DNA prepared from tail or ear clips were partially resistant to digestion by restriction enzyme H*in*-*d*III. Even with two rounds of proteinase K digestion and phenol extraction, the digestion was inconsistent and incomplete. On the other hand, gDNA from brain tissue worked well with the same treatment [24].
2. Not all commercially available membranes work equally. What worked best in our hands is from Roche along with hybridization buffers.
3. Targeting too close to ATG codon, or damaging ATG codon, especially if there is an in-frame, alternative start codon, can lead to the use of the downstream start codon and result a partial protein with a small N-terminal truncation, instead of causing out of frame translation and non-sense-mediated decay of the message RNA. The Rag1 KO rats are an example [34].
4. Premature stop codons do not always trigger non-sense-mediated decay of the mRNA. We have observed no reduction of target mRNA level in multiple KO rats, such as PXR, CAR, and Park2, even though western blot shows no protein, and KO phenotypes are observed (unpublished).

5. We did not observe a correlation between insertion size to efficiency, in contrast to using CRISPR. For example, a 5 kb Mdr1 human cDNA cassette was inserted into the rat Mdr1a locus at comparable efficiency as a restriction site (unpublished). For larger insertions, if active ZFNs are available at the insertion site, they might be more advantageous than CRISPR.
6. The majority of KI founder rats we created using plasmid donors produced around 50% offspring carrying the KI allele, even though the founders can be higher mosaic with various NHEJ alleles, indicating targeted integration occurred before the first cell division or even before pronuclear fusion and ZFNs last into multi cell stages [23, 24]. Similarly, in founders with loxP sites at both loci, the loxP sites were most of the time in the same allele, likely the only one available in the injected pronucleus [24]. Not enough data are available using ZFNs with oligo donors. However, the frequent observation of mutations in the integrated loxP sites by using CRISPR and oligo donors also argues for the advantage of using plasmid donors.
7. Complete linearization is critical to allow T7 RNA polymerase to fall off the template and reinitiate a new transcript, rather than continuing along the circular plasmid and generating long, insoluble RNA molecules.
8. Initially, we were using MegaClear column to purify IVT reactions. However, there were times we observed fiber-like substance in the injection sample that blocked the end of the injection needle. Ammonium sulfate precipitation instead reduced the occurrence needle clogging.
9. Unlike commonly believed, in vitro transcripts without RNase or in the absence of Mg^{2+} are very stable. The stability assay is a simple quality control of the mRNA preparation. Oligo and plasmid donors should be put through the same test to confirm the absence of RNase, which will degrade mRNA in the mixture before the delivery into the cells. Without contamination, the incubated RNA should look the same as RNA without incubation on the gel.
10. If using cuvettes, loosen each cuvette cap and open the package of transfer pipets for all reactions to save time. Cells should be added to the mRNA mixture and then to cuvettes one reaction at a time and immediately nucleofected to minimize the potential exposure of mRNAs to nucleases from culture medium that was not completely removed by washing in PBS and also minimize the toxic effect from the aluminum electroplates in the cuvette.

11. Microinjection of ZFNs into rat zygotes most often results in higher NHEJ rates than in transfected cells, likely because of more efficient delivery of ZFNs. On the other hand, mutation detection assay is likely more accurate at estimating ZFN activity than that of CRISPR because ZFN cleavages tend to result in various NHEJ products, whereas CRISPR often generates predominant, 1 bp insertion or deletion at the cleavage site. ZFN animals tend to have higher mosaicism [19, 23, 24].
12. Up to 200 ng/μl mRNA were tested without adverse effect but also did not increase cutting efficiency compared to around 10 ng/μl range. When two pairs of ZFNs are used, it is helpful to adjust the amount of mRNA based on their relative in vitro activity, i.E., the less active pair should have a higher concentration than the more active one for obtaining large deletion or floxed alleles.
13. Donor plasmid as high as 10 ng/μl has been injected. However, in our hands, plasmid concentration higher than 2 ng/μl lowers birth rates and increased random integration frequency, without significantly increasing efficiency.
14. We injected oligo donors at low concentrations (0.3–0.6 ng/μ l) and obtained some animals with single loxP site insertion but not both [24]. The much higher concentrations up to 100 ng/ ul that is commonly used with CRISPR are likely to work better.
15. Occasionally, the sample can be difficult to inject. For some reason, usually a new preparation from the same stocks results in injectable sample.
16. When it comes to lyse tissue, it is not always the more the better. Too much tissue can cause PCR reactions to fail, possibly due to PCR inhibitors in tissue. When positive controls are not amplifying, dilute the samples and try again.
17. Junction PCRs sometimes can be difficult. Spiking a synthetic template in sample genomic DNA or lysate is an easy positive control for working out conditions: Different Taq polymerase, with and without DMSO, touchdown PCR, nested PCR, etc. partial integration sometimes happens, i.E., one junction is detected but not the other. A positive control is especially helpful under such a situation.
18. If restriction sites are introduced next to loxP site and PCR reaction should do fewer cycles, such as under 25. Too many cycles lead to heteroduplex between with or without insert and REs will not digest. Floxing with two pairs of ZFNs will be competing with individual NHEJ events and deletion between the two target sites. Deletion product will not amplify with junction PCR or Cel-I PCR, leading to false positive for

homozygous floxed. Deletion PCR should be done to differentiate the two possibilities.

19. Digested DNA will be measured more accurately. It is critical to not load too much DNA on the gel for non-radioactive southern blots to avoid high background.
20. The amount of labeled probes should be adjusted, especially if there is too much background on the developed membrane. Increase the number of washes in washing buffer II can help too.

References

1. Robertson E, Bradley A, Kuehn M, Evans M (1986) Germ-line transmission of genes introduced into cultured pluripotential cells by retroviral vector. Nature 323:445–448
2. Thomas KR, Capecchi MR (1987) Site-directed mutagenesis by gene targeting fin mouse embryo-derived stem cells. Cell 51:503–512
3. Doetschman T, Gregg RG, Maeda N et al (1987) Targetted correction of a mutant TPRT gene in mouse embryonic stem cells. Nature 330:576–578
4. Buehr M, Meek S, Blair K et al (2008) Capture of authentic embryonic stem cells from rat blastocysts. Cell 135:1287–1298
5. Li P, Tong C, Mehrian-Shai R et al (2008) Germline competent embryonic stem cells derived from rat blastocysts. Cell 135:1299–1310
6. Rouet P, Smih F, Jasin M (1994) Introduction of double-strand breaks into the genome of mouse cells by expression of a rare-cutting endonuclease. Mol Cell Biol 14:8096–8106
7. Gunn A, Stark JM (2012) I-SceI-based assays to examine distince repair outcomes of mammalian chromosomal double strand breaks. Methods Mol Biol 920:379–391
8. Epinat JC, Arnould S, Chames P et al (2003) A novel engineered meganuclease induces homologous recombination in yeast and mammalian cells. Nucleic Acids Res 31:2952–2962
9. Stoddard BL (2011) Homing endonucleases: from microbial genetic invaders to reagents for targeted DNA modification. Structure 19:7–15
10. Chandrasegaran S, Carroll D (2016) Origins of programmable nucleases for genome engineering. J Mol Biol 428:963–989
11. Carroll D, Moron JJ, Beumer KJ et al (2006) Design, construction and in vitro testing of zinc finger nucleases. Nat Protoc 1:1329–1341
12. Porteus MH, Baltimore D (2003) Chimeric nucleases stimulate gene targeting in human cells. Science 300:763
13. Urnov FD, Miller JC, Lee YL et al (2005) Highly efficient endogenous human gene correction using designed zince-finger nucleases. Nature 435:646–651
14. Miller JC, Holmes MC, Wang J et al (2007) An improved zinc-finger nuclease architecture for highly specific genome editing. Nat Biotechnol 25:778–785
15. Doyon Y, Vo TD, Mendel MC et al (2011) Enhancing zinc-finger-nuclease activity with improved obligate heterodimeric architectures. Nat Methods 8:74–79
16. Santiago Y, Chan E, Liu PQ et al (2008) Targeted gene knockout in mammalian cells by using engineered zinc-finger nucleases. Proc Natl Acad Sci U S A 105:5809–5814
17. Doyon Y, McCammon JM, Miller JC et al (2008) Heritable targeted gene disruption in zebrafish using designed zinc-finger uncleases. Nat Biotechnol 26:702–708
18. Geurts AM, Cost GJ, Freyvert Y et al (2009) Knockout rats via embryo microinjection of zinc-finger nucleases. Science 325:433
19. Carbery ID, Ji D, Harrington A et al (2010) Targeted genome modification in mice using zinc-finger nucleases. Genetics 186:451–459
20. Hauschild J, Petersen B, Santiago Y et al (2011) Efficient generation of a biallelic knockout in pigs using zinc-finger nucleases. Proc Natl Acad Sci U S A 108:12013–12017
21. Flisilowska T, Thorey IS, Offner S et al (2011) Efficient immunoglobulin gene disruption and targeted replacement in rabbit using zinc finger nucleases. PLoS One 6:e21045
22. Dong Z, Ge J, Li K et al (2011) Heritable targeted inactivation of myostatin gene in yellow catfish (Pelteobagrus fulvidraco) using engineered zinc finger nucleases. PLoS One 6: e28897

23. Cui X, Ji D, Fisher DA et al (2011) Targeted integration in rat and mouse embryos with zinc-finger nucleases. Nat Biotechnol 29:64–67
24. Brown AJ, Fisher DA, Kouranova E et al (2013) Whole-rat conditional gene knockout via genome editing. Nat Methods 10:638–640
25. Christian M, Cermak T, Doyle EL et al (2010) Targeting DNA double-strand breaks with TAL effector nucleases. Genetics 186:757–761
26. Miller JC, Tan S, Qiao G et al (2011) A TALE nuclease architecture for efficient genome editing. Nat Biotechnol 29:143–148
27. Tesson L, Usal C, Menoret S et al (2011) Knockout rats generated by embryo microinjection of TALENs. Nat Biotechnol 29:695–696
28. Hsu PD, Lander ES, Zhang F (2014) Development and applications of CRISPR-Cas9 for genome engineering. Cell 157:1262–1278
29. Doudna JA, Charpentier E (2014) Genome editing. The new frontier of genome engineering with CRISPR-Cas9. Science 346:1258096
30. Kaneko T, Sakuma T, Yamamoto T et al (2014) Simple knockout by electroporation of engineered endonucleases into intact rat embryos. Sci Rep 4:6382
31. Qin W, Dion SL, Kutny PM et al (2015) Efficient CRISPR/Cas9-mediated genome editing in mice by zygote electroporation of nuclease. Genetics 200:423–430
32. Kaneko T, Mashimo T (2015) Simple genome editing of rodent intact embryos by electroporation. PLoS One 10:e0142755
33. Chen S, Lee B, Lee AY et al (2016) Highly efficient mouse genome editing by CRISOR ribonucleoprotein electroporation of zygotes. J Biol Chem 291:14457–14467
34. Zschemisch NH, Glage S, Wedekind D (2012) Zinc-finger nuclease mediated disruption of Rag1 in the LEW/Ztm rat. BMC Immunol 13:60

Chapter 18

Organ Generation from Knockedout Rat Blastocysts Complemented with Pluripotent Stem Cells

Masumi Hirabayashi and Shinichi Hochi

Abstract

Regeneration of human organs in domestic animal model would provide enough number of functional donor organs in transplantation therapy. Recent progresses in pluripotent stem cells and nuclease-based genome editing tools have set the stage for investigating the chimeric complementation approach to generate functional organs from embryonic stem (ES) cells or induced pluripotent stem (iPS) cells. In this chapter, protocol for allogeneic or xenogeneic organ generation using knocked-out (KO) rat blastocysts and the rat or mouse ES/iPS cells is described. The protocol includes (1) the preparation of KO rat colony, (2) the preparation of rat or mouse ES/iPS cells, (3) the recovery of rat blastocysts, (4) the stem cell injection into blastocysts, (5) the embryo transfer into pseudopregnant recipient uteri, and (6) the genotyping and organogenetic analysis of chimeric offspring. The accumulation of basic and practical knowledge in the rodent model would be useful in improving therapeutic performance to regenerate 3D organs available for transplantation.

Key words Blastocyst complementation, Gene knocked-out, Organ regeneration, Pluripotent stem cells, Rodent model

1 Introduction

Embryonic stem (ES) cell lines have been established in rats by using a few inhibitors for fibroblast growth factor (FGF) receptor, mitogen activated protein kinase kinase (MEK), and glycogen synthase kinase 3 (GSK3) in differentiation-related signaling pathways [1–3]. Additional supplementation of leukemia inhibitory factor (LIF) and forskolin played a beneficial role in the establishment of "genuine" rat ES cell lines [4]. Induced pluripotent stem (iPS) cell lines also have been established in rats by a forced expression of "Yamanaka factors" (*Oct4*, *Sox2*, *c-Myc*, and *Klf4*) [5–7]. The established rat ES/iPS cell lines were successfully applied to knock out (KO) the endogenous DNA by the homologous recombination (HR)-based gene targeting [8–11].

Chengyu Liu and Yubin Du (eds.), *Microinjection: Methods and Protocols*, Methods in Molecular Biology, vol. 1874, https://doi.org/10.1007/978-1-4939-8831-0_18, © Springer Science+Business Media, LLC, part of Springer Nature 2019

The conventional HR-based genome editing is inefficient [12]. Recently, new genome editing tools, such as zinc finger nucleases (ZFNs) [13], transcription activator-like effector-based nucleases (TALENs) [14], and clustered regularly interspaced short palindromic repeat-CRISPR-associated protein 9 (CRISPR/Cas9) [15], have been developed. These new tools include DNA-binding domains (zinc finger motif, TALE, or guide RNA) and molecular scissors (a pair of FoxI or Cas9 nuclease) to induce double-strand breaks (DSBs) at the desired locations in the genome. The induced DSBs are mis-repaired through nonhomologous end-joining or HR, resulting in the targeted mutations. Since genes such as transcription factors are well known to be critical for lineage specification and organogenesis, knocking out of the genes can produce organogenetic developmental niche in the offspring.

Thus, recent progresses in both the pluripotent stem cells and the nuclease-based genome editing tools have set the stage for investigating the chimeric complementation approach to generate functional organs from ES or iPS cells (Fig. 1). When allogeneic (rat) or xenogeneic (mouse) pluripotent stem cells are microinjected into blastocyst-stage rat embryos that are designed to have the organogenetic niche, the complimented blastocysts can develop to full-term chimeric rat offspring with the stem cell-derived functional 3D organs in the niche.

2 Materials

2.1 Animals

1. Blastocyst donor rats: Gene-modified with conventional HR-based gene targeting or new genome editing tools to have the organogenetic niche. The conventional HR-based gene targeting with ES cells is not only ineffective but also time-consuming, while the new genome editing tools (ZFNs, TALENs, and CRISPR/Cas9) can be applied easily through pronuclear microinjection [16]. Although the CRISPR/Cas9 system is the most convenient, the off-targeting effect by the TALENs is less frequent than the CRISPR/Cas9. A list of genes that can be targeted to induce organogenesis disorder [17–26] is shown in Table 1. While deficiency of thymus and primordial germ cells does not lead to lethal phenotype in the homozygous KO animals, the other organ/tissue deficiencies (kidney, pancreas, liver, heart, or vessel) would induce perinatal death of the KO animals. Therefore, the latter KO phenotype needs to be maintained as a heterozygotic population.
2. Blastocyst recipient rats: Conventional closed-colony strain rats, such as Sprague-Dawley or Wistar rats.

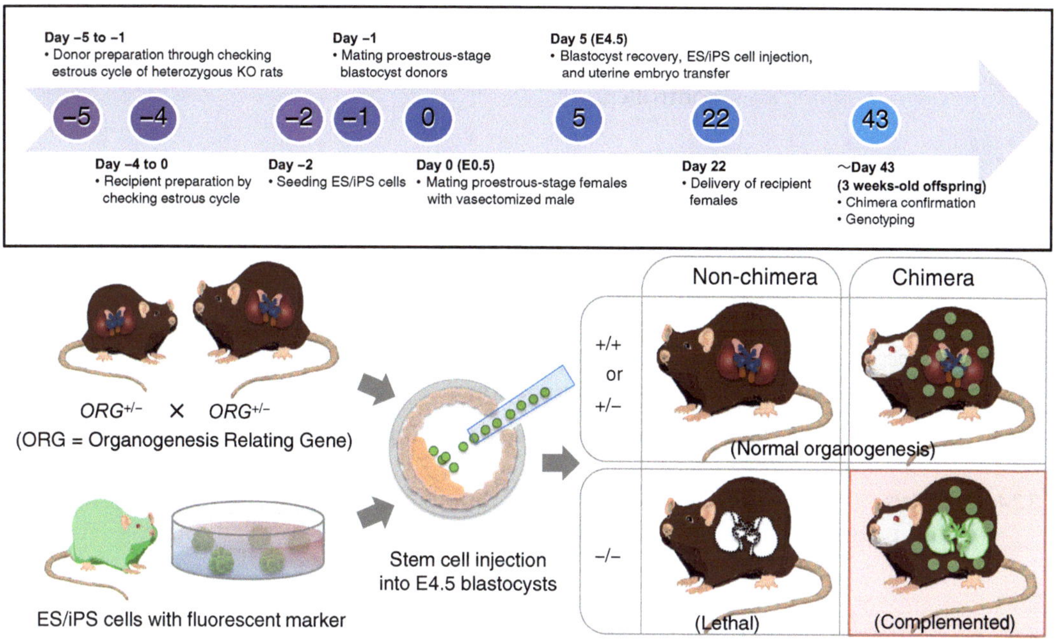

Fig. 1 Outline of organ generation with knocked-out (KO) rats and pluripotent stem cells. By using this blastocyst complementation strategy, not only syngeneic or allogeneic but also xenogeneic organ generation is possible in the rodents (mouse and rat). Homozygous knocking-out *Sall1* and *Pdx1* genes causes kidney- and pancreas-deficiency, respectively, but the KO individuals need to be maintained as heterozygous carriers due to the perinatal lethality of homozygous offspring. In contrast, homozygous KO rats of *Foxn1* gene are not lethal regardless of their thymus-deficient phenotype. An abbreviation ORG is given for organogenesis relating gene to generalize this figure

Table 1
List of genes targeted for organogenesis disorder

Genes	Target organs
Foxn1	Thymus [17]
Sall1, *Kif26b*	Kidney [18, 19]
Prdm14, *Blimp1*	Primordial germ cells [20, 21]
Pdx1	Pancreas [22]
FoxA1, *FoxA2*, *FoxA3*, *Hex*	Liver [23, 24]
Nkx2.5	Heart [25]
Flk1	Vessel [26]

2.2 Equipment

2.2.1 Micromanipulators

1. Narishige micromanipulators and microinjectors.
2. Prime Tech piezo-driven micromanipulator with a pulse controller.

2.2.2 Pipette Puller and Injection Needles

1. Sutter micro pipette puller.
2. Sutter borosilicate tubing for injection needle.
3. Narishige thin-walled glass capillary for holding needle.
4. Narishige microforge.

2.2.3 Microscopes

1. Inverted microscope with 4×, 10×, 20×, and 40× objectives for stem cell injection.
2. Stereomicroscope with halogen fiber illumination systems.

2.2.4 Incubators

1. CO_2 gas incubator. Standard setting: 37 °C, 5% CO_2, and 95% air, humidified by a pan containing distilled water.
2. Water bath.

2.3 Tools and Supplies

1. Handmade injection chamber (50 × 60 × 3 mm acrylic plate), drilled with ø = 25 mm hole and sealed with 30 × 30 × 0.1 mm coverslip by bees wax.
2. Falcon® 3037 organ culture dishes.
3. Falcon® 1029 and 1008 Petri dishes.
4. Falcon® 3001 tissue culture dishes.
5. 5 mL syringe.
6. 26- and 27-G sterile needles.
7. Animal anesthetizer.
8. Autoclip.

2.4 Media and Chemicals

1. MEF medium: DMEM, 10% FBS, 1× penicillin-streptomycin-glutamine.
2. 2iF medium [27]: 1:1 mixture of DMEM/F12 medium and Neurobasal® medium, 0.5× serum-free supplements (B-27® and N-2), 0.5× penicillin-streptomycin-glutamine, 0.1% rat LIF, 3 μM CHIR99021, 10 μM PD0325901, 10 μM forskolin. *See* Table 2 and **Note 1**.
3. E14tg2a-2i medium [28]: GMEM, 10% FBS (ES cell qualified), 1× penicillin-streptomycin-glutamine, 1× MEM nonessential amino acids, 1 mM sodium pyruvate, 0.2% β-mercaptoethanol, 0.01% mouse LIF, 3 μM CHIR99021, 10 μM PD0325901. *See* Table 2 and **Note 1**.
4. Modified rat 1-cell embryo culture medium (modified-R1ECM) [29]: 0.1× Stock solutions A and B, 25 mM $NaHCO_3$, 0.5 mM sodium pyruvate, 1× MEM amino acids/

Table 2
Composition of 2iF medium (for rat ES/iPS cells) and E14tg2a-2i medium (for mouse ES/iPS cells)

Reagents	2iF medium	E14tg2a-2i medium
DMEM/F12 1:1	50 mL	–
Neurobasal® medium	50 mL	–
GMEM ($NaHCO_3$ 2.75 g/L, pH 7.2)	–	86.7 mL
B-27® serum free supplement (50×)	1 mL	–
N-2 supplement (100×)	0.5 mL	–
ES cell qualified fetal bovine serum	–	10 mL
Penicillin-streptomycin-glutamine (100×)	0.5 mL	1 mL
MEM non-essential amino acids (100×)	–	1 mL
Sodium pyruvate (100 mM)	–	1 mL
ß-Mercaptoethanol	–	0.2 mL
Rat LIF (10^6 unit/mL)	0.1 mL	–
Mouse LIF (10^7 unit/mL)	–	0.01 mL
CHIR99021 (30 mM) (GSK3 inhibitor)	0.01 mL	0.01 mL
PD0325901 (10 mM) (MEK inhibitor)	0.1 mL	0.1 mL
Forskolin (5 mM)	0.2 mL	–

MEM nonessential amino acids, 0.1 mM L-glutamine, 0.4% bovine serum albumin (BSA). *See* Table 3 and **Note 2**.

5. Hepes-R1ECM/PVP [30]: 0.1× Stock solutions A and B, 4 mM $NaHCO_3$, 0.5 mM sodium pyruvate, 1× MEM amino acids/MEM nonessential amino acids, 0.1 mM L-glutamine, 22 mM Hepes, 0.1% polyvinyl alcohol (PVA), 12% polyvinyl pyrrolidone (PVP). *See* Table 3 and **Note 3**.
6. Mineral oil, embryo culture-tested.
7. Epinephrine.
8. Isoflurane.
9. Kanamycin.
10. Mitomycin-C (MMC).
11. Trypsin.

3 Methods

3.1 Pluripotent Stem Cell Preparation

Both ES and iPS cell lines that have been established either in rat and mouse can be used for stem cell injection into rat blastocysts.

Table 3
Composition of modified-R1ECM and Hepes-R1ECM/PVP

Reagents		Modified-R1ECM	Hepes-R1ECM/PVP
Stock Solution A	NaCl 4428 mg KCl 239 mg Glucose 1352 mg Penicillin G 75 mg Streptomycin 50 mg Na-lactate (60%) 1.9 mL In 100 mL distilled water	10 mL	10 mL
Stock solution B	$CaCl_2{\cdot}2H_2O$ 294 mg $MgCl_2{\cdot}6H_2O$ 102 mg In 100 mL distilled water	10 mL	10 mL
$NaHCO_3$		210 mg	33.7 mg
Sodium pyruvate		5.5 mg	5.5 mg
MEM amino acids (50×)		2 mL	2 mL
MEM non-essential amino acids (100×)		1 mL	1 mL
L-Glutamine		14.6 mg	14.6 mg
Hepes buffer (1 M)		–	2.2 mL
Bovine serum albumin		400 mg	–
Polyvinyl alcohol		–	100 mg
Distilled water		Mess up to 100 mL	Mess up to 100 mL
Polyvinyl pyrrolidone		–	120 mg/mL of Hepes-R1ECM

Morphology and in vitro growth rate of pluripotent stem cell colonies are obviously different between rat and mouse (*see* Fig. 2). Theoretically, fluorescent marker genes (such as GFP, Venus, tdTomato, or Kusabira-orange) should be equipped in the stem cell lines, as well as the different coat color characteristics against the background strains of host blastocysts. When albino-origin rat ES/iPS cell lines are injected into non-albino blastocysts, the coat color chimerism is often detected on the offspring face only, with very limited white spots [27].

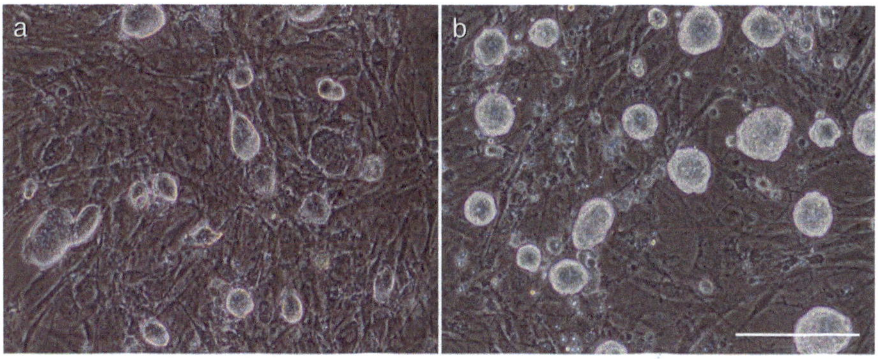

Fig. 2 Morphological feature of pluripotent stem cell colonies on MEF feeders. Scale bar = 200-μm. (a) Mouse ES cells 48 h post-seeding; The ES cells form oval and shiny colonies with clear boundaries. The colonies tightly adhere to the feeder layer. (b) Rat ES cells 48 h post-seeding; The ES cell colonies have compact dome-shape and mildly adhere to the feeder layer. Rat ES cells (doubling time 12.5 hours) can proliferate faster than mouse ES cells (doubling time 15.7 hours), with a shorter lag-phase period

1. MMC-inactivated MEFs are thawed in 37 °C water bath 1 day before ES/iPS cell seeding.
2. The post-thaw MEFs (2×10^6 cells) are suspended in 4.5 mL of the MEF medium, centrifuged at 190 × *g* for 5 min at 4 °C, and resuspended in 3 mL of the MEF medium.
3. An aliquot of the MEF suspension (2 mL) is placed on a 35 mm tissue culture dish, and cultured at 37 °C in 5% CO_2 in air.
4. Cryostock of ES/iPS cells (2×10^5 cells/0.5-mL/vial) is thawed in 37 °C water bath, suspended in 4.5 mL of the MEF medium, centrifuged at 190 × *g* for 5 min at 4 °C, and resuspended with 2 mL of the 2iF or E14tg2a 2i medium.
5. The ES/iPS cells are placed onto the MCC-treated MEFs (after the removal of the MEF medium) and then cultured for 1.5 days (rat) or 2.5 days (mouse).
6. The ES/iPS cells are liberated by 0.25% trypsin/5.3 mM EDTA treatment (5 min, 37 °C), exposed to the MEF medium, and centrifuged at 190 × *g* for 5 min at 4 °C.
7. They are placed in 2 mL of the 2iF or E14tg2a-2i medium on a 35 mm petri dish, and kept for 0.5–1 h at 4 °C until the injection into blastocysts (*see* **Note 4**).

3.2 Blastocyst Recovery

Rat two-cell embryos are capable of developing into blastocysts if modified-R1ECM medium with lowered osmolality is applied [31], but genetic background of donor and recipient rats as well as extent of synchrony between embryos and uteri are among factors influencing in vivo survival of the blastocysts [32]. Therefore, in vivo-derived rat blastocysts may be optimal source for stem cell injection.

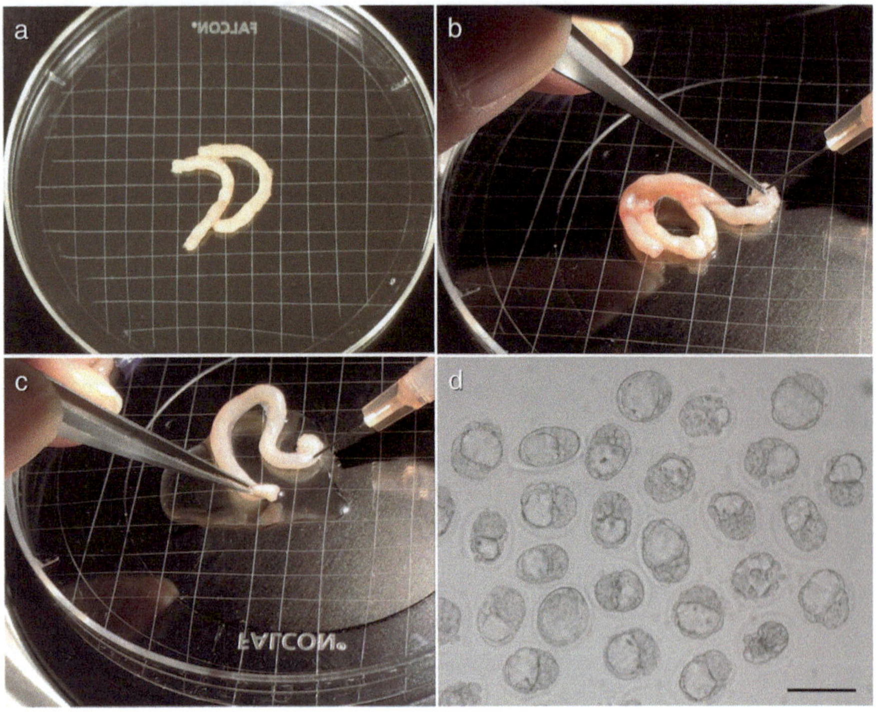

Fig. 3 Recovery of E4.5 blastocysts from KO female rats. (a) A pair of oviduct/uterus is placed on a lid of 100 mm dish with 5- to 10 mm grids. (b) The modified-R1ECM, aspirated into 5 mL syringe connected with a 26-G sterile needle, is injected from oviduct fimbria, and thus the oviduct/uterus is washed by the downstream flushing. (c) Additional uterine downstream flushing is performed by infusing the medium near the utero-tubal junction. During the infusion, the opposite end of uterus (exit) is clipped by fine forceps to induce full swelling of uterine cavity with the medium. Subsequent releasing of the exit may allow recovery of all blastocysts including those hatched from zonae pellucidae. (d) Harvested blastocysts can be temporally cultured for 3 to 4 hours in the modified-R1ECM at 37 °C in 5% CO_2 in air. Scale bar = 100-μm

1. Female KO rats at 7–13 weeks-old are mated with fertile KO males and used as blastocyst donors.
2. After vaginal plug or sperm detection is confirmed next morning.
3. Four days later, the donor females are sacrificed by CO_2 gas euthanasia and cervical dislocation.
4. Oviducts and whole uteri are excised using fine forceps and scissors, and the fats and bloods attached to the oviducts/uteri are removed on a sterilized filter paper.
5. A pair of excised oviduct/uterus (*see* Fig. 3a) is placed on a 100 mm petri dish with 5- to 10-mm grids.
6. The modified-R1ECM, aspirated into 5 mL syringe connected with a 26-G sterile needle, is inserted into the oviductal fimbria, and the oviduct/uterus is washed by the downstream flushing (*see* Fig. 3b).

7. Uterine downstream flushing is added by infusing the medium near the utero-tubal junction (upper part of the uterine horn). During the infusion, the opposite end of the uterine horn (exit) is clipped with fine forceps to induce full swelling of uterine cavity with the medium (*see* Fig. 3c).
8. Subsequent releasing of the exit allows maximal recovery of blastocysts including those hatched from their zonae pellucidae.
9. Harvested blastocysts (*see* Fig. 3d) are washed twice with the modified-R1ECM.
10. The blastocysts are maintained at 37 °C in 5% CO_2 and 95% air until stem cell injection.

3.3 Stem Cell Injection into Blastocysts

Piezo-driven micromanipulator with a pulse controller is used for stem cell injection into blastocysts. Proceeding to the stem cell injection, enough number of glass microtools (blunt-ended injection pipettes and blastocyst holding pipettes) should be prepared (*see* **Note 5**).

1. In a handmade injection chamber, two microdrops of 2iF or E14tg2a-2i medium (30 μL drop-[I] and 10 μL drop-[II]) and one microdrop of hepes-R1ECM/PVP (6 μL drop-[III]) overlaid with the mineral oil are prepared (*see* Fig. 4a).
2. Blastocysts (10–15 per 1 operation) and ES/iPS cells (approximately 10^3) are transferred into drop-[I] and drop-[II], respectively.
3. Inner surface of injecting pipette is washed by repeated blow-off of mercury and aspiration of hepes-R1ECM/PVP in the drop-[III].
4. Ten to thirty ES/iPS cells are aspirated into the injecting pipette in drop-[II], and stand by for injecting in drop-[I].
5. Holding pipette holds a blastocyst as the inner cell mass (ICM) is located at 9:00 o'clock position (*see* Fig. 4b).
6. Zona pellucida is drilled by several piezo-pulses (intensity 3, speed 2).
7. Injection pipette is deeply inserted into the blastocyst cavity through the trophectodermal cells.
8. The ES/iPS cells are deposited near the ICM (*see* Fig. 4c), without touching the pipette tip to the ICM. The injection pipette is slowly withdrawn.
9. The injected blastocysts are transferred into 800 μL of the fresh 2iF or E14tg2a-2i medium in the organ culture dish (*see* Fig. 4d), and kept at 37 °C, 5% CO_2 in air until uterine embryo transfer (*see* **Note 6**).

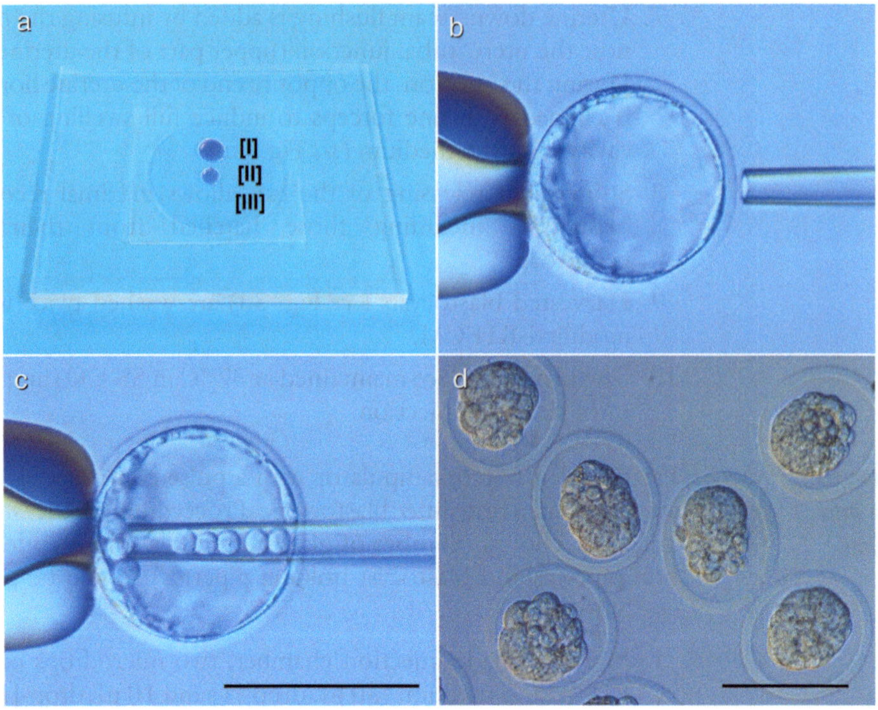

Fig. 4 Stem cell injection into rat blastocysts. (a) The injection chamber contains two microdrops of 2iF or E14tg2a-2i medium ([I] 30-μL and [II] 10-μL) and one microdrop of hepes-R1ECM/PVP ([III] 6-μL). Blastocysts and ES/iPS cells are placed in drop-[I] and -[II], respectively. (b) More than 10 ES/iPS cells are aspirated deep into the injecting pipette. A blastocyst is held by the holding pipette as its inner cell mass (ICM) is located at 9:00 o'clock position. (c) Zona pellucida is drilled by several piezo pulses (intensity: 3, speed: 2), and then the injecting pipette is inserted into the blastocyst cavity through trophectodermal cells. The ES/iPS cells are deposited near the ICM of the blastocyst. (d) The injected blastocysts are moved from drop-[I] to fresh 2iF or E14tg2a-2i medium. The blastocoel cavity often shrinks after the ES/iPS cell injection. However, the injected blastocysts are transferred to recipient uteri without confirming re-expansion of the cavity. Scale bars = 100-μm

3.4 Uterine Embryo Transfer

Rat blastocysts after stem cell injection can be transferred into uteri of surrogate mothers. Mature female rats (>8 weeks-old) are mated with a vasectomized male rat, and embryo transfer is performed 3 days after the vaginal plug is detected. An isoflurane gas at 2–3% using Small Animal Anesthetizer anesthetizes the recipient rats.

1. Uterus is exposed on sterile gauze through a lateral incision under the full anesthesia.
2. One or two drops of epinephrine are dripped on the upper part of the uterus near utero-tubal junction to constrict blood vessels.
3. A 27-G sterile needle is injected into the uterus.
4. When the opening of the uterus becomes visible, a glass capillary is gently inserted into this opening.

5. Blastocysts (usually 6–8 embryos per uterus) aspirated in the capillary with an outer diameter of 140–160 μm are loaded into the uterine horn. It is recommended to place an air bubble into the capillary to mark the position of the last blastocyst to be introduced.
6. Peritoneal wall (muscle layer) is closed using suture-needle and suture.
7. Skin is closed by autoclip.
8. The recipients are allowed to recover on a 37 °C warm plate following subcutaneous administration of 12.5 mg Kanamycin.

3.5 Analysis of Chimeric Offspring

Visualization of stem cell fluorescent marker (immediately after birth) and/or coat color marker (2 weeks after birth) facilitates identification of chimeric rat offspring derived from allogeneic or xenogeneic stem cells. Genotyping of the 3 weeks-old chimeric rat offspring makes clear whether they are homozygous KO rats with the targeted organ deficiency. When the chimeric rats are identified as homozygous KO rats, the presence of 3D organs complemented with the pluripotent stem cells is investigated by autopsy. In addition, their chimerism may be calculated by flowcytometric quantification using lymphocytes prepared from the chimeric rat spleen or peripheral blood. Depending upon the organ type, it is determined what kind of functional analysis should be performed with the regenerated organs. For example, regenerated pancreas is investigated for its ability for glucose-stimulated insulin secretion or glycemic control after islet transplantation into streptozotocin-induced diabetes model rodents [28, 33].

Two of our datasets are briefly shown as follows: *Foxn1*-KO rats were established by CRISPR/Cas9 system [34]. Blastocysts were harvested from homozygous *Foxn1*-KO females × homozygous *Foxn1*-KO males, and injected with seven each GFP-positive mouse ES cells. Total of 25 injected blastocysts were transferred to recipients, resulting in the birth of 23 offspring. Among the offspring, 14 (61%) were GFP-positive xenogeneic chimeras. The GFP-positive thymus was found in autopsied offspring immediately after birth ($n = 2$). Eleven xenogeneic chimeras survived after weaning, with the presence of T-cell population in their peripheral lymphocytes [35]. *Sall1* gene was replaced with tdTomato gene by conventional HR of rat ES cells and the kidney-deficient KO rat model was established through chimeric rat generation [36]. Blastocysts were collected from the *Sall1*-tdTomato heterozygous KO females that had been mated with the heterozygous KO males. Each blastocyst was injected with seven GFP-positive mouse ES cells. Uterine transfer of 168 injected blastocysts resulted in 111 prenatal live fetuses, including 64 interspecies chimeras (58%). Genotyping showed that these interspecies chimeric fetuses were composed from 29 *Sall1*-wild-type, 25 heterozygous *Sall1*-

KO, and 10 homozygous *Sall1*-KO fetuses. Kidneys in six of the ten homozygous *Sall1*-KO fetuses were observed with strong GFP fluorescence [37].

4 Notes

1. The basal media, to which CHIR99021, PD0325901, and ESGRO® have not yet been added, can be stored at 4 °C for 1 month. Once these supplements are added to the basal media, the 2iF medium or E14tg2a-2i medium should not be kept at 4 °C beyond 2 weeks.
2. The two stock solutions can be stored for up to 3 months in a refrigerator (4 °C) after filtration through a 0.22 μm filter unit. When all components except BSA are mixed, osmolarity of the medium is approximately 246 mOsmol/kg. The BSA-free modified R1ECM is sterilized by the filtration, and stored at 4 °C for 1 month. On the day of blastocyst recovery, an aliquot of the medium (10 mL) is placed in a conical tube and BSA (40 mg) is added (no agitation!). The modified R1ECM is sterilized by the filtration, and incubated at 37 °C in 5% CO_2 and 95% air until use.
3. After all components of hepes-R1ECM except PVA/PVP are mixed, the medium is supplemented with PVA and kept for several days at 4 °C until the PVA is completely dissolved. The hepes-R1ECM (10 mL) is then supplemented with PVP (1.2 g) and kept in a room temperature (20 °C) for 2 days until the PVP is completely dissolved. The Hepes-R1ECM/PVP medium (20 μL aliquots after filtration) can be stored at −20 °C, but should not be kept beyond 6 months. The hepes-R1ECM/PVP medium is thawed by placing in a room temperature (20 °C), and should be used within 1 week (ideally within the day of stem cell injection).
4. When rat ES/iPS cells are passaged, attention should be paid not to waste the ES/iPS colonies. Because the attachment of stem cell colonies to the feeder layer is very weak, the supernatant medium must be carefully aspirated preceding the enzymatic digestion.
5. Borosilicate capillaries are kept overnight in 3% HCl suspended in 99% ethanol solution, and rinsed twice with Milli-Q water. The capillaries are dried using the hot air sterilizer at 60 °C, and then sterilized at 160 °C for 2 h. The microtools (pipettes for stem cell injection and blastocyst holding) are processed with equipment such as mechanical puller and microforge. Only holding pipettes are bent to give an angle of approximately 150°.

6. The blastocoel cavity often shrinks after the ES/iPS cell injection. The blastocysts will re-expand 1–3 h later, but they can be transferred to recipient uteri without confirming their re-expansion.

References

1. Buehr M, Meek S, Blair K, Yang J, Ure J, Silva J, McLay R, Hall J, Ying QL, Smith A (2008) Capture of anthentic embryonic stem cells from rat blastocysts. Cell 135:1287–1298
2. Li P, Tong C, Mehrian-Shai R, Jia L, Wu N, Yan Y, Maxson RE, Schulze EN, Song H, Hsieh CL, Pera MF, Ying QL (2008) Germline competent embryonic stem cells derived from rat blastocysts. Cell 135:1299–1310
3. Hirabayashi M, Kato M, Kobayashi T, Sanbo M, Yagi T, Hochi S, Nakauchi H (2010) Establishment of rat embryonic stem cell lines that can participate in germline chimerae at high efficiency. Mol Reprod Dev 77:94
4. Hirabayashi M, Goto T, Tamura C, Sanbo M, Hara H, Hochi S (2014) Effect of leukemia inhibitory factor and forskolin on establishment of rat embryonic stem cell lines. J Reprod Dev 60:78–82
5. Li W, Wei W, Zhu S, Zhu J, Shi Y, Lin T, Hao E, Hayek A, Deng H, Ding S (2009) Generation of rat and human induced pluripotent stem cells by combining genetic reprogramming and chemical inhibitors. Cell Stem Cell 4:16–19
6. Liao J, Cui C, Chen S, Ren J, Chen J, Gao Y, Li H, Jia N, Cheng L, Xiao H, Xiao L (2009) Generation of induced pluripotent stem cell lines from adult rat cells. Cell Stem Cell 4:11–15
7. Hamanaka S, Yamaguchi T, Kobayashi T, Kato-Itoh M, Yamazaki S, Sato H, Umino A, Wakiyama Y, Arai M, Sanbo M, Hirabayashi M, Nakauchi H (2011) Generation of germline-competent rat induced pluripotent stem cells. PLoS One 6:e22008
8. Tong C, Li P, Wu NL, Yan Y, Ying QL (2010) Production of p53 knockout rats by homologous recombination in embryonic stem cells. Nature 467:211–213
9. Kobayashi T, Kato-Itoh M, Yamaguchi T, Tamura C, Sanbo M, Hirabayashi M, Nakauchi H (2012) Identification of rat Rosa26 locus enables generation of knock-in rat lines ubiquitously expressing tdTomato. Stem Cells Dev 21:2981–2986
10. Yamamoto S, Nakata M, Sasada R, Ooshima Y, Yano T, Shinozawa T, Tsukimi Y, Takeyama M, Matsumoto Y, Hashimoto T (2012) Derivation of rat embryonic stem cells and generation of protease-activated receptor-2 knockout rats. Transgenic Res 21:743–755
11. Goto T, Kobayashi T, Hara H, Sanbo M, Hochi S, Nakauchi H, Hirabayashi M (2015) Knock-in of a histone H2B-tdTomato reporter into the Rosa26 locus allows visualization of cell nuclei in rats. Mol Reprod Dev 82:916–917
12. Capecchi MR (1989) Altering the genome by homologous recombination. Science 244:1288–1292
13. Mandell JG, Barbas CF 3rd (2006) Zinc Finger Tools: custom DNA-binding domains for transcription factors and nucleases. Nucleic Acids Res 34:516–523
14. Bogdanove AJ, Voytas DF (2011) TAL effectors: customizable proteins for DNA targeting. Science 333:1843–1846
15. Fu Y, Foden JA, Khayter C, Maeder ML, Reyon D, Joung JK, Sander JD (2013) High-frequency off-target mutagenesis induced by CRISPR-Cas nucleases in human cells. Nat Biotechnol 31:822–826
16. Hirabayashi M, Hochi S (2001) Systematic production of transgenic rats by pronuclear DNA microinjection. In: Miyamoto H, Manabe N (eds) Reproductive biotechnology. Hokuto Shobo, Kyoto, pp 243–250
17. Segre JA, Nemhauser JL, Taylor BA, Nadeau JH, Lander ES (1995) Positional cloning of the nude locus: genetic, physical, and transcription maps of the region and mutations in the mouse and rat. Genomics 28:549–559
18. Nishinakamura R, Matsumoto Y, Nakao K, Nakamura K, Sato A, Copeland NG, Gilbert DJ, Jenkins NA, Scully S, Lacey DL, Katsuki M, Asashima M, Yokota T (2001) Murine homolog of SALL1 is essential for ureteric bud invasion in kidney development. Development 128:3105–3115
19. Uchiyama Y, Sakaguchi M, Terabayashi T, Inenaga T, Inoue S, Kobayashi C, Oshima N, Kiyonari H, Nakagata N, Sato Y, Sekiguchi K, Miki H, Araki E, Fujimura S, Tanaka SS, Nishinakamura R (2010) Kif26b, a kinesin family gene, regulates adhesion of the embryonic

kidney mesenchyme. Proc Natl Acad Sci U S A 107:9240–9245

20. Yamaji M, Seki Y, Kurimoto K, Yabuta Y, Yuasa M, Shigeta M, Yamanaka K, Ohinata Y, Saitou M (2008) Critical function of Prdm14 for the establishment of the germ cell lineage in mice. Nat Genet 40:1016–1022
21. Ohinata Y, Payer B, O'Carroll D, Ancelin K, Ono Y, Sano M, Barton SC, Obukhanych T, Nussenzweig M, Tarakhovsky A, Saitou M, Surani MA (2005) Blimp1 is a critical determinant of the germ cell lineage in mice. Nature 436:207–213
22. Offield MF, Jetton TL, Labosky PA, Ray M, Stein RW, Magnuson MA, Hogan BL, Wright CV (1996) PDX-1 is required for pancreatic outgrowth and differentiation of the tostral duodenum. Development 122:983–995
23. Lee CS, Friedman JR, Fulmer JT, Kaestner KH (2005) The initiation of liver development is dependent on Foxa transcription factors. Nature 435:944–947
24. Martinez Barbera JP, Clements M, Thomas P, Rodriguez T, Meloy D, Kioussis D, Beddington RS (2000) The homeobox gene Hex is required in definitive endodermal tissues for normal forebrain, liver and thyroid formation. Development 127:2433–2445
25. Lyons I, Parsons LM, Hartley L, Li R, Andrews JE, Robb L, Harvey RP (1995) Myogenic and morphogenetic defects in the heart tubes of murine embryos lacking the homeo box gene Nkx2-5. Genes Dev 9:1654–1666
26. Shalaby F, Ho J, Stanford WL, Fischer KD, Schuh AC, Schwartz L, Bernstein A, Rossant J (1997) A requirement for Flk1 in primitive and definitive hematopoiesis and vasculogenesis. Cell 89:981–990
27. Hirabayashi M, Tamura C, Sanbo M, Kato-Itoh M, Kobayashi T, Nakauchi H, Hochi S (2013) A retrospective analysis of germline competence in rat embryonic stem cell lines. Transgenic Res 22:411–416
28. Kobayashi T, Yamaguchi T, Hamanaka S, Kato-Itoh M, Yamazaki Y, Ibata M, Sato H, Lee YS, Usui J, Knisely AS, Hirabayashi M, Nakauchi H (2010) Generation of rat pancreas in mouse by interspecific blastocyst injection of pluripotent stem cells. Cell 142:787–799
29. Oh SH, Miyoshi K, Funahashi H (1998) Rat oocytes fertilized in modified rat 1-cell embryo culture medium containing a high sodium chloride concentration and bovine serum albumin maintain developmental ability to the blastocyst stage. Biol Reprod 59:884–889
30. Kato M, Ishikawa A, Hochi S, Hirabayashi M (2004) Effect of activation regimens for rat oocytes on full-term development after round spermatid injection. Contemp Top Lab Anim Sci 43:13–15
31. Miyoshi K, Abeydeera LR, Okuda K, Niwa K (1995) Effects of osmolarity and amino acids in a chemically defined medium on development of rat one-cell embryos. J Reprod Fertil 103:27–32
32. Kato M, Ishikawa A, Hochi S, Hirabayashi M (2004) Donor and recipient rat strains affect full-term development of one-cell zygotes cultured to morulae/blastocysts. J Reprod Dev 50:191–195
33. Yamaguchi T, Sato H, Kato-Itoh M, Goto T, Hara H, Sanbo M, Kobayashi T, Mizuno N, Yanagida A, Umino A, Ota Y, Hamanaka S, Masaki H, Rashid ST, Hirabayashi M, Nakauchi H (2017) Interspecies organogenesis generates autologous functional islets. Nature 542:191–196
34. Goto T, Hara H, Nakauchi H, Hochi S, Hirabayashi M (2016) Hypomorphic phenotype of Foxn1 gene-modified rats by CRISPR/Cas9 system. Transgenic Res 25:533–544
35. Goto T, Hara H, Hochi S, Nakauchi H, Hirabayashi M (2016) Mouse thymus generated in Foxn1 mutant rats by blastocyst complementation. In: Abstracts of the 49th annual meeting of the Society for the Study of Reproduction, San Diego, CA, USA, 16–20 July 2016
36. Goto T, Hara H, Hochi S, Nakauchi H, Hirabayashi M (2016) Generation of kidney-deficient model rats by knocking-out Sall1 gene. In: Abstracts of the 63rd annual meeting of the Japanese Association for Laboratory Animal Science, Kanagawa, Japan, 18–20 May 2016. Exp Anim 65(Suppl):S114
37. Goto T, Sanbo M, Hochi S, Nakauchi H, Hirabayashi M (2017) Kidney generation from mouse ES cells in homozygous *Sall1*-KO rats by xenogeneic blastocyst complementation approach. In: Abstracts of the 64th annual meeting of the Japanese Association for Laboratory Animal Science, Fukushima, Japan, 25–27 May 2017. Exp Anim 66(Suppl):S94

Chapter 19

Generation of Rabbit Models by Gene Editing Nucleases

Dongshan Yang, Jie Xu, and Y. Eugene Chen

Abstract

Due to the lack of germline transmitting pluripotent stem cells (PSCs) cell lines and the extreme difficulty of somatic cell nuclear transfer (SCNT) in rabbit, the gene targeting technology in rabbit was lagging far behind those in rodents and in farm animals. As a result, the development and application of genetically engineered rabbit model are much limited. With the advent of gene editing nucleases, including ZFN, TALEN, and CRISPR/Cas9, it is now possible to produce gene targeting (i.e., knockout, knockin) rabbits with high success rates. In this chapter, we describe a comprehensive, step-by-step protocol for rabbit genome editing based on gene editing nucleases with specific emphasis of CRISPR/Cas9.

Key words Rabbit, Genome editing, Engineered endonucleases, CRISPR/Cas9, Microinjection

1 Introduction

Rabbit is a classic animal model, and is increasingly becoming a translational model of choice, serving to bridge the gap between rodent models and larger animal models [1–4]. Rabbits are phylogenetically closer to primates than rodents [5, 6]. It is much easier to perform surgical procedures, multiple blood sampling (in larger volumes), tissue and organ sampling and analysis of organ function in vitro (e.g., heart perfusion) in rabbit. Technology or equipment developed for human infants could be easily tested in adult rabbits without modification due to their weight and size. Unlike pigs or rhesus monkeys, rabbits are relatively inexpensive and can be easily adapted to research facilities/institutions. They are easy to breed and handle and are recognized by the scientific and regulatory communities as a well-established model species. In addition, rabbit is a species that can tolerate restraint, which enables imaging technologies like magnetic resonance imaging (MRI) to be applied in an awake and conscious state with minimal preparation and habituation [7].

Since the first transgenic rabbit produced by Hammer and his colleagues [8], transgenic rabbit models have been generated by

Chengyu Liu and Yubin Du (eds.), *Microinjection: Methods and Protocols*, Methods in Molecular Biology, vol. 1874,
https://doi.org/10.1007/978-1-4939-8831-0_19, © Springer Science+Business Media, LLC, part of Springer Nature 2019

different laboratories worldwide for different research purposes, such as studies of lipid metabolism and atherosclerosis [4, 9–13], oncology [14, 15], acquired immunodeficiency syndrome (AIDS) [16–18], and bioreactors for the production of pharmaceutical proteins [19, 20]. Most of these rabbit models are produced by pronuclear microinjection of foreign DNA, which is subjected to position effects, including variations in the expression levels of the transgene caused by variation in the copy number of the transgenes or lack of crucial regulator elements related to the integration site, and potentially disrupting endogenous gene function through insertional mutagenesis. A technology that can insert or delete specific modifications into the rabbit genome more precisely is desired to study human genetic diseases in rabbit model.

The ascent of the mouse as genetic model organism is largely attributed to the development of gene targeting technology in embryonic stem cells (ESCs) that relies on the spontaneous homologous recombination (HR) to introduce defined modifications into sequences of interest in the genome. The gene targeted mice ESCs are capable of integrating into germline in chimeric mouse to produce gene targeted mouse model. In non-rodent animals where germline transmitting PSCs are not available, somatic cell nuclear transfer (SCNT) technology enabled gene targeting in species including pigs [21–23], sheep [24, 25], and cattle [26]. In these cases, gene targeting was performed in the somatic cells, which were then used as nuclear donor for animal cloning.

Unfortunately, germline competent rabbit PSCs are still not available since the first attempt in 1966 [27]. The SCNT efficiency, especially when using fibroblast cells as nucleus donors, is extremely inefficient in rabbit [28, 29]. Thus, production of gene targeted (KO/KI) rabbits has remained an extreme challenge for many years which has limited the application of this invaluable animal model in biomedical research.

Recently, Zinc Finger Nuclease (ZFN), Transcription Activator-Like Effector Nuclease (TALEN), and Clustered Regularly Interspaced Short Palindromic Repeats/CRISPR-associated protein-9 (CRISPR/Cas9) have emerged as powerful means for genome editing [30, 31]. These engineered endonucleases are efficient in generating double-strand breaks (DSB) in the specific genomic loci that can be repaired by error-prone non-homologous end joining (NHEJ) leading to a functional knockout of the targeted gene or used to integrate a DNA sequence at a specific locus through homologous recombination (HR). Thanks to the very high targeted DSB rates, knockout /knockin animals can be now readily derived by direct injection of the engineered endonucleases together with other elements (e.g., donor DNA) into pronuclear stage embryos, bypassing the need for germline transmitting PSCs [32, 33].

The early generations of nucleases, i.e., ZFNs and TALENs, employ custom-designed, locus-specific proteins fused to a sequence-independent nuclease (e.g., FokI) domain to generate targeted mutations. For each new target DNA sequence, the protein motifs of ZFN or TALEN need to be redesigned, synthesized, and validated. Despite these tedious molecular biology works, ZFN and TALEN each has brought breakthroughs over prior technologies in the gene editing field upon its introduction. That being said, there is no doubt that CRISPR/Cas9 has outperformed these two nucleases in many aspects. To name a few, first, CRISPR/cas9 uses so called guide RNA (gRNA) complementation to recognize the target sequence. The only requirement of a targetable sequence is the specific protospacer adjacent motif (PAM) that varies depending on the bacterial species of the Cas9 origin. For example, a NGG PAM is required for SpCas9 target. So for each target sequence, one can almost always find targetable sequence, and only need to synthesize the gRNA (vs protein motifs in ZFN or TALEN). Furthermore, the RNA guided Cas9 nuclease is much more efficient in generating double-strand DNA breaks than ZFN and TALEN. These favorable features have made CRISPR/Cas9 the first choice for gene editing in many applications nowadays [34, 35].

Here we describe comprehensive, step-by-step protocols for producing genetically modified rabbit models using the CRISPR/Cas9 technology. We have developed optimal conditions for superovulation of female New Zealand White (NZW) rabbits, for Cas9 components microinjection, and for embryo transfer into recipient females. The same procedure could also be used in ZFN or TALEN-mediated rabbit gene targeting, with the only difference being the preparation of the nuclease reagents. In general, generation of gene-edited founder rabbits by following this protocol takes as few as 2 months, and homozygous gene-edited rabbit model could be established within 15 months (Figs. 1 and 2, **Note 20**).

2 Materials

2.1 Equipment

1. Inverted microscope with fluorescent light source and digital camera.
2. Vibration Isolation Platform.
3. Thermo Plate for microscope stage.
4. Micromanipulator.
5. Microinjector.
6. Stereomicroscope.
7. Micropipette puller.
8. Microforge.

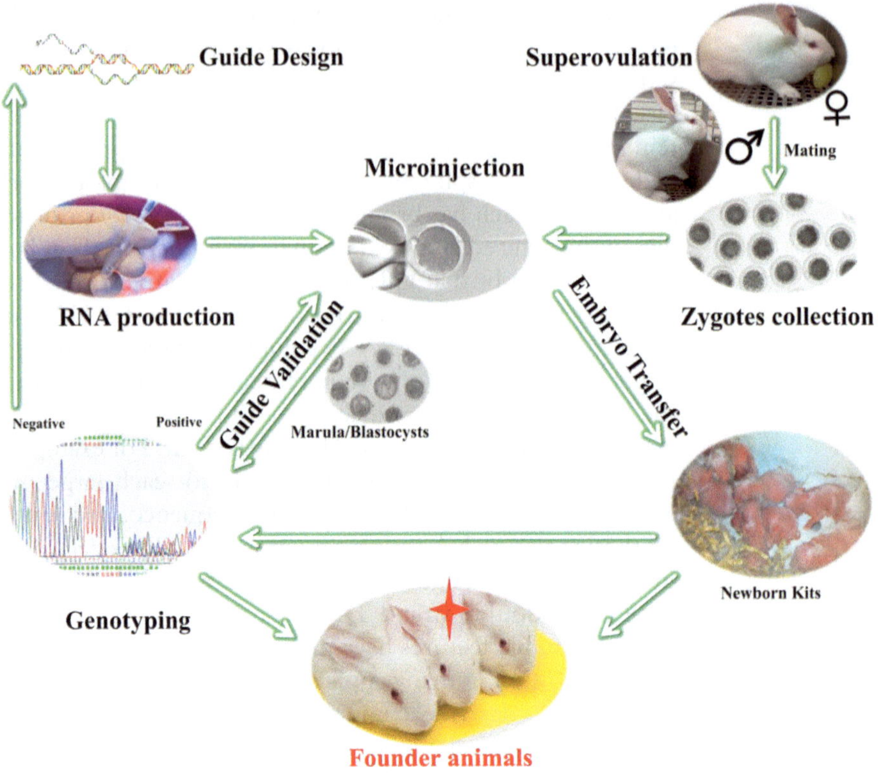

Fig. 1 Procedure of rabbit genome editing using CRISPR/Cas9

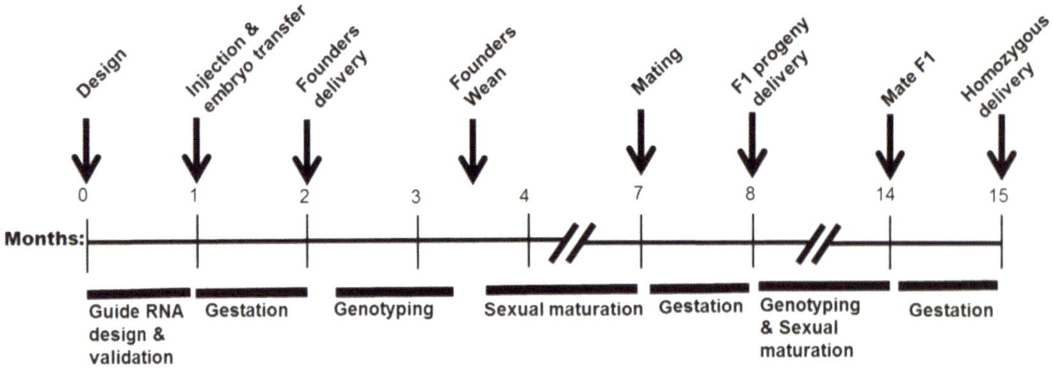

Fig. 2 Timeline for rabbit genome editing

9. Ceramic Tile for scoring glass.
10. CO2 incubator.
11. Alcohol lamp.
12. Embryo moving pipette (pulled Pasteur Pipette) assembled into a hand or mouth operated system.
13. Autoclave-steam sterilizer.

14. Anesthesia Machine.
15. TPR Temperature/Pulse Oximeter/Respiration System.
16. Water heated mat.
17. Vacuum Clipper System: Clipper connected to vacuum.
18. Skin Stapler Handle.
19. Skin Staple Remover.
20. DNA thermal cyclers.
21. Refrigerated Microcentrifuge.
22. Spectrophotometer.
23. Horizontal Gel Electrophoresis System.
24. Ultra-low Freezer.
25. Surgery pack: Thumb Tissue Forceps 5.5″ 1 × 2 teeth, Operating Scissors 6.5″ Sharp/Sharp Straight, Operating Scissors 4.5″ Blunt/Blunt Straight, Forceps Thumb Dressing 5.5″, Adson Dressing Forceps 4.75″, Iris Scissors 3 1/2″ Straight, Scalpel Handle Stainless Steel #3—for #10–15 Blades, Towel Clamps 5.5″, Needle Holder 5.25″, Crile Forceps 5.5″ Straight, Thumb Dressing Forceps 8″.

2.2 Consumables

1. Glass capillary: Thin wall borosilicate tubing with filament, 0.78 mm inner diameter, 1 mm out diameter.
2. Injection slide: Glass Chamber Slide System; 1-well, chamber removed.
3. 4-Well Dish.
4. Petri dish, 35 mm, 60 mm.
5. Centrifuge tubes, sterile, 15 ml, 50 ml.
6. Microcentrifuge Tubes, 0.5 ml, 1.5 ml, Certified RNase- and DNase-free.
7. Syringes, 1 cc, 5 cc Luer Slip.
8. Embryo transfer tube: Tom Catheter 3.5Fr × 4.5″, Sterile.
9. Scalpel Blades #10 Stainless Steel.
10. Powder Free Surgeon Glove, Sterile.
11. Surgery Drape: Poly-lined Sterile 18″ × 26″.
12. Suture #4-0 3/8 Circle Rev. Cut 19 mm/36″.
13. Surgery Gown: XLarge 47″ Non-Reinforced Sterile with Towel.
14. Staple Cartridges Surgi-Close 6 Cartridges × 20 Staples.
15. Soft-E Collar XSmall (Cat/Dog 5-9#).

2.3 Media and Reagents

1. HM: 25 mM HEPES, 10% fetal bovine serum in Medium 199 with Hanks's salts.

2. EBSS-complete medium: 1× MEM nonessential amino acid, 1× BME amino acid, 1 mM L-glutamine, 0.4 mM sodium pyruvate, 10% FBS in Earle's Balanced Salt Solution with calcium, magnesium, and phenol red.
3. Mineral oil: BioReagent, suitable for embryo cell culture.
4. FSH (Follicle-stimulating hormone): 10 mg/ml in saline.
5. HCG (Human chorionic gonadotropin): 200 IU/ml in saline.
6. Euthasol (Euthanize solution): Sodium pentobarbital 390 mg/ml, sodium phenytoin 50 mg/ml.
7. Anesthetic: Isoflurane Inhalant Anesthetic.
8. Analgesic: carprofen Injectable: 50 mg/ml.
9. NP40 LYSIS buffer: 1% Tergitol-type NP-40, 1 μg/ml Proteinase K in Taq Buffer.
10. Digestion buffer for DNA extraction: 5 mM EDTA, pH 8.0, 200 mM NaCl, 100 mM Tris–HCl, pH 8.0, 0.2% sodium dodecyl sulfate (SDS).
11. RNase-Free 0.1× TE Buffer: 1 mM Tris–HCl pH 7.4, 0.1 mM EDTA in DNase/RNase-Free Distilled Water.
12. Molecular biology Kits: T7 Standard RNA IVT Kit, T7 Standard mRNA Production System.

PCR Purification Kit, RNA purification Kit, Whole genome amplification kit.

2.4 Animals

Embryo Donor: 4 months to 2 years old New Zealand White (NZW) does.

Fertile males: 6 months to 2 years old NZW buck rabbits. They should be replaced every 1–2 years.

Recipients: 5 months to 2 year old NZW does.

3 Methods

3.1 Prepare the Engineered Nucleases Reagents

1. Design guide RNA and donor DNA according to project aims (*see* **Notes 1** and **2**).
2. To produce Cas9 mRNA and guide RNA in lab, we use the Cas9 expression plasmid JDS246 and guide RNA expression plasmid PX330 from Addgene as template (*see* **Note 3**).
3. Linearize plasmid JDS246 with PmeI: 5 μg plasmid JDS246 DNA, 10 μl 10× CutSmart® buffer, 1 μl PmeI (10 U/μl), add distilled water to a total volume of 100 μl. Incubate at 37 °C for 2–3 h.
4. Verify the complete digestion by running 2 μl of the cut DNA on a 1% agarose gel.

Table 1
Primers for guide RNA synthesis (*see* Note 1)

Primer	Sequence (5′–3′, NNs stand for customized 20nt guide)
T7-sgRNA_F	TGTAATACGACTCACTATAGGNNNNNNNNNNNNNNNNNNNN gttttagagctagaaatag
T7-sgRNA_R	AAAAGCACCGACTCGGTGCC

5. Clean up the digested DNA using a PCR Purification kit.
6. Measure DNA concentration using a spectrophotometer. Continue to the next step or store the purified DNA at −20 °C.
7. In vitro transcribe, cap, and polyadenylate Cas9 mRNAs from the linearized template using the T7 Standard mRNA Production kit following the manufacturer's protocol.
8. Clean up the resulting mRNA using the RNeasy Mini Kit.
9. Elute the purified mRNA in RNase-free 0.1× TE buffer.
10. Measure Cas9 mRNA concentration using a spectrophotometer.
11. Dilute Cas9 mRNA in RNase-free 0.1× TE Buffer, centrifuge at 20,000 × g at 4 °C for 30 min.
12. Carefully remove the top 2/3 of diluted RNA to a new RNase-free tube and store in −80 °C freezer in 10 μl aliquots.
13. Order primers in Table 1 and use PX330 as PCR template to produce guide RNA in vitro transcription template by PCR amplification. PCR condition: 50 °C, 38 cycle, 2–8 ng plasmid as template in 200 μl PCR reaction. Use Taq DNA polymerase with higher fidelity.
14. Clean up the PCR product using a PCR Purification kit.
15. In vitro transcribe guide RNA by using T7 Standard RNA IVT Kit following the manufacturer's protocol.
16. Clean up guide RNAs using RNeasy Mini Kit. Elute the purified RNA in RNase-free 0.1× TE buffer.
17. Measure the concentration of the guide RNA with Spectrophotometer, and dilute in RNase-free 0.1× TE buffer.
18. Centrifuge at 20,000 × g at 4 °C for 30 min.
19. Carefully remove the top 2/3 of diluted RNA to a new RNase-free tube.
20. Store guide RNA in −80 °C freezer in 10 μl aliquots.
21. Before microinjection, thaw the RNA samples on ice, and keep it on ice.

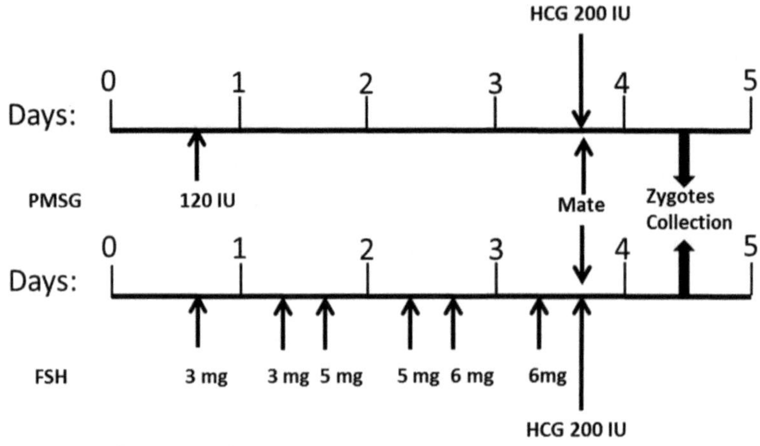

Fig. 3 Time schedule for rabbit superovulation

22. For knockin project, dilute the donor DNA (ssDNA oligo or dsDNA plasmid) in RNase-free 0.1× TE buffer.
23. Centrifuge at 20,000 × *g* for 30 min.
24. Carefully remove the top 2/3 of diluted DNA to a new RNase-free tube.
25. Store DNA in 4 °C fridge.

3.2 Superovulation

1. Subcutaneously inject sexually matured NZW female rabbits with FSH twice/day with a dosage of 3 mg for the first two injections, 5 mg for the next two injections, and 6 mg for the last two injections. Alternatively, PMSG can be used instead of FSH. Intramuscularly inject 150 IU of PMSG in single dosage (*see* Fig. 3 and **Notes 4** and **5**).
2. At 72 h after the first FSH injection, or PMSG injection, inject 200 IU HCG intravenously to induce ovulation (*see* Fig. 3).
3. Put a superovulated doe into a male's cage immediately after hCG injection.
4. Observe and record the successful mating. It will happen within minutes.
5. Optionally, to ensure successful fertilization, put the female to more male's cages to record at least 2–3 successful mating.
6. Put the female back to her cage.

3.3 Zygotes Collection

1. Prepare two 30 mm petri dishes with 2 ml HM medium covered by mineral oil. Keep the dishes on a heated stage at 38.5 °C.
2. Prepare two 50 ml centrifuge tubes with 25 ml HM medium and warm in 37 °C water bath.

3. At 17–19 h after hCG injection, euthanize the superovulated rabbits with an overdose intravenous injection of Euthasol (150 mg/kg). Recover the oviducts with a part of uterus in 25 ml HM medium.
4. Wash the oviducts in a new 25 ml HM medium. Use an iris scissors to remove the fat tissue around the oviduct to make it straight, hold the uterine side with one pair of tissue forceps, insert a blunt needle connected to a syringe filled with 5 ml of HM medium, and flush the oviduct from uterus to infundibulum.
5. Collect the flushed media in a 60 mm dish. Swirl the dish gently to gather eggs in the center.
6. Collect eggs under a stereomicroscope, wash the eggs in the 2 ml pre-warmed HM dish prepared in **step 1**, and kept in the second 2 ml HM dish on a 38.5 °C heated stage (*see* **Note 6**).

3.4 Embryo Microinjection

1. The day before microinjection, prepare a 4-well dish containing 500 μl EBSS-complete medium in each well, cover with 400 μl mineral oil (for embryo culture). Prepare a 30 mm dish with 4 × 80 μl drops of EBSS-complete medium covered with mineral oil (for embryo wash after injection). Place the dishes in 5% CO_2 incubator at 38.5 °C overnight.
2. Prepare holding pipette by hand pulling a piece of the borosilicate glass capillary tubes in the flame of an alcohol lamp. The outside diameter of the drawn-out region of the pipette should be 120–200 μm (i.e., close to the diameter of the zona pellucida). Break the pipette at ~1–2 cm from the shoulder of the pipette by scoring the glass lightly with a ceramic tile at the desired position. Fire-polish the pipette tip with microforge until the pipette shrinks to an inner diameter of ~40 μm. Bend the pipette ~20° angle by approaching the side of the pipette to the heating glass bead on the microforge.
3. Before microinjection, prepare the microinjection platform by placing a 20 μl drop of HM medium on a 1-well chamber slide that has the media chamber removed. Cover with 1.5 ml mineral oil and place on the 38.5 °C heated microscope stage of the inverted microscope (*see* Fig. 4a, b).
4. Place holding pipette into the left arm of the micromanipulator and position into the drop of manipulation medium. Adjust the holding pipette angle to make the tip paralleled to the stage (*see* Fig. 4b).
5. Fabricate injection micropipettes by heating and pulling borosilicate glass capillary tubes in Sutter P-1000 micropipette puller with parameter as following: Heat = 526, Pull = 80, Vel. = 70, Delay = 80, Pressure = 200, Ramp = 516, Delay Mode, Safe heat.

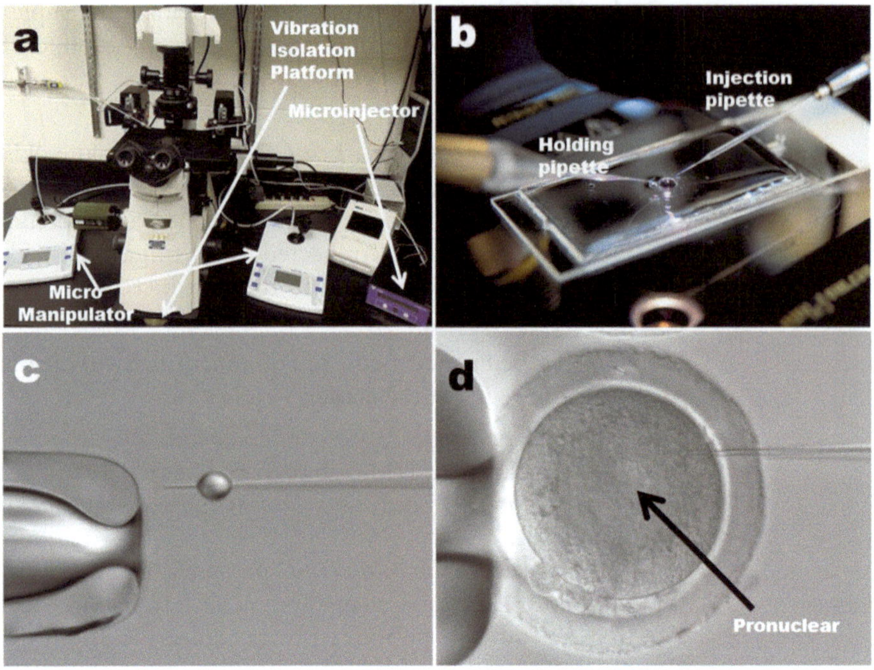

Fig. 4 Rabbit Embryo Microinjection. (**a**) Microinjection setup. Inverted microscope standing on a vibration isolation platform, the microscope was equipped with a Differential Interference Contrast (DIC) device, and a thermoplate. Microinjector was connected with pressured air supply. Holding arm and injection arm were controlled by micromanipulators. (**b**) Microinjection slide: 20 μl drop of HM medium on a 1-well chamber slide that has the media chamber removed, covered with 1.5 ml mineral oil and placed on the 38.5 °C heated microscope stage. (**c**) Opening of the injection pipette. In mineral oil, apply and hold clear pressure (80 psi) on microinjector, open the tip of the injection pipette (right) by gently tapping it against the holding pipette (left). The RNA solution formed a droplet (arrow) on the injection tip once it is opened. (**d**) Injection of rabbit embryo. Avoid pronuclear (arrow) when advancing the injection pipette tip into the embryo

6. Thaw the RNA aliquots on ice, mix Cas9 mRNA and guide RNA in a final concentration of 150/50 ng/μl in Rnase-free 0.1× TE Buffer, and keep on ice. Add donor template DNA in final concentration of 100–200 ng/μl for knockin project (*see* **Note 7**). Drop 1 μl of the injection mixture on the back opening of the injection pipette. A siphon effect will fill the tip of the injection pipette immidiately.
7. Place injection pipette into the right arm of the micromanipulator, and position it into the drop of manipulation medium with ~15° angle to the stage (*see* Fig. 4b).
8. Turn on the compressed air supply that is connected to a microinjector and set up the microinjector. We use the following parameters on our microinjector (Tritech research, MINJ-D): holdP = 2 psi, InjP = 20 psi, ClearP = 80 psi, Injtime = manual.

9. Transfer embryos (30–50) to the micromanipulation drop on the platform.
10. Move the stage to leave the tips out of the medium into oil, apply clear pressure (80 psi), and open the tip of the injection pipette by gently tapping it against the holding pipette. You will see the injection solution forms droplets on the injection tip when it is opened (*see* Fig. 4c).
11. Back to the medium, use the holding pipette to capture an embryo and align the injection needle, embryo and holding pipette along the x-axis.
12. Under 400× magnification, advance the injection pipette through the embryo. Be careful to avoid the nucleus. Once the cell membrane is pierced, press the foot pedal to inject. After injection, withdraw the needle. Release the embryo and repeat till all the embryos in the drop are injected (*see* Fig. 4d).
13. Wash the injected embryos three times in EBSS-complete medium, and incubate at 38.5 °C, 5% CO_2 in air for 1–2 h or overnight before embryo transfer (*see* **Note 7**).
14. For ex vivo validation of guide RNA, culture the injected embryos in EBSS-complete medium in vitro for 3–4 days until they reach morula or blastocyst stage.

3.5 Ex Vivo Validation of Guide RNA

1. For each guide RNA to be validated, collect 5–10 morula or blastocyst stage embryos, put embryos in 10 μl NP40 LYSIS buffer in a PCR tube. Incubate for 0.5–1 h at 55 °C, and 10 min at 95 °C, keep in 4 °C and use as PCR template (*see* **Note 8**).
2. PCR amplify the target sequence using PCR primer pair located upstream and downstream of the target site.
3. PCR products are then purified and Sanger sequenced with a sequencing primer.
4. A successful guide RNA will direct Cas9 to cut the target DNA and results in insertion/deletions (indels) after NHEJ repair, which can be manifested on the sequencing chromatography as overlapping peaks starting around the target site (*see* **Note 9**).
5. More accurate estimation of the efficiency can be calculated with an online software (TIDE https://tide.nki.nl/).
6. Guide RNAs with indels efficiency higher than 30% is recommended for use to produce embryos for embryo transfer (*see* **Note 10**).

3.6 Embryo Transfer

Surgery is performed under sterile conditions, with all surgical tools, media, supplies, etc. sterile. For each animal, a fresh, sterile, surgical pack, containing sterile drapes and tools is used. Personnel involved in the surgical procedures are scrubbed, gloved, gowned,

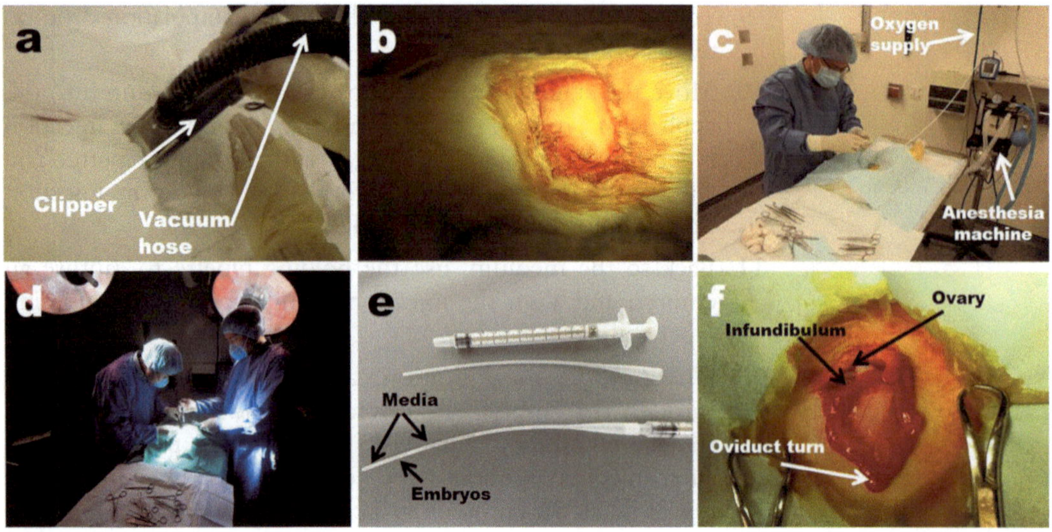

Fig. 5 Rabbit embryo transfer surgery. (**a**) A clipper connected with a vacuum was used to remove the hairs around incision site. (**b**) Incision site was prepared by three cycles of antiseptic scrub, followed by antiseptic solution and wipe with a sterile gauze sponge. (**c**) Surgery room setup. Anesthesia machine connected with oxygen supply and scavenger system. Rabbit was monitored by a TPR System. (**d**) Surgery ongoing. (**e**) Embryo transfer tube assembly. Tom Catheter connected with a 1 ml syringe. The transfer tube is loaded in the following order: 2 cm HM medium (right black arrow), 1–2 cm HM medium with zygotes, and 0.5 cm HM medium (left black arrow) separated by two 0.5 cm air bubbles. (**f**) For embryo transfer, the transfer tube is inserted through the infundibulum into the oviduct. Make sure the tube passed the turn of the oviduct and unload the embryos inside

capped, and masked. The rabbit is placed on the surgical table on a water heated mat maintained at 37 °C, and covered with a sterile drape (*see* Fig. 5).

1. Select female NZW rabbit in good health, weighing approx. 3–4 kg and at least 5 months of age. Check vagina to select does with red or pink vulva color (*see* **Note 11**).
2. To prepare recipients, inject 100 IU of hCG intravenously at same time when donor was mated and injected with hCG (*see* **Notes 12** and **13**).
3. Before surgery, weigh the rabbit and inject 4 mg/kg carprofen subcutaneously as an analgesic (*see* **Note 14**).
4. Connect Anesthesia Machine to oxygen supply and scavenging system (*see* Fig. 5c).
5. Restrain rabbit on its neck and back firmly, put on an anesthesia mask connected with the Anesthesia Machine to the rabbit mouth (*see* **Note 15**).
6. On the Anesthesia machine, adjust flowmeter at 100–200 ml/min and vaporizer at 5% to induce anesthesia. Squeeze the foot

pad to check for lack of pedal reflex which is an indication of adequate anesthesia (*see* **Note 16**).

7. Adjust the evaporator to 2–5% to maintain anesthesia. Monitor temperature, pulse, and respiration during procedure and adjust vaporizer as needed.
8. Clip the rabbit's flanks with a vacuum clipper system (*see* Fig. 5a).
9. Put the rabbit under anesthesia on a water heated mat in lateral recumbence.
10. Wash the shaved flank with antiseptic scrub (Pivodine Scrub) three times, followed by antiseptic solution (70% alcohol), and wipe with a sterile gauze sponge (*see* Fig. 5b).
11. The surgical incision site can be approximated by drawing an imaginary equilateral triangle over the flank with vertices at the wing of the ilium, the greater trochanter, and the ovary.
12. Make a 2–3 cm horizontal (anterior-posterior) skin incision with a scalpel blade over the anticipated location of the ovary. Incise the external and then internal abdominal oblique muscles.
13. Gently elevate the peritoneum with Crile forceps and sharply incise.
14. Grasp the fat pad attached to the oviduct with Thumb Dressing forceps to pull the ovary and oviduct out of the peritoneal cavity.
15. Under a stereomicroscope, 10–15 embryos are loaded into an embryo transfer tube (Tom Catheter) attached to a 1 ml syringe with HM medium (*see* Fig. 5e).
16. Insert the embryo transfer tube through the infundibulum into oviduct of recipients. Make sure the tube passed the big turn of the oviduct and unload the embryos inside. Withdraw the transfer tube gently and carefully (*see* **Note 17** and Fig. 5f).
17. Put the ovary and oviduct back to the peritoneal cavity, close the peritoneum and muscle layer, and then subcutaneous fascia layer with absorbable suture (e.g., Vicryl). Close the skin with Skin Staples.
18. Apply the same procedure to the other side to transfer another batch of 10–15 embryos (*see* **Note 18**).
19. At the end of the surgery, turn vaporizer and gas supply off, place animal in recovery area with thermal support until fully recovered.
20. Put Elizabethan Soft-E Collar on the neck to prevent the rabbit from biting the wound.

21. After surgery, inject carprofen subcutaneously at 4 mg/kg/day as an analgesic for at least 5 days.
22. At 10–14 days post operation, remove the Soft-E Collar and remove the skin staples with a skin staple remover.
23. 1–2 week before the expected delivery date, move the recipient rabbit into a breeding cage with nesting materials (*see* **Note 19**).

3.7 Genotyping Analysis of Founders by PCR and Sequencing

1. For ear skin tissue biopsy from 1-3 week old kits, hold the kit ear firmly with one hand and cut off a tiny piece of ear skin (around 2–5 mm) with a scissors in the other hand.
2. Digest the ear skin tissue with 300 μl digestion buffer with 1 mg/ml proteinase K in a 1.5 ml Eppendorf tube at 55 °C overnight.
3. Add 1 ml of 100% ethanol into each tube, mix completely. Centrifuge at 13,000 × *g* for 12 min, and then pour out the ethanol.
4. Wash the DNA pellets by adding 0.8 ml of 70% ethanol into the tubes, then centrifuge at 13,000 × *g* for 8 min.
5. Pour out the ethanol from the tube, and then suck the remaining ethanol with pipette.
6. Dissolve the DNA pellets with 300 μl TE buffer or distilled water.
7. Store the DNA at 4 °C or −20 °C and use as template for PCR.
8. For KO project, PCR amplify the target locus and sumit for Sanger sequencing using the same procedures as in Subheading 3.5 (*see* **Note 9**).
9. For knockin project using single-stranded oligo donor (ssODN) as template to introduce a point mutation or short tag sequences, identify the knockin kits by deep sequencing or TA cloning and sequencing of the PCR products from **step 8**.
10. For large fragment knockin project, identify the knockin kits by two PCR primer pairs spanning each of the homologous arms.

4 Notes

1. There are dozens of free online tools for guide RNA design. The popular tools include CRISPR Design tool from MIT (http://crispr.mit.edu/), and Breaking Cas from Spain. (http://bioinfogp.cnb.csic.es/tools/breakingcas/?gset=2x1_GENOMES_Ensembl_85). To order T7-sgRNA-F in table 1, please note a 5'-GG was added at the begining to facilitate T7 transcription, so remember to skip the 5' -G or -GG if present in the customized 20nt guide.

2. New Zealand White Rabbit is not a pure inbred line. There are some single-nucleotide polymorphisms (SNPs) in their genome. We recommend PCR sequencing your target region before you design your guide RNA and donor template to exclude SNPs or sequencing mistakes.
3. Cas9 mRNA and guide RNA are also commercially available.
4. NZW rabbits as early as 12 weeks of age are old enough to be stimulated for superovulation with either FSH or PMSG treatments, but the embryo development may be compromised (unpublished data). We recommend using donor rabbit older than 16 weeks.
5. In rabbits, PMSG and FSH have been commonly used for superovulation. Though the PMSG program is simpler and less laborious, FSH program is more reliable [36].
6. It is not necessary to remove cumulus cells using hyaluronidase, since most of the cumulus cells will be detached from the rabbit oocytes at the time zygotes are recovered. For the same reason, the rabbit zygotes are collected by oviduct flush but cannot be released by ampulla tearing as in mice.
7. Donor DNA could be either ssODN or circular plasmid DNA vector. It is not recommended to linearize the plasmid DNA donor due to concerns of higher random insertion risks. All the contents should be injected in cytoplasm, since injection of 100–200 ng/μl DNA in pronuclear will kill the embryos. The survival rate of embryos cultured for 2–3 h after injection is between 90% and 100%. For knock-in project using plasmid donor carrying long homologous arms as template, which depends on homologous recombination, we recommend culture the embryos after injection for 20 h in EBSS-complete medium containing 7.5 μM RS-1 to stimulate Rad51 activity [37].
8. For single-embryo PCR, we recommend using REPLI-g® Mini Kit (Qiagen, Germantown, MD) to amplify genomic DNA before PCR reaction. Briefly, collect embryos individually in 2 μl distilled water, add 1.75 μl Buffer D2 (include in the kit) to each embryo, mix by vortexing, centrifuge briefly, and incubate on ice for 10 min. After that, add 1.75 μl Stop Solution (include in the kit), mix by vortexing, and centrifuge briefly. Add 2 μl of the mixture (denatured DNAs) to 8 μl master mix (include in the kit) and incubate at 30 °C for 10–16 h. Heating at 65 °C for 3 min, keep in 4 °C fridge, and use as PCR template.
9. Since New Zealand White Rabbit is not a pure inbred line, T7E1 and Surveyor mismatch cleavage assays are not always reliable due to potential SNPs. In Sanger sequencing chromatography, the PCR products of the mutant samples will show

different sequencing diagram depending on the type of the mutation. A typical mono-allelic indel mutation will show an "peaks on peaks" starting around the target site on the sequencing diagram, with wildtype sequence still readable. If you cannot read a wild-type sequence from the double curve, it may indicate a hetero double allelic indels. For homo double allelic indels, the sequencing diagram will be as clean as wild-type, but you can read the indels out. Please always keep in mind that these reading from Sanger sequencing data is not accurate. You need a cloning sequencing or targeted deep sequencing to confirm the exact mutant genotype. In practice, many rabbit embryos resulted from microinjection are mosaic, which means there are more than two peaks on the sequence diagram starting around the targeting site.

10. Thanks to the high KO efficiency of CRISPR/Cas9 tools, it is possible to combine 3–4 KO projects in one embryo transfer experiment to reduce the number of recipients. Divide the embryos into 3–4 groups, and inject with different guide RNAs for each group. Mix the embryos from all the groups for embryo transfer. In this way, one or two of the recipients may deliver all the KO kits you need [38].

11. Female rabbits do not have an estrus cycle and are classified as induced ovulators. A doe is considered to be receptive when she accepts service [39]. In recipient preparation, however, we cannot test its receptivity by mating. The color of the vulva is instead used as a strong indication of receptivity. It is reported that higher success rates were achieved using does with red and pink colored vulva in artificial insemination experiments, compared with that using does with white and purple colored vulva [40].

12. Rabbit ovulation is normally induced by the stimuli associated with coitus and occurs 10–12 h after mating. By mimic the natural mating process, ovulation can be induced artificially by mechanical stimulation of the vagina, alternatively, I.V. injection of hCG could also induce ovulation. In our experience, hCG injection is more reliable than mechanical stimulation.

13. Recipients induced the same day as mating the donors, or one day after mating the donors are both acceptable in rabbit embryo transfer, with similar success rates.

14. Pre-operative fasting in rabbit embryo transfer is not recommended, as the risk of aspiration due to vomiting in rabbit is negligible because rabbits lack the vomiting reflex, and are physically almost incapable of regurgitation. Feeding rabbit before surgery helps the gastrointestinal (GI) tract remain

active, which will speed recovery. Rabbit may suffer liver damage when the GI tract is empty [41].

15. Because rabbits lack the vomiting reflex, intubation is not necessary for anesthesia [42].
16. Rabbit can tolerate restraint, it is not necessary to use injectable anesthetics to induce anesthesia in rabbit.
17. Rabbit embryo will form a thick mucin layer around the zona pellucida during their passage through the oviduct. Without proper mucin layer development, their implantation and subsequent viability are adversely affected [43, 44]. It is desirable to transfer early stage embryos into the recipient oviducts, not uterus, to allow the mucin layer formation.
18. In rabbits, the embryos never move from one uterine horn to the other. Bilateral embryo transfer is necessary to fully utilize the recipients. The average rabbit litter size is 7–8 in natural mating. The litter size in gene editing project may vary depending on both the experiences of microinjectionist and the influence of the target gene. A minimum of two implanted embryos seems to be necessary to carry a pregnancy to term [45]. In our experience, when the litter size is larger than 10, the survival rate of the kits will be reduced greatly.
19. The average gestation period in rabbit is 30–32 days. Newborn kits weigh 30–100 g at birth, are hairless, and have closed eyelids. The doe normally nurses her kits once a day, usually in the morning. Kits open their eyes at about 10–12 days, and begin to eat solid food at about 3 weeks. Rabbit kits are weaned at 4–6 weeks.
20. A video demonstration of production of ApoCIII KO rabbits using Zinc Finger Nucleases is available at https://www.jove.com/video/50957/production-apolipoprotein-c-iii-knockout-rabbits-using-zinc-finger [46].

References

1. Fan J, Watanabe T (2003) Transgenic rabbits as therapeutic protein bioreactors and human disease models. Pharmacol Ther 99 (3):261–282
2. Peng X (2012) Transgenic rabbit models for studying human cardiovascular diseases. Comp Med 62(6):472–479
3. Peng X, Knouse JA, Hernon KM (2015) Rabbit models for studying human infectious diseases. Comp Med 65(6):499–507
4. Fan J, Kitajima S, Watanabe T, Xu J, Zhang J, Liu E, Chen YE (2015) Rabbit models for the study of human atherosclerosis: from pathophysiological mechanisms to translational medicine. Pharmacol Ther 146:104–119. https://doi.org/10.1016/j.pharmthera.2014.09.009
5. Li WH, Gouy M, Sharp PM, O'HUigin C, Yang YW (1990) Molecular phylogeny of Rodentia, Lagomorpha, Primates, Artiodactyla, and Carnivora and molecular clocks. Proc Natl Acad Sci U S A 87(17):6703–6707
6. Graur D, Duret L, Gouy M (1996) Phylogenetic position of the order Lagomorpha (rabbits, hares and allies). Nature 379 (6563):333–335. https://doi.org/10.1038/379333a0

7. Miller MJ, Chen NK, Li L, Tom B, Weiss C, Disterhoft JF, Wyrwicz AM (2003) fMRI of the conscious rabbit during unilateral classical eyeblink conditioning reveals bilateral cerebellar activation. J Neurosci 23 (37):11753–11758
8. Hammer RE, Pursel VG, Rexroad CE Jr, Wall RJ, Bolt DJ, Ebert KM, Palmiter RD, Brinster RL (1985) Production of transgenic rabbits, sheep and pigs by microinjection. Nature 315 (6021):680–683
9. Fan J, Ji ZS, Huang Y, de Silva H, Sanan D, Mahley RW, Innerarity TL, Taylor JM (1998) Increased expression of apolipoprotein E in transgenic rabbits results in reduced levels of very low density lipoproteins and an accumulation of low density lipoproteins in plasma. J Clin Invest 101(10):2151–2164. https://doi.org/10.1172/jci1599
10. Fan J, Wang J, Bensadoun A, Lauer SJ, Dang Q, Mahley RW, Taylor JM (1994) Overexpression of hepatic lipase in transgenic rabbits leads to a marked reduction of plasma high density lipoproteins and intermediate density lipoproteins. Proc Natl Acad Sci U S A 91(18):8724–8728
11. Yamanaka S, Balestra ME, Ferrell LD, Fan J, Arnold KS, Taylor S, Taylor JM, Innerarity TL (1995) Apolipoprotein B mRNA-editing protein induces hepatocellular carcinoma and dysplasia in transgenic animals. Proc Natl Acad Sci U S A 92(18):8483–8487
12. Huang Y, Schwendner SW, Rall SC Jr, Sanan DA, Mahley RW (1997) Apolipoprotein E2 transgenic rabbits. Modulation of the type III hyperlipoproteinemic phenotype by estrogen and occurrence of spontaneous atherosclerosis. J Biol Chem 272(36):22685–22694
13. Hoeg JM, Santamarina-Fojo S, Berard AM, Cornhill JF, Herderick EE, Feldman SH, Haudenschild CC, Vaisman BL, Hoyt RF Jr, Demosky SJ Jr, Kauffman RD, Hazel CM, Marcovina SM, Brewer HB Jr (1996) Overexpression of lecithin:cholesterol acyltransferase in transgenic rabbits prevents diet-induced atherosclerosis. Proc Natl Acad Sci U S A 93 (21):11448–11453
14. Knight KL, Spieker-Polet H, Kazdin DS, Oi VT (1988) Transgenic rabbits with lymphocytic leukemia induced by the c-myc oncogene fused with the immunoglobulin heavy chain enhancer. Proc Natl Acad Sci U S A 85 (9):3130–3134
15. Peng X, Olson RO, Christian CB, Lang CM, Kreider JW (1993) Papillomas and carcinomas in transgenic rabbits carrying EJ-ras DNA and cottontail rabbit papillomavirus DNA. J Virol 67(3):1698–1701
16. Yamamura Y, Kotani M, Chowdhury MI, Yamamoto N, Yamaguchi K, Karasuyama H, Katsura Y, Miyasaka M (1991) Infection of human CD4+ rabbit cells with HIV-1: the possibility of the rabbit as a model for HIV-1 infection. Int Immunol 3(11):1183–1187
17. Snyder BW, Vitale J, Milos P, Gosselin J, Gillespie F, Ebert K, Hague BF, Kindt TJ, Wadsworth S, Leibowitz P (1995) Developmental and tissue-specific expression of human CD4 in transgenic rabbits. Mol Reprod Dev 40(4):419–428. https://doi.org/10.1002/mrd.1080400405
18. Dunn CS, Mehtali M, Houdebine LM, Gut JP, Kirn A, Aubertin AM (1995) Human immunodeficiency virus type 1 infection of human CD4-transgenic rabbits. J Gen Virol 76 (Pt 6):1327–1336. https://doi.org/10.1099/0022-1317-76-6-1327
19. Massoud M, Bischoff R, Dalemans W, Pointu H, Attal J, Schultz H, Clesse D, Stinnakre MG, Pavirani A, Houdebine LM (1991) Expression of active recombinant human alpha 1-antitrypsin in transgenic rabbits. J Biotechnol 18(3):193–203
20. McKee C, Gibson A, Dalrymple M, Emslie L, Garner I, Cottingham I (1998) Production of biologically active salmon calcitonin in the milk of transgenic rabbits. Nat Biotechnol 16 (7):647–651. https://doi.org/10.1038/nbt0798-647
21. Prather RS, Lai L (2003) Creating genetically modified pigs by using nuclear transfer. Reprod Biol Endocrinol 1(1):82
22. Ramsoondar JJ, Machaty Z, Costa C, Williams BL, Fodor WL, Bondioli KR (2003) Production of alpha 1,3-galactosyltransferase-knockout cloned pigs expressing human alpha 1,2-fucosylosyltransferase. Biol Reprod 69 (2):437–445
23. Lai L, Park KW, Cheong HT, Kuhholzer B, Samuel M, Bonk A, Im GS, Rieke A, Day BN, Murphy CN, Carter DB, Prather RS (2002) Transgenic pig expressing the enhanced green fluorescent protein produced by nuclear transfer using colchicine-treated fibroblasts as donor cells. Mol Reprod Dev 62(3):300–306
24. Schnieke AE, Kind AJ, Ritchie WA, Mycock K, Scott AR, Ritchie M, Wilmut I, Colman A, Campbell KH (1997) Human factor IX transgenic sheep produced by transfer of nuclei from transfected fetal fibroblasts. Science 278 (5346):2130–2133
25. McCreath KJ, Howcroft J, Campbell KH, Colman A, Schnieke AE, Kind AJ (2000) Production of gene-targeted sheep by nuclear transfer from cultured somatic cells. Nature 405(6790):1066–1069

26. Cibelli JB, Stice SL, Golueke PJ, Kane JJ, Jerry J, Blackwell C, Ponce de Leon FA, Robl JM (1998) Cloned transgenic calves produced from nonquiescent fetal fibroblasts. Science 280(5367):1256–1258
27. Cole RJ, Edwards RG, Paul J (1966) Cytodifferentiation and embryogenesis in cell colonies and tissue cultures derived from ova and blastocysts of the rabbit. Dev Biol 13(3):385–407
28. Chesne P, Adenot PG, Viglietta C, Baratte M, Boulanger L, Renard JP (2002) Cloned rabbits produced by nuclear transfer from adult somatic cells. Nat Biotechnol 20(4):366–369. https://doi.org/10.1038/nbt0402-366
29. Du F, Xu J, Zhang J, Gao S, Carter MG, He C, Sung LY, Chaubal S, Fissore RA, Tian XC, Yang X, Chen YE (2009) Beneficial effect of young oocytes for rabbit somatic cell nuclear transfer. Cloning Stem Cells 11(1):131–140. https://doi.org/10.1089/clo.2008.0042
30. Gaj T, Gersbach CA, Barbas CF 3rd (2013) ZFN, TALEN, and CRISPR/Cas-based methods for genome engineering. Trends Biotechnol 31(7):397–405. https://doi.org/10.1016/j.tibtech.2013.04.004
31. Conklin BR (2013) Sculpting genomes with a hammer and chisel. Nat Methods 10 (9):839–840. https://doi.org/10.1038/nmeth.2608
32. Cui X, Ji D, Fisher DA, Wu Y, Briner DM, Weinstein EJ (2011) Targeted integration in rat and mouse embryos with zinc-finger nucleases. Nat Biotechnol 29(1):64–67. https://doi.org/10.1038/nbt.1731
33. Yang D, Xu J, Zhu T, Fan J, Lai L, Zhang J, Chen YE (2014) Effective gene targeting in rabbits using RNA-guided Cas9 nucleases. J Mol Cell Biol 6(1):97–99. https://doi.org/10.1093/jmcb/mjt047
34. Ran FA, Hsu PD, Wright J, Agarwala V, Scott DA, Zhang F (2013) Genome engineering using the CRISPR-Cas9 system. Nat Protoc 8 (11):2281–2308. https://doi.org/10.1038/nprot.2013.143
35. Hsu Patrick D, Lander Eric S, Zhang F (2014) Development and applications of CRISPR-Cas9 for genome engineering. Cell 157 (6):1262–1278. https://doi.org/10.1016/j.cell.2014.05.010
36. Shuji Kitajima EL, Fan J (2009) Rabbit transgenesis. Rabbit biotechnology: rabbit genomics, transgenesis, cloning and models. Springer, Dordrecht, Heidelberg, London, New York
37. Song J, Yang D, Xu J, Zhu T, Chen YE, Zhang J (2016) RS-1 enhances CRISPR/Cas9- and TALEN-mediated knock-in efficiency. Nat Commun 7. https://doi.org/10.1038/ncomms10548
38. Song J, Yang D, Ruan J, Zhang J, Chen YE, Xu J (2017) Production of immunodeficient rabbits by multiplex embryo transfer and multiplex gene targeting. Sci Rep 7(1):12202. https://doi.org/10.1038/s41598-017-12201-0
39. Lebas F, Coudert P, de Rochambeau H, Thébault RG (1997) The rabbit: husbandry, health and production (new revised version). FAO animal production and health series, no. 21. Food and Agriculture Organization of the United Nations
40. Najjar A, Ben Aicha E, Ben Mrad M (2013) Receptivity of the rabbit doe: which methods that could be predictive for receptivity. Experiment 12(3):786–790
41. Krempels D (2011) Pre- and post-operative care of rabbits. Newsl Buckeye House Rabbit Soc 15(2):1–4
42. Bateman L, Ludders JW, Gleed RD, Erb HN (2005) Comparison between facemask and laryngeal mask airway in rabbits during isoflurane anesthesia. Vet Anaesth Analg 32 (5):280–288. https://doi.org/10.1111/j.1467-2995.2005.00169.x
43. Murakami H, Imai H (1996) Successful implantation of in vitro cultured rabbit embryos after uterine transfer: a role for mucin. Mol Reprod Dev 43(2):167–170. https://doi.org/10.1002/(sici)1098-2795(199602)43:2<167::aid-mrd5>3.0.co;2-p
44. Moore NW, Adams CE, Rowson LE (1968) Developmental potential of single blastomeres of the rabbit egg. J Reprod Fertil 17 (3):527–531
45. Adams CE (1970) Maintenance of pregnancy relative to the presence of few embryos in the rabbit. J Endocrinol 48(2):243–249
46. Yang D, Zhang J, Xu J, Zhu T, Fan Y, Fan J, Chen YE (2013) Production of apolipoprotein C-III knockout rabbits using zinc finger nucleases. J Vis Exp 81:50957. https://doi.org/10.3791/50957

Chapter 20

Production of Genetically Engineered Porcine Embryos by Handmade Cloning

Rong Li, Jinxin Miao, and Zhongde Wang

Abstract

Genetic engineering is essential to realize the full potentials of pigs both as livestock and as animal models of human disease. With the development of new genetic engineering technologies, such as the clustered regularly interspaced short palindromic repeats-associated endonuclease 9 (CRISPR/Cas9) system, the porcine genome can be engineered with high efficiency. In this chapter, we describe a protocol in employing the CRISPR/Cas9 system to genetically engineer the porcine genome in fibroblast cells, the procedures to establish single-cell-derived porcine fibroblast cell colonies carrying the desired genetic modifications, and the handmade cloning (HMC) technique to generate cloned embryos ready for embryo transfer.

Key words Swine, Handmade cloning, Knockout, sgRNA, CRISPR/Cas9, Animal model

1 Introduction

Because of the significance of pigs as livestock and as animal models of human disease, there are great interests and active researches in developing genetically engineered pigs [1]. Genetic engineering has been used to generate several desired animal traits in pigs including more efficient meat production [2], enhanced nutrition values of pork [3], reduced impact on the environment from pig farming [4], and improved disease resistance [5]. Furthermore, genetically engineered pigs may serve as organ donors for human xenotransplantation [6] and as animal models of human disease [7].

Like other livestock species, due to the lack of embryonic stem cells in pigs, genetically engineered pigs are produced either by somatic cell nuclear transfer (SCNT or cloning) from using genetically engineered somatic cells as nuclear donors or by pronuclear injection of foreign DNA or reagents with genetic modification activities (e.g., CRISPR/Cas9 complex) into embryos [1]. Here,

Rong Li and Jinxin Miao contributed equally to this work.

Chengyu Liu and Yubin Du (eds.), *Microinjection: Methods and Protocols*, Methods in Molecular Biology, vol. 1874, https://doi.org/10.1007/978-1-4939-8831-0_20, © Springer Science+Business Media, LLC, part of Springer Nature 2019

we describe a process for generating genetically engineered pigs by SCNT. We divide this protocol into two sections: in Section I, we describe the procedures to engineer the porcine genome with the CRISPR/Cas9 system and to establish single-cell-derived porcine fibroblast cell lines with the desired genetic modifications. In Section II, we describe the handmade cloning (HMC) method to generate cloned embryos using the engineered cells.

The CRISPR/Cas9 system is an adaptive immune system in prokaryotes that protects the host from invading foreign genetic element, e.g., phages and viruses, through a process of recognition and sequence-specific cutting [8]. The functional unit of the CRISPR/Cas9 system is a ribonucleoprotein (RNP) complex comprising a CRISPR RNA (crRNA), a transactivating crRNA (tracrRNA), and the Cas9 endonuclease. The recent understanding of the CRISPR/Cas9 system has allowed scientists to harness its power for use in other organisms to create specific genetic modifications. To achieve site-specific genetic modifications, the CRISPR/Cas9 system has been modified in such a way that a synthetic single-strand guide RNA (sgRNA) can be designed according to a target DNA sequence to direct the Cas9 protein to introduce site-specific double-strand breaks (DSBs) [9, 10]. In mammalian cells, DSBs are repaired either by the non-homologous end joining (NHEJ) pathway or by the homologous directed repair (HDR) pathway [11]. Because the NHEJ path is error prone, nucleotide-insertions or -deletions (indels), as well as nucleotide-conversion mutations, can be introduced to the target gene at the DSBs. If the indels or conversion mutations result in a premature stop codon, the gene will be genetically inactivated or knocked out (KO). Alternatively, it is possible to knock in desired sequences (e.g., gene, gene segment, and point mutations) at the site of a DSB. To accomplish this, a donor DNA template in the form of either oligos or plasmids can be delivered into cells or embryos alongside the RNP. In this situation, the HDR pathway integrates the donor DNA template at or proximal to the DSB site in a sequence-specific manner [12].

The handmade cloning technique was first described by Vajta et al. and Booth et al. in 2001 [13, 14]. This technique circumvents the requirements for expensive equipment by the conventional cloning method. Since then, many animal species have been successfully cloned by HMC, including cattle [15], horses [16], pigs [17], and sheep [18]. Apart from the easiness without using micromanipulators, HMC may be advantageous over conventional cloning in that better quality of cloned embryos can be produced which may lead to higher cloning efficiency [19, 20].

In this chapter, we describe the procedures to establish genetical modified porcine fibroblast cell lines by the CRISPR/Cas9 system carrying indels in the coding sequence for the scavenger receptor cysteine-rich domain 5 (SRCR5) in porcine CD163 (cluster of differentiation 163), the receptor for porcine reproductive and respiratory syndrome virus (PRRSV) entry [21]. We then describe the HMC protocol to generate cloned embryos using the genetically modified porcine fibroblast cell lines.

2 Materials

2.1 General Reagents and Equipment

1. GeneArt Precision gRNA Synthesis Kit: synthesis crRNA.
2. GeneJET RNA Cleanup and Concentration Micro Kit: purify crRNA.
3. E-Gel EX Agarose Gels (2×): detect DNA fragment.
4. Novex TBE-Urea Gels, 10%: detect RNA.
5. Nucleofector: transfect cells.
6. P3 Primary Cell Solution Box: cell transfection supplements.
7. Cas9 nuclease: induce the double-strand break and the mutations.
8. Genomic DNA isolation kit: isolate cells/tissue genomic DNA.
9. Inverted microscope: detect and screen cell colony.
10. Stereomicroscope: embryo in vitro handling, including oocytes/embryos collection, enucleation, fusion, and detection.
11. Incubator: cells/embryos cultivation.
12. Fusion machine: fusion cells and activate embryos.
13. Fusion chamber: fusion cells and activate embryos.
14. Aggregation needle: make aggregation wells for embryonic cultivation.
15. Splitting blade: enucleation of oocytes.
16. Warm plate: keep temperature stable for embryo handling.

2.2 Media

All chemicals are purchased from Sigma-Aldrich unless described otherwise.

1. DPBS.
2. 0.05% Trypsin-EDTA.
3. DMEM supplemented with 10% (v/v) FBS, 1× NEAA, and 1× AA.
4. Aspiration media (AM): M199 supplemented with 2.0 mM $NaHCO_3$, 2.5 mM Pyruvic Acid (Na-Salt), 1.0 mM L-Glutamine, 13.8 mM HEPES (Na-Salt), 11.0 mM HEPES (Acid), 2.5 μg/ml Amphotericin B, and 30 IU/ml Heparin: Adjust pH to 7.2–7.4 and osmolarity to 280 ± 8 mOsm. Filter through a 0.22 μm filter.

Table 1
Formula of porcine zygote media (PZM-3)

Components	Concentration (mM)
NaCl	108.00
$NaHCO_3$	25.07
KCl	10.00
KH_2PO_4	0.35
$MgSO_4$	0.40
Ca-Lactate-5H_2O	2.00
Na-pyruvate	0.20
Myo-Inositol	2.78
Phenol Red	0.27
L-Glutamine	1.00
Hypotaurine	5.00
Gentamicin	0.04 g/L
BSA	3 g/L
NEAA100X	1% (v/v)
EAA50X	1% (v/v)

Note: Adjust pH value to 7.2–7.4; osmolarity to 280 ± 8 mOsm; filter through 0.22 μm filter

5. Maturation media (MM): Use TCM-199 as the base media. Supplement the media with 10% (v/v) cow serum (CS), 10% (v/v) sow follicular fluid (the follicular fluid is collected from normal sow follicles and filtered by 0.45 μm and 0.22 μm filter), 10 IU/ml pregnant mare serum gonadotrophin, and 5 IU/ml human chorionic gonadotrophin.
6. Embryo cultivation medium: Prepare porcine zygote media (PZM-3) as described by Yoshioka K et al. [22]. The formula is listed in Table 1.
7. Fusion media: 0.3 M mannitol, 0.1% (w/v) Polyvinyl alcohol, 99.6 nM $MgSO_4$, and 50 nM $CaCl_2$. Filter and aliquot into 2.5 ml/tube. Fusion media can be stored for 6 months at −80 °C.
8. Activation media: 0.3 M mannitol, 0.1% (w/v) Polyvinyl alcohol, and 99.6 nM $MgSO_4$. Adjust mOsm to 280 ± 8 with water and pH to 7.4–7.8 with trizma-base, then filter and aliquot into tubes with 2.5 ml each. Fusion media can be stored for 6 months at −80 °C.

9. Cytochalasin B (CB), dissolve in DMSO, 5 mg/ml (1000× stock).
10. Cycloheximide (CHX), dissolve in DMSO, 1 mg/ml (100× stock).
11. Hyaluronidase, dissolve in TCM199, 1.0 mg/ml.
12. Phytohemagglutinin (PHA), dissolve in TCM199, 5 mg/ml (12.5× stock).
13. Pronase, dissolve in TCM199, 10 mg/ml. Working concentration is 3.3 mg/ml, diluted in 30% CS of TCM199 medium.
14. T2, Hepes-buffered TCM 199 with 2% CS.
15. T10, Hepes-buffered TCM 199 with 10% CS.
16. T20, Hepes-buffered TCM 199 with 20% CS.
17. Chemical activation media, PZM-3 supplemented with 5 μg/ml CB and 10 μg/ml CHX.

3 Methods

3.1 Establishing Genetically Engineered Porcine Fibroblast Cell Lines

The general processes are described in Fig. 1 with designing a sgRNA for SRCR5 as an example.

3.1.1 Design of Single-Guide RNA (sgRNA) and Genotyping PCR Primers

1. Acquire the gene sequence of interest on the NCBI gene database (Table 2).
2. Design sgRNAs targeting exon 7 (encoding SRCR5) of the porcine CD163 gene using Zhang Lab CRISPR DESIGN (http://crispr.mit.edu:8079/). Select the sgRNAs (Table 2) based on the scores in both on-target and off-target (the overall score is 69). Attention is also given for the availability of restriction enzyme site for genotyping analysis with PCR-RFLP (basic principle described in **Note 1**); in this particular case, a BsaA I site is chosen.
3. Based on the sgRNA target sequence, design a pair of oligonucleotides for the assembly of the crRNA template (following the GeneArt Precision gRNA Synthesis Kit), i.e., the forward (defined as Target F, Table 2) and reverse complement guide sequences (defined as Target R, Table 2).
4. Design PCR primers to amplify the desired DNA sequence (Primer F/R, Table 2) flanking the sgRNA target site (PCR amplicons are normally in the range of 400–600 bp). The PCR fragment can be used for restriction fragment length polymorphism (RFLP).

3.1.2 Synthesis of sgRNAs by the GeneArt Precision gRNA Synthesis Kit

1. Prepare 0.3 μM target F/R oligonucleotide mix working solution by diluting the 10 μM target oligonucleotide mix stock solutions in nuclease-free water.

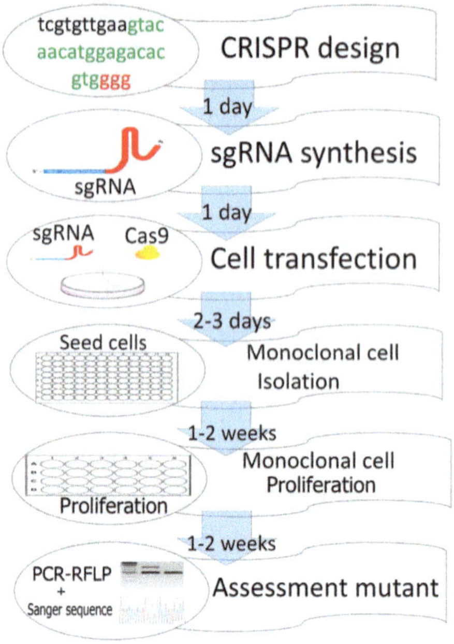

Fig. 1 General workflow for establishing genetically engineered porcine fibroblast cells lines using the CRISPR-Cas9 system. (1) Design and synthesize sgRNA, (2) transfect fibroblast cells to induce genetic modification events, (3) establish and screen monoclonal cell lines, and (4) detect indels with PCR-RFLP, followed by Sanger sequencing

Table 2
sgRNA design for knockout the porcine SRCR5

Name	Description/sequence
SRCR5	NCBI Reference Sequence: NC_010447.5
sgRNA	5′-AAGTACAACATGGAGACACG-3′
Target F	5′-AAGTACAACATGGAGACACG-3′
Target R	5′-CGTGTCTCCATGTTGTACTT-3′
Primer F	5′-ACCTTGATGATTGCGCTCTT-3′
Primer R	5′-TGTCCCAGTGAGAGTTGCAG-3′

2. Set up the reaction solution for the synthesis of sgRNA DNA template at room temperature. Assemble a 25 μl PCR reaction solution with the following components: 1 μl Tracr Fragment + T7 Primer Mix, 1 μl 0.3 μM Target F/R oligonucleotide, 12.5 μl Phusion Flash High-Fidelity PCR Master Mix, and adjust the total reaction solution volume with H_2O to 25 μl. PCR conditions are: 98 °C for 10 s, 32 cycles of (98 °C for 5 s, 55 °C for 15 s), and 72 °C for 1 min.

3. Resolve the PCR product in a 2% E-Gel EX Agarose gel.
4. Perform in vitro transcription (IVT) by the following reaction components and processes: mix 2 μl TranscriptAid Enzyme Mix, 4 μl TranscriptAid Reaction Buffer, 8 μl NTP mix, and 6 μl sgRNA DNA template (synthesized in **step 2**) and incubate the reaction solution at 37 °C for 3 h.
5. Remove the DNA templates by adding 1 μl DNase I to the IVT solution at 37 °C for 15 min.
6. Purify the in vitro transcribed sgRNA product using GeneJET RNA Purification kit as described in the manual. Use 10 μl of nuclease-free water to elute the sgRNA.
7. Analyze IVT product by a 10% Novex TBE-Urea gel to determine IVT sgRNA quality for its size and purity.
8. Measure the concentration and aliquot sgRNA for subsequent use. sgRNA can be kept within 1 week at 4 °C, or can be stored in −80 °C for long time storage (*see* **Note 2**).

3.1.3 Cell Transfection

It is preferable to use early-passage cells for transfection, as they have the greatest potential for the subsequent cloning process.

1. Grow fibroblast cells to 80–90% confluency and change media a day before transfection.
2. Mix 480 ng SRCR5 sgRNA and 2 μg Cas9 protein and incubate for 15 min at room temperature to form RNP. Mix the RNP with the electroporation solution (P3 Nucleofector Solution) by gently tapping the tubes.

   Note: The molar ratio of sgRNA to Cas9 protein should be approximately 1:1 to optimize the production of RNP.
3. Harvest cells by trypsinization and wash cells twice with RNase-free DPBS to remove trypsin and eliminate RNase activity.
4. Resuspend cells in the RNP mixture and transfer the suspension to a Nucleocuvette vessel. For achieving optimal transfection efficiency, 1×10^6 cells should be added per 100 μl of electroporation solution. Transfect cells with the 4D-Nucleofector instrument with the program of EH100.
5. Immediately transfer the transfected cells to a pre-warm 6-well plate with culture media. Incubate for 72 h at 37 °C, 5% CO_2.

3.1.4 Isolation of Single-Cell-Derived Fibroblast Colonies by Limiting Dilution

1. Harvest cells by trypsinization and disaggregate them by gentle pipetting. Quantitate the cell concentration with a hemocytometer.
2. Dilute the cells to a concentration of 5 cells/ml with 15% FBS DMEM. Prepare approximately 10 ml of this cell solution and

seed 100 μl to each well of a 96-well plate, which results an average density of 0.5 cells/well.

3. Visually monitor cell colonies for 1–2 weeks post seeding the cells, select and passage the single-cell-derived colonies that are with sufficient growth (over 2000 cells) (*see* **Note 3**).

3.1.5 Screening CRISPR/Cas9-Targeted Cell Colonies for Desired Indels

1. Isolate genomic DNA from the single-cell-derived colonies using the Gentra Puregene Cell Kit as described in the manual.
2. Assemble a volume of 50 μl PCR reaction with the following components: 0.25 μl Ex Taq DNA Polymerase (TaKaRa, RR001A), 5 μl Ex Taq Buffer (20 mM Mg2+ plus), 4 μl dNTP Mix (2.5 mM each), 1 μl forward/reverse primer (10 μM, Primer F/R, Table 2), 100 ng gDNA, and adjust the reaction volume with H_2O to 50 μl. Run PCR using the following parameters: 95 °C for 3 min, 38 cycles of (98 °C for 10 s, 60 °C for 30 s, 72 °C for 54 s), and 72 °C for 2 min.
3. PCR-RFLP assay: 1 μl BsaA I, 2 μl CutSmart Buffer, 4 μl PCR product, adjust the reaction volume with H_2O to 20 μl, and incubate at 37 °C for 1 h. Run samples on 1% agarose gel at 10 V/cm using 1× Tris-acetate-EDTA (TAE) buffer. Cell lines carrying mutations that interrupt the BsaA I recognition sequence are detected by the presence of bands from undigested PCR fragments.
4. Determine the mutations at nucleotide level by Sanger sequencing the PCR products.

3.2 Production of Handmade Cloned Embryos (see Note 4)

3.2.1 Oocytes Preparation

1. Acquire ovaries from a local slaughterhouse, wash away blood with warm saline solution before delivery. Ovaries are transported in a thermos maintained at 30–35 °C.
2. Wash ovaries twice with warm saline solution (37 °C) before aspiration.
3. Put the ovary on dry filter paper to remove excess liquid.
4. Aspire 3–8 mm follicles with a 21 G needle connected to a negative pressure pump.
5. Collect all cumulus-oocyte complexes (COCs) into AM media. Note: Warm the AM media 2 h before collection on a heating plate (38.5 °C).

3.2.2 Oocytes Maturation

1. Wash COCs twice with AM media and select the COCs with more than two layers of cumulus cells for maturation.
2. Transfer and spread the COCs to maturation dishes with a density of 50–60 COCs/well. The time is recorded as 0 h.
3. Mature the COSs for 41–42 h in a humidified incubator at 38.5 °C, 5% CO_2.

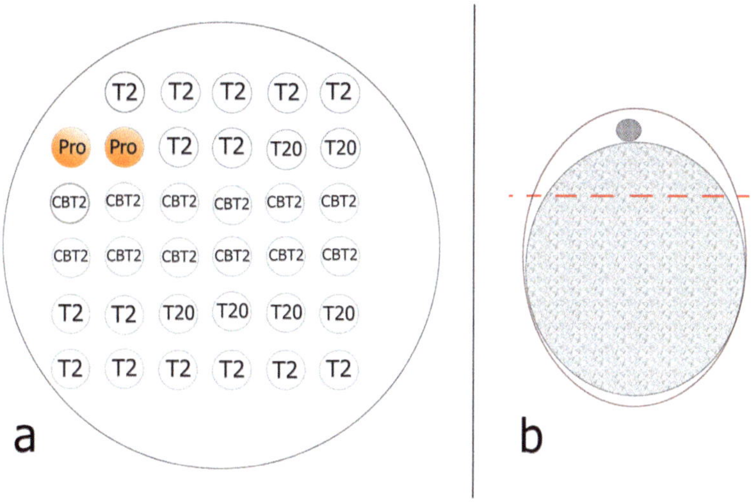

Fig. 2 (**a**) Preparation of bisection dish by using a 60 mm petri dish. The volume of each drop is 20 μl. T2: Hepes-buffered TCM 199 with 2% cow serum (CS); T20: Hepes-buffered TCM 199 with 20% CS; CBT2: T2 supplemented with 25 μg/ml cytochalasin B; Pro: 3.3 mg/ml pronase. (**b**) Diagram of oocyte bisection (enucleation). The position of bisection cut is indicated with the red line (about 1/3 position of the oocyte from the side where the polar body is located)

3.2.3 Oocytes Enucleation

Prepare all manipulation dishes prior to removal of cumulus cells, including enucleation dish, cell fusion dish and activation dishes, which are described in Figs. 2a, 3a and b, respectively.

1. After 41–42 h maturation, denude oocytes from cumulus cells by pipetting them in 1 mg/ml Hyaluronidase.
2. Collect all oocytes into T2 drops (Fig. 2a, line 1) and wash them twice with T2 drops. Discard any dead and abnormal-looking oocytes during washing.
3. Bleach the zona pellucida of oocytes by 3.3 mg/ml pronase (Pro, Fig. 2a, line 2). Wash oocytes in T2 drops once the zona pellucida shows deformation following a wash in T20 (Fig. 2a, line 2).
4. Transfer the oocytes into CBT2 drops and rotate until the polar bodies are positioned at 12 o'clock position; align ~20 oocytes to two lines within a CBT2 drop (Fig. 2a, lines 3 and 4).
5. Bisect oocytes at ~1/3 position of oocytes (Fig. 2b).
6. Collect the enucleated cytoplasm (the low part, Fig. 2b), wash them in T2 and then T20 (Fig. 2a, line 5), and transfer them to T2 drops (Fig. 2a, line 6) (*see* **Note 5**).

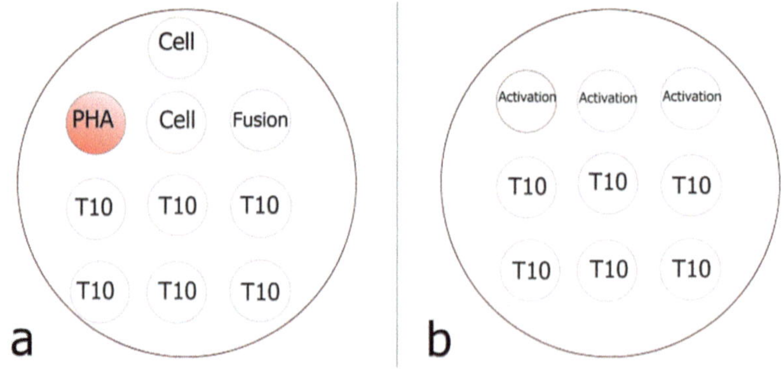

Fig. 3 (**a**) Preparation of cell fusion dish by using a 35 mm petri dish. The volume of each drop is 20 μl. PHA drop: 0.4 mg/ml Phytohemagglutinin; T10 drops: Hepes-buffered TCM 199 with 10% CS; Cell drops: T2 medium; Fusion drop: cell fusion medium. (**b**) Preparation of activation dish by using a 35 mm petri dish. The volume of each drop is 20 μl. T10 drops: Hepes-buffered TCM 199 with 10% CS; Activation drops: activation medium

3.2.4 Preparation of Somatic Cells for Handmade Cloning

1. Grow donor cells to full confluency and change media 1 day prior to cloning.
2. Harvest the cells with 0.05% trypsin meanwhile conducting oocyte enucleation (**step 3**).
3. Resuspend the cells with 500 μl T10, and then keep them at room temperature until use. Normally, the cell density is about 2.5×10^6 cells/ml.

3.2.5 Fusion Donor Cells to Enucleated Oocyte Cytoplasm

1. Transfer the enucleated cytoplasm into T10 drops (Fig. 3a).
2. Transfer 2 μl cells solution prepared in 3.2.4.3 above into cell drops (made by T2, Fig. 3a).
3. Transfer five enucleated cytoplasm into a PHA drop for 5 s, and then transfer them to cell drops to attach a cell for each cytoplasm.
4. Move the cytoplasm-cell complexes (CCCs) into a cell fusion drop, and wait for 10–15 s to balance all CCCs in the fusion media.
5. Place 500 μl cell fusion media in the fusion chamber and transfer the CCCs into the fusion chamber. Apply a 60 V/cm alternating current (AC) to generate an electromagnetic field for the CCCs alignment. Align the cytoplasm part to attach the negative electrode while the CCCs center line is kept vertical to the electrode, and employ a single pulse of 1.6 kV/cm direct current (DC) for 9 μs. Maintain the AC for 1 s and transfer CCCs back into T10 drops.
6. Cover the fusion dish with dark box and keep in the warm plate for 1 h (*see* **Note 6**).

3.2.6 Fusion CCCs with Complemental Cytoplasm and Activation of Reconstructed Embryos

Each CCCs is fused to another enucleated cytoplasm (complemental cytoplasm) in the same procedure of electrical activation to increase the developmental competence.

1. Transfer complemental cytoplasm into the first column of T10 drops in the activation dish.
2. Check the success of cell fusion 1 h after cell fusion and transfer the fused CCCs to the second column T10 drops.
3. Transfer ten complemental cytoplasm into the first activation drop, transfer the corresponding CCCs into another activation drop, and balance them for 10–15 s.
4. Transfer the complemental cytoplasm and CCCs into fusion chamber. Apply a 60 V/cm AC to generate an electromagnetic field for the CCCs alignment. Transfer the CCCs to attach the negative electrode and align the complemental cytoplasm connected to CCCs, then employ a single DC plus of 0.86 kV/cm for 80 μs for electrical activation. Keep the AC for 1 s and transfer the complexes back into T10 drops.
5. Cover the activation dish with dark box and keep in the warm plate for 5 min.
6. Transfer the fused complexes (defined as reconstructed embryos) to chemical activation media and discard the non-fused ones.
7. Incubate the reconstructed embryos in the chemical activation media for 4 h (*see* **Note 7**).

3.2.7 Reconstructed Embryo Cultivation

1. Prepare cultivation dishes and equilibrate the dishes in incubator overnight.
2. Make suitable and enough aggregation wells in a 4-well dish (WOWs, Fig. 4a).
3. Wash the activated embryos twice in new PZM-3 media.
4. Transfer the reconstructed embryo into individual WOWs. Incubate the embryos for 6 days at 38.5 °C, 5% CO_2, 5% O_2, 90% N_2 and 100% humidity.
5. Assess blastocyst formation on Day 6 (Fig. 4b).

4 Notes

1. Basic principle for sgRNA design. (a) Targeting site. To knock out a gene, targeting within the first exons is normally preferred to fully inactivate the gene. (b) Targeting specificity of the sgRNA. A blast search to confirm the uniqueness of sgRNAs to the target sequences in the genome can reduce off-targeting events. (c) Restriction enzyme for conveniently identifying

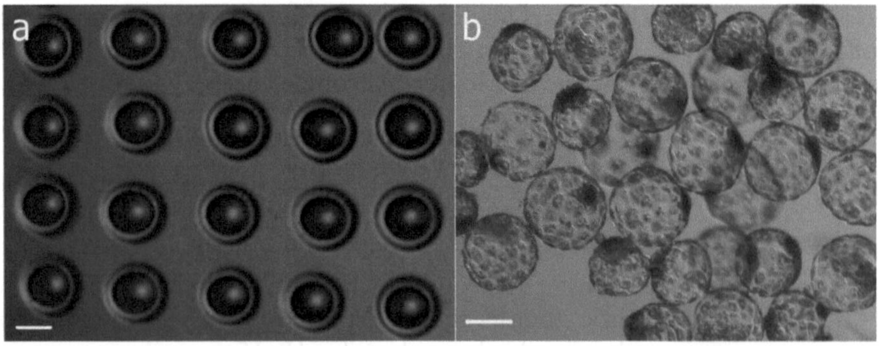

Fig. 4 (**a**) Aggregation wells prepared in the wells of a 4-well plate (well-of-the-well; WOW). Bar = 100 μm; (**b**) Handmade cloned porcine blastocysts. Bar = 100 μm. (Courtesy of Dr. Henrik Callesen's laboratory at Aarhus University, Denmark)

indels. As the Cas9 nuclease tends to cleave 3-nt upstream of the PAM site, a restriction site a few nucleotides upstream of the PAM can be used in the PCR-RFLP assay to detect targeted mutations.

2. Synthesize sgRNA. It is critical to avoid RNase contamination during this procedure. Wear groves and mask during IVT, and make sure to use RNase-free tips/tubes and water/TE buffer for elute/dilute the sgRNA.
3. Single-cell-derived cell colony isolation. Seed an average of 0.5 cells/well to ensure that most of the wells receive a single cell. When screening the cell culture plates, look thoroughly within each well, especially in the corner of a well as cells tend to settle there.
4. Ambient environment. Since porcine oocytes/embryos are sensitive to both light and temperature, reducing light exposure and handling them on a warm plate (38 °C) can improve their survival and developmental competence both in vitro and in vivo. To reduce light exposure: (a) cover all the windows to avoid light exposure, especially to sunlight, (b) shut off all ceiling light, (c) keep a minimum suitable light for embryo handling, (d) cover embryos with a dark box to avoid any extra light exposure except during handling.
5. Keep a thin layer zona pellucida for bisection. A thin zona pellucida can avoid oocyte damage caused by the direct connection with micro blade and the cutting scratches. To keep a thin layer of zona pellucida, the bleach reaction should be closely monitored by the morphology change. Over-digestion can be avoided by washing the oocytes twice in T2 and then twice in T20 once the shape-change of zona pellucida is visibly noticeable.

6. Avoid cytoplasm connection during handling, especially after PHA treatment. Attention should be given in the following two procedures:
   (a) Cell fusion. A suitable diameter of glass pipette can reduce the touch of cytoplasm from each other and reduce the effect of media change to fusion/activation: use ~120 μm and ~140 μm for cell fusion and activation, respectively.
   (b) Zona pellucida residual must be removed, otherwise it can impede the cell attachment and fusion.
   (c) Activation. As the membrane becomes sticky during chemical activation, it is necessary to spread the embryos apart from each other as much as possible during this step; otherwise, the embryos will aggregate into a giant mass.
7. Cytoplasm selection for cell fusion: select the bigger cytoplasms for cell fusion and keep the smaller ones for the activation complemental cytoplasms.

Acknowledgments

The work was supported by a research grant from Utah Science Technology and Research (USTAR) initiative (to Z.W.) and the Next-Generation BioGreen 21 Program, Republic of Korea, grant no. PJ01107704 (to Z.W.). The content is solely the responsibility of the authors and does not necessarily represent the official views of the grantors. We thank Dr. Clay Isom from the Department of Animal, Dairy, and Veterinary Sciences at Utah State University for providing the fetal fibroblast cells used for the procedures described in this protocol. We thank Dr. Nikolas Robl for proofreading the manuscript.

References

1. Wells K, Prather R (2017) Genome-editing technologies to improve research, reproduction, and production in pigs. Mol Reprod Dev 84:1012–1017
2. Ju H, Zhang J, Bai L, Mu Y, Du Y, Yang W, Li Y, Sheng A, Li K (2015) The transgenic cloned pig population with integrated and controllable GH expression that has higher feed efficiency and meat production. Sci Rep 5:10152
3. Lai L, Kang J, Li R, Wang J, Witt W, Yong H, Hao Y, Wax D, Murphy C, Rieke A, Samuel M, Linville M, Korte S, Evans R, Starzl T, Prather R, Dai Y (2006) Generation of cloned transgenic pigs rich in omega-3 fatty acids. Nat Biotechnol 24:435–436
4. Golovan S, Meidinger R, Ajakaiye A, Cottrill M, Wiederkehr M, Barney D, Plante C, Pollard J, Fan M, Hayes M, Laursen J, Hjorth J, Hacker R, Phillips J, Forsberg C (2001) Pigs expressing salivary phytase produce low-phosphorus manure. Nat Biotechnol 19:741–745
5. Burkard C, Lillico S, Reid E, Jackson B, Mileham A, Ait-Ali T, Whitelaw C, Archibald A (2017) Precision engineering for PRRSV resistance in pigs: macrophages from genome edited pigs lacking CD163 SRCR5 domain are fully resistant to both PRRSV genotypes while maintaining biological function. PLoS Pathog 13:e1006206

6. Cooper D, Ekser B, Ramsoondar J, Phelps C, Ayares D (2016) The role of genetically engineered pigs in xenotransplantation research. J Pathol 238:288–299
7. Prather R, Lorson M, Ross J, Whyte J, Walters E (2013) Genetically engineered pig models for human diseases. Annu Rev Anim Biosci 1:203–219
8. Mojica F, Rodriguez-Valera F (2016) The discovery of CRISPR in archaea and bacteria. FEBS J 283:3162–3169
9. Cong L, Zhang F (2015) Genome engineering using CRISPR-Cas9 system. Methods Mol Biol 1239:197–217
10. Paquet D, Kwart D, Chen A, Sproul A, Jacob S, Teo S, Olsen K, Gregg A, Noggle S, Tessier-Lavigne M (2016) Efficient introduction of specific homozygous and heterozygous mutations using CRISPR/Cas9. Nature 533:125–129
11. Chu V, Weber T, Wefers B, Wurst W, Sander S, Rajewsky K, Kühn R (2015) Increasing the efficiency of homology-directed repair for CRISPR-Cas9-induced precise gene editing in mammalian cells. Nat Biotechnol 33:543–548
12. Richardson C, Ray G, DeWitt M, Curie G, Corn J (2016) Enhancing homology-directed genome editing by catalytically active and inactive CRISPR-Cas9 using asymmetric donor DNA. Nat Biotechnol 34:339–344
13. Vajta G, Lewis I, Hyttel P, Thouas G, Trounson A (2001) Somatic cell cloning without micromanipulators. Cloning 3:89–95
14. Booth P, Tan S, Holm P, Callesen H (2001) Application of the zona-free manipulation technique to porcine somatic cell nuclear transfer. Cloning Stem Cells 3:191–197
15. Vajta G, Bartels P, Joubert J, de la Rey M, Treadwell R, Callesen H (2004) Production of a healthy calf by somatic cell nuclear transfer without micromanipulators and carbon dioxide incubators using the Handmade Cloning (HMC) and the Submarine Incubation System (SIS). Theriogenology 62:1465–1472
16. Lagutina I, Lazzari G, Duchi R, Colleoni S, Ponderato N, Turini P, Crotti G, Galli C (2005) Somatic cell nuclear transfer in horses: effect of oocyte morphology, embryo reconstruction method and donor cell type. Reproduction 130:559–567
17. Du Y, Kragh P, Zhang Y, Li J, Schmidt M, Bøgh I, Zhang X, Purup S, Jørgensen A, Pedersen A, Villemoes K, Yang H, Bolund L, Vajta G (2007) Piglets born from handmade cloning, an innovative cloning method without micromanipulation. Theriogenology 68:1104–1110
18. Zhang P, Liu P, Dou H, Chen L, Chen L, Lin L, Tan P, Vajta G, Gao J, Du Y, Ma R (2013) Handmade cloned transgenic sheep rich in omega-3 fatty acids. PLoS One 8: e55941
19. Callesen H, Liu Y, Pedersen H, Li R, Schmidt M (2014) Increasing efficiency in production of cloned piglets. Cell Reprogram 16:407–410
20. Liu Y, Lucas-Hahn A, Petersen B, Li R, Hermann D, Hassel P, Ziegler M, Larsen K, Niemann H, Callesen H (2017) Developmental competence and epigenetic profile of porcine embryos produced by two different cloning methods. Cell Reprogram 19:171–179
21. Van Gorp H, Van BW, Van Doorsselaere J, Delputte P, Nauwynck H (2010) Identification of the CD163 protein domains involved in infection of the porcine reproductive and respiratory syndrome virus. J Virol 84:3101–3105
22. Yoshioka K, Suzuki C, Tanaka A, Anas I, Iwamura S (2002) Birth of piglets derived from porcine zygotes cultured in a chemically defined medium. Biol Reprod 66:112–119

Chapter 21

Electrofusion of 2-Cell Embryos for Porcine Tetraploid Embryo Production

Qingran Kong and Zhonghua Liu

Abstract

The electrofusion of 2-cell embryos proves to be a simple and efficient way of generating mammalian tetraploid embryos. Many factors affect the fusion efficiency, such as fusion medium, electric field intensity, and fusion pulse length. In mice, production of tetraploid embryos by electrofusion has already been investigated; however, the investigation to produce porcine tetraploid embryos is seldom reported. In this chapter, we will describe oocytes in vitro maturation, in vitro fertilization, and the optimum conditions for electrofusion of 2-cell embryos to produce tetraploid embryos in pig.

Key words Pig, Tetraploid embryo, Electrofusion

1 Introduction

Increased discoveries in reproductive biotechnology have enhanced the potential uses of genetically modified pigs in the biomedical field with the purpose of improving human lives [1]. Tetraploid embryos are used to produce chimeras with embryonic stem cells (ESCs) or induced pluripotent stem (iPS) cells and diploid embryos, and it is widely used to create genetically modified mouse [2–6]. However, porcine tetraploid embryos have been seldom reported.

Several approaches have been employed to produce mammalian tetraploid embryos in vitro [7]. The first approach is to transfer an embryonic nucleus into a fertilized egg, which results in the production of a tetraploid embryo [8]. It is also possible to produce parthenogenetic tetraploid embryos by inhibiting the formation of both first (first PB) and second polar bodies (second PB) with cytochalasin B or other cytokinesis-inhibition drugs during parthenogenetic activation of oocytes [9, 10]. The third method is to fuse two diploid blastomeres at the 2-cell stage embryo with either polyethylene glycol (PEG) [11, 12], inactivated Sendai virus [13], or by electrofusion. Among these fusion methods, electrofusion is

Chengyu Liu and Yubin Du (eds.), *Microinjection: Methods and Protocols*, Methods in Molecular Biology, vol. 1874,
https://doi.org/10.1007/978-1-4939-8831-0_21,

the most widely used tool for the production of tetraploid embryos, as it is safer than the chemical- or virus-mediated methods and more convenient and economical than microsurgical injection [14–19].

The mouse tetraploid embryo system has been successfully established by electrofusing 2-cell stage embryos [20]. Mouse 2-cell stage embryos are produced in vivo by mating superovulated females primed with pregnant mare serum gonadotropin (PMSG) and human chorionic gonadotropin (hCG). Since ovaries and follicles are homochronous, mouse embryos develop into the 2-cell stage at almost the same time. In addition, mouse embryos require at least 20 h to develop from the 2-cell to the 3/4-cell stage [21]. In pig, ovaries obtained from an abattoir for in vitro embryo production are variable in estrous cycle stage and follicles size. This heterogeneity of the follicular populations results in retrieval of cumulus-oocyte complexes (COCs) with different developmental potentials that vary in cleavage timing during the first (2-cell stage) and second (4-cell stage) post-fertilization cell cycles [22, 23]. Therefore, compared to mouse, it is much more challenging to obtain synchronous 2-cell stage porcine embryos either in vivo or in vitro.

Many factors affect the fusion efficiency, such as fusion medium, electric field intensity, and fusion pulse length [24–26]. We have explored the optimum condition for electrofusion of 2-cell embryos to produce porcine tetraploid embryos and systematically demonstrated the development of tetraploid embryos [23]. In the chapter, we will describe the method of producing porcine tetraploid embryos by electrofusion in detail.

2 Materials

2.1 Porcine Ovaries and Semen

1. Ovaries are collected from a slaughter house and transported to lab in 0.9% saline at 37 °C.
2. American Large White boars are normally used to collect semen in pig farm.

2.2 Reagents

1. Polyvinyl alcohol (PVA).
2. TCM-199.
3. D-Glucose.
4. Sodium pyruvate.
5. Penicillin G (PG).
6. Streptomycin (SM).
7. Cysteine.
8. Epidermal Growth Factor (EGF).

9. Luteinizing Hormone (LH).
10. Follicle Stimulating Hormone (FSH).
11. $NaHCO_3$.
12. Hepes.
13. NaCl.
14. Bovine serum albumin (BSA).
15. Mannitol.
16. $CaCl_2 \cdot 2H_2O$.
17. $MgCl_2 \cdot 6H_2O$.
18. KCl.
19. KH_2PO_4.
20. $MgSO_4 \cdot 7H_2O$.
21. Ca-(Lactate)$_2 \cdot 5H_2O$.
22. L-Glutaurine.
23. Hypotaurine.
24. BME Amino Acids Solution.
25. MEM Amino Acids Solution.
26. Gentamicin.
27. Paraffin oil.
28. Hyaluronidase.
29. Sorbitol.
30. Tris-base.
31. Hocest 33342.
32. Dulbecco's phosphate-buffered saline (DPBS).
33. Acetic acid.
34. Methanol.
35. 0.9% saline.
36. Hypotonic solution (0.075 M KCl).
37. Alexa 546-dUTP.
38. RNase.
39. 70% paraformaldehyde.
40. 2×SSC solution.
41. Prolong gold antifade reagent.

2.3 Media

Oocyte manipulation medium

Reagents	Concentration
TCM-199	9.5 g/L
$NaHCO_3$	0.05 g/L
Hepes	0.75 g/L
NaCl	1.855 g/L
PG	0.05 g/L
SM	0.06 g/L
BSA	3.0 g/L
Osmolarity (mOsm)	280
pH	7.2–7.4

Maturation medium

Reagents	Concentration
PVA	1.000 g/L
TCM-199	500 g/L
$NaHCO_3$	2.2 g/L
D-Glucose	3.05 mM/L
Sodium pyruvate	0.91 mM/L
Cysteine	0.070 g/L
LH	25 unit/L
FSH	50 unit/L
EGF	10 μg/L
PG	0.070 g/L
SM	0.050 g/L

Note: Cysteine, LH, FSH, EGF are added before use

Modified Tris-buffered medium (mTBM)

Reagents	Concentration
NaCl	6.611 g/L
KCl	0.224 g/L
Tris-base	2.423 g/L
D-Glucose	11 mM/L
Sodium pyruvate	5 mM/L
$CaCl_2{\cdot}2H_2O$	1.102 g/L

Porcine zygote medium-3 (PZM-3)

Reagents	Concentration
NaCl	6.312 g/L
KCl	0.746 g/L
KH_2PO_4	0.048 g/L
$MgSO_4 \cdot 7H_2O$	0.099 g/L
$NaHCO_3$	2.106 g/L
Sodium pyruvate	0.022 g/L
Ca-(lactate)$_2 \cdot 5H_2O$	0.617 g/L
L-Glutaurine	0.146 g/L
Hypotaurine	0.546 g/L
BME	20 ml/L
MEM	10 ml/L
Gentamicin	0.05 g/L
BSA	3 g/L
PH	7.3 ± 0.2

Note: BME, MEM, BSA are added before use

Fusion medium

Reagents	Concentration
Mannitol	54.6 g/L
$CaCl_2 \cdot 2H_2O$	0.15 g/L
$MgCl_2 \cdot 6H_2O$	0.02 g/L
Hepes	0.13 g/L
pH	7.0–7.4

2.4 Instruments

1. OLYMPUS SZ2-LGB stereomicroscope.
2. Thermo HEPACLSS100 water jacketed CO_2 incubator. They are normally set at 38.5 °C in 5% CO_2, 95% air, and saturation humidity by a pan containing distilled water.
3. BTX Elector-Cell Manipulator 2001.
4. Nikon Eclipse 80i epifluorescence microscope.

2.5 Tools

1. 18-Gauge needles.
2. Nunc™ Cell-Culture Treated 4-well dish.
3. Hemocytometer.
4. Slides.
5. 50 ml tubes.

3 Methods

3.1 In Vitro Maturation (IVM) of Oocytes

1. Follicles 3–5 mm in diameter are aspirated using an 18-gauge needle to obtain COCs and follicular fluid and COCs are collected in 50 ml tubes at 36 °C using water bath (*see* **Note 1**).
2. COCs with uniform cytoplasm and several layers of cumulus cells are selected under the stereomicroscope and rinsed three times in oocyte manipulation medium.
3. 40–70 COCs are cultured in a 4-well dish containing 500 μl maturation medium covered by paraffin oil for 42–44 h at 38.5 °C in 5% CO_2, 95% air, and saturation humidity in CO_2 incubator.

3.2 In Vitro Fertilization and Embryo Culture

1. Fresh semen is collected from the donor boars, using the gloved hand technique, and washed three times in DPBS supplemented with 0.1% (w/v) BSA by centrifugation at 1500 × *g* for 3 min.
2. The spermatozoa concentration and the proportion of motile sperm are determined by using a hemocytometer and sperm is diluted with the mTBM.
3. Matured oocytes of IVM are washed three times in mTBM. Approximately 30 oocytes are inseminated in 500 μl mTBM covered by paraffin oil containing sperms at a final concentration of 3×10^5/ml for 4 h at 38.5 °C in 5% CO_2, 95% air, and saturation humidity in CO_2 incubator.
4. 30–50 embryos are transferred to a 4-well dish containing 500 μl PZM-3 covered by paraffin oil and cultured at 38.5 °C in 5% CO_2, 95% air, and saturation humidity in CO_2 incubator. The 2-cell embryos are collected at 48 h.

3.3 Electrofusion of 2-Cell Embryos

1. 2-Cell embryos of 2n are pre-equilibrated in fusion medium for 30 s.
2. Using an AC field of 6 V and 10 s, embryos are aligned in the chamber with their fusion plane parallel to the electrodes. 2DC electro pulses of 0.9 kV/cm electric field intensity for 30 us on BTX Elector-Cell Manipulator 2001 are used to induce blastomere fusion in fusion medium.

3. 30 min after electrofusion, the fusion rate is evaluated and the fused embryos were cultured in PZM-3 as described above (*see* **Note 2**).
4. After 156 h, the tetraploid embryos can develop to the blastocyst stage, and the number of blastocyst cells is evaluated by nuclear staining with 5 μg/ml Hoechst 33342 (*see* **Note 3**).

3.4 FISH Analysis of Ploidy of Porcine Tetraploid Embryos

Tetraploidy is usually confirmed by FISH, and a typical result is shown in Table 3 and Fig. 1 (*see* **Note 4**).

1. Tetraploid blastocysts are washed in TCM199 medium and suspended in hypotonic solution.
2. After being fixed with cold acetic acid/methanol solution (v/v1:3) to dissolve cytoplasm, the embryos are transferred to a slide and frozen in −20 °C for at least 24 h.
3. A 245 bp probe for porcine chromosome 1 is amplified and labeled by PCR using Alexa 546-dUTP with the primer sets 5′GTTGCACTTTCACGGACGCAGC3′ and 5′CTAGCCCATTGCTCGCCATAGC3′. The PCR amplification profiles were as follows: 94 °C initial denaturation for 5 min, 40 cycles of 94 °C denaturation for 30 s, optimal temperature annealing for 30 s, and 72 °C extension for 2 min, followed by a 10 min final extension at 72 °C.
4. RNase is used to treat the embryos on slides, which are then dehydrated in an ethanol series and denatured by incubation in 70% paraformaldehyde in 2×SSC solution at 70 °C for 3 min.
5. After dehydration, the denatured probe is applied onto slides and incubated overnight at 37 °C.
6. Following nuclear staining with 5 μg/ml Hoechst 33342, slides are mounted with Prolong gold antifade reagent and analyzed using Nikon Eclipse 80i epifluorescence microscope.

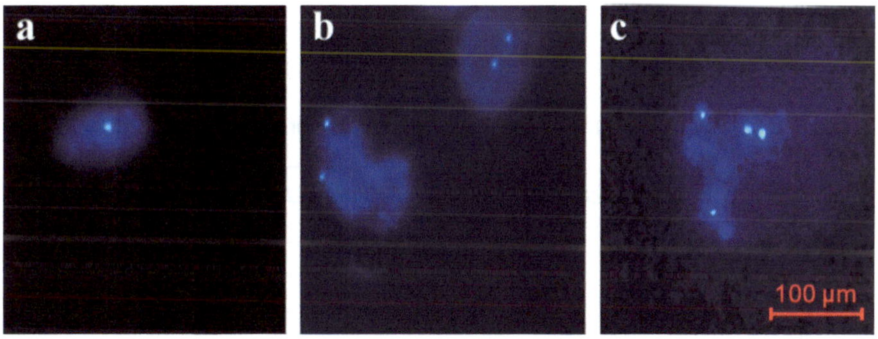

Fig. 1 FISH analysis of ploidy of porcine embryos. (**a**) Sperm; (**b**) diploid blastocyst; (**c**) tetraploid blastocyst. Scale bar, 100 μm

4 Notes

1. Porcine oocytes and embryos are very sensitive to the temperature, so it is very necessary to keep the temperature from 35 to 39 °C.
2. The rates of electrofusion and development to blastocysts are shown in Table 1 [23]. 0.9 kV/cm are used to fuse 2n 2-cell embryos. The fusion rate can reach 85.4%, and the rate of development to blastocyst of presumably 4n embryos is significantly lower than that of 2n embryos, but can reach 28.5%.
3. The developmental characteristics of the 2n and 4n embryos are also compared in vitro (Table 2) [23]. Under the optimum condition, the development of 2n embryos is significantly faster than 4n embryos at the cleavage stage ($P < 0.05$). However, the timing of blastocyst formation of 2n and 4n is very similar and the rates of blastocyst formation between 2n and 4n are not significantly different (23.61 vs. 22.84, respectively; $P > 0.05$). And, the number of cells in 4n blastocysts is significantly lower than that in the 2n embryos ($P < 0.05$). These data suggest that the competence of 4n embryos developing to blastocyst is the same as 2n embryos.
4. The result of ploidy analysis of sperm, diploid and fused embryos at the blastocyst stage by FISH is shown in Table 3 and Fig. 1 [23]. Among the 20 control blastocysts subjected to chromosomal analysis, 19 embryos were diploid and one embryo was a mosaic of diploid and tetraploid cells. In fused presumably 4n embryos, 15 out of 22 embryos were tetraploid, six embryos were diploid, and one embryo was a mosaic of diploid and tetraploid cells. These results indicate that a majority of embryos fused under the optimum condition were tetraploid embryos.

Table 1
Electrofusion of 2-cell embryos and in vitro development of porcine tetraploid embryos

Electric field intensity	No. of 2-cell embryos	No. of embryos fused (%)	No. of blastocysts (%)
Control	200	–	73 (36.5)^{a}
0.9 kV/cm	164	140 (85.4)	40 (28.5)^{b}

Note: Values with different superscripts within columns denote significant difference ($P < 0.05$)

Table 2
In vitro development of 4n and 2n embryos

Hours after electrofusion	Embryos	No. of embryos	No. of embryos developed (%)					No. of blastocyst cells
			2-Cell	4-Cell	8-Cell	Morulae	Blastocyst	
24 h	Diploid	144	73 (50.69)	69 (47.92)				
	Tetraploid	197	30 (15.23)	15 (7.61)				
48 h	Diploid	144	65 (45.14)	53 (36.81)	14 (9.72)			
	Tetraploid	197	31 (15.74)	15 (7.61)	7 (3.55)			
72 h	Diploid	144	42 (29.17)	30 (20.83)	23 (15.97)	34 (23.61)		
	Tetraploid	197	23 (11.68)	19 (9.64)	8 (4.06)	25 (12.69)		
96 h	Diploid	144	40 (27.78)	25 (17.36)	17 (11.80)	33 (22.92)	12 (8.33)	
	Tetraploid	197	21 (10.66)	22 (11.17)	22 (11.17)	18 (9.14)	25 (12.69)	
120 h	Diploid	144	34 (23.61)	27 (18.75)	17 (11.81)	13 (9.03)	34 (23.61)	
	Tetraploid	197	32 (16.24)	20 (10.15)	19 (9.64)	6 (3.04)	38 (19.29)	
132 h	Diploid	144	28 (19.45)	27 (18.75)	18 (12.5)	16 (11.11)	34 (23.61)	45.34 ± 5.00^{a}
	Tetraploid	197	36 (18.27)	19 (9.64)	13 (6.60)	15 (7.61)	45 (22.84)	21.90 ± 4.95^{b}

Note: Values with different superscripts within columns denote significant difference ($P < 0.05$)

Table 3
Ploidy of diploid and fused porcine embryos

Embryos	No. of blastocysts tested	No. of blastocysts with ploidy (%)		
		Diploid	Tetraploid	Diploid/tetraploid
Diploid	20	19 (95.0%)	0 (0)	1 (5.0%)
Fused	22	6 (27.27%)	15 (68.18%)	1 (4.5%)

References

1. Luo YL, Lin L, Bolund L et al (2012) Genetically modified pigs for biomedical research. J Inherit Metab Dis 35:695–713
2. Yang XZ, Smith SL, Tian XC et al (2007) Nuclear reprogramming of cloned embryos and its implications for therapeutic cloning. Nat Genet 39:295–302
3. Eakin GS, Hadjantonakis AK, Papaioannou VE et al (2005) Developmental potential and behavior of tetraploid cells in the mouse embryo. Dev Biol 288:150–159
4. Jouneau A, Zhou Q, Camus A et al (2006) Developmental abnormalities of NT mouse embryos appear early after implantation. Development 133:1597–1607
5. Zhao XY, Lv Z, Li W et al (2010) Production of mice using iPS cells and tetraploid complementation. Nat Protoc 5:963–971
6. Kang L, Gao S (2012) Pluripotency of induced pluripotent stem cells. J Anim Sci Biotechnol 3:5
7. Eakin GS, Behringer RR (2003) Tetraploid development in the mouse. Dev Dyn 228:751–766
8. Modlinski JA (1978) Transfer of embryonic nuclei to fertilised mouse eggs and development of tetraploid blastocysts. Nature 273:466–467
9. Sembon S, Fuchimoto D, Iwamoto M et al (2011) A simple method for producing tetraploid porcine parthenogenetic embryos. Theriogenology 76:598–606
10. Bai C, Liu H, Liu Y et al (2011) Diploid oocyte formation and tetraploid embryo development induced by cytochalasin B in bovine. Cell Reprogram 13:37–45
11. Eglitis MA (1980) Formation of tetraploid mouse blastocysts following blastomere fusion with polyethylene glycol. J Exp Zool 213:309–313
12. Krivokharchenko A, Galat V, Ganten D, Bader M (2002) In vitro formation of tetraploid rat blastocysts after fusion of two-cell embryos. Mol Reprod Dev 61:460–465
13. O'neill GT, Speirs S, Kaufman MH (1990) Sex-chromosome constitution of postimplantation tetraploid mouse embryos. Cytogenet Cell Genet 53:191–195
14. Kaufman MH, Webb S (1990) Postimplantation development of tetraploid mouse embryos produced by electrofusion. Development 110:1121–1132
15. Ozil JP, Modlinski JA (1986) Effects of electric field on fusion rate and survival of 2-cell rabbit embryos. J Embryol Exp 96:211–228
16. Iwasaki S, Kono T, Fukatsu H, Nakahara T (1989) Production of bovine tetraploid embryos by electrofusion and their developmental capability in vitro. Gamete Res 24:261–267
17. Prather RS, Hoffman KE, Schoenbeck RA, Stumpf TT, Li J (1996) Characterization of DNA synthesis during the 2-cell stage and the production of tetraploid chimeric pig embryos. Mol Reprod Dev 45:38–42
18. Prochazka R, Vodicka P, Zudova D et al (2004) Development of in vivo derived diploid and tetraploid pig embryos in a modified medium NCSU 37. Theriogenology 62:155–164
19. Hiriart MI, Bevacqua RJ, Canel NG et al (2013) Production of chimeric embryos by aggregation of bovine egfp eight-cell stage blastomeres with two-cell fused and asynchronic embryos. Theriogenology 80:357–364
20. Kubiak JZ, Tarkowski AK (1985) Electrofusion of mouse blastomeres. Exp Cell Res 157:561–566
21. Sawicki WAJ, Blaton O (1978) DNA synthesis in the second and third cell cycles of mouse preimplantation development. A cytophotometric study. Exp Cell Res 112:7

22. Liu Z, Song J, Wang Z et al (2008) Green fluorescent protein (GFP) transgenic pig produced by somatic cell nuclear transfer. Sci Bull 53:1035–1039
23. He W, Kong Q, Shi Y et al (2013) Generation and developmental characteristics of porcine tetraploid embryos and tetraploid/diploid chimeric embryos. Genomics Proteomics Bioinformatics 11:327–333
24. Suo L, Wang F, Zhou GB, Shi JM, Wang YB, Zeng SM, Tian JH, Zhu SE, Liu GS (2009) Optimal concentration of calcium and electric field levels improve tetraploid embryo production by electrofusion in mice. J Reprod Dev 55:383–385
25. Clement-Sengewald A, Brem G (1989) Electrofusion parameters for mouse two-cell embryos. Theriogenology 32:159–169
26. Cheong HT, Taniguchi T, Hishinuma M, Takahashi Y, Kanagawa H (1991) Effects of various electric fields on the fusion and in vitro development of mouse two-cell embryos. Theriogenology 36:875–885

Chapter 22

Gene Knockouts in Goats Using CRISPR/Cas9 System and Somatic Cell Nuclear Transfer

Zhiqiang Fan, Min Yang, Misha Regouski, and Irina A. Polejaeva

Abstract

The combination of CRISPR/Cas9 and SCNT techniques greatly facilitates the production of gene-edited livestock. Here, we describe the detailed procedure to create gene knockout goats using this strategy starting from the construction of CRISPR/Cas9 targeting vectors to the transfer of cloned embryos into recipient females. In this procedure, the transfection conditions for goat fibroblasts were optimized due to their high sensitivity to electrotransfection, which enabled the isolation of single-cell colonies carrying simultaneous disruption of multiple genes for SCNT with a single co-transfection of pooled CRISPR/Cas9 targeting vectors.

Key words Gene knockout, CRISPR/Cas9, Somatic cell nuclear transfer, Goats

1 Introduction

Genome editing has developed rapidly since the discovery of programmable artificial endonucleases such as ZFN and TALEN [1, 2]. These programmable nucleases produce site-specific DNA double-strand breaks (DSBs) and these DSBs can enhance the efficiency of homology-directed repair (HDR) by at least two orders of magnitude and/or trigger non-homologous end-joining (NHEJ). The NHEJ pathway for DSB repair is often favored and frequently leads to small insertions/deletions (indels) or chromosomal rearrangements, particularly in mammalian cells [3]. Both ZFN and TALEN require complex engineering of highly specific DNA-binding domains for proper targeting. More recently, the type II bacterial clustered, regularly interspaced, short palindromic repeats (CRISPR)-associated protein 9 (Cas9) system has become a highly efficient tool for further advancements of gene targeting strategies due to the simplicity of targeting any locus for cleavage with a single protein and a programmable single-guide RNA (sgRNA) [4]. In addition, because of its high efficiency, CRISPR/Cas9 technology can readily simultaneously target several

Chengyu Liu and Yubin Du (eds.), *Microinjection: Methods and Protocols*, Methods in Molecular Biology, vol. 1874,
https://doi.org/10.1007/978-1-4939-8831-0_22,

loci including both alleles of the same gene. A number of early studies have focused on enhancing the efficiency, specificity, and versatility of the CRISPR/Cas9 system in *C. elegans* [5], zebrafish [6], mouse [4], and human [7] genomes. These findings encouraged researchers to explore possibility of establishing a Cas9/sgRNA-based gene modification platform for large animals.

Genetically modified livestock could be used as an important tool for the production of valuable therapeutic proteins, development of refined biomedical models of human diseases, and improvement of agricultural production traits. For example, recombinant human antithrombin, the first therapeutic protein from genetically altered goats, has been approved by the US Food and Drug Administration (FDA) for the treatment of human diseases. However, it is costly and time consuming to produce genetically modified livestock using routine homologous recombination (HR)-meditated gene targeting. In 2014, Hai et al. reported the efficient generation of monoallelic and biallelic *vWF* knockout pigs in one step by direct cytoplasmic injection of Cas9 mRNA and sgRNA into zygotes with mutation efficiency of 68% (11/16) in newborn piglets [8]. In the same year, Ni et al. proved that Cas9/sgRNAs induced precise mutations at efficiency of 9–70% in goat primary fibroblasts [9]. In this study, the Cas9/sgRNA-modified fibroblasts were used for somatic cell nuclear transfer (SCNT), and live-born goats carrying gene mutations were generated. The combination of CRISPR/Cas9 and SCNT techniques greatly facilitates the production of gene-edited livestock. Researchers from different labs have reported gene knockin in pigs [10] and cattle [11] using this combined strategy.

Here, we describe the detailed procedures in using the CRISPR/Cas9 system together with SCNT to create gene knockout goats, as well as the accompanying steps, starting from the construction of CRISPR/Cas9 targeting vectors to the transfer of cloned embryos into recipient females, for producing live animals. In this protocol, due to the highly sensitive nature of goat fibroblast cells to electrotransfection, we optimized transfection conditions and enabled isolation of single-cell colonies carrying simultaneous disruption of multiple genes with a single co-transfection of pooled Cas9/sgRNAs.

2 Materials

2.1 Tissue and Cell Culture

1. Dulbecco's Phosphate-Buffered Saline (DPBS) without calcium/magnesium.
2. Dulbecco's High Glucose Modified Eagles Medium (DMEM)/High Glucose.
3. TrypLE™ Express.

4. Penicillin Streptomycin Solution.
5. Gentamycin.
6. Fetal Bovine Serum (FBS).
7. Dimethyl sulfoxide (DMSO).
8. Isopropanol (2-Propanol).
9. Nalgene™ Cryo 1 °C Freezing Container.
10. Cell Culture Medium: DMEM supplemented with 15% FBS and 1% Penicillin Streptomycin Solution.
11. Cell Freezing Medium: 90% Cell Culture Medium and 10% DMSO.
12. Chlorhexidine (2% chlorhexidine gluconate). Made to 1:30 diluted solution in DPBS.

2.2 Molecular Biology

1. GoTaq® Green Master Mix, 2×.
2. pGEM®-T Vector System I.
3. JM109 Competent Cells, High Efficiency.
4. T4 DNA Ligase.
5. pX330-U6-Chimeric_BB-CBh-hSpCas9 plasmid (pX330) vector.
6. FastDigest BbsI.
7. T4 Polynucleotide Kinase.
8. GeneJET Plasmid Miniprep Kit.
9. QIAquick® Gel Extraction Kit.
10. Plasmid Plus Midi Kit.
11. DNeasy® Blood & Tissue Kit.
12. TE: 10 mM Tris-HCl, 1 mM EDTA, pH 7.5.

2.3 Cell Transfection

1. Amaxa™ P3 Primary Cell 4D-Nucleofector™ X Kit L.
2. 4D-Nucleofector™ System X Unit.

2.4 SCNT

1. Heparin.
2. Hyaluronidase.
3. Mineral oil.
4. Medium for fibroblast isolation: DMEM supplemented with 15% FBS and 1% Penicillin Streptomycin Solution.
5. Serum Starvation Medium: DMEM supplemented with 0.5% FBS and 1% Penicillin Streptomycin Solution.
6. Tyrode's Lactate HEPES (TL-HEPES) Stock, 1×, for in vitro manipulation of goat oocytes (Table 1).

Table 1
Composition of Tyrode's Lactate HEPES (TL-HEPES) and Synthetic Oviduct Fuid (SOF) medium stocks, 1×

		TL-HEPES		SOF	
Component	Molecular weight	mM	g(or mL)/L	mM	g(or mL)/L
NaCl	58.44	114.00	6.662	107.70	6.294
KCl	74.55	3.20	0.239	7.16	0.534
$CaCl_2 \cdot 2H_2O$	147.01	2.04	0.300	1.71	0.251
$MgCl_2 \cdot 6H_2O$	203.30	0.49	0.100	–	–
$MgSO_4$	120.37	–	–	0.83	0.100
$NaH_2PO_4 \cdot H_2O$	137.99	0.34	0.047	–	–
KH_2PO_4	136.09	–	–	1.19	0.162
$NaHCO_3$	84.01	2.00	0.168	25.07	2.106
HEPES	238.30	10.07	2.400	–	–
Na lactate	112.06	10.00	1.121	7.70	0.863
Na pyruvate	110.04	–	–	0.36	0.040
Na citrate tribasic dihydrate	294.10	–	–	0.34	0.100
Myo-inositol	180.16	–	–	2.78	0.501
Taurine	125.15	–	–	0.50	0.063
Pantothenate	238.27	–	–	0.003	0.0007
Penicillin G potassium salt	372.48	–	–	0.161	0.060
Streptomycin sulfate	728.69			0.069	0.050
Phenol red	–	–	0.005	–	–
BSA fraction V	–	–	3.000	–	–
BSA	–	–	–	–	8.000
FBS	–	–	–	–	25 mL
BME Amino Acid Solution	–	–	–	–	2 mL
MEM NEAA[a]	–	–	–	–	2 mL
GlutaMAX™	–	–	–	–	5 mL
Milli-Q H_2O	–	–	Up to 1 L	–	Up to 1 L

The pH of both two media should be adjusted to 7.2–7.4 and osmolarity 275–280 mosmol. Filter the medium with a 0.22 μM filter. Keep at 4 °C for 15 days

[a]NEAA, non-essential amino acids

7. Ovary Slicing Medium: TL-HEPES Medium supplemented with 1% FBS, 1% Penicillin/Streptomycin, and 30 μg/mL heparin.

8. Oocyte Maturation Medium: TCM-199 supplemented with 10% FBS, 10 μg/mL LH, 5 μg/mL FSH, 1 μg/mL estradiol-17β, and 0.05 g/L gentamycin.
9. Synthetic Oviduct Fuid (SOF) Medium Stock, 1×, for in vitro culture of goat embryos (Table 1).
10. HEPES-buffered SOF (HSOF): SOF containing 20 mM HEPES and 5 mM $NaHCO_3$.
11. Cytochalasin B Medium: SOF supplemented with 7.5 μg/mL cytochalasin B.
12. DMAP/CHX Medium: SOF supplemented with 2 mM 6-dimethylamino purine (DMAP) and 10 μg/mL cyclohexamide (CHX).
13. Ionomycin Medium: SOF supplemented with 5 μM ionomycin.

2.5 Embryo Transfer

1. SOF Medium (detailed previously).
2. Embryo transfer Medium: Dulbecco's Phosphate-Buffered Saline (DPBS) with calcium/magnesium supplemented with 1% FBS.
3. Syncrite Vaginal Sponge.
4. Flunixin Meglumine.
5. Polyflex.
6. Acepromazine.
7. Xylazine.
8. Ketamine.
9. Nunc four-well dishes.
10. Luer-slip 1 mL syringe.
11. Tomcat catheters, open end.
12. Lactated Ringer's.
13. PDS Suture.
14. Ultrasound machine.
15. Rectal Probe.
16. Abdominal Probe.
17. CD&T Vaccine.
18. Bo-Se injectable.

3 Methods

3.1 Preparation of Fetal Fibroblasts

1. Mate a pair of goats 30–40 days prior to harvesting fetuses (*see* **Note 1**).
2. Remove uterus from a euthanized doe and transport it to a laboratory in a sterile container within 30 min from the collection time. Wash fetus(es) in DPBS supplemented with gentamycin at a concentration of 50 μg/mL and transfer it to a class II tissue culture cabinet for evisceration and decapitation. Wash each carcass twice in DPBS/gentamycin.
3. Place the carcass in 1:30 diluted chlorhexidine solution for 30 s. Wash the carcass three times in DPBS/gentamycin.
4. Transfer the carcass to 5 mL TrypLE™ Express and digest for 30 min. During digestion shake the mixture vigorously every 5 min. Facilitate isolation of cells by cutting the carcass to approximately 1 mm^3 cubes with scissors.
5. Stop action of digestion by adding an equal volume of DMEM supplemented with 50 μg/mL gentamycin. Remove larger clumps of tissue by allowing them to sediment for 5 min at room temperature.
6. Transfer the single cells and small clumps contained within the supernatant to a fresh tube and pellet by centrifugation ($200 \times g$ for 5 min). Resuspend the pellet in DMEM/gentamycin medium and place the suspension in a 175 cm^2 culture flask.
7. After 48 h, and every 3–4 days thereafter, replace medium with fresh cell culture medium without gentamycin.
8. Once confluent (usually after 3–7 days of culture), release monolayers with TrypLE™ Express and cryopreserve cells in freezing medium supplemented with 10% DMSO in aliquots of approximately 1×10^6 cells/mL. Place the cryovials into Nalgene™ Cryo 1 °C Freezing Container to ensure slow cooling and store at −80 °C. After 24 h, transfer cryovials to liquid nitrogen for long-term storage.

3.2 Gene Characterization

1. Search NCBI database to find the GenBank Serial Number of the goat gene that you aim to knock out.
2. Search NCBI database to find the GenBank Serial Number of the same genes in other species, such as human, mouse, sheep, pig, cattle, etc.
3. Compare the targeting gene sequence of goat with those from other animals and locate exon and intron regions of the targeting gene.
4. Design specific primers to amplify approximately 500 bp fragments spanning the targeting loci in exons of target genes (Table 2). Purify the PCR products and sequence them.

Table 2
PCR primers for amplification of targeting loci in exons of goat genes

Genes	Primer sequences (5′-3′)	Size of PCR amplicon (bp)
MSTN	Forward: CTGGAAAGGAAGTAGGCTGCTC, Reverse: GTCCTTCTTCTCCTGGTTCTGG	621
NUP155	Forward: TGACTTTCTCCGGTTGCTGTT, Reverse: CTGGAGAAGGGAATGGCAAC	340
BLG	Forward: TGAAGTGCCTCCTGCTTGCCCT, Reverse: AGGAGAAGCCAGCCTGGCCGA	307
PRNP	Forward: TGAAAAGCCACATAGGCAGTT, Reverse: CCACTCCCTCCATTATCTTGA	633
IGMH	Forward: GAAAGTGAATCTCACCCGAAAG, Reverse: GAAGCCAAGGCTCAGGGA	823

3.3 Construction of Gene-Knockout (KO) CRISPR/Cas9 Targeting Vectors

1. Use the pX330 plasmid for the construction of CRISPR/Cas9 targeting vectors.
2. Identify sgRNA targeting sites in each genomic locus by searching for $G(N)_{20}GG$ motifs (Table 3). If the sgRNAs do not begin with a G, an extra G can be added at the 5′ end of sgRNAs. To facilitate mutation analysis, incorporate a restriction enzyme recognition site in each targeting locus (*see* **Note 2**). Predict all the possible off-target sequences using an online software (Benchling) based on the on-target sgRNAs.
3. Synthesize the DNA oligos for each targeting site.
4. Digest the pX330 plasmid with BbsI and purify the linearized plasmid using the Gel Extraction Kit. Anneal a pair of oligos for each targeting site and insert them to the linearized pX330 vector as described by Zhang's lab (http://www.addgene.org/crispr/zhang/).
5. Confirm the final constructs by sequencing.

3.4 Transfection of Goat Fibroblasts

1. Seed goat fetal fibroblasts in a T25 flask to achieve 70–80% confluence on the day of transfection (*see* **Note 3**).
2. Remove media from the cultured cells and wash cells once with DPBS.
3. Harvest the cells by trypsinization using TrypLE™ Express (*see* **Note 4**).
4. Count an aliquot of cells and determine cell density.

Table 3
sgRNAs for gene knockout in goat genome

Target genes	sgRNA sequence (5′-3′)[a]	Restriction enzymes	Mutation efficiency (%)	Exons	GenBank No.
MSTN-1	CTTGACA TGAACCCAGGCAC TGG	Van91 I	59	2	KU991727.1
MSTN-2	AAAGACGGTACAAGG TATACTGG	Bst1107 I	9	2	KU991727.1
NUP155–1	GGCAGCCC TGCAGGAAGC TTTGG	Hind III	57	1	XM_005694783.3
NUP155–2	GCAAGAAGACCGCA TGTACCCGG	Fat I	50	1	XM_005694783.3
BLG	ATCGTCACCCAGACCA TGAAAGG	Nla III	19	1	KF992214.1
PRNP	AACCGCTATCCACC TCAGGGAGG	Bsu36 I	70	1	NM_001314247.1
IGMH	GAAAGGCGC TTGAGGAATGCAGG	Bsm I	20	1	EU182621.1

[a]The underlined 'NGG' is protospacer adjacent motif (PAM) sequence recognized by Cas9 for target binding

5. Centrifuge approximately 5×10^5 cells at $200 \times g$ for 2.5 min at room temperature. Remove the supernatant completely.
6. Resuspend the cell pellet carefully in 100 μL 4D-Nucleofector™ Solution. Add 2 μg of targeting vector or control plasmid and gently pipette the solution approximately 10 times.
7. Transfer mastermixes into the 100 μL Nucleocuvette™ Vessels.
8. Transfect cells using 4D-Nucleofector™ System X Unit according to the manufacturer's protocol. Use the Amaxa™ P3 Primary Cell 4D-Nucleofector™ X Kit L and Program EN150 for goat fibroblast transfection with a single targeting vector.
9. When transfecting cells with multiple vectors simultaneously, pool all plasmids (0.5–1 μg of each) together and mix them with cells and transfection solution. Use optimized program for multiple vector transfection (*see* **Note 5**).
10. After running Nucleofection™ process, carefully remove the Nucleocuvette™ Vessel from the retainer and incubate Nucleocuvette™ 10 min at room temperature.
11. Resuspend cells with pre-warmed media and transfer cells into 3 mL cell culture medium. Culture cells at 38.5 °C, 5% CO_2, and 5% O_2 in an incubator (*see* **Note 6**).

3.5 Selection of Gene-KO Single Cell-derived Colonies

1. 72 h after transfection, harvest the cells by trypsinization using TrypLE™ Express.
2. Separate cells in two 15 mL tubes with approximately 1/3 of cells placed in one and the rest (2/3) in the other tube.
3. Extract genomic DNA from 1/3 of cells using the DNeasy® Blood & Tissue Kit. Cryopreserve the rest of the cells.
4. Detect the mutant efficiency by the PCR-Restriction Fragment Length Polymorphism (RFLP) assay (Fig. 1a).
5. After the cells are confirmed to contain the knockouts of interest by PCR-RFLP and sequence analysis, thaw the previously cryopreserved cells and seed them individually into 96-well plates by serial dilution for isolation of single-cell colonies.
6. Culture the cells in 96-well plates for 5–7 days until most of them reach 70–90% confluence, and then transfer single-cell-derived colonies to 24-well plates.
7. After culture for 3–7 days, when most of the colonies in 24-well plates reach 70–90% confluence, harvest 1/3 of cells from each of the wells by trypsinization.
8. Extract genomic DNA individually from each of the harvested colonies.
9. Identify gene-KO single-cell-derived colonies using the PCR-RFLP assay (Fig. 1b) and sequence uncut PCR products (Table 4 and Fig. 1c).
10. Design specific primers flanking each off-target site (target score ≥ 0.9) and amplify approximately 500 bp of DNA fragments spanning the off-target sites. Analyze the PCR products by Sanger sequencing. Discard the gene-KO cell colonies that have mutations in the analyzed off-target sites.
11. Cryopreserve the gene-KO single-cell-derived colonies without off-target mutations in liquid nitrogen.

3.6 SCNT Procedures

3.6.1 Donor Cell Preparation

1. Thaw the gene-KO cells 3–5 days before nuclear transfer. Rapidly thaw one vial of cells taken out of liquid nitrogen cryostorage in a water bath at 37 °C.
2. Transfer cells to a 15 mL tube and add 10 mL of cell culture medium.
3. Centrifuge the tube at 200 × *g* for 5 min to remove DMSO and resuspend the cell pellet in 2 mL of cell culture medium.
4. Pipette 0.5 mL of cell suspension into each well of a 4-well plate and culture at 38.5 °C, 5% CO_2.
5. Replace culture medium with Serum Starvation Medium 24 h before SCNT procedure.

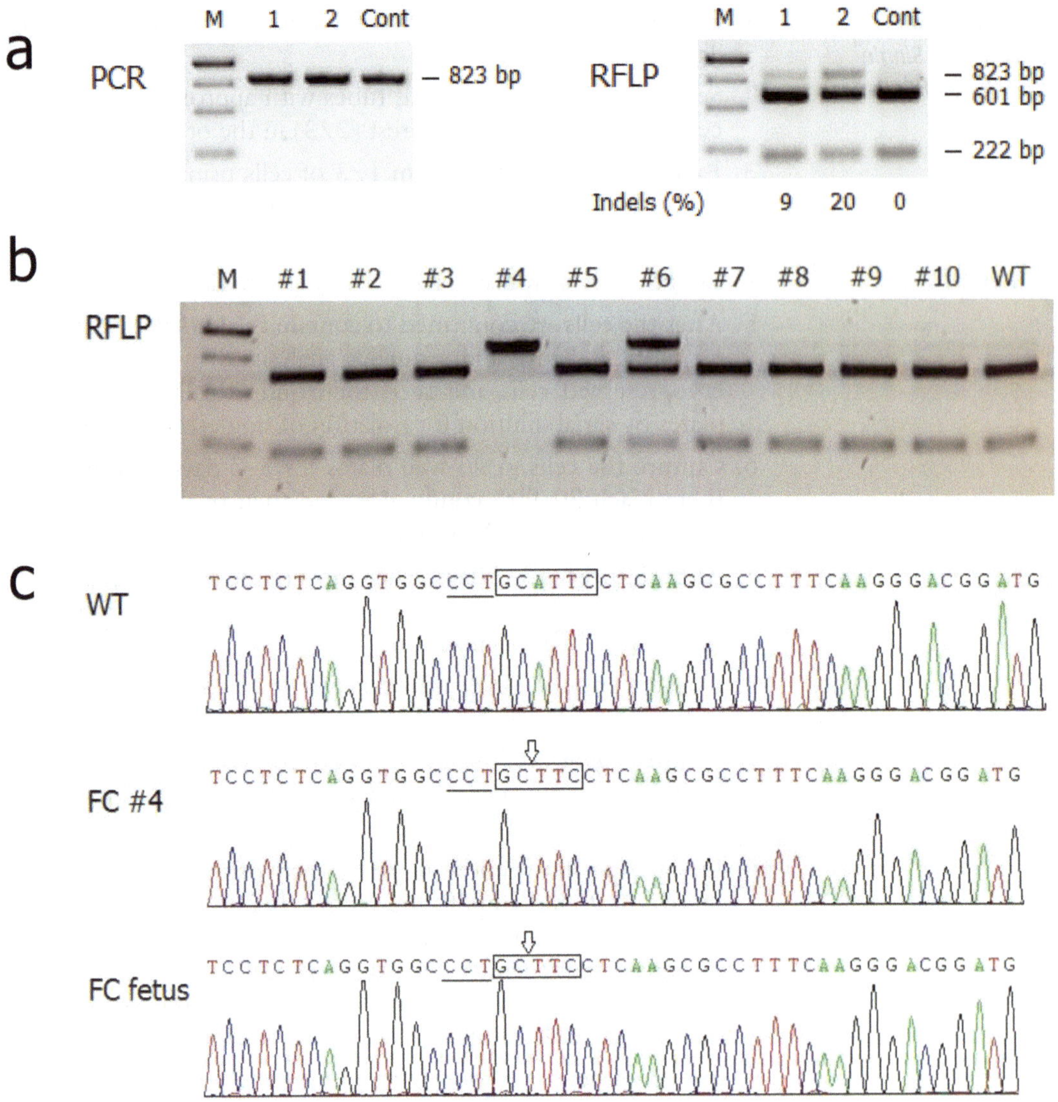

Fig. 1 Detection of mutation efficiency by PCR-RFLP and identification of goat IGMH-knockout fibroblast colonies. (**a**) Mutation efficiency analysis by PCR-RFLP. The PCR products in lane 1 and 2 are derived from fibroblast cells transfected with CRISPR/Cas9 targeting vectors using transfection program EH100 and EN150, respectively. The PCR products spanning the IGMH target site are shown in the left panel. The mutation efficiencies are shown in the right panel. The PCR products are digested by Bsm I. DNA repair through NHEJ pathway after double-strand breaks led to the loss of restriction enzyme site at targeting locus. (**b**) Detection of mutant fibroblast colonies by PCR-RFLP after selection of single-cell colonies. The results show that #4 and #6 are biallelic and monoallelic knockout colonies, respectively. (**c**) Sequencing analysis of fibroblast colonies. The colony #4 is derived from single-cell selection after transfection of the CRISPR/Cas9 targeting vector. The PAM sequences are underlined. Restriction enzyme sites are circled by the frame. The sequence of a fibroblast colony isolated from a cloned fetus using colony #4 as cell donor is shown at the bottom panel. The arrow shows an "A" deletion at the target locus of colony #4 or fetus fibroblast cells (FC). WT, wild type. RFLP, restriction fragment length polymorphism

Table 4
Cas9/sgRNA-mediated multiple gene disruptions in goat fibroblasts

		No. of mutant colonies (%)			
Target genes	**Colonies tested**	**Single**	**Double**	**Triple**	**Quadruple**
MSTN	22	7 (32)	–	–	–
MSTN/PRNP	45	19 (42)	9 (20)	–	–
MSTN/PRNP/BLG/NUP155	107	35 (33)	30 (28)	9 (8)	3 (3)

6. Approximately 1 h before SCNT, rinse the cells with DPBS, then add 150 μL of TrypLE™ Express, and incubate at 38.5 °C in 5% CO_2 in the air for 5 min until the cells round up and become detached.
7. Add 500 μL of cell culture medium and gently pipet five or six times across the well.
8. Transfer all the cells in to a 15 mL tube containing 10 mL of cell culture medium and collect by centrifugation at 200 × *g* for 5 min.
9. Remove the supernatant and gently resuspend the pellet in 100 μL of HSOF medium.

3.6.2 Oocyte Collection and In Vitro Maturation

1. Collect goat ovaries from a local abattoir and transport them in saline containing 100 U/mL penicillin/streptomycin at 20–27 °C to a laboratory within 4 h after collection (*see* **Note 7**).
2. Upon arrival to the laboratory, wash the ovaries three times in saline and then three times in DPBS containing 100 U/mL penicillin/streptomycin.
3. Recover cumulus-oocyte complexes (COCs) from ovaries by slicing the surface of the ovaries with a scalpel blade in a square Petri dish with grid in the ovary slicing medium.
4. Pour the COC containing medium into a 50 mL tube and allow it to sediment for 5 min.
5. Transfer the sediment to a new Petri dish containing 20 mL of fresh slicing medium. Select oocytes with three or more layers of compact cumulus cells and homogeneous cytoplasm.
6. Wash the COCs three times in maturation medium.
7. Place 50 oocytes in each well of a 4-well Petri dish containing 500 μL of maturation medium. Incubate the COCs in the maturation medium under a humidified 5% CO_2 in air at 38.5 °C for 22 h.

8. After 22 h of culture, remove cumulus cells mechanically from oocytes by pipetting in TL-HEPES medium containing 1 mg/mL hyaluronidase.
9. Select matured oocytes and use them as the donor cytoplasts for SCNT (*see* **Note 8**).

3.6.3 Enucleation and Cell Transfer

1. Place the oocytes in HSOF medium supplemented with 10 μg/mL cytochalasin B under a mineral oil overlay.
2. After 10 min, immobilize an oocyte with a holding pipette.
3. Use a sharpened beveled micropipette to pierce the zona pellucida, but not the ooplasmic membrane. Aspirate the first polar body and a small portion of cytoplasm under it. Avoid aspirating excessive amount of surrounding cytoplasm. Transfer enucleated oocytes in maturation medium before cell transfer.
4. Place a group of 30–50 enucleated oocytes and dispersed donor cells in a drop of HSOF medium under oil overlay for cell transfer.
5. Pick up a donor cell using the sharpened beveled manipulation pipette and insert it into the perivitelline space of enucleated oocyte to form a couplet. Ensure that there are good contacts between donor cells and enucleated oocytes.
6. Conduct fusion immediately after SCNT is completed for a group of 30–50 oocytes.

3.6.4 Fusion

1. Prepare a 4-well dish for fusion. Pipette 500 μL of TL-HEPES in well #1, 250 μL of TL-HEPES and 250 μL of fusion medium in well #2, 500 μL of fusion medium in well #3, and 500 μL of TL-HEPES in well #4.
2. Place the reconstructed couplets in TL-HEPES in well #1 of the fusion dish. Move the couplets from well #1 to well #2, and allow them to equilibrate for approximately 1–2 min, and then move them to well #3.
3. Place a few of the couplets (two to four) between the electrodes of the fusion chamber BTX 45–105 in fusion medium. Position the couplets the way that the cell-ooplast contact area is in parallel to the electrodes. Apply a single DC electric fusion pulse of 1.75 kV/cm for 15 μs.
4. Transfer the embryos into the well #4, and when a group of 30–50 oocytes is fused, place them in an incubator until activation.
5. Evaluate fusion of the donor cell with the oocyte cytoplasm under a stereo microscope 30 min after the pulse.

3.6.5 Activation

1. Move the fused couplets to pre-warmed cytochalasin B holding medium droplets under mineral oil and incubate at 38.5 °C, 5% CO_2 in air for 50 min.
2. Wash the couplets three times in pre-warmed SOF medium and incubate for 5 min in 1 mL of Ionomycin Medium.
3. Wash the couplets three times in DMAP/CHX Medium and incubate at 38.5 °C, 5% CO_2 in air for 4 h in DMAP/CHX medium droplets under mineral oil for activation.
4. Wash embryos six times in pre-warmed SOF medium following DMAP/CHX incubation.
5. Culture the embryos under oil in 20 μL droplets of SOF medium for 8–12 h prior to transfer into synchronized recipient females.

3.7 Embryo Transfer

3.7.1 Estrous Synchronization

1. Place a Syncrite sponge containing 45 mg of flugestone acetate into the vagina of a recipient for 10 days (*see* **Note 9**).
2. Pull the sponges and weigh the recipients on the day oocytes are collected or 48 h prior to embryo transfer surgery.
3. Remove food and water access from recipients 24 h prior to surgery.
4. Monitor recipients closely for the next 48 h for signs of estrus at 12 h intervals (*see* **Note 10**). Select does exhibiting signs of estrus for surgery.

3.7.2 Surgery Preparation

1. Give does pre-surgical drugs: flunixin meglumine at a concentration of 2.2 mg/kg intravenously (IV), Polyflex (ampicillin) at 10 mg/kg intramuscularly (IM), and a mixture of acepromazine and xylazine at 0.05 mg/kg and 0.03 mg/kg IV, respectively.
2. Wait 10–15 min, and then give a dose of ketamine at a concentration of 3 mg/kg IV. Place a recipient in a gurney for surgical preparation. Move the recipient to the surgical table, place rope restraints on each limb, and brace across the carpals or hocks. Perform a surgical scrub. Cover the recipient with a sterile drape with a precut opening (fenestration) over the prepared surgical site.

3.7.3 Surgery

1. Make a linear ventral midline incision, 10–13 cm, along the linea alba starting 2–3 cm cranial to the udder and moving forward, and avoid damaging mammary veins (*see* **Note 11**).
2. Expose the uterus and ovaries with great care by grasping the surrounding ligaments as much as possible without gripping the actual tract. If tract must be handled, do so with a gentle touch.

3. Visualize the uterus and ovaries and identify the presence of ovulation, or follicles that have at least a 7 mm diameter (Fig. 2a). If ovulation only occurs on one side transfer the embryos to the ipsilateral side.
4. Load embryos into a tomcat catheter with a luer-slip 1 mL syringe attached. Load approximately 3 μL of embryo transfer medium followed by a 1–2 mm air bubble, then, load all embryos to be transferred in 3–5 μL of embryo transfer medium, again followed by a 1–2 mm air bubble and 3 μL of embryo transfer medium.

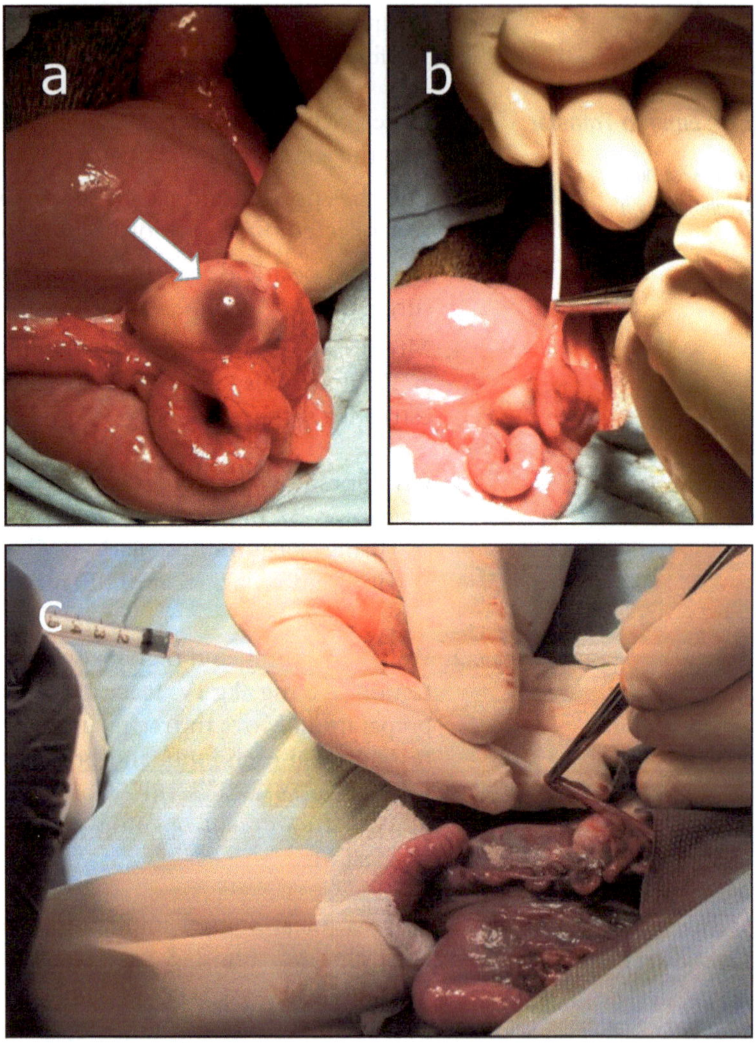

Fig. 2 Surgical embryo transfer. (**a**) Visualize the uterus and ovaries and identify the presence of follicles that have at least a 7 mm diameter. The arrow indicates a preovulatory follicle. (**b**, **c**) Place embryos into the ostium of the infundibulum via a tomcat catheter and deposit them at least 3.5 cm into the oviduct

5. Introduce embryos into the ostium of the infundibulum via a tomcat catheter and deposit them at least 3.5 cm into the oviduct (Fig. 2b, c). Remove the catheter gently and check for the presence of embryos.
6. Pour 500 mL of warm lactated ringer's (or sterile 0.9% saline) into the abdominal cavity and place the reproductive tract back into the abdominal cavity.
7. Close the cavity by using size one coated, braided, absorbable suture (PDS II) in an appositional pattern (*see* **Note 12**).
8. Close the skin with size one non-absorbable suture and tapered or reversed cutting needle (*see* **Note 13**).
9. Allow recipients to recover until normal functions resume and does can get up and walk normally on their own. Reinstate access to water and regular feed (typically a total mixed ration of alfalfa and corn silage) ad libitum. Observe recipients closely for any signs of pain or distress.

3.7.4 Pregnancy Monitoring

1. Assess the pregnancy status of recipients via ultrasonography (rectal probe) at day 45 of gestation. If not pregnant (open), use recipients for another embryo transfer surgery if approved by an IACUC.
2. If pregnant, ultrasound recipients (abdominal probe) at day 90 and day 120 of gestation.
3. If the recipient has not kidded by day 152 of gestation, induce into parturition via an IM injection of 15 mg prostaglandin F2α and 12 mg dexamethasone. If the doe has still not kidded by day 154 of gestation after induction, perform a cesarean section.
4. When performing a cesarean section, give a dorsal epidural anesthesia using lidocaine at a concentration of 6 mg/kg.
5. Make a midline incision through the linea alba and exteriorize the uterus, and make another incision in the greater curvature of the uterus and remove the fetus. Close the uterine incision with a monofilament absorbable suture in an inverting pattern. Close the cavity and skin with the suture methods described above.

3.7.5 Offspring Delivery

1. Observe the neonate for suckling behavior. If the surrogate has milk, continue monitoring daily. If no milk or the surrogate rejects the neonate, provide fresh or thawed (from a herdmate) colostrum to the neonate (*see* **Note 14**).
2. Weigh, tag, and vaccinate the neonate, then dip the neonate's navel in an iodine solution directly after birth. If a large umbilicus is present, place a clip as close to the body wall as possible.

3. Vaccinate routinely throughout gestation and kidding. Give 2 mL of clostridium perfringens types C & D Tetani Bacterin-Toxoid (CD&T) vaccine to the recipient at day 120 of gestation. At birth, give the neonate 2 mL of CD&T vaccine along with 0.5 mL Bo-Se (a selenium supplement containing 1 mg selenium and 50 mg vitamin E per mL) if in a selenium deficient soil area. Booster the CD&T vaccine at 1 month of age.

4 Notes

1. Several different types of cells have been successfully used for cloning including fetal fibroblasts, adult fibroblasts, cumulus cells, epithelial cells, etc. We describe a method for the isolation of fetal fibroblasts, which are the most commonly used cell type for SCNT.
2. Site-specific mutations make the target locus resistant to the restriction enzyme treatment (uncut), which can be detected by the PCR-RFLP assay (Fig. 1a, b).
3. The suitable number of cells for a 100 μL single Nucleocuvette™ is 5×10^5 cells and lower or higher cell numbers may influence transfection results. The optimal confluency for goat fibroblasts before Nucleofection™ is 70–80%. Do not use cells after passage 4 because this may result in relatively lower gene transfer efficiency and viability.
4. Goat fibroblast cells are sensitive to trypsin treatment. In order to enhance the survival rate of transfected cells, we optimized the procedure for transfection. We diluted TrypLE™ Express 1: 5 in 1× DPBS for cell harvesting and shortened the digestion time to 3–5 min. Do not incubate the cells in TrypLE™ Express longer than 10 min. When a single plasmid is used, a relatively low survival rate of fibroblasts is acceptable because the cells survived from transfection are sufficient for single-cell colony selection.
5. Transfection with a higher amount of plasmids greatly decreases the survival rate of transfected cells. We optimized transfection programs using the control pmaxGFP™ plasmid included in the kit according to the manufacturer's protocol (http://bio.lonza.com/6.html).
6. The goat fibroblast cells can be cultured in standard tissue culture conditions (20% oxygen); however, low oxygen stimulates proliferation of fibroblasts when seeded as a single cell.
7. All ovary collections for cloning were performed during the goat breeding season. The COC collections from ovaries were conducted at 30 °C.

8. Maturation status was assessed by the presence of a first polar body under a stereo-microscope. Oocytes at this stage are termed as metaphase II oocytes.
9. The number of recipients used depends on the number of embryos to be transferred. We usually transfer 10–20 embryos to one recipient.
10. When monitoring the recipients for the signs of estrus, we use a vasectomized male in an adjoining pen. The signs for estrus include wagging tail, standing for mounting, and attempting contact with buck.
11. The incision must be large enough to allow the uterus and ovaries to be exteriorized.
12. Suture placement should be 0.5–0.75 cm from the edge of the incision and 1 cm apart from skin. We test closure tightness by trying to insert a finger or butt-end of scalpel handle between or through sutures. If a finger or butt-end of scalpel handle can be inserted the suture should be replaced, pulled tighter, or more sutures added. Sutures should not be over-tightened in order to avoid tissue necrosis.
13. For goats, a tapered needle can easily penetrate the skin. Cutaneous sutures should not be any tighter than bringing the incision into apposition.
14. If no suckling behavior or general listlessness is observed, we tube the neonate with colostrum. In the case of rejection by mother, or no milk production, we continue to bottle-feed (or tube) the neonate with colostrum every 4 h for the first 3 days. Then transition to thawed milk or milk replacer at 4 h intervals four times daily for 2 weeks. After that, transition to an ad libitum system of feeding. Monitor milk consumption and adjust the volume based on size of neonate. Wean the neonate at 3 months of age.

References

1. Meyer M, de Angelis MH, Wurst W, Kuhn R (2010) Gene targeting by homologous recombination in mouse zygotes mediated by zinc-finger nucleases. Proc Natl Acad Sci U S A 107:15022–15026
2. Hockemeyer D, Wang HY, Kiani S, Lai CS, Gao Q, Cassady JP et al (2011) Genetic engineering of human pluripotent cells using TALE nucleases. Nat Biotechnol 29:731–734
3. Lieber MR (2010) The mechanism of double-strand DNA break repair by the nonhomologous DNA end-joining pathway. Annu Rev Biochem 79:181–211
4. Cong L, Ran FA, Cox D, Lin SL, Barretto R, Habib N et al (2013) Multiplex genome engineering using CRISPR/Cas systems. Science 339:819–823
5. Chen C, Fenk LA, de Bono M (2013) Efficient genome editing in Caenorhabditis elegans by CRISPR-targeted homologous recombination. Nucleic Acids Res 41:e193
6. Hwang WY, Fu Y, Reyon D, Maeder ML, Kaini P, Sander JD et al (2013) Heritable and precise zebrafish genome editing using a CRISPR-Cas system. PLoS One 8:e68708
7. Mali P, Yang LH, Esvelt KM, Aach J, Guell M, DiCarlo JE et al (2013) RNA-guided human

genome engineering via Cas9. Science 339:823–826

8. Hai T, Teng F, Guo R, Li W, Zhou Q (2014) One-step generation of knockout pigs by zygote injection of CRISPR/Cas system. Cell Res 24:372–375
9. Ni W, Qiao J, Hu S, Zhao X, Regouski M, Yang M et al (2014) Efficient gene knockout in goats using CRISPR/Cas9 system. PLoS One 9: e106718
10. Ruan JX, Li HG, Xu K, Wu TW, Wei JL, Zhou R et al (2015) Highly efficient CRISPR/Cas9-mediated transgene knockin at the H11 locus in pigs. Sci Rep 5:14253
11. Gao YP, Wu HB, Wang YS, Liu X, Chen LL, Li Q et al (2017) Single Cas9 nickase induced generation of NRAMP1 knockin cattle with reduced off-target effects. Genome Biol 18:13

Chapter 23

Generating Goat Mammary Gland Bioreactors for Producing Recombinant Proteins by Gene Targeting

Liyou An, Lan Yang, Yuejin Huang, Yong Cheng, and Fuliang Du

Abstract

Exogenous genes can be site-specifically integrated into the genomic DNA of animals by homologous recombination, generating transgenic animals. These animals have a clear hereditary background, although position effects of the exogenous genes and potential functional disruption of host genes can be caused by the genetic inserts. Therefore, the generation of mammary gland bioreactors via gene-targeting methods is a great asset for producing recombinant proteins in milk. Here, we describe a method of generating gene-targeted goats with the human alpha-lactalbumin gene (*hα-LA*) integrated into the beta-lactoglobulin gene (*BLG*) locus. The milk from these goats will be less allergenic and will be enriched with components of human milk protein.

Key words Mammary gland bioreactors, Gene targeting, Recombinant proteins, Goat, Human alpha-lactalbumin

1 Introduction

Homologous recombination (HR) is a type of genetic recombination in which nucleotide sequences are exchanged between two identical or similar molecules of DNA. HR functions to accurately repair harmful double-strand DNA breaks in cells and is also utilized for gene targeting for site-specific modification of genomic DNA [1]. Gene targeting is a genetic technique that is used to alter or replace an endogenous gene. Exogenous genes can be site-specifically integrated into the genomic DNA of animals via HR via the gene-targeting process. To target genes in mice, a gene-targeting construct is transfected into mouse embryonic stem cells in culture. After cells with the correct site-specific modification in the genomic DNA have been selected, these cells can then be injected into a mouse embryo to facilitate contribution to the genetic makeup of the animal [2]. Finally, chimeric mice, in which modified cells have made up the reproductive organs, are selected for breeding and propagation of the genetic modification. As a

Chengyu Liu and Yubin Du (eds.), *Microinjection: Methods and Protocols*, Methods in Molecular Biology, vol. 1874,
https://doi.org/10.1007/978-1-4939-8831-0_23,

result, the entire body of the offspring will reflect the gene-targeted genetic modification based on the previously selected embryonic stem cells.

In livestock, germline competent embryonic stem cells are not currently available. Thus, gene-targeted offspring of livestock cannot be generated as a chimeric animal as described above for mice. Therefore, somatic cell nuclear transfer (SCNT) has been found to offer an efficient genetic engineering strategy for the production of genetically modified animals [3]. In this procedure, the nuclear donor cells—primary cultured somatic cells—are modified via HR to target the gene of interest. Then, cloned livestock are generated by SCNT using the gene-targeted modified cells [4, 5]. Gene-targeted animals are not affected by any positional effects of the exogenous genes in the host genome or by an potential functional disruption of the host genes due to genetic inserts [6]. Additionally, these animals with site-specific gene insertions have an advantage over breeding selection since the hereditary background of the gene-targeted animals is known [7]. Gene-targeted cattle [8–10], pigs [11–13], sheep [14], and goats [15, 16] have been generated; however, it remains difficult to produce transgenic domestic animals with site-specific gene modifications. A gene-targeting strategy can also be used to produce a mammary gland bioreactor for producing recombinant proteins by these animals [17]. Important applications of this gene-targeted mammary gland bioreactor include the modification of milk proteins to eliminate allergens and/or enhance nutritional value and the expression of various recombinant proteins in the mammary gland. In our previous work, we successfully produced gene-targeted goats that the beta-lactoglobulin gene (*BLG*) replaced with the human alpha-lactalbumin gene (*hα-LA*) so that the goat mammary glands will produce *hα-LA* instead of *BLG*. Here, we describe a procedure for the production of goat mammary gland bioreactors via gene targeting.

2 Materials

2.1 Reagents

1. pGEM-T vector.
2. Porf-hav1tk.
3. Pyrobest DNA polymerase.
4. Restriction endonucleases *Sal*I, *Not*I, *Xho*I, *Dra*I.
5. T4 DNA ligase.
6. QIAquick gel extraction kit.
7. DMEM/F12.
8. Ham F-12 medium.

9. FBS.
10. Collagenase I.
11. Transferrin.
12. Hydrocortisone.
13. Sodium acetate.
14. Insulin.
15. Hyaluronidase.
16. Ethanolamine.
17. Prolactin.
18. Trypsin.
19. Mineral oil.
20. EDTA.
21. Hoechst 33342.
22. Cytochalasin B.
23. Ionomycin.
24. N-6 dimethylaminoprine.
25. M2 medium.
26. M16 medium.
27. BSA.
28. Geneticin® Selective Antibiotic (G418).
29. Penicillin-Streptomycin.
30. Cloprostenol.
31. FSH.
32. LHRH.
33. Rabbit anti human hα-LA monoclonal antibody.
34. Goat anti rabbit IgG polyclonal antibody.
35. DIG-high prime DNA labeling and detection starter kit I.

2.2 Preparation of the Medium and Solution

1. Goat mammary gland cell culture medium: DMEM/F12 containing 10% FBS, 1.2 mg/mL $NaHCO_3$, 5 mmol sodium acetate, 5 μg/mL transferrin, 0.5 mmol ethanolamine, 10 μg/mL insulin, 5 μg/mL hyaluronidase, 100 U/mL penicillin, and 100 μg/mL streptomycin (pH 7.2–7.4).
2. Goat fetal fibroblast culture medium: DMEM/F12 containing 15% FBS, 100 U/mL penicillin, and 100 μg/mL streptomycin (pH 7.2–7.4).
3. D-Hank's solution: 8 g/L NaCl, 0.4 g/L KCl, 0.05 g/L Na_2HPO_4, 0.06 g/L KH_2PO_4, 0.35 g/L $NaHCO_3$ (PH 7.2~7.4).

4. Collagenase digestion solution:DMEM/F12 containing 200 U/mL collagenase I, 100 U/mL hyaluronidase, 100 U/mL penicillin, and 100 μg/mL streptomycin (pH 7.2–7.4).
5. Trypsin digestion solution: Hank's solution supplemented with 0.5 g/L trypsin and 0.4 g/L ethylenediamine tetraacetic acid (EDTA) (pH 7.2–7.4).
6. Cell freezing medium: DMEM/F-12 supplemented with 10% dimethyl sulfoxide (DMSO) and 20% FBS.
7. Expression induction medium: goat mammary gland cell culture medium supplemented with 5 μmol prolactin.
8. Cell lysis buffer: 40 mM Tris/HCl pH 8.9, 0.9% Triton X-100, 0.9% Nonidet P-40, and 0.4 mg/mL proteinase K.
9. Fusion buffer: 0.3 M mannitol, 0.1 mM $MgSO_4$, 0.05 mM $CaCl_2$, 0.5 mM HEPES, and 3 mg/mL BSA.

2.3 Supplies and Instruments

1. 200 μL PCR tube.
2. 1.5 mL Eppendorf tube.
3. 15 mL centrifuge tube.
4. 50 mL centrifuge tube.
5. 48-well cell culture plate.
6. 24-well cell culture plate.
7. 6-well cell culture plate.
8. Water bath.
9. 2 mm gap electroporation cuvettes.
10. Multiporator (Eppendorf AG.).
11. Eppendorf Mastercycler.
12. ECM-2001 (BTX).
13. Fluorescence microscopy with micromanipulator apparatus.

3 Methods

3.1 Construction of the Gene-Targeting Vector

3.1.1 Amplification of hα-LA Genomic DNA by PCR

1. The genomic coding sequence of human *α-LA*, including the signal peptide (+26 to +82), four exons and introns (0–2361), and the polyA (+2339 to +2344) sequences were amplified by PCR. The primers LA1 and LA2 (Table 1) were designed according to sequence information obtained from GenBank (*hα-LA*, accession no: X05153.1), and an *Xho*I restriction endonuclease site was added to the 5′-end of the primers to allow for easy tracking of the cloned gene. The 2351-bp DNA fragments were amplified using Pyrobest DNA polymerase.

Table 1
PCR primers for gene amplification and detection

Primer name	Primer sequences	TM (°C)	PCR product size (bp)
LA1	GCTCGAGATGAGGTTCTTTGTCCCTC	67.8	2351
LA2	GCTCGAGTGACTTCAAAGTGGGACC	67.4	
pA F	TGTTCCCATAGTAACGCCAAT	57.5	843
pA R	AGCGAAACTCCACCTCAAA	55.4	
tA F	CTTGCTCCTGCCGAGAAAGTATCCATC	67.5	3538
tA R	GGAGGGTGTTCCCAGGTCACAGTCTAC	67.2	
BIN	GGTCACTGAATCTAACAGCCGTTGT	65.4	3868[a]
Neo 1	GCTATTCCAGAAGTAGTGAGG	51.2	896
Neo 2	CAGAAGAACTCGTCAAGAAG	49.6	

[a]Primers tA R and BIN were used to amplify wild-type *BLG3′* sequences to confirm gene targeting only at monoallelic sites

2. PCR products were cloned into the pGEM-T vector to yield the plasmid pGEM-LA.
3. Cloned sequence was confirmed by DNA sequencing.

3.1.2 Construction of the Gene-Targeting Vector pBla5-TK

1. The hα-LA gene was cloned into pgBLC/LF14 vector (available from our lab) at the *Xho*I site. This vector is a mammary gland-specific expression vector under the control of the *BLG5′* regulatory region, cytomegalovirus (CMV) promoter and enhancer, and *BLG3′* regulatory region [18]. This step yielded the expression vector pBlac.
2. The vector containing the Porf-hav1tk-encoded negative screening gene HSV-TK was recombined with pBlac to create the replacement targeting vector pBla5-TK (Fig. 1), which can be screened by a positive-negative selection strategy.

3.2 Analysis of the Plasmid Expression in Goat Mammary Gland Cells (GMGECs)

3.2.1 Culture of Primary GMGECs

1. GMGECs were isolated from lactating Sannen goats. Approximately 5 g of secretory tissue was removed from one of the two udders and was trimmed of visible fat and connective tissue.
2. After washing with Ca^{2+} and Mg^{2+}-free D-Hank's solution, the tissue sections were minced finely and digested in collagenase digestion solution with agitation at 37 °C for 1 h.
3. The digestion was stopped by the addition of one-fourth volume of FBS. Samples were subjected to centrifugation at 800 × *g* for 5 min to recover CMGECs.
4. The cells were then washed three times with D-Hank's solution. The isolated CMGECs were then cultured in goat mammary gland cell culture medium at 37 °C in a humidified atmosphere containing 5% CO_2.

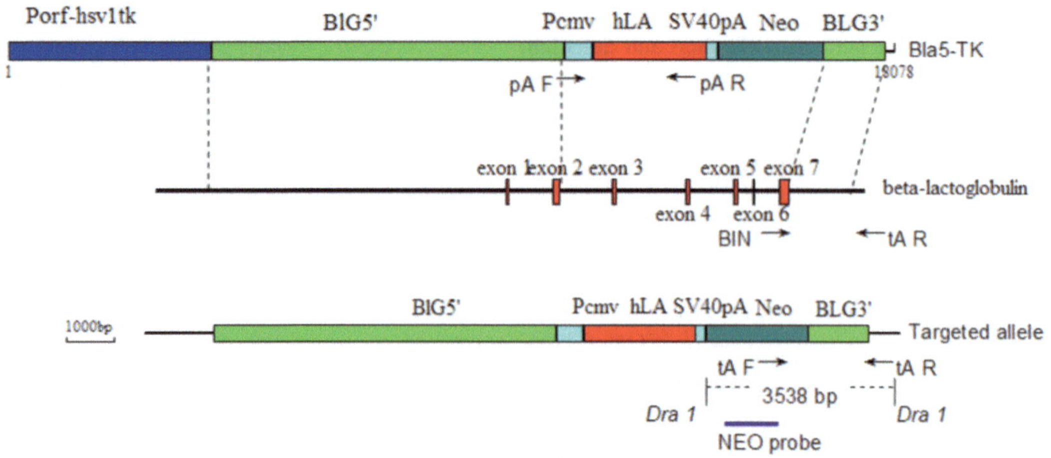

Fig. 1 pBla5-tk gene-targeting vector and detection strategy. The primers pA F and pA R were used to detect *hα-LA* integration, and primers tA F, tA R, and BIN were used to detect homologous recombination. The amplified 3538 bp region is indicated. The targeted allele contains sites for *Dra*I digestion and can be detected with a *NEO* probe

5. CMGECs that became attached to the culture dish were washed several times with D-Hank's solution after 24 h to remove dead cells. The cells were continuously cultured in goat mammary gland cell culture medium for 4–7 days.
6. For subculturing, the CMGECs were digested in trypsin digestion solution at 37 °C until the cells detached from the plastic dish. Then, the cells were recovered by centrifugation at 800 × *g* for 5 min and resuspended in fresh medium. The cell suspension was plated in culture dishes.
7. The epithelial cell population was enriched by selective detachment and attachment methods.
8. The purified cells were frozen in freezing medium or used for transfection.

3.2.2 Transfection of GMGECs

1. When GMGECs reached 80% confluence, the cells were transfected with pBlac plasmid, which was linearized by *Sal*I and *Not*I digestion. Transfection was performed by electroporation. The cells (2×10^6 cells) were mixed with 15 μg/mL pBlac DNA in hypoosmolar buffer (Eppendorf AG., Hamburg, Germany) and transferred into 2 mm gap electroporation cuvettes (Eppendorf AG.).
2. Electroporation was completed using a Multiporator (Eppendorf AG.) at 270 V for 100 μs.
3. Transfected cells were plated in 48-well culture plates at extra-low cell densities that each well plated no more than ten cells.

4. Antibiotic drug selection in mammary gland cell culture medium with 400 μg/mL G418 began 48 h after transfection and continued for 8–12 days.
5. Cell colonies were expanded into 6-well culture plates via subculture in mammary gland cell culture medium with 200 μg/mL G418.

3.2.3 Induction and Detection of hα-LA Expression in GMGECs

1. Transfected GMGECs were treated with 5 μmol prolactin to induce hα-LA expression. Cell culture medium was collected after 48 and 72 h for the detection of hα-LA expression.
2. 50 μL cell culture medium were used in an ELISA assay using standard laboratory western blotting protocols. The primary antibody was rabbit anti-human hα-LA monoclonal antibody (1:1000, Santa Cruz), and the secondary antibody was goat anti-rabbit IgG polyclonal antibody (1:2000, Santa Cruz).
3. To further assess the hα-LA expression, 24 μL cell cultured medium that was positive for hα-LA in the ELISA assay was analyzed by western blotting. Western blotting was performed using standard laboratory western blotting protocols. Recombinant hα-LA protein, human whey, induction medium from non-transfected GMGECs, and culture medium without prolactin from transfected GMGECs were used as controls (*see* **Note 1**).

3.3 Generation of Gene-Targeted Cloned Goats

3.3.1 Transfection and Selection of Goat Fetal Fibroblasts

1. The gene-targeting vector pBla5-TK was linearized with *Not*I. DNA fragments were purified using the QIAquick gel extraction kit.
2. Goat fetal fibroblasts were isolated from a 35-day-old female fetus recovered surgically from a Sannen goat [19]. Fetal fibroblasts were separated and propagated in goat fetal fibroblast culture medium.
3. When cultured fetal fibroblasts reached 80% confluency, the cells (2×10^6 cells) were mixed with 15 μg/mL pBla5-TK DNA fragments in hypoosmolar buffer (Eppendorf AG., Hamburg, Germany) and transferred into 2 mm gap electroporation cuvettes (Eppendorf AG.).
4. Electroporation was performed using a multiporator (Eppendorf AG.) at 270 V for 100 μs. Transfected cells were plated in 60 mm cell culture dishes.
5. DMEM/F12 supplemented with 10% FBS and 600 μg/mL G418 was begun 36 h after transfection and continued for 8–12 days in order to select for recombinant cells.
6. Healthy stable colonies were collected using a cloning ring and expanded via subculture in DMEM/F12 supplemented with 10% FBS and 400 μg/mL G418.

7. A portion of the cells were collected for cryopreservation, and the remaining cells were used for PCR screening for gene recombination at the BLG locus (*see* **Note 2**).

3.3.2 Screening Gene Targeting in Transfected Cells by PCR

1. Drug-resistant cells (1000 cells per 10 μL) were lysed in cell lysis buffer at 65 °C for 30 min, and then proteinase K was inactivated by incubation at 96 °C for 10 min.
2. 4 μL cell lysate were used as the DNA template in PCR. First, integration of exogenous genes was determined using primers pA F and pA R (Table 1). Next, PCR amplification was performed to screen for HR events using primer tA F and tA R, which amplified the sequence from the *NEO*^{r} gene to the outside of the BLG3′ homologous arm (Table 1). PCR were performed according to the following protocol: pre-denaturing at 95 °C for 5 min, 32 cycles of denaturing at 94 °C for 40 s and annealing and extension at 68 °C for 7 min, and a final extension at 72 °C for 5 min.
3. PCR products were analyzed by agarose gel electrophoresis, and product sequences were confirmed by direct sequencing.

3.4 Production of Cloned Fetuses by SCNT

3.4.1 SCNT and Activating Reconstructed Embryos

1. The cells containing the genomic-targeted modification were used as donor cells for SCNT. Donor cells were cultured with low-serum medium (DMEM/F12 + 0.5% FBS) in 24-well cell culture plates and incubated at 38 °C in an atmosphere with 5% CO_2 for 48 h. Prior to SCNT, donor cells were collected by trypsin digestion and resuspended in M2 medium plus 10% FBS.
2. Oocyte donor goats (Yangtze River Delta white goats) were synchronized, and superovulation was induced [20]. Oocytes were recovered surgically by flushing the oviducts with Ham F-12 medium containing 1% FBS.
3. Cumulus cells were removed from the oocytes by vortexing the cumulus-oocyte complexes in M2 medium containing 1 mg/mL hyaluronidase.
4. SCNT was performed as follows. Oocytes were incubated in M2 medium containing 2 mg/mL Hoechst 33342 and 7.5 mg/mL cytochalasin B at 38 °C with an atmosphere of 5% CO_2 for 20 min. A group of oocytes (20–30) were enucleated in M2 medium containing 7.5 mg/mL cytochalasin B by aspiration of the first polar body and the MII chromosomes lying beneath the polar body under fluorescence microscopy. After each individual oocyte was enucleated, a donor cell was injected directly into the perivitelline space using a glass pipette.
5. The karyoplast-cytoplast couplets were equilibrated briefly in M16 medium for 30 min before fusion. The karyoplast-

cytoplast couplets were washed briefly in fusion buffer and fused by the application of two direct current pulses (1.7 kV/cm for 40 ms) delivered by ECM-2001 (BTX). Cell fusion was evaluated after 30 min, and a second fusion pulse was administered to unfused couplets.

6. Fused couplets were washed extensively in M16 medium and then incubated for 5 h in M16 medium, before incubation in M16 medium containing 7.5 mg/mL cytochalasin B and 5 mM ionomycin for 5 min.
7. The constructed embryos were washed and then cultured in M16 medium containing 7.5 mg/mL cytochalasin B and 2 mM N-6 dimethylaminopurine for 5 h.
8. Finally, the reconstructed embryos were washed and cultured in M16 medium until embryo transfer (*see* **Note 3**).

3.4.2 Embryo Transfer and Ultrasonography

1. Reconstructed embryos were surgically transferred into synchronized recipient dairy goats. Each oviduct received five to seven cloned embryos.
2. Recipients were subjected to a transvaginal ultrasonographic evaluation on days 30, 60, and 90 of gestation.
3. Pregnant recipients were separated and placed in groups of two females per pen. For pregnancies that continued beyond 149 days, parturition was induced with prostaglandin F2a (Lutalyse, 4 mg im). Cotyledon number and birth weights were recorded.

3.5 Identification of Cloned Offspring

3.5.1 PCR Analysis of the Genomic DNA of Cloned Offspring

1. Genomic DNA was extracted from the umbilical cords or ear tissue of newborn cloned offspring 30 days after birth (Fig. 2).
2. For the detection of the integration of the exogenous gene, PCR was performed with primers pA F and pA R. For determination of HR events, the primers tA F, tA R, and BIN were used to amplify the genomic-targeted modification and wild-type *BLG3′* sequences (Fig. 1).

3.5.2 Southern Blot Analysis of the Genomic DNA of Cloned Offspring

1. To further confirm the site-specific gene modification, the genomic DNA of cloned offspring was subjected to southern blot analysis. The genomic DNA was digested with *Dra*I. The digested DNA was separated by 0.8% agarose gel electrophoresis, capillary transferred to a positively charged nylon membrane, and hybridized with a DIG-labeled probe for the NEO gene, which had been amplified from the pBlac-TK construct with primers neo 1 and neo 2 (Fig. 1).
2. Hybridization results were visualized with NBT/BCIP (DIG-high prime DNA labeling and detection starter kit I; Roche diagnostics, Indianapolis, IN, USA).

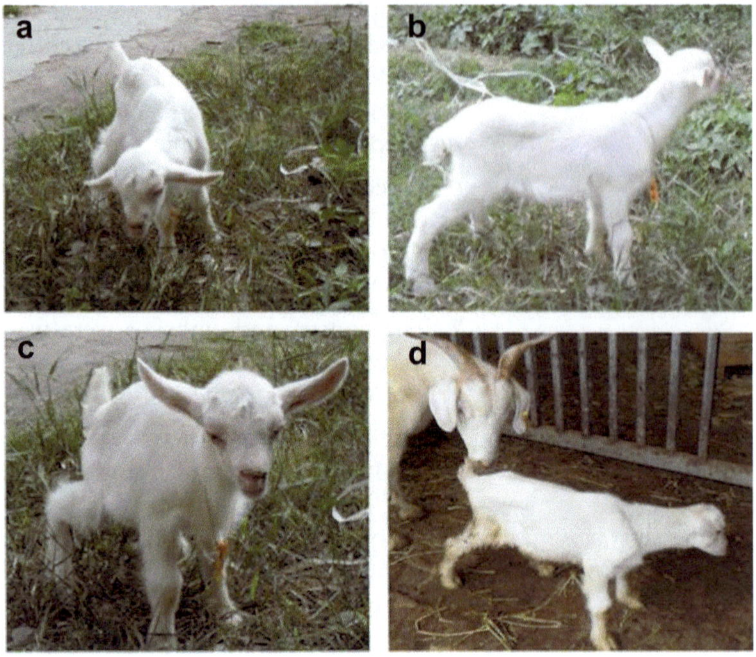

Fig. 2 Four transgenic cloned goats (**a**: TC-102, **b**: TC-108, **c**: TC-113, **d**: TC-114)

4 Notes

1. This procedure includes the methods for gene-targeting vector construction, analysis of the expression from the vector, SCNT to generate gene-targeted cloned goats, and detection of the expression of the recombinant protein. It is important to confirm the constructed vector has the ability to express the recombinant protein. Analysis of the expression function of the vector in GMGCs is a valid method to achieve this purpose.
2. The efficiency of gene targeting in somatic cells is low, usually $<10^{-4}$. Thus, in order to obtain a sufficient number of cell lines of gene-targeting modified cells for SCNT, transfection and selection of cells must be performed on a larger scale, e.g., 1×10^7 cells or more.
3. Gene-targeted cells undergo lengthy (>30 days) selection and culture expansion, potentially inducing cell differentiation and lack of activity. The efficiency of SCNT may be decreased if these cells are used. To overcome this barrier, a second round of cloning can be performed, in which the donor cells are separated from the 30-days cloned fetus. With this approach, the cloned fetus can be assessed to determine whether it carries the gene-targeted genotype, its karyotype, and the growth activity of the cells. This method improves the efficiency of generation of healthy cloned offspring.

References

1. Capecchi MR (1989) Altering the genome by homologous recombination. Science 244 (4910):1288–1292
2. Anastassiadis K, Schnutgen F, von Melchner H, Stewart AF (2013) Gene targeting and site-specific recombination in mouse ES cells. Methods Enzymol 533:133–155
3. Proudfoot C, Carlson DF, Huddart R, Long CR, Pryor JH, King TJ, Lillico SG, Mileham AJ, McLaren DG, Whitelaw CB, Fahrenkrug SC (2015) Genome edited sheep and cattle. Transgenic Res 24(1):147–153
4. Laible G, Alonso-Gonzalez L (2009) Gene targeting from laboratory to livestock: current status and emerging concepts. Biotechnol J 4 (9):1278–1292
5. Wang X, Niu Y, Zhou J, Yu H, Kou Q, Lei A, Zhao X, Yan H, Cai B, Shen Q, Zhou S, Zhu H, Zhou G, Niu W, Hua J, Jiang Y, Huang X, Ma B, Chen Y (2016) Multiplex gene editing via CRISPR/Cas9 exhibits desirable muscle hypertrophy without detectable off-target effects in sheep. Sci Rep 6:32271
6. Samiec M, Skrzyszowska M (2011) Transgenic mammalian species, generated by somatic cell cloning, in biomedicine, biopharmaceutical industry and human nutrition/dietetics--recent achievements. Pol J Vet Sci 14 (2):317–328
7. Murray JD, Maga EA (2016) Genetically engineered livestock for agriculture: a generation after the first transgenic animal research conference. Transgenic Res 25(3):321–327
8. Kuroiwa Y, Kasinathan P, Matsushita H, Sathiyaselan J, Sullivan EJ, Kakitani M, Tomizuka K, Ishida I, Robl JM (2004) Sequential targeting of the genes encoding immunoglobulin-mu and prion protein in cattle. Nat Genet 36(7):775–780
9. Richt JA, Kasinathan P, Hamir AN, Castilla J, Sathiyaseelan T, Vargas F, Sathiyaseelan J, Wu H, Matsushita H, Koster J, Kato S, Ishida I, Soto C, Robl JM, Kuroiwa Y (2007) Production of cattle lacking prion protein. Nat Biotechnol 25(1):132–138
10. Zhao LH, Zhao YH, Liang H, Yun T, Han XJ, Zhang ML, Zhou X, Hou DX, Li RF, Li XL (2015) A promoter trap vector for knocking out bovine myostatin gene with high targeting efficiency. Genet Mol Res 14(1):2750–2761
11. Takahagi Y, Fujimura T, Miyagawa S, Nagashima H, Shigehisa T, Shirakura R, Murakami H (2005) Production of alpha 1,3-galactosyltransferase gene knockout pigs expressing both human decay-accelerating factor and N-acetylglucosaminyltransferase III. Mol Reprod Dev 71(3):331–338
12. Rogers CS, Hao Y, Rokhlina T, Samuel M, Stoltz DA, Li Y, Petroff E, Vermeer DW, Kabel AC, Yan Z, Spate L, Wax D, Murphy CN, Rieke A, Whitworth K, Linville ML, Korte SW, Engelhardt JF, Welsh MJ, Prather RS (2008) Production of CFTR-null and CFTR-DeltaF508 heterozygous pigs by adeno-associated virus-mediated gene targeting and somatic cell nuclear transfer. J Clin Invest 118(4):1571–1577
13. Klymiuk N, Mundhenk L, Kraehe K, Wuensch A, Plog S, Emrich D, Langenmayer MC, Stehr M, Holzinger A, Kroner C, Richter A, Kessler B, Kurome M, Eddicks M, Nagashima H, Heinritzi K, Gruber AD, Wolf E (2012) Sequential targeting of CFTR by BAC vectors generates a novel pig model of cystic fibrosis. J Mol Med 90(5):597–608 Berl
14. McCreath KJ, Howcroft J, Campbell KH, Colman A, Schnieke AE, Kind AJ (2000) Production of gene-targeted sheep by nuclear transfer from cultured somatic cells. Nature 405(6790):1066–1069
15. Yu G, Chen J, Yu H, Liu S, Xu X, Sha H, Zhang X, Wu G, Xu S, Cheng G (2006) Functional disruption of the prion protein gene in cloned goats. J Gen Virol 87(Pt 4):1019–1027
16. Zhu C, Li B, Yu G, Chen J, Yu H, Xu X, Wu Y, Zhang A, Cheng G (2009) Production of Prnp −/− goats by gene targeting in adult fibroblasts. Transgenic Res 18(2):163–171
17. Luo Y, Wang Y, Liu J, Cui C, Wu Y, Lan H, Chen Q, Liu X, Quan F, Guo Z, Zhang Y (2016) Generation of TALE nickase-mediated gene-targeted cows expressing human serum albumin in mammary glands. Sci Rep 6:20657
18. Cheng Y, An LY, Yuan YG, Wang Y, Du FL, Yu BL, Zhang ZH, Huang YZ, Yang TJ (2012) Hybrid expression cassettes consisting of a milk protein promoter and a cytomegalovirus enhancer significantly increase mammary-specific expression of human lactoferrin in transgenic mice. Mol Reprod Dev 79 (8):573–585
19. An LY, Yuan YG, Yu BL, Yang TJ, Cheng Y (2012) Generation of human lactoferrin transgenic cloned goats using donor cells with dual markers and a modified selection procedure. Theriogenology 78(6):1303–1311
20. Zou X, Wang Y, Cheng Y, Yang Y, Ju H, Tang H, Shen Y, Mu Z, Xu S, Du M (2002) Generation of cloned goats (Capra hircus) from transfected foetal fibroblast cells, the effect of donor cell cycle. Mol Reprod Dev 61 (2):164–172

Chapter 24

Production of Transgenic Chickens Using Cultured Primordial Germ Cells and Gonocytes

Ellen J. Collarini, Philip A. Leighton, and Marie-Cecile Van de Lavoir

Abstract

The unique characteristics of the avian embryo, with its large opaque yolk, have necessitated the development of different approaches to transgenesis from those that have been successful in mammalian species. Genetic modification of birds was greatly advanced by the ability to grow long-term cultures of primordial germ cells (PGCs). These cells are obtained from embryos, established in culture, and can be propagated without losing the ability to contribute to the germline when reintroduced into a host animal. PGCs can be genetically modified in culture using traditional transfection and selection techniques, including gene targeting and site-specific nuclease approaches. Here, we describe our methods for deriving cell lines, long-term culture, genetic modification, production of germline chimeras and obtaining fully transgenic birds with the desired genetic modifications.

Key words Primordial germ cell, Gonocyte, Transgenic chicken, Avian transgenesis, Germline transmission, Genetic modification

1 Introduction

In order to enable the widest range of possible genetic modifications, a cell-based system for transgenesis is required. In mammalian species, embryonic stem cells have most often been used. In birds, initial attempts to replicate the success of embryonic stem cells resulted in stem cell lines that can be cultured and reintroduced into host embryos where they differentiate into somatic lineages of all three germ layers. However, after extensive effort it was concluded that these cells do not contribute to the germline [1]. The development of methods for culturing germline-specific, primordial germ cells (PGCs) and gonocytes provided a system in which genetic modifications could be completed in vitro and fully characterized prior to producing transgenic animals [2, 3]. Clonal selection allows for stable insertion of transgenes, knockouts, knockins, and any other genetic modifications that can be envisaged.

Chengyu Liu and Yubin Du (eds.), *Microinjection: Methods and Protocols*, Methods in Molecular Biology, vol. 1874, https://doi.org/10.1007/978-1-4939-8831-0_24,

It is still unclear when PGCs in the developing chicken embryo arise or even if the germline is predetermined. We do know, however, that primordial germ cells are present in several locations in the developing embryo. The earliest identification of PGCs during development is in the area pellucida, followed by the germinal crescent, and finally in the blood during their migration to the gonad [4, 5]. After being transported in the vasculature, PGCs migrate into the gonads, at which time they are referred to as gonocytes. Primordial germ cells and gonocytes are functionally very similar. Both populations can be grown and expanded in vitro and retain the potential to contribute to the germline when reintroduced into the vasculature.

We have successfully derived cell lines from germinal crescent, blood, and gonads, and have shown germline transmission from these cell lines [2, 3, 6]. The protocol described here is for the derivation of gonocytes, but the general method can be used for all three sources. The gonad is the richest source of germ cells, and is the easiest source with which to work, because the cells are visible from the start and it is relatively easy to assess their growth. PGCs and gonocytes require a feeder layer of irradiated Buffalo rat liver (BRL) cells, as well as BRL-conditioned medium. Subheading 3.1 contains the protocols for culturing BRL cells, preparing conditioned medium, preparing irradiated feeders, and obtaining drug-resistant feeders for transfections. Subheading 3.2 contains protocols for deriving and expanding the germ cells, and Subheading 3.3 describes how to transfect the cells. Subheading 3.4 contains protocols for injecting PGCs and gonocytes into embryos, hatching the embryos, and determining germline transmission.

2 Materials

2.1 General Equipment

1. Biosafety cabinet for sterile cell culture work.
2. Laminar flow hood that can accommodate a stereomicroscope to isolate gonadal tissue and to perform injections.
3. Inverted microscope with fluorescence capability for observing cell cultures.
4. Stereomicroscope with gooseneck lights, preferably equipped with a blue filter. The blue light increases the contrast between the blood vessels and the yolk.
5. CO_2 humidified incubator, CO_2 at 5%.
6. Table-top centrifuge with swing-out rotor.
7. Microcentrifuge.
8. 37 °C water bath to warm medium.

9. X-irradiation or gamma-irradiation source to irradiate BRL cells.
10. BTX ECM 830 square wave pulse electroporator.
11. Hemacytometer cell counting chamber.
12. Sterile scissors and forceps.
13. 70% isopropanol for cleaning scissors and forceps.
14. Cryogenic liquid N_2 storage tank.
15. Fluorescent head lamp (*see* **Note 1**).
16. High-speed rotary tool (e.g., Dremel MultiPro) fitted with a 220 mm diameter, two-sided, diamond-coated disc.

2.2 Cell Culture Reagents

1. Buffalo Rat Liver cells (BRL; ATCC, CRL-1442).
2. BRL medium: DMEM (with 4.5 g/L glucose and sodium pyruvate, without glutamine), 5% Fetal Bovine Serum (FBS; *see* **Note 2**), 1× glutamax or equivalent, 1× penicillin/streptomycin.
3. Dulbecco's PBS, Ca^{2+}/Mg^{2+}-free.
4. 0.25% Trypsin/EDTA, store at −20 °C.
5. Conditioning medium: KO-DMEM (ThermoFisher; *see* **Note 3**), 5% FBS, 1× glutamax or equivalent.
6. Manipulation Medium: CO_2-independent medium (ThermoFisher; *see* **Note 3**), 10% FBS, 1× glutamax or equivalent, 1× penicillin/streptomycin.
7. Freezing Medium: Manipulation Medium supplemented with 20% dimethyl sulfoxide (DMSO).
8. PGC medium: 47.5% KO-DMEM, 40% BRL-conditioned medium (BCM, collected as described below), 7.5% FBS, 2.5% chicken serum (*see* **Note 2**), 1× glutamax or equivalent, 1× nonessential amino acids, 0.4% sodium pyruvate, and 0.1% beta-mercaptoethanol. Filter sterilize and aliquot into sterile 50 mL centrifuge tubes. Store at 4 °C and use within 1 week. On the day of use, add recombinant human basic fibroblast growth factor (bFGF) to 4 ng/mL and recombinant murine stem cell factor (SCF) to 6 ng/mL. Make 2500× stock solutions of bFGF and SCF in PBS supplemented with 0.1% Bovine Serum Albumin (BSA) and store in aliquots at −80 °C. Once an aliquot is thawed, keep at 4 °C and use for up to 1 week.
9. Selection reagents: 5 mg/mL puromycin or 10 mg/mL blasticidin, filter sterilize, aliquot, and store at −20 °C; 250 mg/mL G418 in 100 mM HEPES pH 7.5, filter sterilize, aliquot, and store at −20 °C; 50 mg/mL hygromycin, store at 4 °C.

Table 1
List of cell culture vessels along with the recommended amounts of irradiated BRLs and total volume

Well/flask size	Area	Volume of BRL cells	Total volume
48	1 cm^2	50 μL	300 μL
24	2 cm^2	100 μL	500 μL
12	5 cm^2	200 μL	1 mL
6	10 cm^2	400 μL	2 mL
T25	25 cm^2	1 mL	5 mL
T75	75 cm^2	3 mL	15 mL
T150	150 cm^2	6 mL	30 mL

Total volumes can be slightly higher than that listed; when passing in smaller wells, it is difficult to keep the volumes exact

10. Nucleofector buffer V kit (Lonza, VCA-1003; *see* **Note 4**; includes 2 mm cuvettes and transfer pipets). Bring to room temperature before using.
11. PBS/BSA: Dulbecco's PBS, Ca^{2+}/Mg^{2+}-free, 0.1% BSA, filter sterilize, store at 4 °C.
12. Irradiated BRLs (or drug-resistant irradiated BRLs, prepared and cryopreserved as described in Subheading 3.1.3); thaw a 0.5 or 1.0 mL vial quickly in a 37 °C water bath, transfer to a 50 mL centrifuge tube, and slowly dilute with 5.5 mL (0.5 mL vial) or 11 mL (1.0 mL vial) PGC medium (final density 2.5×10^6 cells/mL), for use only on the day they are thawed. Table 1 indicates how much of this cell suspension is to be used with a given cell culture vessel.

2.3 Cell Culture Supplies

1. 50 mL conical centrifuge tubes.
2. 24- and 48-well cell culture plates, T75 and T150 cell culture flasks, and cellSTACK™ culture chambers (Corning; *see* **Note 5**).
3. Flat neck (NOT canted neck) T75 cell culture flask for X-irradiation.
4. Controlled-rate cell freezing container.
5. Sterile 500 mL polycarbonate bottles.
6. Gelatin-coated 48-well cell culture plates: using a 10 mL pipet, add three drops (~200 μL) of 0.1% gelatin solution to each well, aspirate solution, and leave the dish open in biosafety cabinet to dry under sterile conditions. If not using the same day, wrap in plastic bag and store at 4 °C.
7. Cryovials.
8. Sterile 1.5 mL microcentrifuge tubes.

2.4 Molecular Biology Reagents

1. Lysis buffer: 7 mM SDS, 10 μM EDTA, 100 mM Tris-HCL, 200 mM NaCl, with 0.8 mg/mL Proteinase K.
2. 3 M Sodium Acetate, pH 5.2.
3. Ethanol, 100 and 70%.
4. Standard PCR reagents (polymerase, buffer, dNTPs).
5. Sexing primers for W-chromosome:

   USP1: 5′-CTATGCCTACCACMTTCCTATTTGC-3′ and

   USP3: 5′-AGCTGGAYTTCAGWSCATCTTCT-3′. Control primers for actin: Actin-F3, 5′-ATGCGCATAAAACAAG ACGAGATT-3′,

   Actin-R3, 5′-GGGGACTGTAAAGCCTTCATTCAC-3′.
6. For transfections of BRLs and PGCs: DNA of interest containing drug-resistant gene (prepare using an endotoxin-free kit; Qiagen-prepared DNA works well). For random integration (~10 μg/transfection) and homologous recombination (10–15 μg/transfection), linearize 100 μg overnight in a 500 μL reaction. Check a small amount on a gel. If digest is complete, precipitate with 1 mL ethanol, pellet, wash with 70% ethanol, dry, and resuspend in water to at least 1 μg/mL. For integrase-mediated transfection use circular construct (3–15 μg of insert construct, 3–15 μg of integrase construct).

2.5 Embryology Supplies

1. Weigh boats (3 × 3 in.).
2. Mold to hold the weigh boat when transferring. Using mold-making material from a craft store, make an impression mold of a weigh boat, and shape a funnel at one of the corners to facilitate pouring the egg into the surrogate shell (*see* Fig. 3g).
3. Injection needles: 40 μm I.D. with a 35°bevel.
4. Aspirator tube assembly (product A5177-Sigma) to connect the needle to the injection tubing.
5. Fertile eggs incubated for 6–7 days (to derive gonocytes).
6. Chicken embryos at Stage 14–16 H&H for injection. All eggs need to be weighed before incubation to be matched with a surrogate shell (*see* **Note 6**). Set 50% more eggs than needed in case of infertile eggs and breakage.
7. Surrogate shells. Eggs should be clean. Weigh all eggs and write weight on pointy side. Since most fertile eggs will be between 50 and 65 g, the preferred weight of the surrogate shells is between 80 and 95 g (*see* **Note 6**). Sort eggs by weight and put them at 4 °C until use.
8. Saran Cling Plus Wrap. Small squares of plastic film, 20–25 mm^2, to seal the openings in the egg. Make in advance and wrap in foil packages containing ~100 squares, interleaved

with filter paper. Other plastic film can be used provided it is similar to the shell in permeability to water vapor, oxygen, and carbon dioxide.

9. Egg white collected from several eggs with 20% Penicillin/Streptomycin to seal the plastic to the eggshell.
10. Cotton applicator sticks.
11. Scalpel with scalpel blade.
12. Kraft boxes (4 × 4 × 4 in.) with perforations to allow airflow.

3 Methods

3.1 BRL Cell Culture, Conditioned Medium, Irradiation, and Generation of Drug-Resistant BRL Cells

Buffalo Rat Liver (BRL) cells produce factors that promote the growth of PGCs and are used in two ways in our culture system: as a source of conditioned medium, and as a feeder layer [2] (*see* **Note** 7). For the feeder layer, the BRL cells need to be sufficiently irradiated to stop growth so they do not overtake the PGC cultures, but viable enough to attach and form a monolayer. Guidelines for irradiation are given, but every irradiation source will need to be tested at various doses to find the best setting.

3.1.1 Initial Expansion of BRL Cells and Creation of Frozen Stock

1. All work to be carried out under sterile conditions in a biosafety cabinet.
2. Thaw vial of cells from ATCC by gently swirling vial in a 37 °C water bath until thawed, transfer cells to a 50 mL centrifuge tube, and slowly add 10 mL of BRL medium.
3. Centrifuge at 300 × *g* for 5–7 min.
4. Aspirate supernatant, resuspend cells gently with a P1000 tip, and transfer to a T75 culture flask with 15 mL final volume of BRL medium.
5. Incubate in CO_2 incubator at 37 °C until confluent.
6. Wash flask 2× with 10 mL PBS.
7. Add 3–5 mL trypsin, let it wet the surface, then remove most of the liquid. Place flask in an incubator for 5–10 min until cells are detached from surface and from each other.
8. Use 10 mL of BRL medium to collect the cells, and transfer to a 50 mL centrifuge tube. Count cells, centrifuge at 300 × *g* for 5–7 min.
9. Resuspend the cells in medium and transfer to multiple flasks, plating approximately 2.5×10^6 cells/T75 flask or 5×10^6 cells/T150 flask. Incubate until confluent, about 3 days.
10. Trypsinize as above, using proportionately more volumes of PBS and trypsin for a T150 flask. Replate cells into multiple T150s or into a cellSTACK (4.5×10^7 cells in 225 mL;

see **Note 5** for instructions on cellSTACK use), and incubate until confluent, about 3 days.

11. Trypsinize, count cells, centrifuge, and resuspend in Manipulation Medium at 9×10^7 cells/mL. Slowly add an equal volume freezing medium and aliquot 1 mL of cell suspension (4.5×10^7 cells/mL) to cryovials. Put tubes in controlled-rate cell freezing container at −80 °C for 24 h. Transfer to a LN_2 tank for long-term storage.

3.1.2 Collection of BRL-Conditioned Medium

1. *Day 1*: Thaw one vial (4.5×10^7 cells) of cryopreserved BRL cells, add medium as described above.
2. Spin at 300 × *g* for 5–7 min, and resuspend in 5 mL medium.
3. Add 220 mL warm growth medium to *bottom of stack* (this can be done earlier so the medium can warm up in the incubator; otherwise use warm medium).
4. Transfer the cells to the bottom of the cellSTACK. Gently swirl stack to evenly distribute the cells, and divide the cell suspension equally over both levels of the cellSTACK (*see* **Note 5**).
5. Incubate for 3 days. Make sure incubator shelf is level.
6. *Day 4*: Wash cellSTACK 2× with 40 mL of PBS (*see* **Note 5**).
7. Add 50 mL of trypsin, making sure both layers are covered by the trypsin solution, then aspirate the trypsin. Incubate for 5–10 min until the cells are detached from the surface and from each other.
8. Add 40 mL of medium to stack, distributing over both levels. Pipet up and down a few times in the stack to resuspend cells, then split the cell suspension evenly into 2 × 50 mL tubes.
9. Wash with 40 mL medium and split evenly into the two 50 mL tubes.
10. Count cells (make a 1:10 dilution of cell suspension for counting).
11. Add 220 mL growth medium to bottom of desired number of cellSTACKs. Add 1.3×10^7 cells to the bottom of each stack, swirl to mix thoroughly, and distribute the suspension over the two layers.
12. Put stacks in the incubator and leave for 4 days.
13. *Day 8*: Cells should be at or near confluence.
14. Use the aspirator to remove the medium.
15. Add 225 mL of pre-warmed conditioning medium (point the pipet toward the side of the stack, not the bottom surface, so as to not disturb the attached cells), and distribute over both layers.
16. Return to incubator for 3 days.

17. *Day 11*: Label cellSTACKs A, B, C, etc. Label sterile 500 mL bottles with date and cell stack reference (e.g., 160215A1, 160215A2—bottles A1 and A2 will be collected from cellSTACK A).
18. Remove 110 mL of the BRL-conditioned medium and place in bottle A1. Remove 110 mL and place in bottle A2.
19. Add 225 mL of fresh conditioning medium to cellSTACK and return to the incubator.
20. Place collection bottles at −20 °C.
21. Repeat this procedure with all the cellSTACKs, and incubate the cells another 3 days.
22. *Day 14*: Retrieve bottles A1 and A2 from freezer, add 110 mL medium from cellSTACK A to bottles A1 and A2, add 225 mL fresh conditioning medium to stack, return to incubator for another 3 days, and return bottles to −20 °C.
23. Do the same for the remaining stacks.
24. *Day 17*: Repeat the same procedure as for day 14. Each bottle will now have 330 mL of conditioned medium (*see* **Note 8**).
25. Return the bottles to −20 °C.

3.1.3 Preparation of Irradiated Feeder Cells

1. *Day 1*: Thaw and plate BRL cells as described above, grow for 3 days.
2. *Day 4*: Trypsinize as above and count cells. Plate 2.65×10^7 cells into each of 4–6 cellSTACKs. Incubate for 4 days.
3. *Day 7*: Trypsinize cells—it is easiest to do this with two people, one for aspirating, and one for adding PBS/trypsin. Wash and add trypsin (freshly thawed) as above. Remove trypsin with pipet and transfer to the next washed cellSTACK, then aspirate off trypsin. 50 mL trypsin can be used for 2–3 cellSTACKs. Incubate to detach the cells.
4. Add 40 mL of media to the cellSTACK, recover cells as above, and transfer to one 50 mL tube.
5. Wash the same cellSTACK with another 40 mL medium, retrieve the medium, and use this medium for the first wash of the second cellSTACK.
6. Repeat this procedure with all stacks; put the second wash of the last stack in a separate tube.
7. Centrifuge all tubes for 5–7 min at $300 \times g$.
8. Aspirate supernatants and resuspend each pellet using a P1000 tip and Manipulation Medium. Combine all pellets into one tube, wash the rest of the tubes with Manipulation Medium, and add to the combined cells. Final volume should be 10–20 mL of cell suspension.

9. Make a 200× dilution and count cells. Recovery is usually 3–4.3 × 10^8 cells per cellSTACK.
10. Transfer cells at a density of ~4 × 10^7 cells/mL in multiple 50 mL tubes for gamma irradiation, or 3–3.4 × 10^7 cells/mL in a flat-neck (NOT canted) T75 flask for X-irradiation. We irradiate using a gamma source at 12,000 rads, or using an X-ray source at 17,500 rads on a level that accommodates a T75 flask within the 95% irradiation uniformity range (e.g., on level 4 of a RS-2000 Irradiator).
11. After irradiation, pellet cells by centrifugation for 5–7 min at 300 × *g*, take off the supernatant, and resuspend pellet in Manipulation Medium (try to keep the cell density above 6 × 10^7 cells/mL). Make a 200× dilution and count. Cell recovery should be between 65 and 100%.
12. Add Manipulation Medium to a final cell density of 6 × 10^7 cells/mL. Split into aliquots of no more than 10 mL if volume is over 10 mL. Add an equal volume of freezing medium to the cell suspension in each tube, adding 1–3 drops at a time, and swirling the tube after each addition (if the final volume is over 20 mL, mixing by swirling the tube is difficult). Final density is 3 × 10^7 cells/mL.
13. Add 0.5 or 1 mL to each cryotube and freeze using controlled-rate cell freezing containers.
14. To check quality of irradiated cells, thaw a 0.5 mL vial quickly in a 37 °C water bath, transfer to a 50 mL centrifuge tube, slowly add 5.5 mL BRL or PGC medium, and use 1 mL of this suspension to seed a T25 with 5 mL BRL or PGC medium.
15. Check 24 h later that there is a monolayer. If cells are not dense enough, resuspend subsequent vials of the batch in a slightly smaller volume of medium so that 1 mL in a T25 flask will result in a monolayer.
16. Properly-irradiated cells will form a monolayer that will last for 3 days, and then the cells will slowly start lifting. Change the medium every few days and keep in culture for at least 2 weeks to make sure there are no BRL cells that are actively growing. If so, the whole batch should be discarded and a new batch made.

3.1.4 Creation of Drug-Resistant BRL Cells

The standard irradiated BRL cells can be used for PGC transfections in which puromycin is the selection agent. PGCs can be selected on 0.5 μg/mL puromycin, which will not kill BRL cells. When using other selection agents, drug-resistant BRL feeders are needed.

1. Thaw and expand BRL cells as above.
2. For each transfection, fill a sterile microcentrifuge tube or cryovial with 0.5 mL BRL medium, and place in incubator.

3. Trypsinize, count, and pellet 5×10^6 cells (multiple transfections can be done to generate a larger resistant cell population faster). Use an aspirator fitted with a P1000 unfiltered tip to remove as much medium as possible from the pellet so as not to dilute the transfection buffer. It helps to hold the tube horizontally so the medium will collect where the conical section of the tube starts. Alternatively, the pellet can be transferred with 0.5–1 mL medium to a microcentrifuge tube, and pelleted again. The residual medium is easier to remove in a smaller tube.
4. Add approximately 50 μL of Nucleofector transfection buffer (*see* **Note 4**) for each 5×10^6 cells, resuspend cells, and measure volume. Add more transfection buffer to obtain a final volume of 100 μL/transfection, minus the volume of DNA to be added.
5. Add 10 μg of linearized DNA for each transfection, mix, and transfer 100 μL to a 2 mm cuvette with a P200. Cap the cuvette, transfer the cuvette to the electroporator, and pulse at 250 V, 8 pulses, 100 μs pulse length. *See* **Notes 9** and **10** for more transfection information.
6. Recover cells with a transfer pipet, add to 0.5 mL medium in a microcentrifuge tube or cryotube, place in incubator for 15 min to recover, and then transfer all cells to a T150 flask with 30 mL of medium.
7. 2 days later, start selection using 800 μg/mL G418, 600 μg/mL hygromycin, and/or 10 μg/mL blastocidin.
8. Change the medium every few days to remove cell debris. You should be able to see small colonies of resistant cells growing in the flask after a week or so.
9. After 10–14 days, trypsinize, collect, and count the remaining live cells, and replate in a suitable size flask. Expand as above for regular BRL cells, make frozen stock, and irradiate. We irradiate at slightly lower doses; 10,000 rads on a gamma source, 17,000 rads on an X-irradiation source. We do *not* use these cells as a source of conditioned medium.

3.2 Gonocyte Derivation and Expansion

Here we describe deriving germ cell cultures from gonadal tissue, but with slight modifications this protocol can be used for PGC derivations from the germinal crescent and the blood (*see* **Note 11**).

Although germline transmission of PGCs through the opposite sex has been reported [7], we have never seen any germline transmission of male cells through the female germline or vice versa in many test breedings. Germline transmission of our PGC/gonocyte lines in the chicken has always been gender-specific; male PGCs transmitted through the male germline and female PGCs transmitted through the female germline. We therefore suggest that when deriving cell lines one should keep the final application in mind. For our purposes, we focus exclusively on male cell lines. Male chimeras become sexually mature earlier than female chimeras, and many more offspring can be obtained from a male chimera than from a female chimera. Therefore, we determine sex shortly after setting up the cultures and only continue with the male cultures. Plan to start with 20 cultures. Half will be female, and depending on your research goals, can be dropped. Most of the remaining cultures will start growing. Generally, growth is robust for the first 7–10 days. After that time, some cells adjust to the culture environment and continue to grow well, and some cultures slow down and never recover. Some cultures will grow well and expand to 2–3 $\times$ 10^6 cells, then stop expanding. In our hands we usually obtain 1–2 cell lines that keep growing and can be expanded indefinitely. Be aware that these are primary cell cultures; there is an art to the process that cannot easily be described, so do not be discouraged if no cell line is obtained in your first set of derivations.

PGCs and gonocytes grow in suspension, but often lightly stick to the feeder layer. They have a unique morphology; they are large, round and have distinctive vesicles in the cytoplasm. One defining feature is that even after cytokinesis the cells remain attached and these doublets (*see* Fig. 1a) are often the first indication that PGC/gonocytes are present.

3.2.1 Derivation of Gonocyte Cell Lines

1. Label two 1.5 mL microcentrifuge tubes for each embryo: one empty and one with 400 μL of sterile PBS/BSA.
2. In a laminar-flow hood, crack open an incubated egg and transfer the embryo to a petri dish.
3. Decapitate, cut open embryo, and remove overlying organs to reveal the mesonephros. The gonads are located on the medial side of the mesonephros and are slightly more opaque than the mesonephros.
4. With the aid of a stereomicroscope, use curved forceps to remove the gonads from the underlying mesonephros and place gonads from each embryo into a tube with PBS/BSA. Some find it easier to remove the mesonephros from the embryo first, and then remove the gonads.
5. Collect a small piece of any soft tissue (~2 mm^3) from each embryo and place in the matching empty microcentrifuge tube. This sample may be kept at −20 °C until it is convenient to perform sex determination PCR (*see* Subheading 3.2.2).
6. Wipe all utensils with 70% isopropanol before moving on to the next embryo.
7. All following work is done in a sterile biosafety cabinet.
8. When all gonads have been collected, add 100 μL 0.25% Trypsin/EDTA (freshly thawed) to each tube.

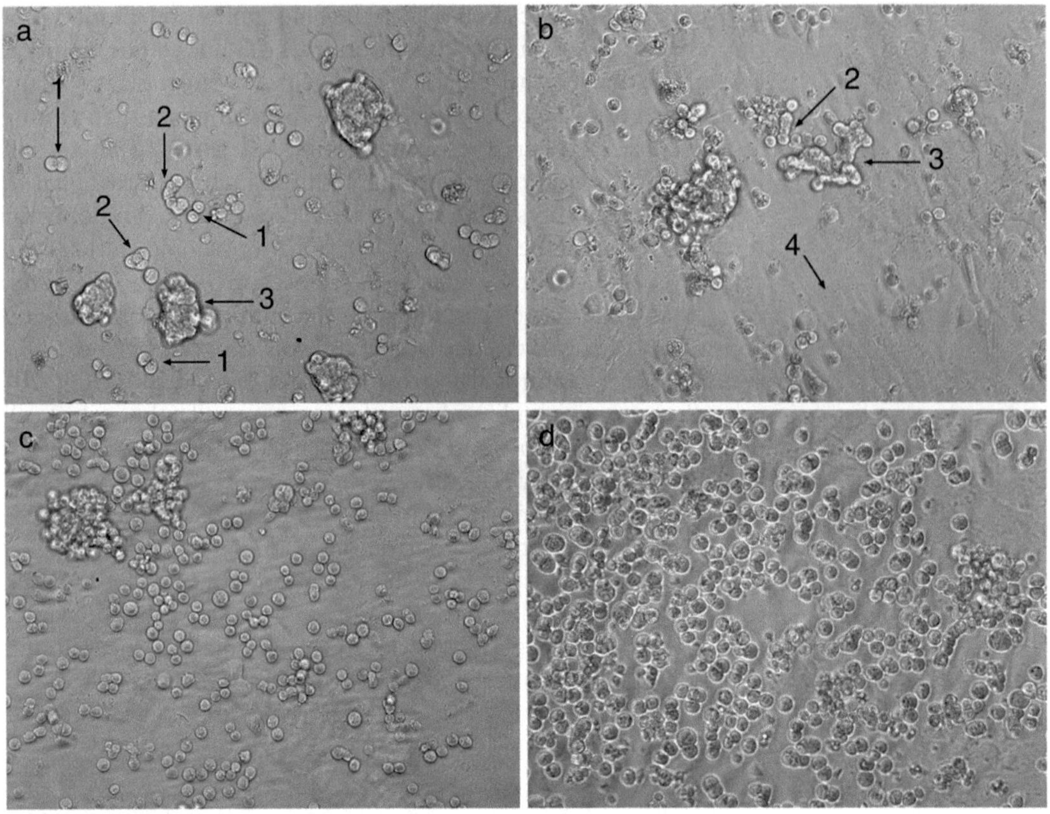

Fig. 1 (**a**, **b**) Images of early stage gonadal cultures, showing gonocyte doublets (1), chains and small clusters (2), and clumps (3). The irradiated BRL cells form a monolayer underneath the gonocytes (4). (**c**, **d**) Later stage cultures

9. Incubate in a 37 °C water bath for 12 min.
10. Add 200 μL PGC medium, and triturate the tissue with a P1000 tip until tissue is completely broken up (do not add more volume than 200 μL, since it will be hard to triturate without the medium spilling out of the tube).
11. Add 500 μL more medium to dilute/inactivate the trypsin and centrifuge for 3–4 min at 300 × *g*.
12. Resuspend pellet in 350 μL PGC medium and add each cell suspension to a well in a 24-well plate. Incubate for 3–4 h to allow any adherent cells to attach to the plastic surface.
13. Swirl the dish to dislodge non-adherent cells, and transfer the contents of each 24 well to a well in a 48-well plate. To prevent evaporation of medium in the cultures, do not use the outside edge wells on the 48-well plate, only the inner 6 × 4 wells.
14. Wash each 24 well with 100 μL of medium and add to the medium in the 48 wells.

15. Add 50 μL of irradiated BRL cell suspension (thawed and diluted as described in **item 12** of Subheading 2.2) to each well, and place in the incubator.
16. Carry out PCR for sex determination (*see* Subheading 3.2.2) and, depending on research goals, eliminate the female cultures.
17. Gonocytes should be visible in the next day or two. They will be loosely attached to the BRL cell monolayer (*see* Fig. 1a, b for examples of early stage germ cell cultures). Contaminating adherent fibroblast-like cells can take over the culture quickly, so pass cells within 1–2 days.
18. To pass cells: remove and discard ~350 μL medium (the empty wells around the edges can be used for the waste medium; these wells can be checked the next day to confirm that no gonocytes were discarded). Use a P200 tip to gently dislodge the germ cells with remaining medium, transfer volume to a new well below the original well, use 150 μL of PGC medium to very gently wash the original well, and add that medium to the new well. Add 50 μL freshly thawed and diluted BRL cells to the new well, and add 300 μL PGC medium to the original well. The original well (now called a "leftover well") can still yield gonocytes. *See* Fig. 2 for an example of how to set up the cultures on the plate.
19. From this point on, cells are evaluated and either passed or have a medium change on a Monday, Wednesday, and Friday schedule. During the first week of culture they will probably need passing on every culture day (*see* **Note 12**).
20. At the next passage, remove 150 μL medium from the well, use the remaining 150 μL to transfer the germ cells to a new well, wash the original well with fresh medium and add that to the new well, add BRL cells, and add medium to the old well (a new leftover well). If germ cells are visible in the first leftover well, collect and plate into a new well with BRL cells (*see* **Note 13**). This well can be combined with the main well at the next passage. *See* Fig. 2 for passaging scheme.
21. At the next passage, remove almost all the medium from the replated leftover well and gently wash that well when washing the main well, and combine with the main culture.
22. If there is no sign of contaminating adherent cells, but the cells are still not very dense (*see* Fig. 1d for an example of a dense culture), remove most of the medium and replace with fresh medium instead of passing the cells.
23. Once the germ cells have become dense, they can be passed to a 24-well plate. Transfer the entire contents of the 48-well, gently wash the well with fresh PGC medium and add to the

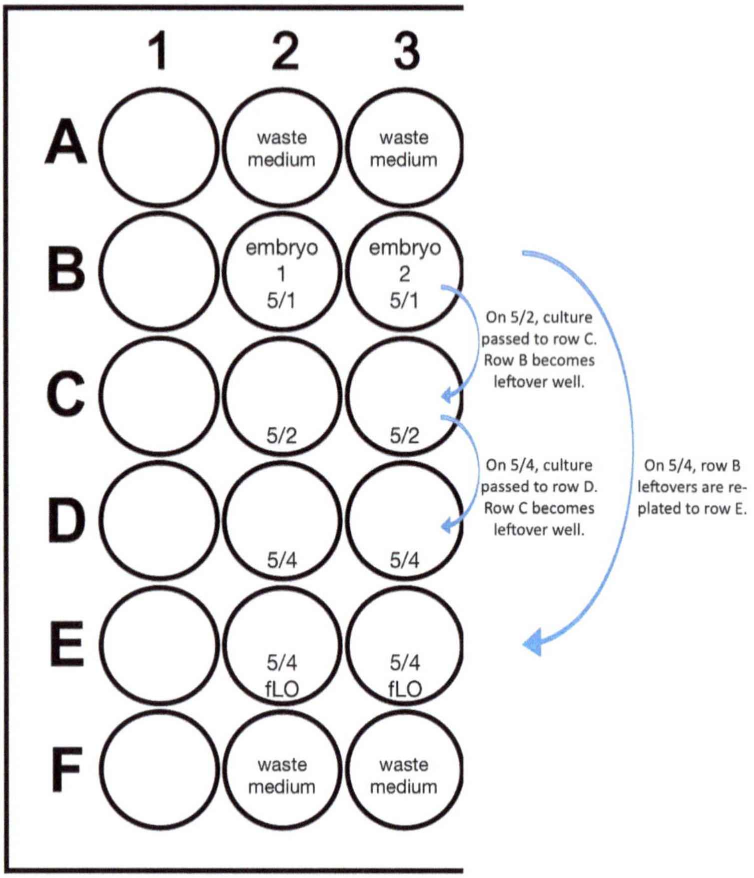

Fig. 2 Diagram of initial culture plate. Each gonadal culture is set up in a separate well and dated (e.g., 5/1). When passing cells, transfer to the well below to keep all wells from the same embryo in a column. In this example, the main culture was passed on 5/2 to row C. On 5/4, the main culture was passed to row D, and the leftovers from row B were passed to row E (fLO indicates "from leftovers"). At the next passage, a new plate will be needed, but keep each embryo in the same column on the new plate to avoid accidental cross-contamination

same 24 well, and add 100 μL irradiated BRL cells to the 24-well. Add fresh medium back to the leftover well.

24. Once in 24-well plates, the cells are passed by transferring the suspension to a microcentrifuge tube, pelleting at 300 × *g* for 4 min, and replating. If just changing the medium, pellet the cells and return to the same well. Continue to examine leftover wells for germ cells and replate with BRL cells, then collect, pellet, and combine with the main culture at the next passage.
25. Continue to expand the cells to 12-well plates and then 6-well plates, continuing a Monday, Wednesday, Friday schedule. *See* Table 1 for guidelines on culture vessel size, volume, and BRL cell additions.

26. Once cells have expanded to multiple six wells, collect and count cells; a good culture will expand to at least 2×10^6 cells in 25–30 days.
27. Pellet the cells in microcentrifuge tubes at $300 \times g$ for 4 min or, for larger volumes, in 15 or 50 mL conical centrifuge tubes at $300 \times g$ for 5–7 min.
28. Replate in fresh PGC medium at 1.25×10^5 cells/mL for a 2-day culture (Monday to Wednesday, for example), or 1×10^5 cells/mL for 3-day culture (Friday to Monday).
29. On each culture day collect, count, pellet and replate the cells, to expand the cells into larger flasks (*see* **Note 14**). Keep the leftover flasks, and when replating move them to a smaller vessel (e.g., replate leftovers from a T75 into a T25). Up to 30% of the medium of the replated leftover flask can consist of supernatant recovered from spinning down the main culture (*see* **Note 15**).
30. When cells have expanded enough, cryopreserve multiple vials of cells (at least 2×10^6/vial) as described for freezing BRL cells. Record the number of days the cells were cultured.

3.2.2 Sexing of Embryos by W-Chromosome PCR

1. Add 0.5 mL lysis buffer with Proteinase K to each tube containing an embryo tissue fragment (~2 mm^3).
2. Incubate at 56 °C overnight, with shaking or turning.
3. Tissue should be completely dissolved.
4. Centrifuge lysate at $13{,}000 \times g$ for 5 min.
5. Remove a 50 μL aliquot into a new tube.
6. Add 5 μL 3 M NaOAc, pH 5.2 and 150 μL ethanol to the 50 μL lysate and vortex.
7. Centrifuge to pellet genomic DNA, $13{,}000 \times g$ for 10 min.
8. Wash pellet once with 70% ethanol.
9. Carefully remove the supernatant.
10. Resuspend DNA in 200 μL water.
11. Perform PCR in a 20 μL reaction using 2 μL genomic DNA with 0.25 μM primers for W and a separate reaction for actin.
12. Cycling parameters for both W and actin: 95 °C 3 min to activate polymerase, followed by 40 cycles of: 95 °C for 5 s, 56 °C for 30 s, 72 °C for 1 min.
13. Primers for the W-chromosome [8]: USP1, 5′-CTATGCCTACCACMTTCCTATTTGC-3′ and USP3, 5′-AGCTGGAYTTCAGWSCATCTTCT-3′. Control primers for actin: Actin-F3, 5′-ATGCGCATAAAACAAGACGAGATT-3′, Actin-R3, 5′-GGGGACTGTAAAGCCTTCATTCAC-3′.

14. Run samples on an agarose gel. W PCR product is ~370 bp; actin is 446 bp. Presence of a W-specific band indicates a female embryo.

3.3 PGC and Gonocyte Transfection

3.3.1 Strategies for Genetic Modification

1. General considerations.

   PGCs and gonocytes can be stably transfected with a variety of constructs. In our experience, stable transfection with standard linearized constructs bearing a selectable marker does not produce drug-resistant colonies, unless the selectable marker is flanked by the insulator element from the chicken β-globin gene (HS4) [9]. We have successfully used the following selectable marker genes in PGCs and gonocytes: neomycin, puromycin, hygromycin, and blasticidin. Even with the insulator element, however, random insertion of linearized constructs is not efficient in PGCs and gonocytes, so for non-targeted insertion it is preferable to facilitate integration with a system such as an integrase or transposon. For targeted insertions (see below), the efficiency of obtaining colonies is also low, however the frequency of correctly targeted clones is high, and a sufficient number of correctly targeted clones is readily obtainable. The use of a site-specific nuclease such as CRISPR/Cas9 with targeting vectors greatly increases the frequency of drug-resistant colonies and most of these clones are correctly targeted [10].

   For easy screening of germline progeny derived from the genetically modified germ cells, it is useful to include a fluorescence marker such as GFP on the transgene construct. When germline chimeras are bred to wild-type hens, chicks can be quickly screened after hatch using UV light and goggles to find those that are GFP-positive (Fig. 4c). GFP-fluorescence can also be used to screen semen samples from chimeras by flow cytometry, as there is a small amount of GFP present in the sperm. Alternatively, feather pigmentation markers can be used to screen visually for germline transmission, followed by PCR genotyping for the presence of the transgene.

2. Random integration.

   (a) phiC31 integrase-mediated transfection.

      For random integration of a transgene, the frequency of obtaining stably transfected clones is increased by the use of phiC31 integrase or a transposon. In the case of phiC31 integrase, the plasmid to be inserted must carry a copy of the attB site recognized by integrase, which will recombine with either a pseudo-attP site in the chicken genome, or a previously integrated authentic attP site placed in the genome, and mediate insertion of the plasmid into the genome. The insertion plasmid is co-transfected with an integrase-expression construct, both as circular DNAs, and drug selection is used for the insertion of the transgene construct. The integrase-bearing plasmid should not insert into the genome and will be lost during subsequent cell divisions.

(b) Transposon-mediated transfection.

To use a transposon-based system for random integration, a transposase expression vector is co-transfected with the transgene construct to be integrated, which contains inverted terminal repeats (ITRs) recognized by the transposase. We have used the *piggyBac*-based transposon system [11], which generates very high efficiency insertion.

3. Gene targeting by homologous recombination

For classical gene targeting using homologous recombination, it is essential that the homology regions be isogenic to the cell line used for targeting. In our targeting of the immunoglobulin light chain, a construct with five single base pair changes in one homology region failed to target, whereas the isogenic targeting vector targeted with an efficiency of 23% [12]. Because most chicken strains are not inbred, it is likely that the two alleles of a given locus contain polymorphisms relative to each other in any given individual or cell line. Different cell lines derived from that same breeding could also contain different alleles, depending on the inheritance from the parents. This means that each cell line could have a different complement of alleles, and homology regions must be prepared using genomic DNA from the cell line to be used for targeting. The homology regions (5′ and 3′, flanking the selectable marker cassette) must also be from the same allele, to ensure that the sequence will completely match one of the chromosomes. If using PCR to clone the homology regions from genomic DNA, overlapping amplicons should be cloned and sequenced to confirm that the amplicons are derived from the same allele. We have used homology regions totaling 8 kb (IgL), 4.4 kb (IgH), or 8 kb (IgH) with success. Shorter homology regions may be used with CRISPR/Cas9 (see below).

Homology regions are cloned on either side of a selectable marker cassette, including a drug-selection cassette for selection of stable transfection, and a GFP-expression cassette (if desired for germline testing). The targeting construct must be linearized, so the construct must include a unique restriction site outside of the homology regions.

4. CRISPR/Cas9 method

We have used CRISPR/Cas9 in combination with a donor vector, to increase the targeting frequency of the donor [10]. We design guide RNAs on the MIT server (http://crispr.mit.edu/). Guide RNAs are cloned into a single vector that expresses the gRNA from the U6 promoter, and expresses the wild-type Cas9 from the CMV promoter. To make our donor vector, we use homology regions (2 kb total) from either side of the Cas9 cut site, with a small gap of ~100 bp between the two homology regions; this 100 bp region is where the gRNAs direct Cas9 cutting. Larger deletions may be possible [13]. The homology regions are cloned with a drug-selectable marker between them.

In theory, CRISPR/Cas9 could be used in PGCs and gonocytes without drug selection, to induce small indels or in combination with non-selectable donor DNAs to introduce specific nucleotide changes. We have not pursued this approach. Because the modified PGC population would not be clonally derived, but rather a pool of potentially modified and un-modified cells, the frequency of the desired modification in the cell population would need to be high enough to make it feasible to screen for germline transmission of the mutation in progeny. The genetic modification could be combined with an unlinked GFP transgene for germline screening, if germline transmission rates are low in general (to identify progeny derived from the injected cells), followed by screening for the presence of the desired modification.

5. Other transgenic methods

TALENs have also been used to modify chicken PGCs and generate birds carrying the modifications [14, 15].

We have produced transgenic chickens expressing the Cre recombinase ubiquitously [16]. These birds were made by inserting a Cre transgene, driven by the β-actin promoter, into PGCs using phiC31 integrase as described above. These birds grow normally and are healthy, and the Cre catalyzes excision of sequences flanked by loxP sites ("floxed"). We routinely breed our chimeras to Cre-recombinase expressing hens to remove the selectable marker and GFP expression cassettes present in the transgenes. The Cre gene does not have to be inherited for the floxed region to be excised [16].

3.3.2 Transfection Protocol

We use 4–5 $\times$ 10^6 cells for one transfection. To obtain clonal cell lines, the cells are transfected by electroporation, plated with BRL cells (drug-resistant if necessary) in a 48-well plate, and selection started 4–5 days later. After 10–14 days, the wells are examined for surviving cells. If every well has drug-resistant cells, then there is a high probability that the cells arose from more than one colony, and the transfection should be repeated with less DNA to obtain fewer colonies. If only 1–20 wells contain resistant cells, the chance that these arose from single colonies is relatively high. For targeted insertions, the frequency of resistant colonies is low, and multiple transfections may be needed to obtain several colonies.

1. It is best to start transfections on a Monday, if possible, so that the cells can be monitored during the first week.
2. Fill as many cryotubes or microcentrifuge tubes as there will be cuvettes with 0.5 mL PGC medium and place them in the incubator.
3. Turn on the electroporator and set the required pulse.
4. Calculate volumes of buffer/DNA solution needed: 100 μL total volume, including DNA, per transfection.
5. Cells will be in multiple flasks; collect into 50 mL centrifuge tubes and count. Use 5×10^6 cells/transfection. If the same conditions are to be used, multiple aliquots can be pelleted together. Do not include any replated leftover cells when transfecting; the cell debris from these cultures can reduce the transfection efficiency. Spin 5–7 min at $300 \times g$.
6. Remove some of the supernatant with a pipet and transfer to another tube; this will be used when replating the transfected cells (referred to as "PGC-conditioned medium").
7. Use an aspirator fitted with a P1000 unfiltered tip to remove as much medium as possible from the pellet so as not to dilute the transfection buffer. It helps to hold the tube horizontally so the medium will collect where the conical section of the tube starts. Alternatively, the pellet can be transferred with 0.5–1 mL medium to a microcentrifuge tube, and pelleted again. The residual medium is easier to remove in a smaller tube.
8. Add approximately 50 μL of Nucleofector transfection buffer (*see* **Note 4**) for each 5×10^6 cells, resuspend cells, and measure volume. Add more transfection buffer to obtain a final volume of 100 μL/transfection, minus the volume of DNA to be added.
9. Add the DNA, mix using a P200 tip, transfer 100 μL to a 2 mm cuvette, and cap the cuvette.
10. Transfer the cuvette to the electroporator and pulse at 350 V, 8 pulses, 100 μs pulse length.
11. Use a transfer pipet to quickly transfer cells into 500 μL of pre-warmed medium. Place in incubator for 20 min to allow the cells to recover.
12. In the meantime, for each transfection, add the following to a 50 mL centrifuge tube; 8 mL of PGC-conditioned medium, 2.5 mL of irradiated BRL cells (drug-resistant if necessary, thawed and diluted as described in **item 12** of Subheading 2.2), and 13 mL of fresh PGC medium.
13. After 20 min add the transfected cells (600 μL) to the medium/BRL mixture, and add 0.5 mL of the suspension to each well of a 48-well plate. There will be 50 μL of BRL cells/well.

14. After 2 days remove 350 μL from each well and add 150 μL fresh PGC medium. The medium can be added with a repeat pipettor with sterile tips.
15. Start selection when the cells are growing well, generally 4–5 days after transfection. Add 300 μL of a 2× concentration of the selective agent in fresh PGC medium; puromycin 1 μg/mL (0.5 μg/mL final), G418 700 μg/mL (350 μg/mL final), hygromycin 80 μg/mL (40 μg/mL final), or blasticidin 4 μg/mL (2 μg/mL final). The final volume in the wells will be approximately 600 μL.
16. Change medium on the next culture day by removing 450 μL and adding 150 μL fresh PGC medium containing the selection reagent at 1×.
17. Continue to change the medium on each culture day (Mon-Wed-Fri), keeping approximately 300 μL in each well. Add BRL cells when necessary (usually once every 5–7 days); use approximately 1/2 normal seeding density. For example, when changing medium, add 1.3 mL BRL cells to 6.7 mL PGC medium (with selection reagent) and add 150 μL to each well.
18. Colonies will take approximately 2 weeks to become visible. If a GFP construct has been used, the plate can easily be screened on fluorescence for wells with GFP-positive growing cells.
19. Continue changing the medium on culture days until the cells have expanded, then pass them to a new 48 well. At this point, selection medium is not required, but continue using drug-resistant BRL cells for at least another passage, until the remaining selection agent has been diluted out. Expand cells as for derivation, collecting leftovers when necessary. When the cells reach the 24–12-well stage, a portion of the cell population should be collected for PCR confirmation of the modification.
20. Expand the positive clones until there are enough cells to inject and cryopreserve.
21. We generally inject four clonal cell lines for any given genetic modification. We will expand at least six PCR-confirmed colonies in case some clones grow too slowly to be useful.

3.4 Production of Germline Chimeras

3.4.1 Injection of Cells

1. Three days before injection, incubate fertile eggs (*see* **Note 16**). Before incubation be sure to write the weight of the egg on the shell (*see* **Note 6**).
2. On the day of injection: collect and count cells to be injected. Determine total volume necessary for injecting the embryos, and calculate cells needed for a density of 3000 cells/μL. Spin cells at 300 × *g*, and resuspend in appropriate volume of

Manipulation Medium (e.g., $4.5 \times 10^5/150\ \mu L$). Transfer to a 0.7 mL microcentrifuge tube for injection.

3. Select the surrogate shells and use a template (38 mm diameter) to draw a circle on the blunt side of the shell. This can be done the day before injection. Match the weight of the surrogate shell with the weight of the fertile eggs. The surrogate shell needs to be 30 g heavier (*see* **Notes 6** and **17**).
4. Follow Fig. 3 for the next steps. On the day of injection, using a Dremel tool, cut off the blunt side of the surrogate egg, cleanly and accurately, following the circle that was drawn. Remove any

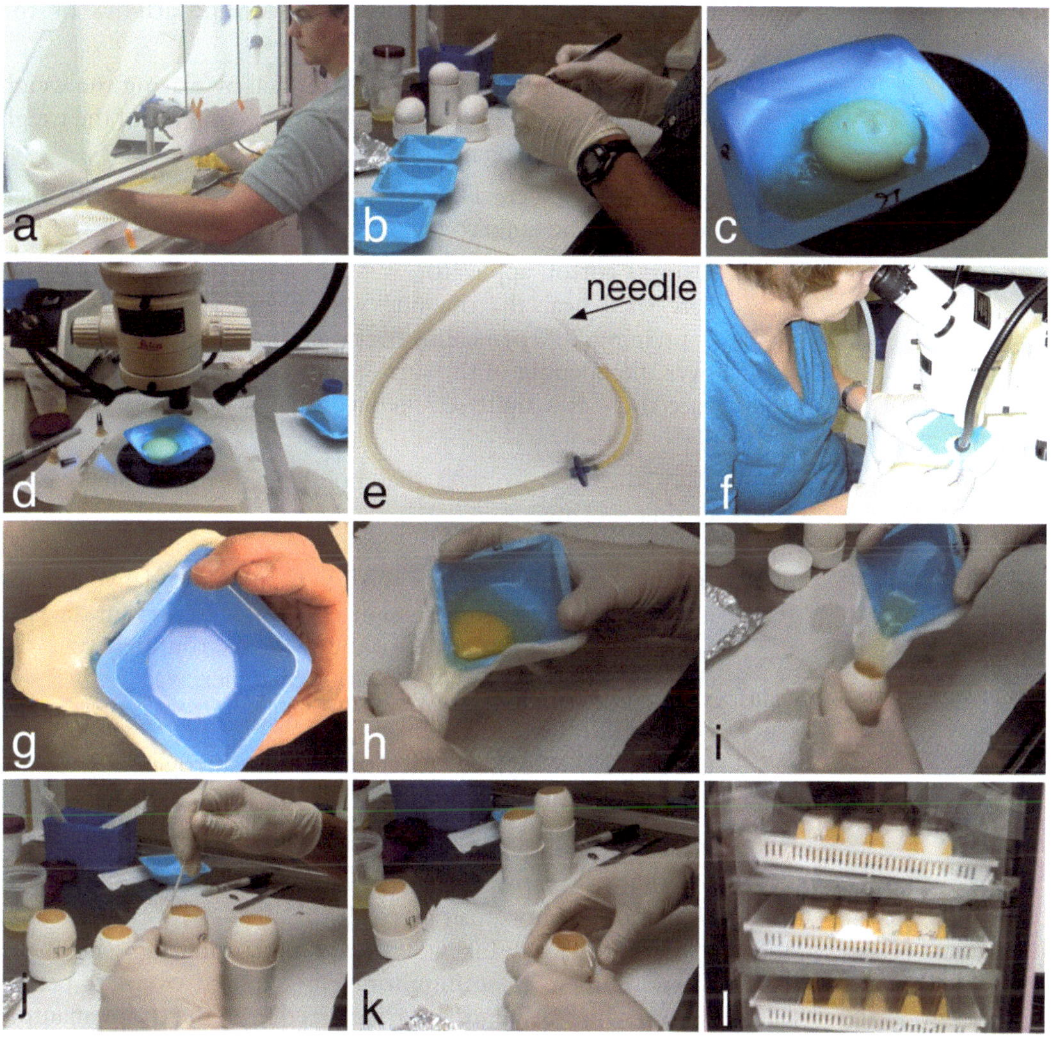

Fig. 3 (**a**) Cut surrogate eggs and score eggs. (**b**) Crack eggs with the help of a scalpel and pour into a weigh boat (**c**). (**d**) visualize embryo under the microscope. (**e**) Tubing and needle used to inject the cells (**f**). (**g**) Mold holding a weigh boat. (**h**, **i**) Pour embryo into surrogate shell. (**j**) Wet the edge of the shell with albumin (with penicillin/streptomycin) and fold the plastic square over the egg (**k**). (**l**) Incubate and rock eggs

surrogate shells that display cracks. Discard the contents and place the egg shell upside down on bench paper.

5. "Score" (make little indentations around the egg at its widest part) the incubated eggs (stage 14–15 H&H; *see* **Note 16**) with the Dremel tool. This will facilitate cracking the contents of the egg cleanly into a weigh boat. The yolk of a Stage 14–15 H&H embryo is fragile and breaks easily.
6. To crack a fertile egg, wiggle a scalpel blade into the indentations in the egg shell, dislodge the two halves of the egg shell, and deposit the contents of the egg into a weigh boat.
7. Inject 3000 germ cells into the vasculature. Cells can be injected into the dorsal aorta, the vitelline arteries or the marginal vein.
8. Transfer the embryo to a surrogate shell by placing the weigh boat in the casting mold and pouring the contents of the weigh boat into the surrogate shell.
9. Add 1 mL of Penicillin/Streptomycin to the surrogate shell containing the embryo.
10. Using a cotton applicator, wet the outer surface of the surrogate shell near the opening with a thin layer of egg white containing 20% penicillin/streptomycin. Place a plastic square over the opening of the surrogate shell, smoothing it to make sure there is a tight seal between the plastic square and the shell.
11. Place the eggs in racks that hold the surrogate shells tight during rocking.

3.4.2 Incubation of Injected Embryos

1. Incubate the injected eggs at 37.5 °C and 60% relative humidity with hourly rocking through a 60° angle to prevent the yolk from sticking to the inside of the shell. We routinely rock the eggs until they are transferred to the hatcher.
2. Evaluate embryos daily and remove dead embryos.
3. On days 17–19 of incubation transfer the surrogate eggs to the hatcher.
4. Maintain a temperature of 37.5 °C and 60% relative humidity.
5. Hatching in surrogate shells requires extra attention as the chimera does not push through the shell as an embryo does in the intact egg. Likewise, the chimera cannot push off the plastic film, so close monitoring is required over the period during which respiration is transferred from the chorioallantoic membrane (CAM) to the lungs and the chimera is ready to hatch (*see* Fig. 4a).

Fig. 4 (**a**) Chicks in surrogate shells at the time of hatch. Plastic covering has been removed from the three chicks at the left. The other embryos are still covered. (**b**) Chimeras being grown to sexual maturity. (**c**) Offspring from PGCs expressing GFP

3.4.3 Hatching of Surrogate Embryos

1. When the blood has begun to drain from the vessels in the CAM and the chimera has pipped into the air space and is breathing consistently, perforate the plastic film by making two small holes with a sterile hypodermic needle. At this time the air space has a high concentration of CO_2 and a low concentration of O_2. Ambient air will enter through the perforations and gradually equilibrate the air space with the surrounding air.
2. Once the head of the chimera is up and it is looking around, take off the plastic and lay the shell in a 4 × 4 × 4 carton box to allow the chimera to push itself free. This energetic process aids the final stages of yolk sac retraction and closing of the navel sphincter.

On average 50% of the injected embryos will hatch. To determine if a cell line transmits through the germline it is sufficient in our experience to evaluate 6–10 chimeras per cell line. Since we breed only with male chimeras, twice as many chimeras need to be hatched. We inject 30 embryos per cell line and hatch on average 15 (50%) male and female chicks. Be aware that chicks can be less robust after hatching out of a surrogate shell and might need more attention post-hatch.

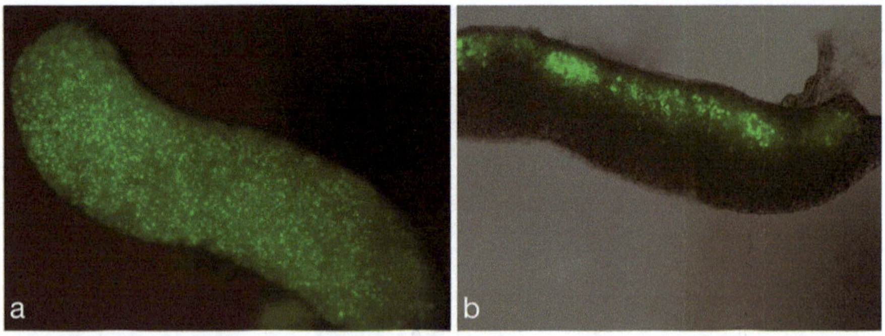

Fig. 5 (**a**) Testis colonized with GFP-positive PGCs. The PGCs are dispersed throughout the seminiferous cords. (**b**) Ovary colonized with GFP-positive PGCs. The PGCs are only present in the ovarian cortex

3.4.4 Germline Testing

All injected cells contain a β-actin-GFP construct. This feature allows us to visualize the cells after injection into the recipient embryo. Since we focus our breeding on male chimeras we routinely euthanize the female chicks after hatch and evaluate their gonads for colonization. Male PGCs/gonocytes will colonize the female gonad and this should be evaluated within a few days after hatch. If there is poor or no colonization, the chance for germline transmission through the male chimeras is very low and euthanization of the chimeras should be considered.

1. Collect the combs of all putative chimeras at hatch and determine sex as described in Subheading 3.2.2.
2. Euthanize female chicks and, using an inverted fluorescent microscope, evaluate the gonads for GFP expression as depicted in Fig. 5 (*see* **Note 18**).
3. Grow male chimeras to sexual maturity (Fig. 4b), collect sperm and evaluate germline transmission by performing test matings (*see* **Note 19**).
4. Inseminate several hens with semen of a putative chimera.
5. Collect eggs for 1 week and incubate the eggs for 7 days.
6. Open the eggs and evaluate the embryos for GFP expression with the fluorescent head lamp. It is not possible to observe green fluorescence through the intact eggshell so it is necessary to open the eggs.

4 Notes

1. Any fluorescent head lamp will be ok. We like the one that can be obtained from BLS, Budapest, Hungary (www.bls-ltd.com).
2. It is not necessary to use the highest quality fetal bovine serum that is produced, such as that produced specifically for mouse

ES cell culture. We have had good results with Hyclone Characterized serum. Chicken serum is less standardized compared to fetal bovine serum and there can be major batch differences. Once PGC/gonocyte cultures are established, serum manufacturers and lots can be tested for optimal growth.

3. KO-DMEM and CO_2-independent media are proprietary media made by Life Technologies. KO-DMEM has a lower osmolarity than standard DMEM, which is optimal for PGC growth; we do not know of any other commercially available medium that can be used as a substitute. CO_2-independent medium is used when a CO_2 atmosphere cannot be maintained. Other CO_2-independent media may be available, but we have not tested them.
4. The type of electroporation buffer used has a major effect on viability of the cells and this should always be evaluated. In our hands the Amaxa V-buffer has consistently performed well during germ cell transfections.
5. CellSTACKs (Corning, #3269) are convenient for large-scale cell culture because of the large surface area (1272 cm^2). As far as we are aware, there are no equivalent cell culture chambers with such a large surface area. Multiple T150 flasks can be used instead, but will be less efficient. Using a cellSTACK: Each cellSTACK has two caps. Use the same cap for all medium additions and collections; tighten the other cap and keep closed. Add growth medium and cells to *bottom of stack*, and mix back and forth gently to achieve a uniform cell suspension. To divide the medium equally over both levels of the cellSTACK, set the cell stack vertically on its long side OPPOSITE to the side you added the medium; allow liquid to equilibrate between the two levels. Rotate to the short end opposite the caps, then bring the cellSTACK to a horizontal position; the media should now evenly cover both layers. To remove medium, tilt flask at approximately 30° toward the open cap; the liquid from the top layer will flow to the bottom and the total volume can be collected with a pipet. Make sure when washing with PBS and trypsin that the volume is equilibrated over both layers.
6. When using surrogate shells, a large opening, or window, must be made in the surrogate shell to permit transfer of the recipient embryo. The window is cut at the blunt end of the egg, removing the natural air space. An artificial air space is generated above the embryo after transfer. Cutting the window reduces the volume of the surrogate shell. In order to accommodate the recipient embryo, the surrogate shell must not only be large enough to provide an artificial air space into which the embryo will pip and begin lung ventilation prior to hatching, it

also has to be large enough to prevent albumin touching the plastic window during rocking of the eggs. Consequently, the egg supplying this shell needs to be 30 g heavier than the fertile egg. For example, if the fertile recipient egg is 55 g, it is matched with a surrogate egg that is 85 g.

7. BRL cells are not the only feeder cells that can be used; STO cells also support the derivation and growth of PGCs/gonocytes. However, PGCs and gonocytes are much easier to remove from the BRL feeder layer than from the STO feeder layer.
8. A 330 mL bottle of conditioned medium can be thawed and re-distributed to smaller aliquots if smaller batches of PGC medium are preferred. Further freezing and thawing is not recommended.
9. BRL cells may be able to be transfected using chemically mediated methods, but we have not tried this.
10. To generate universal drug-resistant BRL cells, we used one construct to obtain G418-resistant BRLs, then later transfected these cells with two constructs, one expressing the hygromycin-resistance gene and one the blasticidin-resistance gene, to obtain BRL cells that could be used for all possible selections. All three constructs could probably be used at one time, but we have not done this ourselves.
11. The method to derive PGCs from germinal crescent is very similar to the one described above, except that the tissue is transferred to a tube with PGC medium, triturated, and transferred to a 48-well plate; trypsinization is not necessary [3]. When blood is taken to start a PGC culture, red blood cells block the observation of the PGCs in the first week of culture [2]. However, since there are very few somatic cells present, the cells can stay in the same well with just medium changes for the first 7–10 days.
12. If the gonadal cultures are started on a Tuesday they can be passed on Wednesday and Friday, which will reduce a lot of the adherent cell contamination before the 3-day culture over the weekend.
13. In 48-and 24-well plates, the PGCs and gonocytes will be loosely stuck to the BRL layer and may be somewhat clumpy, and need to be gently washed off. It is difficult to get all the cells, which is why we add medium to the original well (now called the leftover well) and continue to monitor the well. Often the remaining cells will grow and detach enough so that by the next culture day there will be a good number of cells that can be combined with the main culture. Tightly attached PGCs and gonocytes will have a tendency to differentiate, and should be left behind. As the culture expands up to

6-well plates, the cells usually become less adherent and clumpy. Once the cells are growing routinely in flasks, just collect the cells in suspension, or at most gently tap the flask if there are cells loosely attached. The strongly attached cells should be left behind in the leftover flask.

14. Often when moving from 6-well plates to T25 or T75 flasks PGC/gonocyte growth slows down. In wells, the cells tend to stay in the center and are therefore at a higher cell density than the actual plating density. With larger surface areas the cells are more spread out, so the cells experience an apparent drop in cell density. We sometimes use a slightly higher cell density when first plating in flasks to compensate.
15. If cell growth seems slow when culturing in flasks, you can save some of the supernatant from the main culture after centrifugation and use up to 25–30% of this "PGC-conditioned medium" when replating. We generally use 25–30% PGC-conditioned medium on leftover flasks to save on medium.
16. According to the publications of Hamburger and Hamilton (1952) [17], fertile eggs reach Stage 14 (H&H) after 50–53 h of incubation. This has not been our experience; we need to incubate eggs 60–64 h before they reach Stage 14. Not only is the pace of embryo development dependent on the conditions of the incubator, the percentage of embryos that hatch can be even more affected. Smaller incubators often have large temperature differentials between the top and bottom of the incubator. To monitor the incubation parameters it is helpful to use a remote temperature and monitor system that can be accessed in real time and that sends daily reports.
17. Although more labor intensive, we routinely culture the injected embryos in surrogate shells due to the unimpeded visibility allowing accurate PGC injection and subsequent observation of embryo development. Embryos can be evaluated easily, and viability and progress of development can be recorded daily. Alternatively, the embryo can be injected through a small hole in the pointed side of the egg, which is then re-sealed [3].
18. The morphological evaluation of colonization in the testis and ovary is different (*see* Fig. 5). In the male, PGCs disperse among the seminiferous tubules that are present throughout the testis. In the ovary, the PGCs will be restricted to the ovarian cortex. This means that when evaluating colonization in the ovaries, care has to be taken that the whole ovary is scanned. It is possible for the cortex to be completely hidden behind the medulla, in which case no GFP-positive cells will be seen. Colonization of ovaries has to be evaluated within a few days after hatch since the formation of primordial follicles

makes it difficult to evaluate GFP expression in the ovary a week after hatch. In the testis, GFP-positive germ cells will remain visible until sexual maturity. In general, GFP expression can be very bright and thus lead to an overestimation of colonization.

19. Semen can be evaluated for GFP-expressing sperm by flow cytometry, or by quantitative PCR for the presence of the transgene, to give some idea of which chimeras might be the best transmitters. However, a test mating still needs to be done to demonstrate germline transmission.

References

1. van de Lavoir M-C, Mather-Love C, Leighton P et al (2006) High-grade transgenic somatic chimeras from chicken embryonic stem cells. Mech Dev 123:31–41. https://doi.org/10.1016/j.mod.2005.10.002
2. van de Lavoir M-C, Diamond JH, Leighton PA et al (2006) Germline transmission of genetically modified primordial germ cells. Nature 441:766–769. https://doi.org/10.1038/nature04831
3. Song Y, Duraisamy S, Ali J et al (2014) Characteristics of long-term cultures of avian primordial germ cells and gonocytes. Biol Reprod 90:15. https://doi.org/10.1095/biolreprod.113.113381
4. Fujimoto T, Ukeshima A, Kiyofuji R (1976) The origin, migration and morphology of the primordial germ cells in the chick embryo. Anat Rec 185:139–145. https://doi.org/10.1002/ar.1091850203
5. Nakamura Y, Kagami H, Tagami T (2013) Development, differentiation and manipulation of chicken germ cells. Develop Growth Differ 55:20–40. https://doi.org/10.1111/dgd.12026
6. Collarini E, Leighton P, Pedersen D et al (2014) Inserting random and site-specific changes into the genome of chickens. Poult Sci. https://doi.org/10.3382/ps.2014-04372
7. Liu C, Chang I-K, Khazanehdari KA et al (2017) Uniparental chicken offsprings derived from oogenesis of chicken primordial germ cells (ZZ). Biol Reprod 96:686–693. https://doi.org/10.1095/biolreprod.116.144253
8. Itoh Y, Suzuki M, Ogawa A et al (2001) Identification of the sex of a wide range of Carinatae birds by PCR using primer sets selected from chicken EE0.6 and its related sequences. J Hered 92:315
9. Leighton PA, van de Lavoir M-C, Diamond JH et al (2008) Genetic modification of primordial germ cells by gene trapping, gene targeting, and ϕC31 integrase. Mol Reprod Dev 75:1163–1175. https://doi.org/10.1002/mrd.20859
10. Dimitrov L, Pedersen D, Ching KH et al (2016) Germline gene editing in chickens by efficient CRISPR-mediated homologous recombination in primordial germ cells. PLoS One 11:e0154303. https://doi.org/10.1371/journal.pone.0154303
11. Sanders TA, Llagostera E, Barna M (2013) Specialized filopodia direct long-range transport of SHH during vertebrate tissue patterning. Nature 497:628–632. https://doi.org/10.1038/nature12157
12. Schusser B, Collarini EJ, Pedersen D et al (2016) Expression of heavy chain-only antibodies can support B-cell development in light chain knockout chickens. Eur J Immunol. https://doi.org/10.1002/eji.201546171
13. Wang L, Yang L, Guo Y et al (2017) Enhancing targeted genomic DNA editing in chicken cells using the CRISPR/Cas9 system. PLoS One 12:e0169768. https://doi.org/10.1371/journal.pone.0169768
14. Taylor L, Carlson DF, Nandi S et al (2017) Efficient TALEN-mediated gene targeting of chicken primordial germ cells. Development 144:928–934. https://doi.org/10.1242/dev.145367
15. Park TS, Lee HJ, Kim KH et al (2014) Targeted gene knockout in chickens mediated by TALENs. Proc Natl Acad Sci U S A 111:12716–12721. https://doi.org/10.1073/pnas.1410555111
16. Leighton PA, Pedersen D, Ching K et al (2016) Generation of chickens expressing Cre recombinase. Transgenic Res. https://doi.org/10.1007/s11248-016-9952-6
17. Hamburger, V., & Hamilton, H. L. (1952). A series of normal stages in the development of the chick embryo. 1951. Developmental Dynamics, 195: 231–272.

Chapter 25

Using Microinjection to Generate Genetically Modified *Caenorhabditis elegans* by CRISPR/Cas9 Editing

Jyoti Iyer, Nicole DeVaul, Tyler Hansen, and Ben Nebenfuehr

Abstract

In this chapter, we describe the procedure for generating genetically modified *Caenorhabditis elegans* using microinjection via the Cas9-mediated Clustered Regularly Interspaced Short Palindromic Repeats (CRISPR) editing technique. Specifically, we describe the detailed method of performing CRISPR editing by microinjection using the Cloning-free Co-CRISPR method described by the Seydoux lab. This microinjection protocol can also be used for CRISPR editing with protocols from other labs as well as for a variety of other editing techniques including *Mos1*-mediated single-copy transgene insertions (MosSCI), transcriptional activator-like nucleases (TALENs), and zinc-finger nucleases (ZFNs). Further, this microinjection protocol can also be used for injecting plasmid DNA to generate heritable extrachromosomal arrays for gene expression and mosaic analysis, performing RNAi by injection and delivering RNA, dyes or other molecules into the *C. elegans* germline.

Key words Microinjection, CRISPR, Cas9, *C. elegans*, Gene editing, CRISPR screening

1 Introduction

The soil-dwelling nematode *Caenorhabditis elegans* has characteristics that make it an attractive model system for investigating various biological processes [1–4]. It is transparent, has a short developmental time of ~3 days to develop from an embryo to a fertile gravid adult, and is easy to cultivate in the laboratory [5]. Further, *C. elegans* is a multicellular eukaryotic organism that is amenable to both forward and reverse genetics [1, 6, 7]. Its entire genome has been sequenced (*C. elegans* Sequencing Consortium 1998) and the fate of each of its cells has been thoroughly mapped and characterized [8–11]. The *C. elegans* distal gonad is a syncytium [12] and is amenable to microinjection with several different kinds of molecules including RNA, DNA, nanoparticles, etc. [6, 13–16].

Nicole DeVaul and Tyler Hansen contributed equally to this chapter.

Chengyu Liu and Yubin Du (eds.), *Microinjection: Methods and Protocols*, Methods in Molecular Biology, vol. 1874,
https://doi.org/10.1007/978-1-4939-8831-0_25,

One of the most important applications of performing *C. elegans* microinjections is its utility to edit the worm genome. Microinjected molecules such as DNA may be incorporated into the germ cells of an injected animal and consequently, generate a heritable change in the progeny of the injected animals [14, 17, 18]. Cas9-mediated Clustered Regularly Interspaced Short Palindromic Repeats (CRISPR) editing and *Mos1*-mediated single-copy transgene insertions (MosSCI) are two of the most popular current *C. elegans* genome editing techniques. The CRISPR technique facilitates precise editing at any endogenous locus within the genome [19]. In contrast, MosSCI enables insertion of a single extra copy of a gene or a DNA fragment at specific locations within the genome [20]. The success of both these techniques relies heavily on the ability of the investigator to microinject accurately into the correct location (syncytial gonad of the worm) and with minimum damage to the injected worm. Since this is the most technically challenging aspect of these editing techniques, it is critical to master the technique of microinjection to be able to successfully edit the *C. elegans* genome. This chapter will describe in detail how to perform microinjection in *C. elegans* and will also provide helpful tips at each step to improve the success of the microinjection procedure. As one important application of microinjection, this book chapter will focus on utilizing CRISPR/Cas9 to edit the *C. elegans* genome.

The establishment of the CRISPR/Cas9 system has revolutionized gene editing in all organisms including *C. elegans* [19, 21]. This is a powerful technique that enables the generation of precise mutations at any location in the genome. The CRISPR/Cas9 system relies on the ability of the bacterial Cas9 enzyme to recognize NGG protospacer-adjacent motifs (PAM motifs) within the genome [22–24]. The Cas9 enzyme in association with a CRISPR RNA (crRNA) containing a 20–25 base-pair specific guide sequence (with homology to the target of interest) and a *trans*-activating RNA (tracrRNA) generates a double-stranded DNA break 3 base-pairs upstream of a PAM motif within the genome [24–26]. This double-stranded break can be repaired by either homology directed repair (HDR) or by non-homologous end joining (NHEJ). Scientists can synthesize crRNAs with different guide sequences and use these to target the Cas9 enzyme to the specific region(s) of the genome that they wish to edit. With continued improvements in the CRISPR technique, performing CRISPR editing in *C. elegans* has become increasingly easier. There are several different established protocols for performing CRISPR editing in *C. elegans* (Reviewed in [27–29]). The advantages and disadvantages of some of the commonly used CRISPR protocols for *C. elegans* are summarized in Table 1 [30–34]. In this

Table 1
Cloning-Free Co-CRISPR should be considered for all basic CRISPR edits due to easy preparation and screening

	Common CRISPR methods compared			
Method	Cloning-Free Co-CRISPR, *Short range HDR*	Self-Excising Cassette, *Long Range HDR*	SapTrap, *Long Range HDR*	Dual Marker Cassettes, *Long Range HDR*
Use for	Point mutation(s), deletions, small insertions (GFP, 3xFLAG, HA, etc.)	Null mutations, large insertions, transcriptional reporters	Insertions and tissue specific promoters	Gene knockouts/ deletions
Pros	High efficiency, clean edits (silent mutation of the PAM site), short homology arms, no cloning, on rare occasions homozygously edited worms are directly obtained upon injection, can be used to easily insert short epitope tags (1–2 kb) using a single crRNA and make relatively large gene deletions (up to 5 kb) by using two crRNAs	Simple and easy selection, SEC plasmids available, reduced PCR screening, can be used to screen for rare edits, can obtain a loss-of-function mutant, GFP- and 3xFlag-tagged gene and transcriptional reporter in a single experiment	Kit with 26 unique modular components, short homology arms, tissue specific tags, long range edits	Easy visual screening, generate gene knockouts and tag simultaneously
Cons	Limited insert size and limited distance from cut site (edit site must be less than 10 bp from the cut site for highest efficiency)	Cloning required, leaves a single LoxP site, requires heat shock, SEC spontaneously excises in some cells even in the absence of heat shock, low efficiency	Must inject *unc* animals, assembly of repair template is simplified yet cumbersome, secondary Cre injection required, outcrossing required, variable efficiency	Low efficiency, knockouts only
Maximum insertion size	1–2 kb	Up to 12 kb	<10 kb	n/a
Cloning?	N	Y	Y	Y
Maximum efficiency[a]	>50%[b]	0–12.5%	0–36.9%	5–10%

(continued)

Table 1 (continued)

Common CRISPR methods compared				
Screening approach	Co-Conversion (*dpy-10, pha-1, sqt-1*)	Hygromycin resistance and *sqt-1* Roller phenotype	*unc-119* rescue	G418 resistance and F1 fluorescence patterning (dependent on injected plasmids)
Preparation time	<1 h	1–2 days	1–2 days	1–2 days
Screening time	6–8 h	1.5 h	1st screen: 2 h; second injection: 1 h; second screen: 2 h	1–2 h
Waiting time	3–5 days	~2.5 weeks	4 weeks	3–5 days
Cost	Relatively expensive. ~$350 for a typical CRISPR experiment using a single crRNA, short oligonucleotide repair template and home-made Cas9 protein	Low cost	Low cost (Kit costs $350)	Low cost
Kit Available?	N	N	Y	Y
Relevant papers	Arribere et al. [35] *Genetics;* Paix et al. [30] *Genetics;* Kim et al. [36] *Genetics;* Ward et al. [37] *Genetics;* Paix et al. [31] *NAR*	Dickinson et al. [32] *Genetics;* Dickinson and Goldstein, [27] *Genetics*	Schwartz and Jorgensen, [33] *Genetics*	Norris, Kim et al. [34] *Genetics*

Self-Excising Cassette, SapTrap, and Dual Marker Cassette approaches are more involved with a cloning step but allow for larger insertions (SEC and SapTrap) or easy gene knockouts (Dual Marker Cassettes)

[a]Efficiency is the frequency of Cas9-directed modification per number of successfully injected worms

[b]Efficiency dependent on distance between edit and cut site

chapter, we will describe in detail the method for utilizing microinjection to perform CRISPR editing using the Cloning-free Co-CRISPR protocol by Paix et al. [30, 31]. Notably, this method takes less than 2 weeks to obtain edited homozygotes.

2 Materials

2.1 Microinjection

1. 100, 60, and 35 mm diameter MYOB or NGM plates seeded with OP50 bacteria-MYOB plates: Dissolve 1.1 g of Tris-Cl, 0.48 g Tris-OH, 40 g Bacto-peptone, 4 g NaCl, 16 mg cholesterol in 2 L of water and add 60 g Bacto-agar. Autoclave and pour into petri plates.
2. Synchronized young adult worms of the desired strain (with ten or less embryos in the uterus).
3. Worm pick: Made of platinum wire (28 gauge 99.9% platinum wire) fused to a broken glass Pasteur pipette or mounted on a metal holder. To make a worm pick fused to a glass Pasteur pipette, partially break the tip of a glass Pasteur pipette. Insert a ~3 cm piece of the platinum wire into the broken tip of the Pasteur pipette such that ~0.5–1 cm of the wire is inside the tip of the Pasteur pipette. Heat the glass tip enclosing the platinum wire over the flame of a Bunsen burner. Upon heating, the glass will fuse to seal the Pasteur pipette tip leaving the platinum wire attached to the Pasteur pipette. The tip of the platinum wire may be flattened using Chain nose pliers or a hammer and bent slightly to allow for efficient worm transfer.
4. Glass Coverslips: Dimensions = 48 × 60 mm, Thickness = 0.13–0.17 mm.
5. 2% agarose solution: 0.2 g agarose in 10 mL of water. Make the 2% agarose solution in water by heating in a microwave oven.
6. Halocarbon 700 oil or Heavy Paraffin Oil.
7. Borosilicate glass capillary tubes for injection needles: 100 mm in length with a 1.2 mm outer diameter and a 0.68 mm inner diameter.
8. Refrigerated table-top centrifuge.
9. 20 μL microloader pipette tips.
10. Needle loaders (optional): The center of a glass capillary is heated over a Bunsen burner. Following this, the two ends of the capillary are immediately pulled and snapped apart, thereby generating two pieces of the capillary each with a long and thin extension. These can be used to load the injection mix into the needle. For long-term storage, needle loaders can be stored by mounting vertically (with the tip at the top) on a piece of modeling clay or in a microfuge tube rack (refer to [35] for more details).
11. Modeling clay (optional) for storing pulled needles and/or needle loaders.
12. Micropipette storage jar (optional) for storing pulled needles.

13. M9 buffer: 3 g KH_2PO_4, 6 g Na_2HPO_4 (or 11.3 g $Na_2HPO_4 \cdot 7H_2O$), 5 g NaCl, 1 mL 1 M $MgSO_4$ to 1 L with distilled water. Note: Mix all *but* $MgSO_4$. Autoclave and then add $MgSO_4$ under sterile conditions.
14. Recovery buffer (optional): 0.1% salmon sperm DNA, 4% Glucose, 2.4 mM KCl, 66 mM NaCl, 3 mM $CaCl_2$, 3 mM HEPES, pH 7.2 [18, 36].

2.2 Microinjection Setup

The injection setup can be tailored according to the needs of the laboratory. There are several microscopes, injectors, micromanipulators, and needle pullers that are commercially available to generate a functional injection setup. A single-injection setup can be shared between multiple labs. The injection setup for our laboratory is depicted in Fig. 1.

1. Needle puller (*see* **Note 1**).
2. Dissecting microscope to aid in picking and transferring worms to and from injection pads.
3. Inverted microscope with DIC/Normaski optics. The microscope should have a low power 5× or 10× objective to locate the worm and 40× air objective to visualize the gonad while performing the injection.
4. Pressurized injection system: Less expensive injectors are manual in nature and usually involve a pressure source such as a nitrogen tank, which is connected to the needle. These may or may not involve a foot pedal for injection and/or a monitor for visualizing the injection (*see* **Note 2**).
5. Micromanipulator: Micromanipulators enable fine control of the needle motion in *X*, *Y*, and *Z* directions (*see* **Note 3**).
6. Recovery apparatus/Mouth pipette system/Aspirator: Used to transfer injected worms to fresh plates. Materials used for construction: Pipette tip, 15″ aspirator tube assembly (*see* **Note 4**), needle after pulling with its tip broken slightly. *See* Fig. 3 for a visual description.

2.3 CRISPR/Cas9 Editing (Cloning-Free Co-CRISPR Protocol by Paix et al. [30, 31])

1. *S. pyogenes* Cas9 protein: Can be commercially purchased or purified in the laboratory according to the protocol from Paix et al. [30].
2. crRNAs: Lyophilized crRNAs are reconstituted in 5 mM Tris-HCl pH 7.5 to a concentration of 8 μg/μL, aliquoted and stored at −80 °C until use.
3. Single-stranded oligonucleotide (ssODN) repair templates: 1 μg/μL (up to 2.2 μg total in a final volume of 20 μL); PCR repair templates: 500 ng/μL (up to 10 μg total in a final volume of 20 μL).

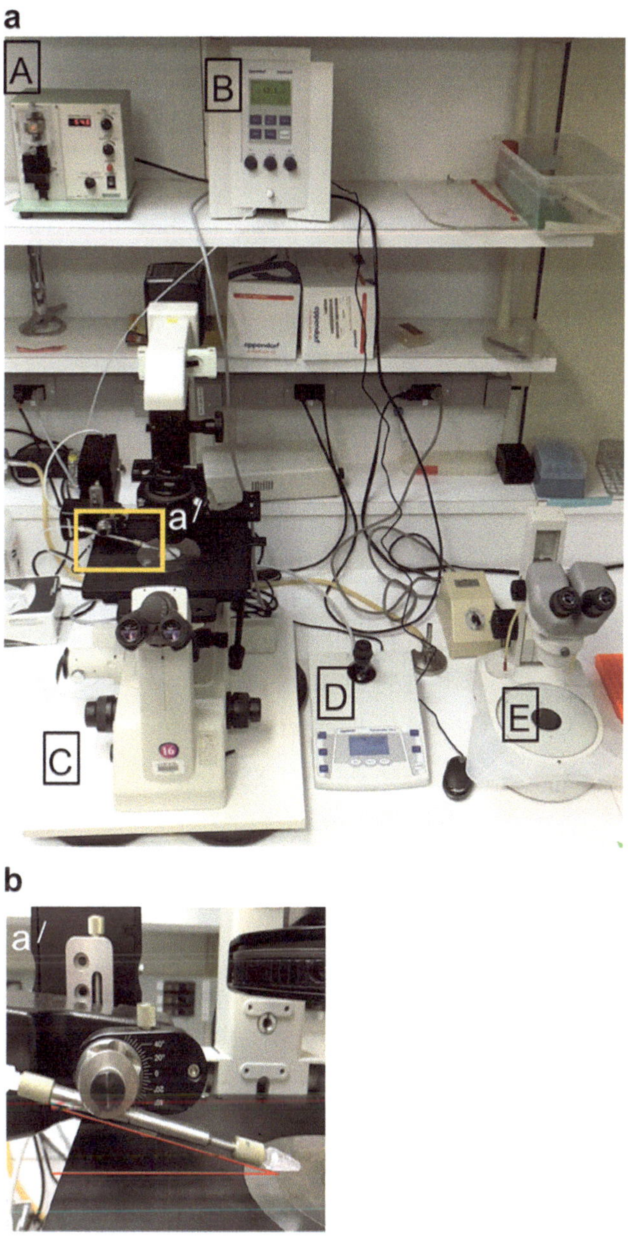

Fig. 1 (**a** and **b**) Injection equipment setup. The injection setup is depicted with labeled equipment (A) needle puller, (B) Injector, (C) Inverted microscope with Normaski optics (Injection microscope): (a′) A blow-up of region from (C) showing the needle holder is aligned to a 15°–45° angle, (D) Micromanipulator keypad, and (E) Dissecting scope

4. tracrRNA: Lyophilized tracrRNA is reconstituted in 5 mM Tris-HCl pH 7.5 to a concentration of 4 μg/μL, aliquoted and stored at −80 °C until use.
5. 200 mM HEPES pH 7.5.
6. 1 M KCl.
7. 5 mM Tris-HCl pH 7.5.
8. Nuclease-free water.
9. Refrigerated table-top centrifuge.
10. −80 °C freezer.
11. Worm lysis buffer: 5 mL 1 M KCl, 1 mL 1 M Tris–HCl pH 8.3, 0.25 mL 1 M $MgCl_2$, 0.45 mL NP-40 (or IGEPAL CA-630), 0.45 mL Tween-20, 0.5 mL 2% Gelatin to 100 mL with distilled water. Autoclave or filter sterilize.
12. Proteinase K (Recombinant, PCR grade, ~20 mg/mL).
13. Lysis buffer + Proteinase K: Add 1 μL of ~20 mg/mL Proteinase K to 99 μL of lysis buffer.
14. Thermocycler.
15. PCR reagents.
16. Restriction enzymes (optional).
17. Agarose gel reagents and apparatus.

3 Methods

3.1 Preparing Worms for Injection

Ensure that you have worms of the desired genotype and developmental stage ready on the day of the injection. Injecting young adult worms with up to ten embryos inside the uterus yields the most favorable results. Synchronize the worms so that they are at the young adult stage on the day of the injection procedure:

1. Wash out a 60 mm starved MYOB plate containing mostly L1 larvae with 1 mL of M9 buffer.
2. Using a 1 mL pipette, dispense about five drops of this solution onto a 10 cm MYOB plate.
3. Incubate the 10 cm MYOB plate for the appropriate time and temperature so that the worms are young adults on the day of injection.
4. Alternatively, pick L4 worms on the day before the injection and incubate them at 16 °C overnight.

3.2 Making Agarose Injection Pads

1. Arrange 48 × 60 mm coverslips horizontally in a few rows on a benchtop.

2. Near each arranged coverslip, place another coverslip of the same dimensions adjacent to it in a perpendicular orientation.
3. Use a glass Pasteur pipette attached to a rubber bulb to draw up molten 2% agarose solution and dispense one drop (~100 μL) of it in the center of the 48 × 60 mm coverslip.
4. Immediately place the perpendicularly placed coverslip on the top of the dispensed drop (*see* **Note 5**).
5. Repeat this procedure for all the remaining coverslips.
6. After >5 s, carefully slide the top coverslip off the bottom coverslip.
7. Determine which coverslip the agarose pad has stuck to by blowing on the two coverslips that were in contact with the drop of agarose.
8. Mark the outline of the area of the agar pad using a permanent marker for easy visualization (*see* **Note 6**).
9. The coverslips can either be left out on the benchtop overnight at room temperature (our lab does this) or they can be placed in a box lined with aluminum foil and baked in an oven at 80 °C for 1 h or longer [36, 37].
10. Place the agarose injection pads in the box that the 48 × 60 mm coverslips came in. The agarose injection pads can be stored in this box for several months at room temperature (*see* **Note 7**). Each coverslip may be used for performing multiple worm injections.

3.3 Preparing the Injection Mix

Make the injection mix that you desire to inject with according to the purpose of your experiment (containing DNA/RNA/nanoparticles, etc.). For CRISPR editing, the injection mix needs to be made differently depending on the type of CRISPR editing protocol that is employed. Refer to Table 1 to determine which CRISPR editing protocol best suits the requirements of your experiment.

Our lab utilizes the Cloning-free Co-CRISPR genome editing protocol from the Seydoux lab where purified Cas9 protein in complex with the crRNA(s) is directly delivered with the appropriate repair template(s) into the *C. elegans* gonad by microinjection. This protocol employs the co-conversion strategy in which the probability of obtaining successfully edited worms is increased by screening for simultaneous repair at the *dpy-10* locus. Worms that have been edited successfully at the *dpy-10* locus will present either a roller or dumpy (Dpy) phenotype.

Using this protocol, genetically modified worms can be generated by CRISPR editing in a short period of time (<2 weeks to obtain homozygously edited worms), the protocol is cloning-free, and is easy to perform. In spite of some shortcomings as described in Table 1, this is an efficient protocol to use for adding short epitope and fluorescent tags to genes, making point mutations, short insertions (<2 kb), gene deletions (<6 kb), and gene replacements. Here, we describe the procedure for making the injection mix for the Cloning-free Co-CRISPR genome editing protocol according to Paix et al. [30, 31].

The following protocol for making injection mixes for Cloning-free Co-CRISPR has been adopted from the publication from the Seydoux lab [30].

Below is the injection mix recipe for editing using the Cloning-free Co-CRISPR protocol. This protocol utilizes Cas9 protein that is purified in the laboratory (*see* **Note 8**):

Cas9 prep (10 μg/μL): 5 μL.

1 M KCl: 0.5 μL.

200 mM HEPES pH 7.4 (or pH 7.3): 0.75 μL.

dpy-10 crRNA (8 μg/μL): 0.4 μL.

Targeted gene crRNA (8 μg/μL): 1 μL (for editing a single locus and *dpy-10*) or 0.75 μL crRNA#1 + 0.75 μL crRNA#2 (for editing two loci simultaneously and *dpy-10*).

tracrRNA (4 μg/μL): 5 μL (for editing a single locus and *dpy-10*) or 6.7 μL (for editing two loci simultaneously and *dpy-10*).

dpy-10 ssODN (500 ng/μL): 0.55 μL.

ssODN (several ssODN can be mixed) (1 μg/μL): 2.2 μL total.

OR PCR template(s) (several can be mixed): Up to 500 ng/μL final in the mix.

H_2O: add if necessary to reach a final volume of 20 μL.

1. Make the injection mix in the following order: Cas9 protein, KCl, HEPES pH 7.5, crRNAs, tracrRNA, ssODNs, H_2O, and finally, PCR fragments if used. If a PCR fragment is used as a repair template, its concentration must be close to 1 μg/μL (*see* **Note 9**). A highly pure and concentrated PCR template increases editing efficiency.
2. Spin the injection mix at 13,000 rpm for 2–5 min prior to performing the injection (*see* **Note 10**). If the injection mix is not going to be used immediately, it should be spun and stored at −80 °C until further use. The stored injection mix may be used 1–2 additional times for subsequent injections on different days. However, repeated thawing and freezing beyond this point in our experience decreases editing efficiency (*see* **Note 11**).
3. Keep the injection mix on ice while the microinjection procedure is being performed.

3.4 Preparing Needles and Microinjection Setup

3.4.1 Pulling Needles for Injection

Needles are generated from borosilicate glass capillary tubes. Most needle pullers consist of a filament which when heated to a specific temperature results in pulling of the needle. The settings for each needle puller must be optimized to yield a high-quality needle. These settings may depend upon factors such as the temperature and the humidity of the room in which the needle puller is housed as well as the age of the needle puller filament. A good needle will taper quickly and smoothly to a sharp open tip [36, 37] and will dispense a single drop of the injection mix (~1–10 pL [38]) when the injection pressure is applied. Upon applying the injection pressure, a good needle should dispense enough injection mix to fill the worm gonad (whose maximum capacity is estimated to be between 1 and 50 pL [17, 38]) with minimum physical damage to the worm. Once the appropriate settings are identified, they can be maintained constant for pulling all subsequent needles.

A needle with a broken tip can be used for picking up the injected worms and transferring them to fresh MYOB plates for recovery (Fig. 3). The pulled needles can either be used immediately for performing microinjection or stored in a jar or in a petri dish with clay for a few days until needed. Be very careful while handling the pulled needles as their tips are fragile and will break upon even slight contact.

Some microinjection protocols recommend breaking the tip of the needle prior to using it. However, we do not find this necessary with the settings on our Narishige PC-10 needle puller. To break the needle tip, gently graze the tip of the needle against the side of a glass capillary or on a coverslip shard while visualizing this under 40× magnification on the injection microscope [35].

3.4.2 Loading Needles

1. Use 20 μL microloader pipette tips (or needle loaders—*see* Subheading 2.1 for procedure to make needle loaders) to withdraw 1.5–2 μL of the injection mix from the top of the injection mix solution (*see* **Note 12**).
2. Insert the microloader pipette tip through the back open end of the injection needle and push it down all the way into the bottom of the needle.
3. Dispense the injection mix into the needle. Avoid the formation of bubbles in the injection mix (*see* **Note 13**).

3.4.3 Assembling Needles on the Needle Holder and Preparing the Setup for Injection

The angle of the needle is critical for a good injection. The settings for the needle holder and the injector will differ depending upon which injection setup is used. In general, the needle holder is fixed at a 15°–45° angle with respect to the injection stage (Fig. 1ca′). There are three important pressure parameters that need to be taken into consideration while adjusting the settings of the injector or the pressure source: (1) the injection pressure (pressure used to inject), (2) the cleaning pressure (higher pressure used to flush out the needle), and (3) the compensation pressure (constant pressure used to hold the injection mix inside the needle). The injection pressure usually ranges between 5 and 40 psi (*see* **Note 14**). For our Eppendorf Femtojet Injection setup, we set the injection pressure to 18.1 psi and the compensation pressure to 4.35 psi. The pressure settings on each injection setup must be optimized such that ~1–10 pL of the injection mix are dispensed from the needle when the injection pressure is applied [38]. Test each needle to ensure that it dispenses the right amount of injection mix when the injection pressure is applied prior to mounting worms on the injection pad. To assemble the needle on the needle holder and get it ready for injecting, perform the following steps:

1. Carefully insert the back end of the loaded needle into the needle holder (*see* **Note 15**).
2. On the injection microscope (Fig. 1c), change to a 10× objective and set to brightfield.
3. Using the micromanipulator, move the needle in the *X*, *Y*, and *Z* directions to bring the tip of the needle into the center of the bright light.
4. Once this position is located, switch to the 40× objective.
5. Subtly move the needle in the *X*-*Y* directions to bring the tip of the needle into the center of the 40× field.
6. After completing this step, refrain from moving the needle in the *X*-*Y* direction using the joystick. Only move the needle in the *Z*-direction.
7. Move the needle upward in the *Z*-direction so that it is away from the injection microscope stage.
8. Place an agarose pad with a drop of halocarbon 700 oil on the stage of the injection microscope.
9. Focus the drop of oil first under 10× and then under 40× magnification.
10. Under the 40× magnification, bring the needle down slowly in the *Z*-direction so that its tip just penetrates the drop of oil on the agarose pad. Once the needle is closer to the drop of oil, only use the fine adjustment to move the needle.
11. Dispense the injection mix through the needle by applying the injection pressure to ensure that the needle functions as expected.
12. If the needle is not able to dispense the injection mix, flush the needle using the cleaning pressure to flush out debris that may be clogging the needle. Alternatively, gently graze the needle against a glass capillary or a glass coverslip to break its tip slightly under the 40× magnification (*see* **Note 16**).

13. If this still does not resolve the problem, the needle might need to be changed as it may be clogged. This is a common and one of the most frequently encountered problems while performing *C. elegans* microinjections (*see* **Note 17**).
14. Repeat **steps 1–12** until a good needle for injection is obtained. Once this is done, the needle is ready for performing injections. A single loaded needed may be used for performing multiple worm injections.

3.5 Mounting Worms on Agarose Injection Pads

After fixing the needle in the correct position, start mounting worms on the agarose injection pad. The worm mounting procedure is performed on a dissecting scope located adjacent to the main injection microscope (Fig. 1E). Perform the following steps to mount worms on the injection pads and prepare them for injection:

1. Place the agarose injection pad on an inverted lid of a 60 mm petri dish with the agarose pad facing upward.
2. Put a drop of halocarbon 700 oil on the part of the glass coverslip that is outside the agarose pad.
3. Dip a worm pick in the oil and gently tap it on the agarose pad to dispense one tiny droplet of oil on the agarose pad.
4. Dip the worm pick in the oil again and use this oil-coated pick to pick up one to five healthy young adult worms. Beginners should start with injecting one worm at a time.
5. Ensure that worms are picked from a bacteria-free region of the plate (*see* **Note 18**).
6. Using the pick, transfer the worms to the oil droplet on the agarose pad and allow the worms to adhere to the pad. If the worms do not adhere on their own, gentle pressure may be applied using the pick to make the worms stick. Do not apply too much pressure on the worms, otherwise they will rupture.
7. Ensure that the vulvas of the mounted worms are located on the same side so that the coverslip need not be moved considerably between injections.
8. Pay attention to the dorsal side of the worms so that the injection pads are placed on the microscope with the dorsal side of the worms facing the needle.
9. Once the worms are mounted, quickly inject them within 5 min; otherwise, the agarose pad will desiccate the worms [18, 36].
10. After mounting the worms under the dissection microscope (Fig. 1E), quickly place the agarose pad with the mounted worms on the stage of the injection microscope (Fig. 1C).

11. Focus the worms under the 10× objective and lower the injection needle in the *Z*-direction until the needle is just above the drop of oil.
12. Switch to the 40× objective and use the micromanipulator to lower the needle gently in the *Z*-direction so that the tip of the needle just penetrates the oil drop.

3.6 Microinjection

Inject the distal gonad so that upon injection, the injected material fills the entire U-shaped germline (Fig. 2a). Injections are performed with the needle at about a 45° angle with respect to the body wall of the worm (Fig. 2b–d). Many labs inject both arms of the worm gonad. Although this increases the success rate of obtaining edited transgenic worms, it also increases the frequency of worm death after the injection. Therefore, we recommend beginners to start with injecting a single-gonad arm. Locating the gonad and injecting in the right place takes some practice and experience. The following is a link to a helpful youtube video demonstrating how to perform injections in *C. elegans*: https://www.youtube.com/watch?v=DLMvbJqlz4c (courtesy of Dr. Amy Fabritius). Perform the following steps for microinjection:

1. Locate the distal gonad which is situated across from the oocytes on the dorsal side of the worm, as shown in Fig. 2a.
2. Focus the distal gonad so that only the outer and innermost rows of circular germ cells are in focus (Fig. 2b–d, f). Do not inject when the focused germline appears as in Fig. 2g. This is the wrong plane to inject.
3. Ensure that the tip of the needle is focused in the same plane as the gonad of the worm to be injected by subtly moving the needle in the *Z*-direction.
4. Move the stage to move the distal gonad of the worm toward the needle (Fig. 2b) (*see* **Note 19**).
5. Move the stage further to prick the needle into the distal gonad (Fig. 2c) and insert it into the syncytial space within the gonad as shown in Fig. 2d–f. If you are finding it difficult to penetrate the worm cuticle, gently tap the microscope stage to aid in piercing of the cuticle.
6. Once the needle is inside the rachis (the central canal that connects the germ cells) (Fig. 2d–f), apply the injection pressure to inject. If the injection is successful, the injection mix will fill the entire U-shaped gonad. This will cause the gonad to swell and result in the appearance of cobblestone-like nuclei in the syncytial gonad (*see* **Note 20**).
7. After injecting, retract the needle carefully by moving the stage.
8. (Optional step) Once outside the injected worm, flush the needle by applying the cleaning pressure and dispense the injection mix on the injected worm. This ensures that the

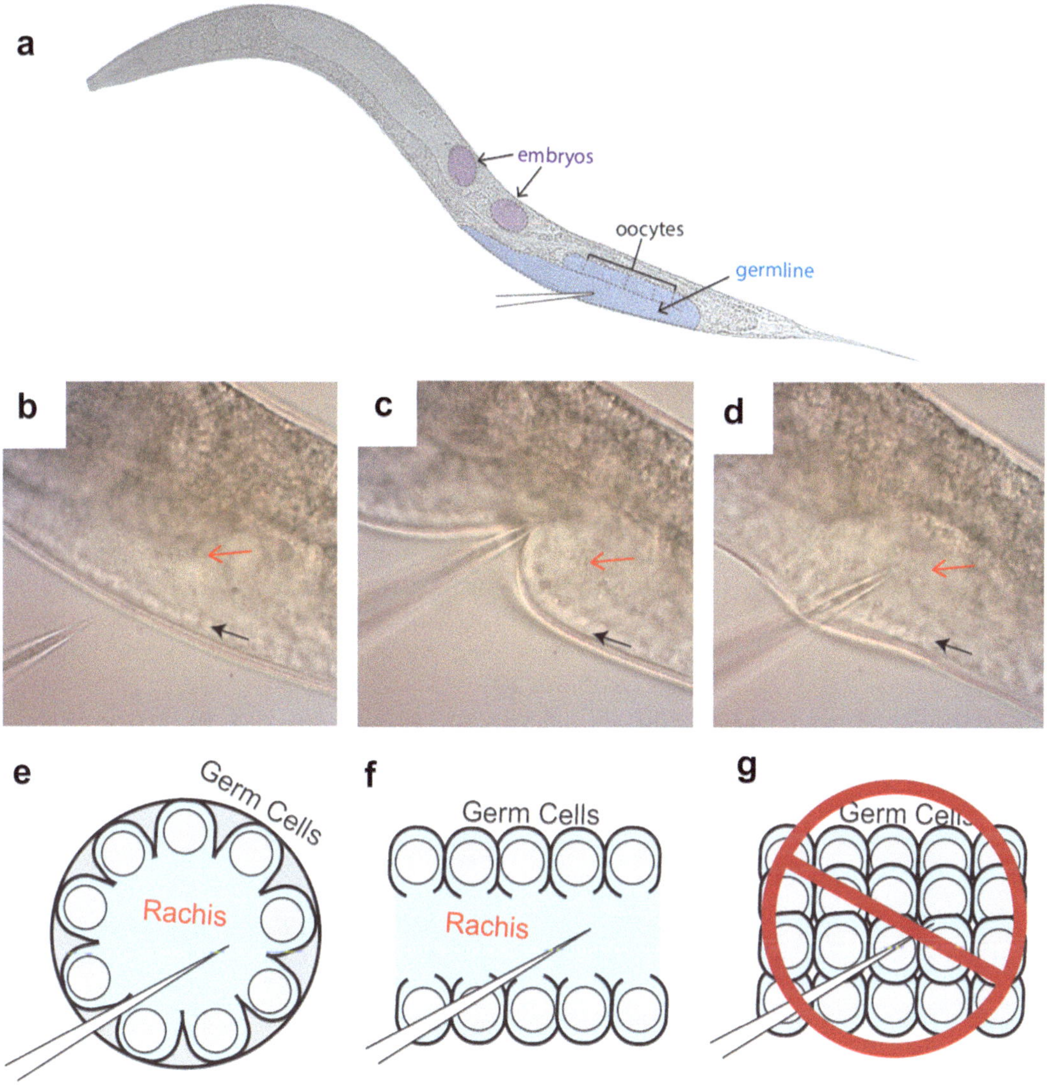

Fig. 2 Injection of the *C. elegans* gonad. (**a**) A *C. elegans* hermaphrodite and injection needle positioned against the site of injection. Position the needle on the side with the distal gonad at about a 45° angle with respect to the body wall. (**b** and **c**) Three-step injection procedure. Red arrows indicate the rachis of the germline; Black arrows indicate germ cells. The first step (**b**) is to position the tip of the needle at the correct focal plane and next to the germline. The second step (**c**) is to push the needle tip through the cuticle and into the gonad. The final step (**d**) is to inject the mix into the gonad, causing it to swell. A good injection will fill the entire germline (not depicted). (**e**) A cartoon cross-section of the *C. elegans* gonad. The syncytial gonad contains many germ cells that share a common cytoplasm called the rachis. It is important to inject directly into the rachis. (**e**–**g**) Cartoon views of the *C. elegans* gonad. (**f**) Depicts the microscopic view of the middle of the gonad. This is the correct focal plane for injection. (**g**) Depicts the microscopic view of the top or bottom of the gonad. Do not inject at this focal plane since the rachis is inaccessible. You may be outside the gonad at this point

injected worm is coated with a thin layer of the injection mix and prevents its desiccation (*see* **Note 21**).

9. Move to injecting the next worm.

3.7 Recovering Worms

1. Following microinjection, quickly remove the agarose pad from the stage of the injection microscope (Fig. 1C).
2. Place the agarose pad on the cover of a 60 mm petri dish with the agarose pad facing upward and place this assembly on the stage of the dissecting microscope (Fig. 1E).
3. Locate the injected worms on the agarose pad by viewing them through the dissecting microscope.
4. Use the recovery apparatus/mouth pipette system/aspirator (Fig. 3) to withdraw M9 buffer from an eppendorf tube using capillary action.
5. Place the capillary tube filled with M9 buffer above the oil drop with the injected worms and blow gently into the recovery apparatus through the pipette tip to cover the oil drop containing the injected worms with M9 buffer. After a few seconds, the worms will release from the agarose pad and float into the M9 buffer (*see* **Note 22**).
6. Gently mouth pipette the worms floating in the M9 buffer into the glass capillary tube of the recovery apparatus and release the

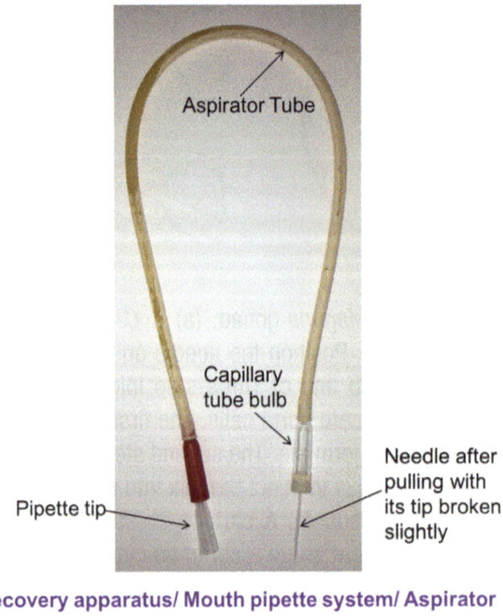

Fig. 3 Components of the recovery apparatus/mouth pipette system/Aspirator. The materials used for making the recovery apparatus and their arrangement are depicted. M9 is drawn into the needle tip by capillary action. Light pressure through the pipette tip is used to expel the liquid out of the needle tip onto the worms

worms on the bacterial lawn of a fresh MYOB plate (*see* **Note 23**).

7. Let the worms recover on the MYOB plates at room temperature for about 1 h. Once the worms recover, they will be able to move freely on the plates. A good sign that the worm was not harmed is that it will start thrashing at this point.
8. Transfer each recovered injected worm to a new 35 mm MYOB plate using a worm pick.
9. Place the plates in an incubator set at the optimal temperature for the growth of the injected strain.

   If the worms are severely desiccated, using a recovery buffer improves survival of the injected worms [18, 36]. Briefly, in this method, the worms are suspended in recovery buffer and the osmotic strength is gradually reduced by adding M9 buffer in a dropwise manner over a course of an hour (at a rate of 20 drops of M9 per hour). The worms are then picked and transferred onto fresh plates [18, 36].

3.8 Screening for Edits

If microinjection was performed for genome editing purposes, depending on which technique was used for genome editing, successfully edited worms can be screened using a variety of methods. These include acquisition of a particular visible phenotype, fluorescence, drug resistance, PCR, or PCR followed by restriction digestion. For CRISPR screening using the Cloning-free Co-CRISPR protocol, please refer to Subheading 3.8.1.

3.8.1 Screening Procedure for the Cloning-Free Co-CRISPR Protocol

Figure 4 briefly outlines the screening procedure for the Cloning-free Co-CRISPR protocol [30, 31]. For information on screening worms for some common CRISPR editing methods from other labs, please refer to the respective papers for these methods listed in Table 1. The screening method for the Cloning-free Co-CRISPR protocol will vary depending upon the type of repair template that has been used for making the desired edit. For fluorescent tag fusions to abundant proteins, screening by fluorescence microscopy may be sufficient. In all other cases, PCR is used to screen for positively edited worms.

A diagnostic PCR could be used for screening edited worms when a fairly large gene deletion or insertion is being made using CRISPR. The deletion or insertion must be large enough to be able to distinguish between wild-type and mutant bands obtained after PCR upon agarose gel electrophoresis. Alternatively, a diagnostic PCR could also be used for screening successfully edited worms in a scenario when only a mutant worm but not a wild-type worm would be expected to produce a band of the appropriate size upon PCR amplification. For example, a diagnostic PCR could be used for screening worms in which GFP has been fused to an endogenous gene using CRISPR. PCR can be performed using one primer which anneals to the endogenous gene locus and

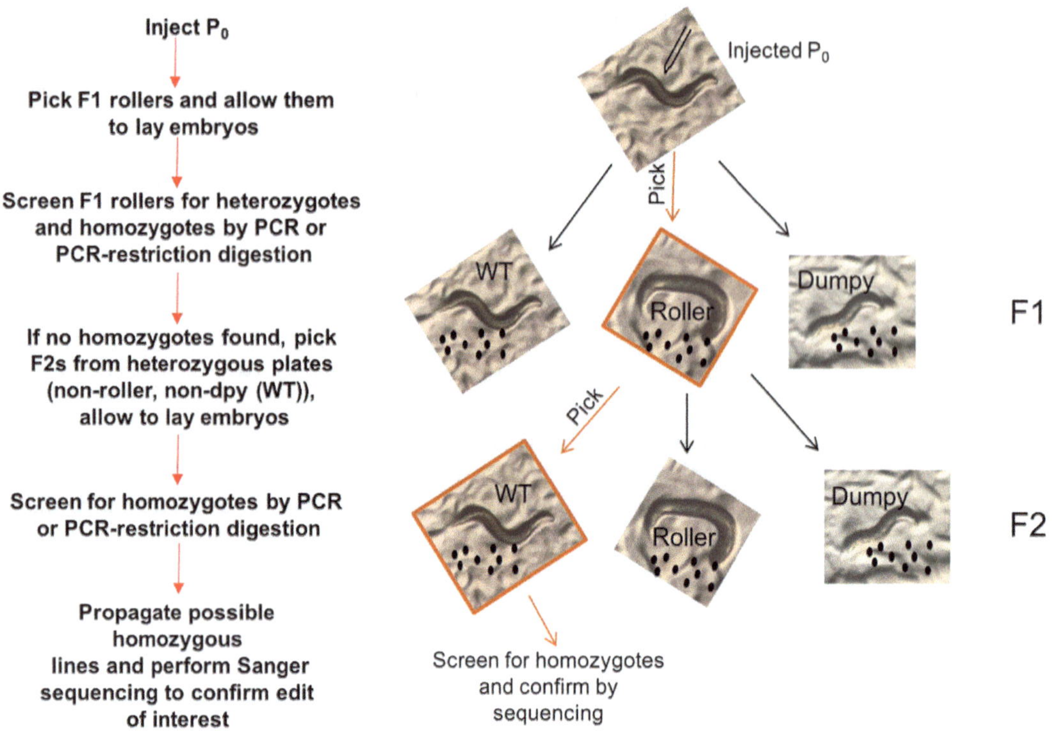

Fig. 4 Procedure for screening injected animals for the Cloning-free Co-CRISPR method. Left panel: Flowchart of the screening procedure. Right panel: A visual guide for the screening process

another primer that anneals to GFP. In such a case, only successfully edited worms will produce a band of the appropriate size upon PCR amplification (wild-type worms will not produce any band) (*see* **Note 24**).

When a smaller edit is to be made using CRISPR editing such as a point mutation or a short insertion or deletion, a restriction site is introduced or abrogated by silent mutation of the repair template to facilitate screening. In such a case, candidate F1 roller and/or Dpy worms are subjected to PCR to amplify a short band (~500 bp) spanning the edit. This PCR product is then digested with a restriction enzyme and run on an agarose gel. Lanes that show digested bands of expected sizes are selected for further analysis. Figure 5 shows a cartoon representation of the experimental scheme and possible outcomes for introducing a silent mutation by Cloning-free Co-CRISPR.

The step-by-step screening procedure for screening using the Cloning-free Co-CRISPR protocol is described below:

1. Let the injected worms (P_0) lay eggs for 24 h and then transfer them to fresh 35 mm MYOB plates.
2. Monitor the plates after L4s and young adults begin to appear on the plates.

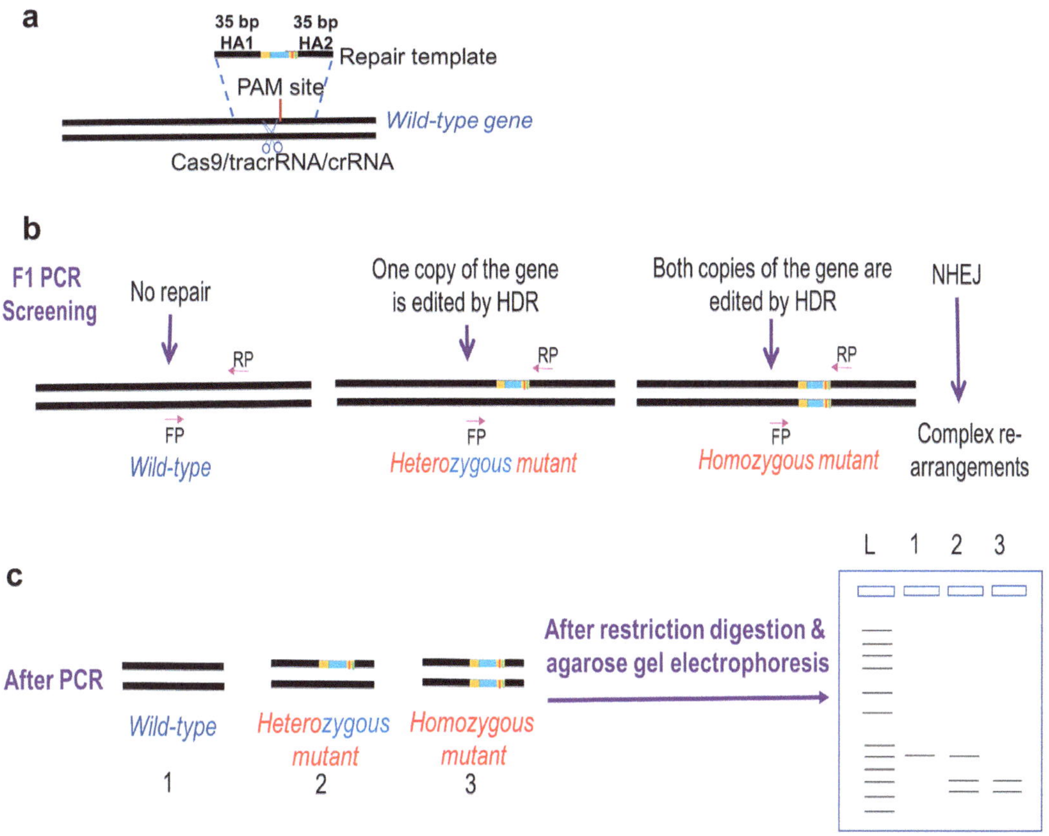

Fig. 5 Cartoon explaining the process of introducing silent mutations by the Cloning-free Co-CRISPR method. (**a**) In the presence of the Cas9-crRNA-tracrRNA complex, Cas9 introduces a double-stranded break 3 base pairs upstream of the PAM site. A generic repair template for making point mutations by Cloning-free Co-CRISPR is presented. The repair template for introducing point mutations consists of the following components: (1) homology arms (HA1 and HA2), (2) region of the repair template that does not have homology to the target gene (yellow bar), restriction site introduced by silent mutation (blue bar), desired point mutation to be introduced into the target gene (red line), silent mutation of the PAM site (green line). (**b**) Four possible genotypes that may be seen in the F1 generation during PCR screening after microinjecting with CRISPR editing reagents: (1) No repair occurs. This results in the target gene remaining unedited (wild-type) in the F1 animals. (2) Only one strand of the target gene of F1 animals is edited by Homology-Directed Repair (HDR) resulting in the F1 animals being heterozygotes. This scenario is commonly observed. (3) HDR occurs on both strands of the target gene of the F1 animals yielding homozygotes. Although rare, we have observed this scenario occurring in a few instances. (4) Non-homologous end joining (NHEJ) may also occur resulting in complex re-arrangements of the target gene. *FP* forward primer, *RP* reverse primer. (**c**) Cartoon representing three genotypes that may arise upon either no repair or upon HDR and an example of how each of these genotypes may appear on an agarose gel after PCR and restriction digestion. L = DNA ladder

3. Identify plates that have roller and Dpy worms on them (*see* **Notes 25** and **26**).
4. Identify the top 3–5 plates that have the greatest number of roller worms [jackpot broods] [39] and screen these further for the presence of your edit-of-interest.

5. Single out 30–50 candidate F1 roller and dumpy worms that are suspected to carry the edit-of-interest onto fresh 35 mm MYOB plates.
6. Let these worms lay eggs overnight.
7. Once there are sufficient eggs on the plate and some hatched F2 L1 larvae appear, perform single-worm lysis for each of the F1 mothers:

   Pick the single F1 mother from each plate into a PCR tube on ice containing 2.5 μL of lysis buffer + Proteinase K. Try to avoid picking bacteria as doing so might prevent efficient lysis and PCR (*see* **Note 27**). Transfer the PCR tubes to dry ice for 10–15 min. Although not mandatory, you may overlay the lysis reaction with 15 μL of mineral oil to prevent its evaporation. Perform the lysis protocol in a thermocycler with the following conditions: 60 °C for 1 h, 95 °C for 15 min and hold at 4 °C. The samples (worms in lysis buffer) can be stored either before or after lysis at −20 °C or −80 °C until further analysis.
8. After the lysis procedure is completed, the screening can be performed either by setting up a diagnostic PCR or by setting up a PCR reaction followed by a restriction enzyme digestion.
9. For setting up PCRs with the single-worm lysis reaction:

   Add the PCR reagents and primers to each entire single worm lysis reaction (*see* **Note 28**). Spin each PCR tube containing these components briefly (at maximum speed for about 10 s) to ensure that the PCR reagents mix with the single worm lysis reaction and the mineral oil floats to the top. The mineral oil is inert and will not interfere with any of the downstream screening steps. Perform the PCR reaction in a thermocycler.
10. Perform restriction enzyme digestion if necessary.
11. Perform agarose gel electrophoresis for PCR/PCR-digestion reactions to identify edited homozygotes and heterozygotes. Most of the identified positives that are obtained upon screening will likely be heterozygotes. However, we have also observed homozygously edited F1 worms on a few occasions.
12. If only edited heterozygotes are obtained, perform the following steps to identify homozygously edited worms:

    Pick 8–12 wild-type looking non-dumpy-non-roller F2 progeny from each of the plates where the F1 mother was found to be positive for the edit (*see* **Note 29**). Repeat **steps 6** through **11** to identify homozygotes.
13. After identifying homozygously edited animals, verify the desired edit by Sanger sequencing.

We have included a troubleshooting guide (Table 2) that discusses some of the commonly encountered problems while performing CRISPR editing in *C. elegans* using the Cloning-free

Table 2
Troubleshooting table for CRISPR editing using the Cloning-free Co-CRISPR protocol

Problem	Possible reasons and solution(s)
Only Dpy worms appeared after CRISPR but no rollers obtained	If the *dpy-10* repair template has been frozen and thawed repeatedly, it gets degraded and does not work properly. If this happens, the *dpy-10* crRNA still works resulting in dsDNA breaks. However, because the *dpy-10* repair template is degraded, the homology directed repair does not work. In such a case, non-homologous end joining causes the appearance of dumpy worms but no rollers are obtained. Get a fresh aliquot of the *dpy-10* repair template and try injecting again. You could try screening the Dpy worms for your edit-of-interest first before attempting re-injecting.
No roller or dumpy worms obtained	Several possibilities: (1) If the injector is inexperienced, it is possible that the mix was not injected in the correct place. Try injecting more worms to increase the chance of the injection mix being delivered to the right place. (2) Get a fresh aliquot of the *dpy-10* crRNA and try re-injecting. The *dpy-10* crRNA is a robust co-CRISPR marker. Therefore, one would expect to at least see a few Dpy worms appear if an injection is successful. (3) It is also possible that the Cas9 protein was not functional. Test your Cas9 protein in an in vitro reaction with a short PCR product spanning your region of interest in the presence of the crRNA and tracrRNA. If the Cas9 protein is able to cleave the PCR product, this indicates that the Cas9 protein was functional.
Heterozygously edited roller worms were obtained but none of their progeny were homozygotes	(1) If the gene-of-interest is an essential gene, it might not be possible to obtain homozygotes since interfering with the function of both copies of the gene could be lethal. This may result in the F2 progeny all being wild-type and heterozygotes. (2) Check to ensure that your gene-of interest is not linked to *dpy-10*. If this is the case, you will not be able to obtain non-Dpy-non-roller homozygotes and almost all of the non-Dpy-non-roller F2 progeny will exhibit a wild-type genotype.
While tagging a gene with an epitope tag, homozygotes were obtained but they exhibited significant embryonic lethality	In our experience, many genes do not tolerate fusions with large epitope tags (such as GFP) very well. Tagging some genes with these tags may interfere with their function, resulting in embryonic lethality or sterility in the homozygous state. If this happens, try switching the position of the tag. For instance, if you initially tagged the N-terminus of the gene, now try tagging the

(continued)

Table 2 (continued)

Problem	Possible reasons and solution(s)
	C-terminus instead. Alternatively, try tagging the gene-of-interest with smaller tags such as HA, His, Ollas or 3xFlag. Even in the case of gene fusions to smaller epitope tags, some tags are tolerated better than others. This is unique to each edited gene.
The editing efficiency at the gene-of-interest was very low	You obtained a lot of rollers and Dpys but still obtained a low editing efficiency at your gene-of-interest. (1) It is possible that the crRNA targeted against the gene-of-interest did not function optimally. In our experience, certain crRNAs work very well and yield a high editing efficiency. However, some crRNAs are not very efficient and yield a low editing efficiency. Ensure that your crRNAs have a high GC content (>50%). (2) Design a new crRNA to the region-of-interest and try injecting again. The direct delivery CRISPR method requires that the edit site be very close to the cut site otherwise the editing efficiency drops dramatically. Check to make sure that the edit site is less than 10 base pairs away from the Cas9 cut site. (3) Ensure that the PAM motif within the supplied repair template was designed to carry a silent mutation otherwise, the repair template will be cut by the Cas9 enzyme and HDR will not occur effectively.

Co-CRISPR protocol and have also suggested potential solutions to these problems.

4 Notes

1. Needle pullers may be purchased from Narishige, Sutter Instrument Company, Warner Instruments, or others. Alternatively, needles may be commercially purchased from companies such as Tritech Research (~$54 for ten needles).
2. Injectors may be purchased from companies like Eppendorf, Narishige, and Tritech Research. Tritech research sells a simple preassembled analog microinjector system (MINJ-1) for ~ $650. Some more expensive injection setups involve a self-sufficient injector with a built-in pressure controller that

enables a precise and reproducible dispensing of the injection mix into the worms.

3. Micromanipulators are sold by several companies including Eppendorf, Narishige, Zeiss, and Leica.
4. The 15″ aspirator tube assembly components may be purchased from Drummond Scientific Company (Catalog # 2-000-000) or Sigma Aldrich (Catalog # A5177-5EA).
5. This needs to be done quickly, otherwise the 2% agarose drop will solidify.
6. The coverslip to which the agarose pad did not stick can be saved and used as the bottom coverslip for the next time.
7. Test an agarose pad from each new batch first with a control worm to ensure that the pad functions as expected. If the mounted worm does not stick to the agarose pad, try making a thicker pad by increasing the agarose concentration. On the other hand, if the mounted worm desiccates quickly when mounted on the agarose pad, try making a thinner pad by slightly decreasing the agarose concentration. You could also re-bake the pad prior to use to make the worms stick better.
8. The Cas9 protein used in this protocol [30] can be purified in the laboratory using the method described in the above publication.
9. Good sterile technique must be utilized while making the injection mix such as using fresh gloves and RNase-free water and filter tips. We recommend using the Qiagen minelute PCR kit for this process. However, any such alternative kit(s) may be used.
10. The direct delivery protocol recommends incubating the injection mix at 37 °C for 10–15 min after spinning to assemble the Cas9-ribonucleoprotein complexes. However, we omit this step and find that this does not affect the editing efficiency significantly.
11. Alternatively, the injection mix may also be stored at 4 °C for 1–2 weeks before use (Andy Golden, personal communication).
12. If the injection mix is drawn from a region near the bottom of the injection mix, it will cause the needle to clog and the injection will not work.
13. If bubbles do form, hold the loaded needle vertically and gently tap the side of the needle with your fingers until the bubbles disappear. Do not tap the needle too hard, otherwise the needle will break.
14. Very high injection pressures (>80 psi) can cause the needle to shoot out of the needle holder and rupture gaskets [36].

15. Ensure that the tip of the needle does not touch anything while mounting the needle because the tip of the needle is very fragile and can break easily upon contact.
16. Note that if the entire tip of the needle is broken, the needle will not penetrate the worm body correctly and will need to be replaced. If the bore of the needle is too wide, it will injure the worm and may cause it to rupture.
17. Spin the injection mix for a longer time (~30 min) at ~13,000 rpm in a table-top centrifuge if every new needle that is pulled keeps getting clogged.
18. Bacteria can clog the injection needle and prevent the worms from adhering efficiently to the agarose injection pad. Alternatively, to prevent the carry-over of bacteria, you could transfer several worms to an MYOB plate without bacteria prior to picking the worms.
19. We always move the stage to inject. We do not move the needle. However, this step is highly dependent on the type of injection setup that is used. Some injection setups require movement of the needle to perform the injection. Check your injection setup to determine if you should move the stage or the needle to inject.
20. If the volume of the injection mix in a successful injection is high, it forces the expulsion of embryos from the uterus through the vulva. Thus, the expulsion of embryos upon injection can also be used to gauge if an injection was successful. Note that it is not necessary to inject such a high amount of injection mix for the injection to be successful.
21. If you are having difficulty with dispensing the injection mix on the worm, ensure that the needle is not clogged. To do this, move the stage to remove the needle from the worm body and flush it by applying the cleaning pressure. If the needle does not dispense any injection mix, the needle might need to be replaced. To do this, repeat procedures from Subheadings 3.4.1–3.4.3.
22. If the worms are still stuck to the agarose pad, gently blow on the top of the worms through the recovery apparatus to aid in their release.
23. Sometimes, a few of the worms stick inside the recovery apparatus. If this is the case, suck up fresh M9 buffer into the capillary tube and blow into the recovery apparatus to try and dispense the worms on to the MYOB plate. Some worms will not survive the injection due to physical damage by the injection and hence will not recover.
24. Note that in this scenario, the original primer set that was used for the initial screening process (in which one primer anneals

within the gene and the other primer anneals to GFP) cannot be used for identifying homozygotes. For detecting homozygotes, design new primers such that the two primers now flank either side of the GFP insertion. In this case, there will be a size difference between the wild-type and GFP-inserted bands that can be easily detected after agarose gel electrophoresis.

25. These plates are selected because the P_0 mother on these plates must be germline-edited at the *dpy-10* locus for roller and Dpy worms to appear. This suggests that this P_0 worm must also have a high probability of being simultaneously edited at your region-of-interest.
26. Although we prefer to screen roller worms for edits, we have observed on several occasions that Dpy F1s that segregate all dumpy F2s have a good chance of being edited at our gene-of-interest. However, one must be cautious with screening Dpy worms and ensure that all their progeny also exhibit the Dpy phenotype. The reason is that sometimes, F1 worms that are only edited in the soma and not the germline can also present a Dpy phenotype. In such a case, the progeny of these F1 animals will all have a wild-type phenotype. We do not analyze such F1 Dpy animals that segregate only wild-type F2s.
27. We usually move the worm to a region outside of the bacterial lawn, let it leave behind the bacteria as it crawls away and then scoop the worm up using a clean worm pick.
28. We use the MyTaq RedMix from Bioline (Catalog # BIO-25043) to perform PCRs for Cloning-free Co-CRISPR screening. Since this already has premixed PCR reagents as well as a loading dye added, it is convenient to use this reagent to screen several worms by PCR with relatively little effort. However, other less expensive PCR reagents may also be used for PCR screening.
29. The reason why we prefer picking wild-type looking worms for screening is that they would have already lost the *dpy-10* mutation and hence be wild-type at the *dpy-10* locus. This will prevent the need for further crosses to get rid of the *dpy-10* mutation.

Acknowledgments

We would like to thank Kevin O'Connell, Andy Golden, Aimee Jaramillo-Lambert, Tetsu Fukushige, Prabhu Sankaralingam, and Daman Kumari from the National Institutes of Health for their critical feedback and comments on this manuscript. Nicole DeVaul and Tyler Hansen contributed equally to this chapter.

References

1. Brenner S (1974) The genetics of *Caenorhabditis elegans*. Genetics 77(1):71–94
2. Baumeister R, Ge LM (2002) The worm in us - *Caenorhabditis elegans* as a model of human disease. Trends Biotechnol 20(4):147–148
3. Poulin G, Nandakumar R, Ahringer J (2004) Genome-wide RNAi screens in Caenorhabditis elegans: impact on cancer research. Oncogene 23(51):8340–8345
4. Kenyon C (2005) The plasticity of aging: insights from long-lived mutants. Cell 120 (4):449–460
5. Corsi A, Wightman B, Chalfie M (2015) A transparent window into biology: a primer on Caenorhabditis elegans. Genetics 200 (2):387–407
6. Fire A, Xu SQ, Montgomery MK et al (1998) Potent and specific genetic interference by double-stranded RNA in *Caenorhabditis elegans*. Nature 391(6669):806–811
7. Kutscher L, Shaham S (2014) Forward and reverse mutagenesis in *C. elegans*. WormBook:1–26
8. Sulston J, Horvitz HR (1977) Post-embryonic cell lineages of the nematode, *Caenorhabditis elegans*. Dev Biol 56(1):110–156
9. Kimble J, Hirsh D (1979) The postembryonic cell lineages of the hermaphrodite and male gonads in *Caenorhabditis elegans*. Dev Biol 70 (2):396–417
10. Sulston J, Schierenberg E, White JG et al (1983) The embryonic cell lineage of the nematode *Caenorhabditis elegans*. Dev Biol 100 (1):64–119
11. Deppe U, Schierenberg E, Cole T et al (1978) Cell lineages of the embryo of the nematode Caenorhabditis elegans. Proc Natl Acad Sci U S A 75(1):376–380
12. Hirsh D, Oppenheim D, Klass M (1976) Development of the reproductive system of *Caenorhabditis elegans*. Dev Biol 49 (1):200–219
13. Kimble J, Hodgkin J, Smith T et al (1982) Suppression of an amber mutation by microinjection of suppressor tRNA in *C. elegans*. Nature 299(5882):456–458
14. Stinchcomb D, Shaw JE, Carr SH et al (1985) Extrachromosomal DNA transformation of *Caenorhabditis elegans*. Mol Cell Biol 5 (12):3484–3496
15. Mohan N, Chen CS, Hsieh HH et al (2010) In vivo imaging and toxicity assessments of fluorescent nanodiamonds in *Caenorhabditis elegans*. Nano Lett 10(9):3692–3699
16. Daniels B, Masi BC, Wirtz D (2006) Probing single-cell micromechanics in vivo: the microrheology of *C. elegans* developing embryos. Biophys J 90(12):4712–4719
17. Mello C, Kramer JM, Stinchcomb D et al (1991) Efficient gene transfer in *C.elegans*: extrachromosomal maintenance and integration of transforming sequences. EMBO J 10 (12):3959–3970
18. Fire A (1986) Integrative transformation of *Caenorhabditis elegans*. EMBO J 5 (10):2673–2680
19. Hsu PD, Lander ES, Zhang F (2014) Development and applications of CRISPR-Cas9 for genome engineering. Cell 157(6):1262–1278
20. Frøkjær-Jensen C, Davis W, Hopkins CE et al (2008) Single-copy insertion of transgenes in *Caenorhabditis elegans*. Nat Genet 40 (11):1375–1383
21. Friedland A, Tzur YB, Esvelt KM et al (2013) Heritable genome editing in *C. elegans* via a CRISPR-Cas9 system. Nat Methods 10 (8):741–743
22. Sapranauskas R, Gasiunas G, Fremaux C et al (2011) The *Streptococcus thermophilus* CRISPR/Cas system provides immunity in *Escherichia coli*. Nucleic Acids Res 39 (21):9275–9282
23. Deveau H, Barrangou R, Garneau JE et al (2008) Phage response to CRISPR-encoded resistance in *Streptococcus thermophilus*. J Bacteriol 190(4):1390–1400
24. Garneau JE, Dupuis ME, Villion M et al (2010) The CRISPR/Cas bacterial immune system cleaves bacteriophage and plasmid DNA. Nature 468(7320):67–71
25. Gasiunas G, Barrangou R, Horvath P et al (2012) Cas9-crRNA ribonucleoprotein complex mediates specific DNA cleavage for adaptive immunity in bacteria. Proc Natl Acad Sci U S A 109(39):E2579–E2586
26. Jinek M, Chylinski K, Fonfara I et al (2012) A programmable dual-RNA-guided DNA endonuclease in adaptive bacterial immunity. Science 337(6096):816–821
27. Dickinson D, Goldstein B (2016) CRISPR-based methods for *Caenorhabditis elegans* genome engineering. Genetics 202 (3):885–901
28. Chen X, Feng XZ, Guang SH (2016) Targeted genome engineering in *Caenorhabditis elegans*. Cell Biosci 6(1):60
29. Frøkjær-Jensen C (2013) Exciting prospects for precise engineering of *Caenorhabditis*

elegans genomes with CRISPR/Cas9. Genetics 195(3):635–642
30. Paix A, Folkmann A, Rasoloson D et al (2015) High efficiency, homology-directed genome editing in *Caenorhabditis elegans* using CRISPR-Cas9 ribonucleoprotein complexes. Genetics 201(1):47–54
31. Paix A, Schmidt H, Seydoux G (2016) Cas9-assisted recombineering in *C. elegans*: genome editing using in vivo assembly of linear DNAs. Nucleic Acids Res 44(15):e128
32. Dickinson D, Pani AM, Heppert JK et al (2015) Streamlined genome engineering with a self-excising drug selection cassette. Genetics 200(4):1035–1049
33. Schwartz ML, Jorgensen EM (2016) SapTrap, a Toolkit for High-Throughput CRISPR/Cas9 Gene Modification in *Caenorhabditis elegans*. Genetics 202(4):1277–1288
34. Norris A, Kim HM, Colaiacovo MP et al (2015) Efficient genome editing in *Caenorhabditis elegans* with a toolkit of dual-marker selection cassettes. Genetics 201(2):449–458
35. Arribere JA, Bell RT, Fu BX, et al (2014) Efficient Marker-Free Recovery of Custom Genetic Modifications with CRISPR/Cas9 in *Caenorhabditis elegans*. Genetics 198 (3):837–846
36. Kim H, Ishidate T, Ghanta KS, et al (2014) A Co-CRISPR Strategy for Efficient Genome Editing in *Caenorhabditis elegans*. Genetics (197)4:1069–1080
37. Ward JD (2015) Rapid and Precise Engineering of the Caenorhabditis elegans Genome with Lethal Mutation Co-Conversion andInactivation of NHEJ Repair. Genetics 199 (2):363–377
38. Berkowitz LA, Knight AL, Caldwell GA et al (2008) Generation of stable transgenic C. elegans using microinjection. J Vis Exp 18 (833):e833
39. Mello C, Fire A (1995) DNA transformation. In: Epstein HF, Shakes DC (eds) Methods in cell biology, vol 48. Academic, San Diego, pp 451–482
40. Jin Y (1999) Transformation. In: Hope IA (ed) C. elegans: a practical approach, vol 213. OUP, Oxford, pp 69–96
41. Rieckher M, Kourtis N, Pasparaki A et al (2009) Transgenesis in *Caenorhabditis elegans*. In: Cartwright EJ (ed) Transgenesis techniques, vol 561. Humana Press, a part of Springer Science, New York, pp 21–39
42. Paix A, Wang YM, Smith HE et al (2014) Scalable and versatile genome editing using linear DNAs with microhomology to Cas9 sites in *Caenorhabditis elegans*. Genetics 198 (4):1347–1356

Chapter 26

Microinjection in Zebrafish for Genome Editing and Functional Studies

Wuhong Pei and Shawn M. Burgess

Abstract

A major strength of zebrafish as a model organism is their rapid, in vitro development. The easy access to embryos compared to mammals, allows larval molecular and cellular composition to be manipulated by microinjection, providing a powerful avenue for biological and translational studies. Here, we describe the essential steps and different applications of microinjection in zebrafish for genome editing and functional studies, along with some experimental tips that are critical for microinjection success.

Key words Zebrafish, Microinjection, CRISPR, Mutagenesis, Transgenesis, Functional studies

1 Introduction

Zebrafish has enjoyed increasing popularity over the past two decades. Several features of zebrafish make it an attractive model for understanding basic biology and modeling human genetic diseases. External fertilization, rapid development, and embryo transparency make it an ideal model for studying vertebrate developmental biology. A fully annotated genome sequence coupled with simple and cheap husbandry extend its utilities to large-scale mutagenesis, high-throughput drug screening, and modeling a variety of human genetic diseases ranging from cancer, cardiovascular disease, to congenital and age-related diseases [1–3].

Microinjection is an essential technique used in zebrafish research. For example, microinjection can be used to alter the genomic DNA sequence, thus permanently changing the function of the targeted gene. Targeted genomic lesions can be obtained by injection of zinc finger nucleases (ZFN), transcription activator-like effector nucleases (TALEN), or clustered regulatory interspaced short palindromic repeat (CRISPR)-associated protein-9 nuclease (Cas9) (CRISPR/Cas9) [4]. Random genomic mutagenesis can be obtained by injection of retroviruses to create retroviral DNA

Chengyu Liu and Yubin Du (eds.), *Microinjection: Methods and Protocols*, Methods in Molecular Biology, vol. 1874, https://doi.org/10.1007/978-1-4939-8831-0_26, © Springer Science+Business Media, LLC, part of Springer Nature 2019

integrations [5], or transposable elements containing gene trap constructs [6].

Microinjection has also been used to study gene function by transiently altering the levels of protein translation. Loss-of-function can be achieved by injection of morpholino (MO), an antisense oligonucleotide analog that blocks the translation of mRNA or interrupts the splicing of pre-mRNA [7]. Gain-of-function or ectopic expression can be achieved by injection of synthetic mRNAs that will be translated to protein in vivo [8].

In addition, microinjection has been used in zebrafish to monitor the activity of regulatory elements such as promoters and enhancers, by delivering a DNA cassette into genome to create gene- or enhancer-traps [9]. The DNA cassette usually contains a reporter gene encoding a fluorescent protein, whose spatiotemporal expression mirrors the activity and tissue specificity of the neighboring promoter and/or enhancer. The expression of the fluorescent protein under the control of a tissue-specific promoter is also a powerful tool for in vivo studies. These transgenes allow the researcher to carefully monitor the responses of cells or tissues under different experimental conditions. Again, microinjection can play a role as it can be used to inject a biological stimulus into the fluorescent-reporter system [10]. For example, a zebrafish expressing green fluorescent protein (GFP) in neutrophils could be used to monitor an inflammatory response when bacteria are injected into a zebrafish larva.

Microinjection has been used to generate a wide-ranging set of zebrafish mutant and transgenic lines currently being used by the zebrafish research community. The majority of these lines were created by microinjection of embryos at very early stages (i.e., 1–4 cell stages). However, recent years have witnessed successful uses of injection into later stage zebrafish larvae for various functional studies. For example, 1 day post fertilization (dpf) embryos were injected with clodronate to kill macrophages [11]. 2 dpf embryos were injected with human leukemia cells to study cancer cell proliferation and chemotherapy responses [12]. 1–3 dpf embryos were injected with bacteria to study their interaction with host innate immune cells [10]. In addition, immune-compromised adult zebrafish were transplanted with hematopoietic, muscle stem cells, and cancer cells to determine the optimal conditions for long-term cell engraftment [13].

This chapter will briefly describe our laboratory protocols for using early stage embryos to deliver DNA and/or RNA to create genetically engineered zebrafish, as well as our recent experiments using embryos at later stages to locally deliver molecules such as lipopolysaccharide (LPS) or proteins like heat shock protein 60 (HSP60) to study immune cell migration and tissue regeneration.

2 Materials

1. World Precision PV830 microinjector.
2. Leica MZ12.5 stereo microscope.
3. World Precision M3301R micromanipulator.
4. Zeiss Axio Observer A1 fluorescent microscope.
5. Injection needles. Glass or quartz 1 mm capillary.
6. Needle puller. KOPF 730 needle/pipette puller for glass needles. Sutter P-2000 Puller for quartz needles.
7. Razor blade.
8. Fisher Scientific Micro-spatula.
9. (Optional) Parafilm.
10. Zebrafish: wild-type, Tg(*mpx:EGFP*)/Tg(*mpx:EGFP*), or other preferred transgenic lines.
11. Embryo medium 1× Holtfreiter's buffer. We make Holtfreiter's buffer as a 20× stock solution. For 1 L of 20× stock solution, mix 70 g NaCl, 2.6 g $CaCl_2{\cdot}(2H_2O)$, and 1 g KCl, filter-sterilize or autoclave. For making 1× Holtfreiter's buffer, dilute the 20× stock into 1× and add HEPES (pH 7.6) to a final concentration of 5 mM.
12. Becton Dickinson Labware 3 mL plastic transfer pipet.
13. Plastic petri dishes, 4 and 10 cm.
14. Agarose injection bed with depression grooves. Made by pouring 30 mL of 2% agarose in 1× Holtfreiter's buffer into a 10 cm petri dish, and then placing custom-designed plastic mold to create grooves on the agarose surface.
15. Gilson pipettes and tips.
16. Eppendorf Microloader tips.
17. Injection buffer (1× Danieau solution): 58 mM NaCl, 0.7 mM KCl, 0.4 mM $MgSO_4$, 0.6 mM $Ca(NO_3)_2$, 5 mM HEPES, pH 7.6. Sterilize by 0.2-μm filter. We make 10× stock solution, aliquot and store at room temperature for up to 1 year.
18. Phenol red, 0.5%.
19. Mineral oil.
20. Micro-ruler. 1 × 0.01 mm stage micrometer for droplet calibration.
21. Ambion mMessage mMachine Transcription Kit.
22. NEB HiScribe T7 Quick High Yield RNA Synthesis Kit.
23. Qiagen Endofree Plasmid Maxi Kit.
24. 1× Tricaine methanesulfonate (MS-222). 250 mg/L, buffered with sodium bicarbonate to pH = 7.0–7.5.

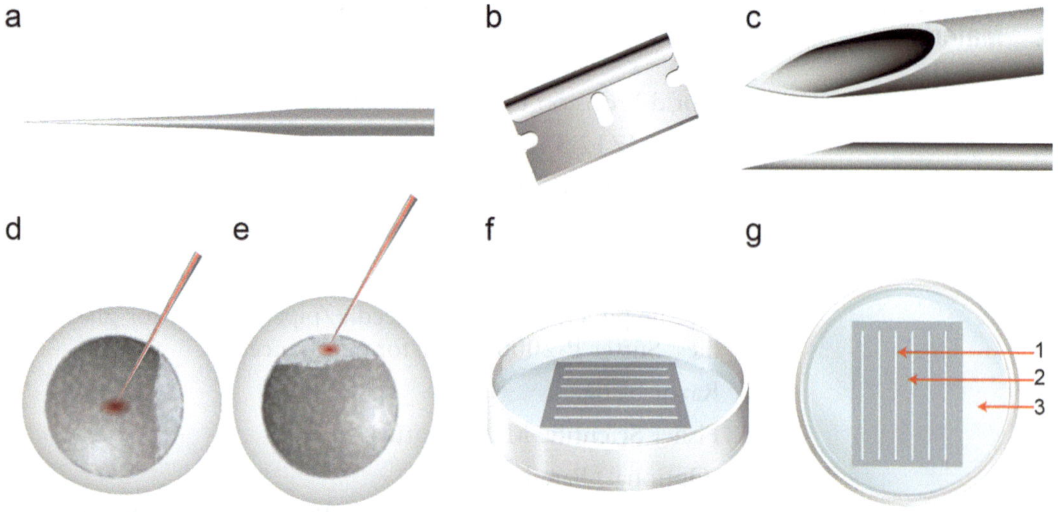

Fig. 1 Microinjection into 1-cell stage embryos. Schematic images are used in all panels. (**a**) A needle after the capillary has been pulled but before cutting. The taper can be adjusted to injector's preference by altering pulling speed and temperature. (**b**) A razor blade is used for cutting the end of the needle. (**c**) A needle after cutting should have a bevel that allows better penetration into the cell or tissue. The two images show the wedge-shaped opening from two different angles. (**d**) Yolk injections are fast and relatively easy. (**e**) Cytoplasmic injection is slower but increases efficiency of cutting and DNA integration. (**f**) Injection agarose bed is made by laying the silicone mold containing the grooves on the surface of liquid 2% agarose. (**g**) Different areas can be used for different types of injections depending on need. Area 1 in the groove is best for yolk and cytoplasmic injections. Area 2 is better for brain injections where orienting the dorsal view requires more working room. Area 3 is ideal for injections into the trunk

25. Lipopolysaccharide (LPS).
26. Recombinant human GAPDH.
27. Recombinant human HSP60.
28. Kimwipe tissue paper.

3 Methods

3.1 Injection Needle Preparation

Whether a glass or quartz capillary needle should be used is based on experimental design (details *see* **Note 1**). An illustrative example for creating an opening in a needle is shown in Fig. 1a–c. For detailed information see steps below.

1. Clean the microscope stage with 70% ethanol.
2. Hold a needle stationary either with a needle holder or between thumb and forefinger. Place the needle tip at the center of the dissecting microscope's field.
3. Set the microscope to the highest magnification.
4. Move the razor blade, blade side downward, slowly to the center of the microscope field until it is close to the tip of the needle.

5. Tilt razor blade to approximately a 30° angle to the needle. Move the blade to cut the very tip of the needle. The resulting needle should have a small, beveled opening (Fig. 1c).
6. To avoid the needle being crushed at the cutting site, place a sheet of parafilm under the needle as cushion before cutting.

3.2 Needle Loading and Injection Volume Calibration

1. Injection solution preparation. To deliver genetic materials (DNA, RNA, and/or protein), dilute the materials into 1× injection buffer to the desired concentration. A small amount of phenol red can be included at a concentration of 0.05% when there is a need to visualize the injection volume and distribution in the animal. Extra care should be taken for preparing RNA-containing solutions (*see* **Note 2**).
2. Needle loading. The injection solution can be loaded into the back of the needle using a MicroLoader pipet tip. For needles with an internal filament, the filament will guide the solution to the tip of the needle. For needles without an internal filament, try to load the solution as close as possible to the tip of the needle. MicroLoader tips are long and fine enough to reach very close to the very tip of the needle.
3. Injection volume calibration. Place a micro-ruler in the center of the microscope's field. Place a 4 cm petri dish filled with 5 mL of mineral oil on the top of the micro-ruler. Set the microscope to a low magnification. Submerge the loaded needle into the mineral oil. Press foot pedal to get a liquid sphere to calculate the volume of the injection (more information in **Note 3**).

3.3 Preparing 1-Cell Stage Embryos for Injections

Most microinjections performed in zebrafish are done with embryos at 1-cell stage, either into the cytoplasm or yolk. Cytoplasmic injection offers higher delivery efficiency than yolk injection. Microinjection at this stage will distribute the injection solution uniformly in all cells as development proceeds. Since 1-cell stage lasts approximately 20 min, it is essential to plan ahead and make sure all things are ready prior to injection (for details *see* **Note 4**).

1. In the late afternoon the day before injection, set up a breeding pair of male and female zebrafish in a small breeding tank with a divider to separate the two fish.
2. The next morning, pull the divider and leave the female and male fish together, check on the spawning every 15 min to collect freshly laid eggs.
3. Needles can be pulled at any time and kept in a 10 cm petri dish held in place embedded in a wedge of clay or a loop of tape. Before setting up the fish to breed, place the needles on a

precleaned microscope stage, and cut as described in Subheading 3.1. Fill the needle with injection solution and hold in a petri dish set on ice until time for injection.

4. Collect the freshly laid eggs, and rinse the eggs gently with 1× Holtfreiter's buffer.
5. To temporarily extend the time of embryo at the 1-cell stage, place the embryo-containing dish on ice or in ice-water. However, if done too long the eggs will lyse and die.
6. Use the 3 mL plastic transfer pipet to transfer eggs into a 10 cm Petri dish containing the 1× Holtfreiter's buffer. Remove any fish waste or milky-looking (dead) eggs. Transfer normal-looking eggs to the agarose injection bed with injection grooves and use a micro-spatula to align the embryos in the grooves (Fig. 1f, g). Check all areas of the injection bed and make sure no embryo is missed from injection.
7. Remove excess medium from the injection bed without letting the eggs dry. This is to help prevent embryos from drifting or sticking to the injection needle.
8. For cytoplasmic injections (Fig. 1e), set the microscope to a low magnification and orient each of the embryos so the embryonic cell faces the direction of injection needle. For yolk injections (Fig. 1d), there is no need to orient the eggs.
9. Calibrate the injection needle and start the injection. The injections should be done quickly during the 1-cell stage. Inject at least 50 embryos for each injection condition.
10. If multiple injection conditions need to be done in the same morning, arrange multiple tanks of breeding pairs, and set up crosses at 30 min intervals to produce a steady supply of freshly laid eggs. To ensure high quality injections, always kill poorly injected embryos. This can be done by injecting multiple times to deliver an injection volume larger than 5 nL.
11. Save approximately 30 eggs as un-injected control to determine egg quality. It is necessary to use eggs from the same clutch as the un-injected control to monitor egg quality for each injection condition.
12. Eggs laid late may not be suitable for injections. These eggs usually have a reduced fertilization rate and poor quality. If parental fish are maintained in a light cycle of 8 am–10 pm, it is recommended to inject eggs laid before 12 pm.
13. Some useful tips for high quality injections are listed in **Note 5**.

3.4 Injecting Embryos at 1-Cell Stage for Creating Targeted Genomic Lesions by CRISPR/Cas9

Three major approaches have been used in zebrafish for targeted genome editing: ZFN, TALEN, and CRISPR [4]. The high efficiency of the CRISPR/Cas9 system has greatly accelerated the speed of targeted genome editing, making it a technique available to any lab capable of basic molecular biology. Below are the key steps for creating CRISPR/Cas9-mediated targeted mutations.

1. Synthesize Cas9 mRNA using the mMessage mMachine (or comparable) Transcription Kit.
2. Synthesize short guide RNA by HiScribe T7 Quick High Yield RNA Synthesis Kit.
3. Make sure the synthesized RNAs with an OD260 nm/OD280 nm ratio of ~2. Freshly synthesized Cas9 mRNA and guide RNA generate higher mutation efficiencies. It is recommended to make small aliquots of freshly synthesized Cas9 mRNA and guide RNA and store them in −80° freezer. An aliquot is thawed and used only once. Avoid freeze-thaw cycles. Commercially available Cas9 protein can be substituted for Cas9 mRNA.
4. Prepare injection solution by mixing Cas9 mRNA and guide RNA with 1× injection buffer. For multiplexing, up to six guide RNAs can be included in a single injection needle. We usually inject 150–300 pg of Cas9 mRNA and 25–50 pg of each guide RNA. A calculator for determining injection concentrations and dilution called CRISRPR-Calculator was described previously [14].
5. Inject Cas9 mRNA and guide RNA mixture into the yolk or cytoplasm of 1-cell stage embryos.
6. Collect some of the injected embryos at 2–5 dpf to measure guide RNA activity. This measurement can be done using the CRISPR-STAT method described previously [14]. Since the CRISPR/Cas9 system creates mosaic mutations and majority of the mutations are small insertions/deletions, fluorescent-PCR-based size separation methods are more efficient in detecting mutations in the injection embryos, than conventional sequencing analysis.
7. If the guide RNA shows an acceptable activity, raise the remaining injected embryos to adults, and use these adults as the founder fish to transmit germline mutations.
8. Incross or outcross the adult founder fish to get offspring with inherited mutations in the next generation.
9. Useful resources for creating zebrafish CRISPR mutations are listed in **Note 6**.

3.5 Injecting Embryos at 1-Cell Stage for Creating Random Genomic Integrations by Tol2 Transposon

Random genomic integrations in zebrafish can be made by several transgenesis approaches such as pseudotyped retroviruses [15], sleeping beauty transposon [16], or Tol2 transposon [17]. The Tol2 system is the most popular in zebrafish, due to its high integration efficiency and its capacity for accommodating large inserts. Below are the essential steps for the generation of Tol2 transgenic zebrafish.

1. Construct a Tol2 plasmid containing the desired expression cassette.
2. Transform the constructed plasmid into host bacteria to amplify the recombinant plasmid.
3. Use the Qiagen Endofree Plasmid Maxi Kit to purify the plasmid. Purification of the Tol2 plasmid by the Endofree plasmid kit can eliminate endotoxins released from the host bacteria, thus reducing the toxicity to the injected embryos.
4. Synthesize Tol2 transposase mRNA in vitro using the mMessage mMachine Transcription Kit.
5. The morning of the injections, prepare the injection solution by diluting Tol2 transposase mRNA and Tol2 plasmid DNA with 1× injection buffer. A recommended concentration is 18 ng/μL for each.
6. Place the injection solution on ice until embryos are ready for the injection.
7. Inject 1.4 nL of the injection solution into the cytoplasm of 1-cell stage embryos. Injection quality determines the efficiency of Tol2 integration. It is recommended to add phenol red to the injection solution to monitor the volume of the injection. Injecting directly into the cell's cytoplasm (instead of the yolk) is recommended for maximum efficiency.
8. If the construct has a reporter gene, examine the embryos for fluorescent protein expression (e.g., GFP) in the injected embryos at 1–5 dpf. The timing of GFP expression depends on the activity of the neighboring promoter.
9. Sort out the GFP positive embryos and raise them to adult founder fish. Raising embryos that demonstrate a high degree of transgenic somatic insertion increases the chance of germline transgenesis and obtaining construct-positive offspring.
10. Outcross the adult founder fish to generate stable transgenic lines.

3.6 Preparation of 5 dpf Embryos for Injections

Later-stage injection provides an opportunity for in vivo manipulation (for more information *see* **Note** 7). As an illustrative example, here are the steps for preparing later-stage embryos for brain or trunk injections to assay the neutrophil-mediated inflammation response.

1. Select healthy parental fish with Tg(*mpx:EGFP*)/Tg(*mpx:EGFP*) background for brain ventricle injection, with Tg(*mpx:EGFP*)/Tg(*mpx:EGFP*) or wild-type background for trunk vessel injection.
2. In the late afternoon, set up pairwise crosses in small breeding tanks with dividers to separate the fish.
3. It is recommended to use 5 dpf embryos from a pairwise cross for injection to ensure a consistent genetic background. If a large quantity of embryos is needed, embryos can be obtained by a group cross with multiple fish.
4. The next morning shortly after the lights turn on, pool the female and male fish together. Check on the spawning every 15 min to collect the freshly laid eggs.
5. Incubate the eggs in 1× Holtfreiter's buffer. Avoid exposing embryos with fluorescent transgenes to blue egg water (containing methylene blue) to reduce autofluorescence.
6. Remove the dead and unfertilized eggs at the end of day. Place up to 50 embryos in a 10 cm petri dish containing 30 mL of 1× Holtfreiter's buffer.
7. Optional. Treat the embryos with 0.003% PTU to inhibit pigmentation formation.
8. Grow the embryos to 5 dpf.
9. Select healthy-looking embryos for injection.
10. Anesthetize the selected embryos immediately before the injection, by placing embryos in a 1× MS-222 solution.
11. Prepare the injection needle. A quartz needle is stiffer and more durable than glass so is preferred for injections of 5 dpf embryos.

3.7 Injecting LPS into the Brain Ventricle

Below is an example of using LPS to induce neutrophil migration. LPS is injected at a concentration of 250 ng/μL. BSA is injected at the same concentration as a control. Injection volume is 1 nL. Neutrophils are labeled by Tg(*mpx:EGFP*) transgenic line. Neutrophil migration is examined at 3 h post injection. **Note 8** includes more information on the reasons for choosing brain ventricle and the keys to a successful injection.

1. On the day of injection, dilute LPS with 1× injection solution containing phenol red to a final concentration of 250 ng/μL. BSA at the same concentration is prepared to use as a control.
2. Pull a quartz needle. Cut the tip in a bevel to improve penetration.
3. Load the needle with injection solution. Calibrate the needle so each press delivers an injection volume of ≈1 nL.

4. Allocate thirty 5 dpf embryos generated from an incross of Tg (*mpx:EGFP*)/Tg(*mpx:EGFP*) fish for injection. If injections have a poor survival rate, more than 30 embryos should be injected.
5. Anesthetize the embryos with 1× MS-222.
6. Transfer the anesthetized embryos into the grooves of the agarose bed (Fig. 1f, g). Place several embryos in each groove. Remove excessive liquid. Orient them in the same direction.
7. Use the following steps to orient each embryo: place a metal pin under the middle of the embryonic trunk, gently lift the pin to transfer the embryo from the groove to the flat area of agarose, move the tail close to one side of the head, and then orient the head in a dorsal view with the brain ventricle up (Fig. 2a).
8. Adjust the injection needle parallel to the midline of the brain.
9. Pierce the needle through the skin of the posterior dorsal brain ventricle, and then push the needle to the center of the brain ventricle.
10. Press foot pedal once to inject, injection solution should fill the brain ventricle. To examine the location of brain injection, roll the injected embryo to a lateral view to see the brain ventricle filled with phenol red injection dye. It is normal to see a small fraction of the injection solution leaking into blood circulation.
11. If the injection is not successful, use the injection needle to gently push the embryo back into the grooves.
12. Gently move the injection bed to place the next embryo in the center of microscope's field.
13. Proceed with injections until all the embryos have been injected.
14. Remove all the poorly injected embryos from the grooves.
15. Gently add 1× Holtfreiter's buffer to the injection bed, let the buffer spread, and slowly cover the injected embryos.
16. Use a metal pin to gently detach the embryos from the injection bed, and let the injected embryos rehydrate for 1 min.
17. Fill a 10 cm plastic petri dish with 30 mL of 1× Holtfreiter's buffer.
18. Gently transfer all the injected embryos into the plastic dish. No shaking or swirling of the dish. Filling the plastic petri dish with 1× Holtfreiter's buffer before transferring the injected embryos can avoid embryos getting struck in a plastic petri dish.
19. Inject an equal number of embryos for LPS and for BSA.
20. Incubate the injected embryos at 28 °C for 3 h.

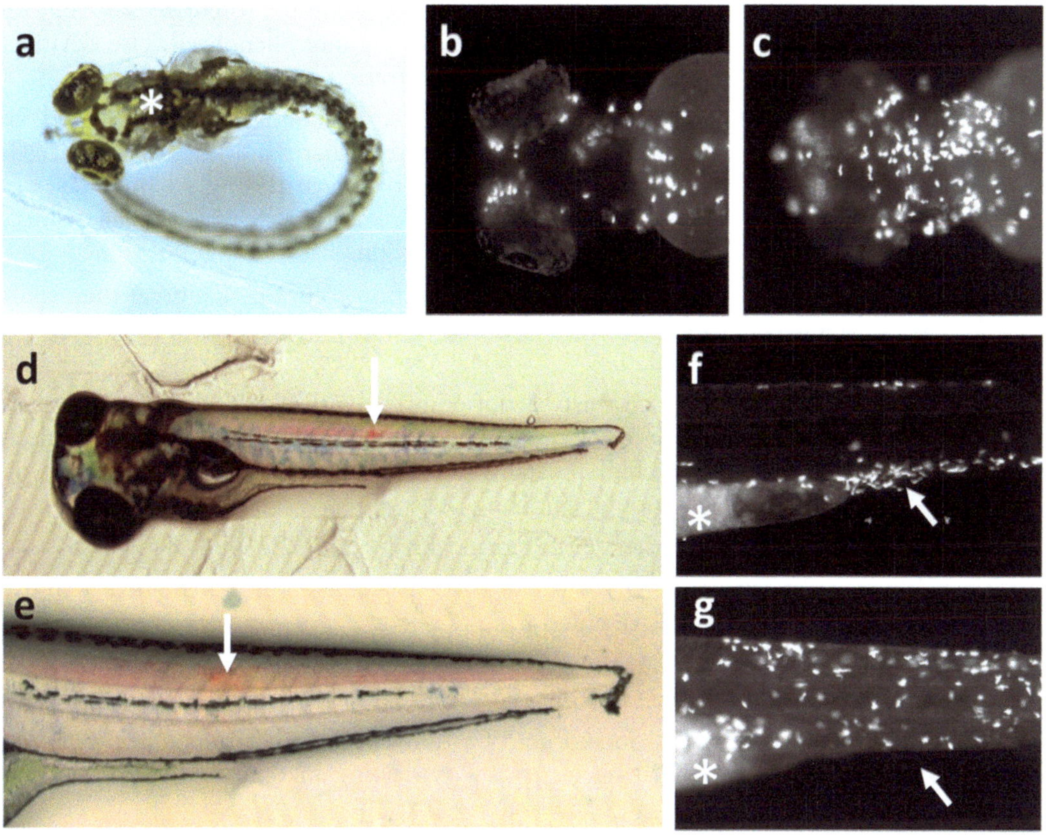

Fig. 2 Microinjection into the brain or trunk of 5 dpf embryos. All images were taken with live animals. (**a**) An embryo oriented for brain injection. The white asterisk indicates the location of the brain ventricle. (**b** and **c**) Neutrophils accumulated in the brain after injection of control protein BSA (**b**) or immunoreactive LPS (**c**). Neutrophils are labeled by Tg(*mpx:EGFP*). More neutrophils are present in the brain area after LPS injection compared to the control BSA injection. (**d**) A larva oriented for the trunk injection. The white arrow points to the injection site. (**e**) A higher magnification of (**d**), showing how the injected solution diffuses throughout the trunk. (**f** and **g**) Neutrophils accumulating in the trunk after injection of control GAPDH (**f**) or HSP60 (**g**) proteins. The white asterisks indicate the yolk extension. White arrows point to the neutrophil distribution in the trunk after injection. Neutrophils are mostly localized in the ventral side of the trunk after GAPDH injection (where they usually reside) (**f**), however, the neutrophils migrate dorsally after injection of HSP60 protein (**g**)

21. Mount the injected embryos with 3% methylcellulose with a dorsal view.
22. Image the dorsal brain area to analyze the neutrophils migrated to the injection site (Fig. 2b, c).

3.8 Injecting HSP60 Protein into the Dorsal Flank

Recombinant human HSP60 is injected into the dorsal vessel of the middle trunk so it can be distributed along anterior-posterior axis of the lateral line. Injection can be performed with neutrophil labeled Tg(*mpx:EGFP*) larvae (for immune cell migration analysis), or wild-type larvae (for hair cell regeneration analysis). For detailed information on the experimental design, *see* **Note 9**.

1. The day of injection, dilute HSP60 protein to 250 ng/μL using 1× injection buffer including phenol red. Use GAPDH protein at the same concentration as a control.
2. Use quartz needles for these injections. The needle used for trunk injection should be very sharp and fine, optimal if it is made from quartz. It has to pierce through a thick layer of tissue to reach the dorsal vessel to deliver a small volume (≈0.5 nL) of injection solution.
3. Load the needle with injection solution. Calibrate the needle opening so the injection volume for each press is approximately 0.5 nL.
4. Allocate thirty 5 dpf embryos with Tg(*mpx:EGFP*)/Tg(*mpx: EGFP*) or wild-type background for injection. Trunk injections have a high mortality rate. Injecting more larvae will ensure you get enough surviving larvae.
5. Anesthetize the larvae with 1× MS-222.
6. Transfer the embryos to the flat area of agarose bed. Remove excessive liquid from the bed.
7. Arrange the embryos in a lateral view, with all embryos oriented in the same direction (Fig. 2d).
8. Twist one corner of a Kimwipe or other tissue paper, use the twisted tip to dry the remaining medium off the larvae, this helps make embryos stick to the agarose bed.
9. Orient the injection needle parallel to the dorsal trunk.
10. Pierce the needle through skin and somite tissue, and advance it into dorsal vessel.
11. Press foot pedal once to inject. Injection solution should diffuse in a straight line inside the dorsal vessel along anterior-posterior axis (Fig. 2e). Needle clogging can cause considerable variation in injection volume. Whether the right volume is injected can be estimated by the length of the phenol red spread in the dorsal vessel along anterior-posterior axis.
12. If injection solution forms a V-shaped line across dorsal-ventral axis, it indicates the needle was in the somite junction and failed to deliver injection material into the dorsal vessel. These poorly injected embryos should be removed or killed.
13. After injection of all the embryos, gently add 1× Holtfreiter's buffer onto the injection bed. Let the buffer spread over the injected larvae.
14. Gently detach the larvae from the agarose bed with a metal probe, and let them rehydrate for 1 min.
15. Fill a 10 cm plastic dish with 30 mL of 1× Holtfreiter's buffer.

16. Gently transfer all the injected larvae into the plastic dish. Avoid shaking or swirling the dish.
17. Inject an equal number of larvae for HSP60 and for GAPDH.
18. Incubate the injected larvae at 28 °C for 3 h.
19. Use the healthy-looking Tg(*mpx:EGFP*) larvae for an immune cell migration assay (Fig. 2f, g).
20. For hair cell regeneration, use wild-type larvae at 5 dpf for injections, ablate hair cells at 3 h post injection, and measure the regeneration at 48 h post hair cell ablation.

4 Notes

1. Glass or quartz capillaries are the two major types of needles used for microinjection in zebrafish. We typically use glass needles to inject solutions (such as DNA, RNA, or protein) into embryos at the 1-cell stage. We use quartz needles to inject into tissues of larvae at later stages. We pull glass needles with a KOPF needle/pipette puller. Glass or quartz needles can be pulled with a Sutter P-2000 (https://www.youtube.com/watch?v=JGN8aeaZPoo). The machine settings are based on manufacturer's recommendation, the researcher can test various parameters to produce needles shaped the way he or she prefers.
2. For preparing injection solutions containing RNA that is prone to degradation, use DEPC-treated water, clean chemicals, and RNAse-free consumables. Also, try to use fewer components in the injection solution if possible. Phenol red can be omitted when there is no need to carefully visualize the injection process. To eliminate degradation of RNA in the injection solution, place the injection solution on ice, use a freshly pulled needle, and load the injection solution into the needle immediately before the injection.
3. For measuring the needle opening and calculating injection volume, we usually press the foot pedal ten times to observe a liquid sphere forming in the mineral oil. Calculate the volume of the sphere using the micro-ruler to measure the diameter of the sphere and divide by 10. To obtain the desired volume for each press, adjust the time and pressure settings of the microinjector apparatus. To reduce the chance of needle clogging, a quick spin of the injection solution before needle loading is recommended. A clogged needle could be cleared by increasing injection time or pressure to push out the clog. If that fails, recut the needle or preferably have several prepared needles ready for each injection session.

4. Necessary preparations before injecting embryos at 1-cell stage: needles should be pulled, embryos should be laid out in the injection bed, and oriented properly if cytoplasmic injection is planned. If delivery efficiency and even distribution are not the top priorities, injections can be done until the 8-cell stage. It is worth noting that some large molecules such as DNA do not diffuse evenly throughout organism when injected later than the 1-cell stage.
5. Useful tips for high quality injections. A very fine, sharp needle is the key to successful injections. Do not try to use a needle with a poorly cut or too large a hole, otherwise you will not achieve optimal injections. If possible, always add phenol red to the injection solution to increase the visibility of injection location and injection volume. Align all the embryos neatly before injection. Check all areas of the injection bed and make sure no embryo is missed from injection. Kill all the poorly injected embryos. This can be done by injecting multiple times at the same site. Handle all freshly injected embryos very gently.
6. Useful resources for creating zebrafish CRISPR/Cas9 mutations: (1) www.zfin.org, for information of zebrafish genes and existing mutations. (2) https://www.addgene.org/CRISPR/, for information of zebrafish Cas9 plasmid map and sequence. (3) www.crisprscan.org, for CRISPRscan target design [18]. (4) Zebrafish NHGRI Data Hub: http://genome.ucsc.edu/cgi-bin/hgHubConnect, for CRISPR target design and NHGRI-1 zebrafish genome database [19]. (5) CRISPRz: http://research.nhgri.nih.gov/crisprz/, a database documenting the sequences and activities of previous analyzed CRISPR guide RNA [20]. (6) Carrington et al., Nucleic Acid Research 2016. https://www.ncbi.nlm.nih.gov/pubmed/26253739, a method article using CRISPR-STAT to evaluate the target-specific activity of CRISPR guide RNA [14]. (7) Varshney et al., Nature Protocol 2016. https://www.ncbi.nlm.nih.gov/pubmed/27809318, an experimental protocol documenting all the essential steps and tips for generating zebrafish CRISPR mutations [21].
7. In contrast to a systemic delivery of injected material achieved from 1-cell stage injection, later-stage injection is intended to deliver the injected material to local areas or specific tissues. Injection into later stage embryos is not as time restrained as the 1-cell stage injections. The major challenge of later-stage injection is how to deliver the injection material into the desired tissue without causing significant tissue damage.
8. 5 dpf embryos have been used to inject lipopolysaccharide (LPS) into dorsal brain ventricle to study the migration of

innate immune cells [22]. There are three reasons that dorsal brain ventricle was chosen for the immune cell migration assay: (1) the brain ventricle is easily accessible by injection needle and has space to accommodate the injection solution, (2) the brain ventricle can retain the injected material for a period of time, (3) the dorsal brain area normally has very few resident immune cells. When compared to injection into 1-cell stage embryos, injection into 5 dpf embryos is more time-consuming, more technically challenging, often accompanied with a higher mortality rate. To improve the survival of injected embryos, researchers who perform the injection should have a tight control on injection volume, needle entry point, and depth of needle invasion.

9. HSP60 is classically a chaperone protein functioning in protein folding. It also has a moon-lighting function acting as an immune antigen. Study from our laboratory has shown HSP60's immune activity promotes tissue regeneration [22]. To examine the effect of HSP60 on lateral line hair cell regeneration, we injected HSP60 into the dorsal vessel of the middle trunk to deliver it along the anterior-posterior axis of the lateral line. Since the dorsal vessel is thin in diameter and deep inside the midline, the injection volume has to be very small and injection needle has to be very fine to ensure minimal damage to the injected embryos. We usually inject ≈0.5 nL. It is important to note that trunk vessel injection has a high mortality rate, there is a need to prepare and inject more embryos to ensure the results for each data point have a strong statistical power.

Acknowledgments

This work was supported by the Intramural Research Program of the National Human Genome Research Institute (ZIAHG200386-05). We thank Dr. Benjamin Feldman from NICHD for technical advice on microinjections; Dr. Gaurav K. Varshney and other members of the Burgess laboratory and NHGRI Zebrafish Core for improving the protocols.

References

1. Santoriello C, Zon LI (2012) Hooked! Modeling human disease in zebrafish. J Clin Invest 122(7):2337–2343. https://doi.org/10.1172/JCI60434
2. Van Houcke J, De Groef L, Dekeyster E, Moons L (2015) The zebrafish as a gerontology model in nervous system aging, disease, and repair. Ageing Res Rev 24 (Pt B):358–368. https://doi.org/10.1016/j.arr.2015.10.004
3. White R, Rose K, Zon L (2013) Zebrafish cancer: the state of the art and the path forward. Nat Rev Cancer 13(9):624–636. https://doi.org/10.1038/nrc3589
4. Varshney GK, Sood R, Burgess SM (2015) Understanding and editing the zebrafish

genome. Adv Genet 92:1–52. https://doi.org/10.1016/bs.adgen.2015.09.002

5. Jao LE, Maddison L, Chen W, Burgess SM (2008) Using retroviruses as a mutagenesis tool to explore the zebrafish genome. Brief Funct Genomic Proteomic 7(6):427–443. https://doi.org/10.1093/bfgp/eln038
6. Kawakami K (2005) Transposon tools and methods in zebrafish. Dev Dyn 234(2):244–254. https://doi.org/10.1002/dvdy.20516
7. Nasevicius A, Ekker SC (2000) Effective targeted gene 'knockdown' in zebrafish. Nat Genet 26(2):216–220. https://doi.org/10.1038/79951
8. Feldman B, Gates MA, Egan ES, Dougan ST, Rennebeck G, Sirotkin HI, Schier AF, Talbot WS (1998) Zebrafish organizer development and germ-layer formation require nodal-related signals. Nature 395(6698):181–185. https://doi.org/10.1038/26013
9. Kawakami K, Asakawa K, Hibi M, Itoh M, Muto A, Wada H (2016) Gal4 driver transgenic zebrafish: powerful tools to study developmental biology, organogenesis, and neuroscience. Adv Genet 95:65–87. https://doi.org/10.1016/bs.adgen.2016.04.002
10. Bhuiyan MS, Ellett F, Murray GL, Kostoulias X, Cerqueira GM, Schulze KE, Mahamad Maifiah MH, Li J, Creek DJ, Lieschke GJ, Peleg AY (2016) Acinetobacter baumannii phenylacetic acid metabolism influences infection outcome through a direct effect on neutrophil chemotaxis. Proc Natl Acad Sci U S A 113(34):9599–9604. https://doi.org/10.1073/pnas.1523116113
11. Bernut A, Herrmann JL, Kissa K, Dubremetz JF, Gaillard JL, Lutfalla G, Kremer L (2014) Mycobacterium abscessus cording prevents phagocytosis and promotes abscess formation. Proc Natl Acad Sci U S A 111(10):E943–E952. https://doi.org/10.1073/pnas.1321390111
12. Corkery DP, Dellaire G, Berman JN (2011) Leukaemia xenotransplantation in zebrafish--chemotherapy response assay in vivo. Br J Haematol 153(6):786–789. https://doi.org/10.1111/j.1365-2141.2011.08661.x
13. Tang Q, Abdelfattah NS, Blackburn JS, Moore JC, Martinez SA, Moore FE, Lobbardi R, Tenente IM, Ignatius MS, Berman JN, Liwski RS, Houvras Y, Langenau DM (2014) Optimized cell transplantation using adult rag2 mutant zebrafish. Nat Methods 11(8):821–824. https://doi.org/10.1038/nmeth.3031
14. Carrington B, Varshney GK, Burgess SM, Sood R (2015) CRISPR-STAT: an easy and reliable PCR-based method to evaluate target-specific sgRNA activity. Nucleic Acids Res 43(22):e157. https://doi.org/10.1093/nar/gkv802
15. Lin S, Gaiano N, Culp P, Burns JC, Friedmann T, Yee JK, Hopkins N (1994) Integration and germ-line transmission of a pseudotyped retroviral vector in zebrafish. Science 265(5172):666–669
16. Davidson AE, Balciunas D, Mohn D, Shaffer J, Hermanson S, Sivasubbu S, Cliff MP, Hackett PB, Ekker SC (2003) Efficient gene delivery and gene expression in zebrafish using the Sleeping Beauty transposon. Dev Biol 263(2):191–202
17. Kawakami K, Shima A, Kawakami N (2000) Identification of a functional transposase of the Tol2 element, an Ac-like element from the Japanese medaka fish, and its transposition in the zebrafish germ lineage. Proc Natl Acad Sci U S A 97(21):11403–11408. https://doi.org/10.1073/pnas.97.21.11403
18. Moreno-Mateos MA, Vejnar CE, Beaudoin JD, Fernandez JP, Mis EK, Khokha MK, Giraldez AJ (2015) CRISPRscan: designing highly efficient sgRNAs for CRISPR-Cas9 targeting in vivo. Nat Methods 12(10):982–988. https://doi.org/10.1038/nmeth.3543
19. LaFave MC, Varshney GK, Vemulapalli M, Mullikin JC, Burgess SM (2014) A defined zebrafish line for high-throughput genetics and genomics: NHGRI-1. Genetics 198(1):167–170. https://doi.org/10.1534/genetics.114.166769
20. Varshney GK, Zhang S, Pei W, Adomako-Ankomah A, Fohtung J, Schaffer K, Carrington B, Maskeri A, Slevin C, Wolfsberg T, Ledin J, Sood R, Burgess SM (2016) CRISPRz: a database of zebrafish validated sgRNAs. Nucleic Acids Res 44(D1):D822–D826. https://doi.org/10.1093/nar/gkv998
21. Varshney GK, Carrington B, Pei W, Bishop K, Chen Z, Fan C, Xu L, Jones M, LaFave MC, Ledin J, Sood R, Burgess SM (2016) A high-throughput functional genomics workflow based on CRISPR/Cas9-mediated targeted mutagenesis in zebrafish. Nat Protoc 11(12):2357–2375. https://doi.org/10.1038/nprot.2016.141
22. Pei W, Tanaka K, Huang SC, Xu L, Liu B, Sinclair J, Idol J, Varshney GK, Huang H, Lin S, Nussenblatt RB, Mori R, Burgess SM (2016) Extracellular HSP60 triggers tissue regeneration and wound healing by regulating inflammation and cell proliferation. NPJ Regen Med 1. https://doi.org/10.1038/npjregenmed.2016.13

Chapter 27

Microinjection of Marine Fish Eggs

Rie Goto, Taiju Saito, Takahiro Matsubara, and Etsuro Yamaha

Abstract

Microinjection is a powerful tool for studying embryonic development and analyzing gene functions in fish. This technique was first applied to model species of fish such as zebrafish and medaka whose egg chorions could be removed or softened before microinjection. Recent progress in genome editing using TALEN and CRISPR has opened the opportunity to analyze gene functions in a much wider range of fish including those important to marine aquaculture. Therefore, application of the microinjection technique is also required in these species. However, the characteristics of fish eggs vary widely between species and several technical difficulties need to be overcome in order to use microinjection in a wider range of species. To obtain consistent results with microinjection, an optimal method has to be developed for each target species. In this chapter, we describe the physical characteristics of the eggs of fish species that have been used in microinjection experiments in our laboratory and detail the microinjection system we developed for fish eggs with a hard chorion, such as those of marine species.

Key words Microinjection, Fish egg, Marine fish species, Glass needle, Yolk, Blastodisc, Aquaculture species

1 Introduction

Application of the microinjection technique to fish eggs, including those of species used for aquaculture, has become more important with the development of genome editing technologies such as TALEN and CRISPR. The microinjection method was initially developed in the model fish species, zebrafish and medaka [1, 2]. However, the conventional method is not always suitable for the eggs of other species of fish. The main factors that affect the outcome of microinjection are (1) the nature of the yolk cell, (2) the buoyancy of the eggs, (3) the width of the perivitelline space, (4) the hardness of the chorion, (5) the difficulty of removing the egg chorion, and (6) the stickiness of the chorion.

In most bony fish species, the eggs undergo meroblastic cleavage and, as a consequence, a thin region of the cytoplasm at the animal pole forms the embryo. After fertilization, the widely distributed cytoplasm moves toward the animal pole to form the

Chengyu Liu and Yubin Du (eds.), *Microinjection: Methods and Protocols*, Methods in Molecular Biology, vol. 1874, https://doi.org/10.1007/978-1-4939-8831-0_27,

Table 1
Fish species listed according to the yolk type and physical characteristics of their eggs

Yolk type	Common name[a] and an approximate diameter of egg[b]	Type of egg and physical character of egg chorion[c]
Single yolk mass 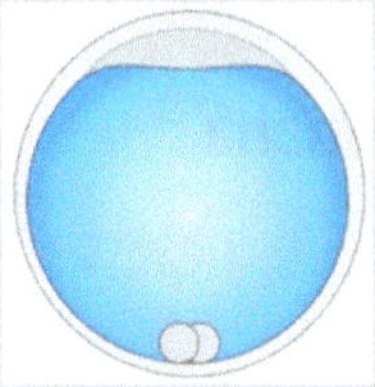	Kawakawa, bluefin tuna (1.0 mm), (1.0 mm)	Marine sp., pelagic, round, unremovable
	Barfin flounder (1.6–1.9 mm)	Marine sp., pelagic, round, unremovable
	Spotted halibut (1.5 mm)	Marine sp., pelagic, round, unremovable
	Thread-sail filefish (0.6 mm)	Marine sp., adherent demersal, round, unremovable
	Medaka (1.5 mm)	Fresh water sp., adherent demersal, round, soften
Yolk granules 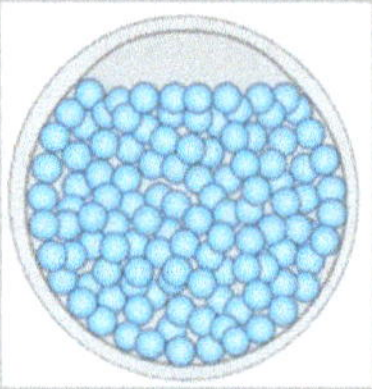	Japanese anchovy (L: 1.0 mm, S: 0.5 mm)	Marine sp., pelagic, spherical, unremovable
	Amberjack (1.2–1.4 mm)	Marine sp., pelagic, round, unremovable
	Pacific herring (L: 1.3–1.6 mm, S: 1.1–1.4 mm)	Marine sp., adherent demersal, round, soften
	Japanese eel (1.6 mm)	Marine sp., pelagic, round, removable
	Japanese smelt (1.0 mm)	Anadromous sp., adherent demersal, round, removable
	Arctic rainbow smelt (1.0 mm)	Anadromous sp., adherent demersal, round, removable
	Zebrafish (0.6 mm)	Fresh water sp., demersal, round, removable
	Goldfish, loach (2.0 mm), (1.1–1.4 mm)	Fresh water sp., adherent demersal, round, removable
	Rosy bitterling (L: 2.7 mm, S: 1.5 mm)	Fresh water sp., demersal, tear drop shaped, removable

[a]Species in the upper column possess multiple yolk granules and in the lower column possess a single yolk mass
[b]Egg size varies among environments, strains, maternal nutrients, etc.
[c]Chorion condition after enzymatic treatment after fertilization. The chorion of Pacific herring and medaka can be removed after a few hours treatment

blastodisc. Fish eggs vary widely among species in size, shape, and physical characteristics (Table 1) and can be categorized into two types using features of the yolk. Some fish have a single yolk mass in their eggs, others have multiple yolk granules. In the former group, the cytoplasm moves toward the animal pole on the surface of the unitary yolk to form the blastodisc. In the latter group, the cytoplasm moves between the yolk granules to the animal pole. In eggs with a single-yolk mass, an injectant needs to be directly introduced into the blastodisc since molecules cannot pass through the yolk-cytoplasm boundary (Fig. 1a–c). By contrast, in eggs with multiple yolk granules, such as the Japanese anchovy, an injectant can be introduced either directly into the blastodisc or into the yolk region; in both cases, the injectant is held in the cytoplasm and is able to move to the blastoderm with the cytoplasm during embryonic development (Fig. 1d–f). Therefore, the nature of the egg

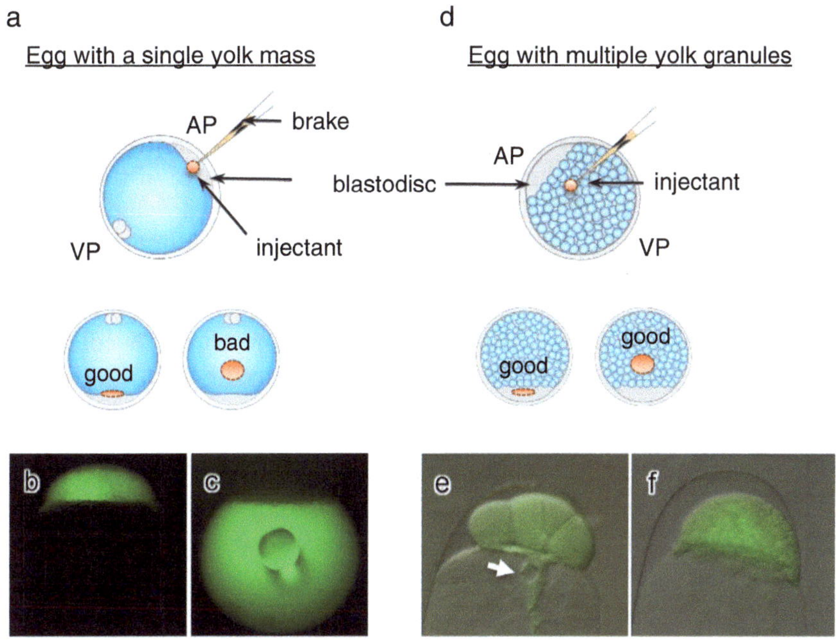

Fig. 1 Schematic illustration of microinjection into fish eggs. (**a**) Microinjection into the blastodisc of an egg with a single yolk mass. The injectant needs to be introduced directly into the blastodisc to achieve a good result, otherwise the injectant remains in the yolk of embryo which produces a bad outcome. (**b**) Blastoderm of a blastula stage embryo of kawakawa, *Euthynnus affinis,* labeled by injecting FITC-dextran at the 1-cell stage. (**c**) Injection of FITC-dextran into the yolk of a 1-cell stage embryo results in the retention of the label in the yolk at the blastula stage in kawakawa embryos. (**d**) The injectant can be introduced into either the blastodisc or the space between yolk granules. (**e**) Japanese anchovy embryo at the 8-cell stage after injecting *bucky ball-GFP* mRNA into the center of the fertilized egg. Arrow shows the cytoplasmic movement toward the blastoderm as the embryo develops. (**f**) Blastula stage Japanese anchovy embryo showing GFP-labeled cytoplasm that has completely moved to the blastoderm

determines the position of injection and is of primary importance for the success of microinjection.

After fertilization, the egg chorion gradually hardens as development proceeds [3]. In some species, treatment immediately after fertilization with commercially available proteolytic enzymes or hatching enzyme can soften or remove the egg chorion. These eggs or embryos can then be microinjected much more easily and reliably. However, softening or removing the chorion is much more problematic in many marine fish species. The eggs are buoyant (pelagic eggs) and stay on the surface of sea water with the animal pole facing downward as the blastodisc is heavier than the yolk. For microinjection, the egg needs to be fixed to a base material and positioned so that the blastodisc faces the microinjection needle. However, in pelagic eggs, the embryo can roll to the bottom even when the egg is fixed in place. This effect is particularly obvious in eggs with a large perivitelline space such as those of tuna and Japanese eel. Instability in the positions of the eggs, a hard chorion, and repeated attempts at injection can result in the breakage of the tip of the glass needle and make it difficult to control the pressure of injection. In addition, the eggs may have high internal pressure. In this case, the use of the needles normally employed for microinjection into eggs of model species may result in back flow of egg material due to the high internal pressure of the egg; this loss of material may disturb normal embryonic development.

Eggs of some fish species have an adhesive chorion that aids attachment to various surfaces [4, 5, 7]. This sticky nature can make it difficult to remove an egg from a microinjection glass needle after injection as the chorion can adhere to the needle. The adhesive layer of the eggs needs to be removed before injection.

To overcome these difficulties of microinjection into marine fish eggs, we have developed a modified microinjection system. A constriction, which we term the "brake," is included at the tip of the glass needle to allow a constant flow of injectant into the egg but prevent back flow of cytoplasm after puncture of the egg. Even if the tip of the glass needle is damaged, it does not affect the injection pressure. This type of needle was first reported in 1986 by Kishimoto who used it for microinjection into starfish oocytes [6]. It was then employed in a modified method for microinjecting into the eggs of some marine fish species [7] and to the pelagic eggs of flatfish [8]. In addition to a modified microinjection needle, skill at using a microscope is also important for microinjection of marine fish eggs. Most fish eggs possess a transparent embryonic body and yolk except for the eggs of species such as rosy bitterling that have nontransparent solid yolk. Use of a stereomicroscope with light transmitted throughout the base of the stage is essential to visualize transparent eggs and the injected substances in treated eggs.

2 Materials

2.1 Collecting Pelagic Eggs from the Aquarium Tank or the Net Cage

1. Fish net with a circular stainless steel frame attached to a rope 15–20 m long (*see* **Note 1**).
2. A 15–20 L container in which to transfer eggs to the injection site.
3. Glass beakers of different sizes (200–500 mL).
4. A flashlight.

2.2 Preparation of the Microinjection Workspace

1. A stereomicroscope with light transmitted throughout the base of the stage (*see* **Note 2**).
2. Manual microinjector.
3. Three-dimensional coarse manual manipulator with a simple magnetic stand and an iron plate.

4. A dual-stage glass micropipette puller.
5. Precision glass capillary tube.
6. Forceps and prepared glass pipettes with rubber nipples (*see* Subheading 3.3 and **Note 3**).
7. Extra-long tips for filling of microcapillaries.
8. Artificial or fresh seawater sterilized by passing through a 0.45 μm filter.
9. Incubator set at the optimal temperature for embryo development.
10. 1% penicillin/streptomycin solution in distilled water kept at −20 °C.
11. Agar plate, such as 100 mm glass dish, coated with a 0.5–0.8 cm thick layer of 1–2% agar.
12. Ethyl 3-aminobenzoate methanesulfonate salt (Tricaine/MS222) at a final concentration of 0.025%. Tricaine can be stored in the freezer as a 2.5% stock solution in water.

2.3 Instrument for Making Microinjection Needles (See Fig. 2)

1. Binocular compound microscope with a ×4 objective lens.
2. Voltage regulatory type power supply (0–6 V, 5 amp, A.C. 100 V) (*see* **Note 4**).
3. Two three-dimensional micromanipulators (*see* **Note 5**).
4. Two pipette holders (*see* **Note 6**).
5. Looped platinum wire, 0.3 mm (0.012 in.) diameter, 99.9% (metal content) (*see* **Note 7**).
6. Supporting glass capillary tube attached to a slide glass (*see* **Note 8**).
7. 2 cm alligator clips and copper wire.

3 Methods

3.1 Assembling the Instrument for Making the Constriction

1. Fix the alligator clips to the plate and add the pipette holder (Fig. 2a).
2. Cut the platinum wire into 3 cm lengths, and make a loop of 1 mm diameter at the middle of each segment of platinum wire (Figs. 2b and 3a).
3. Attach the micromanipulators to each side of the microscope to control the position of the looped platinum wire and the needle, respectively (Fig. 2a).
4. Connect alligator clips to the wire from the power supply connector (Fig. 2a).

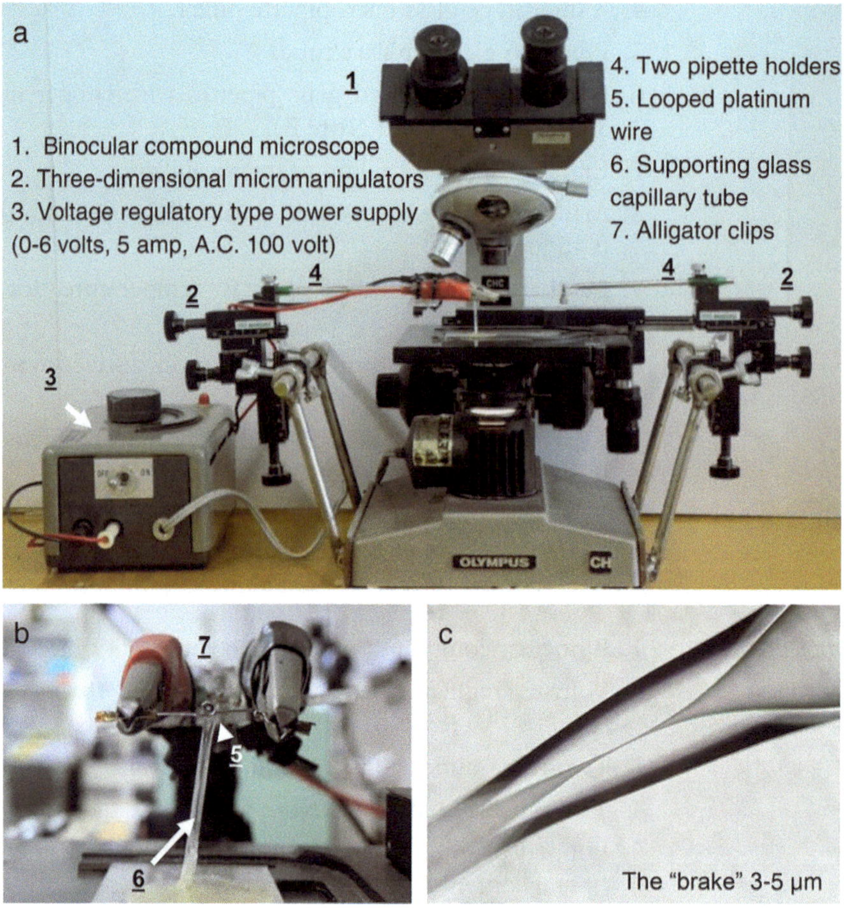

Fig. 2 The apparatus for adding the constriction "brake" to a microinjection needle. (**a**) The binocular compound microscope equipped with a unit to add the brake to the microinjection needle and a power supply for heating a looped platinum wire. (**b**) An image of the looped platinum wire captured from the side of the apparatus. (**c**) The constriction brake of a microinjection needle

5. Attach the alligator clips to both ends of the looped platinum wire (Figs. 2b and 3b, *see* **Note 9**).
6. Place a slide on the stage with the tip of the supporting glass facing the loop of the platinum wire (Figs. 2b and 3a).

3.2 Preparation of Microinjection Needle

1. Smooth both ends of the glass capillary tube by heating in order to protect the rubber ring of the microinjector.
2. The capillary tube is pulled to make two needles using a Dual-Stage Glass Micropipette Puller (*see* **Note 10**).
3. A constriction (termed a "brake") is formed in the upper part of the tip (approx. 1–2 mm from the tip) with a 3–5 μm inner diameter (Fig. 2c and *see* **Note 11**).

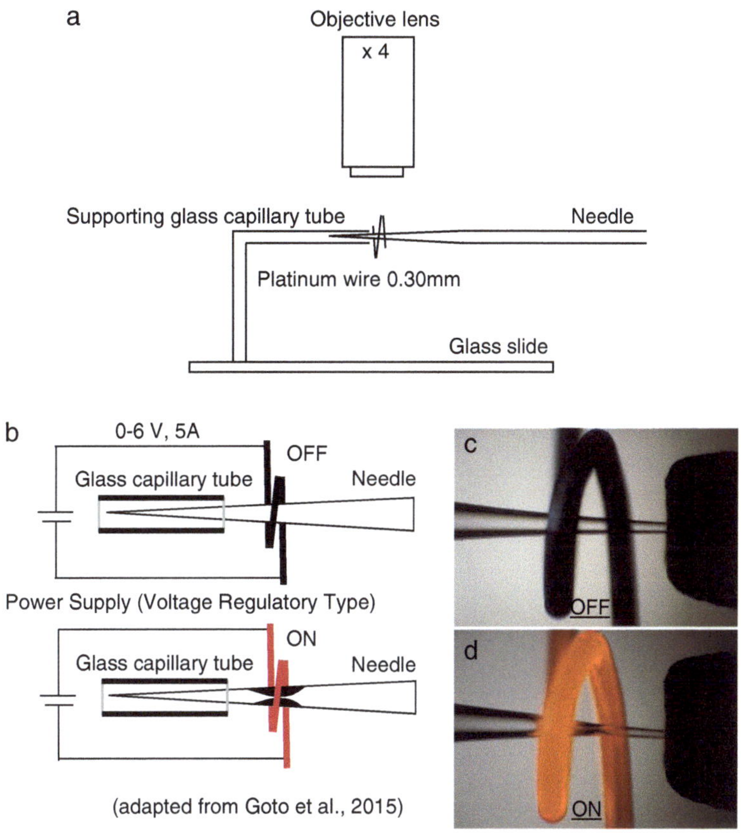

Fig. 3 Illustration of the setup for preparing glass needles for microinjection. (**a**) Schematic drawing of the front view showing the position of the microscope objective and the needle. (**b**) Schematic drawing of the image seen using the microscope as in (**a**). The microscopic views of a microinjection needle set for adding the brake before heating the looped platinum wire (**c**) and during heating to form the brake (**d**). Reproduced from [8] with permission

4. The relative positions of the microinjection needle and the platinum wire loop are controlled using the micromanipulators under the microscope (Fig. 3c).
5. The tip of the microinjection needle is inserted through the loop into the supporting glass capillary tube to hold the microinjection needle in a horizontal position (Fig. 3a).
6. The brake is formed by applying an electric current to the loop of platinum wire (Fig. 3b–d) (*see* **Note 12**).

3.3 Preparation of Glass Pipettes

1. Cut the tip of a glass pipette with a diameter larger than that of the eggs.
2. Heat the tip of the glass pipette to make it smooth in order to protect the eggs from rough surfaces.
3. Place a rubber nipple onto the glass pipette (*see* **Note 13**).

3.4 Preparation of RNA or DNA Solution for Injection

1. Dissolve RNA or DNA in 0.2 M KCl that has been sterilized by passing through a 0.22 μm filter.

   For example, 300–600 ng/μL of RNA for labeling primordial germ cells [7, 8], 50–200 ng/μL of Platinum TALEN RNA (unpublished data), 50–100 μM of morpholino oligos against the *dead end* gene [9], etc. (*see* **Note 14**).
2. Add dye to the injection solution to make it easier to judge the success or failure of injection (*see* **Note 15**).
3. Centrifuge injectant solution at maximum speed for 5 min just before use (*see* **Note 16**).

3.5 Loading Microinjection Needle and Attaching to the Needle Holder

1. Load a few μL of injectant solution into the needle above the brake.
2. Fill the needle holder with mineral oil, then insert the needle into the holder.
3. Set the plate coated with agar on the center of the stage and fill with one-third or half-concentration seawater containing 0.01% streptomycin/penicillin.
4. Slowly bring the needle down and dip the tip into the diluted seawater.
5. Apply a slight pressure to the needle using the microinjector. Then, touch the tip of the needle gently using the loading tip or forceps. The air between the tip and the brake will come out as fine bubbles and the injectant solution will pass through the brake. Check that the injectant solution continues to flow slowly from the tip of the needle.

3.6 Prepare the Injection Tray

1. After filling the agar plate with diluted seawater, make a small (egg-size related) hole on the agar surface into which the egg is positioned for injection (Fig. 5b). The depth of the hole should be about one-third that of the egg.

3.7 Collection of Fertilized eggs

3.7.1 Collection of Pelagic Eggs of Japanese Anchovy from a Rearing Tank

1. A few hundred adult Japanese anchovies can be kept in a 1 ton tank at a temperature of 20–26 °C and 14L:10D photoperiod with flow-through water. The fish are fed artificial pellets 3–4 times a day.
2. A few hours after light-off, Japanese anchovies start spawning in the tank and fertilized eggs rise to the surface of the seawater.
3. Through use of a relatively strong circulation in the seawater, fertilized eggs can be drawn down through a drain and discharged into a collection net.
4. The fertilized eggs are collected into a container, taking care to avoid exposure to air (*see* **Note 17**).

3.7.2 Collection of Pelagic Eggs of Kawakawa from a Net Cage

1. Several dozen mature kawakawa can be kept in a 600–1000 ton net cage (8–10 m depth). A few hours before sunset, males will chase females as part of the mating behavior.
2. After visual confirmation of spawning behavior, slowly pull up the net from different depths in the net cage.
3. Collected eggs are immediately transferred to a container with clean seawater and brought to the laboratory.

3.8 Microinjection into the Egg

1. Scoop eggs from the surface of the seawater and put in a glass container.
2. Shine a light onto the glass container from one side to make the floating eggs visible, then transfer the eggs to a dish filled with filtered seawater to remove other organisms (Fig. 4, STEP 1).
3. Transfer the eggs to the dish filled with diluted seawater, then rotate the glass dish horizontally by a hand to gather the eggs to the center (Fig. 4, STEP 2 and *see* **Note 18**).

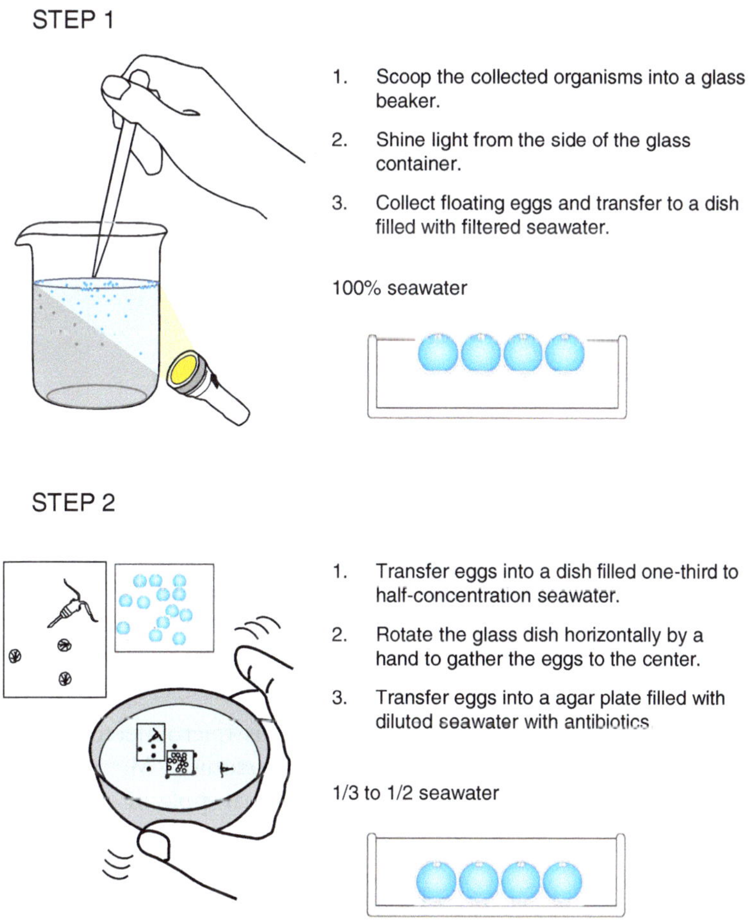

Fig. 4 Illustration of the preparation of pelagic eggs for microinjection

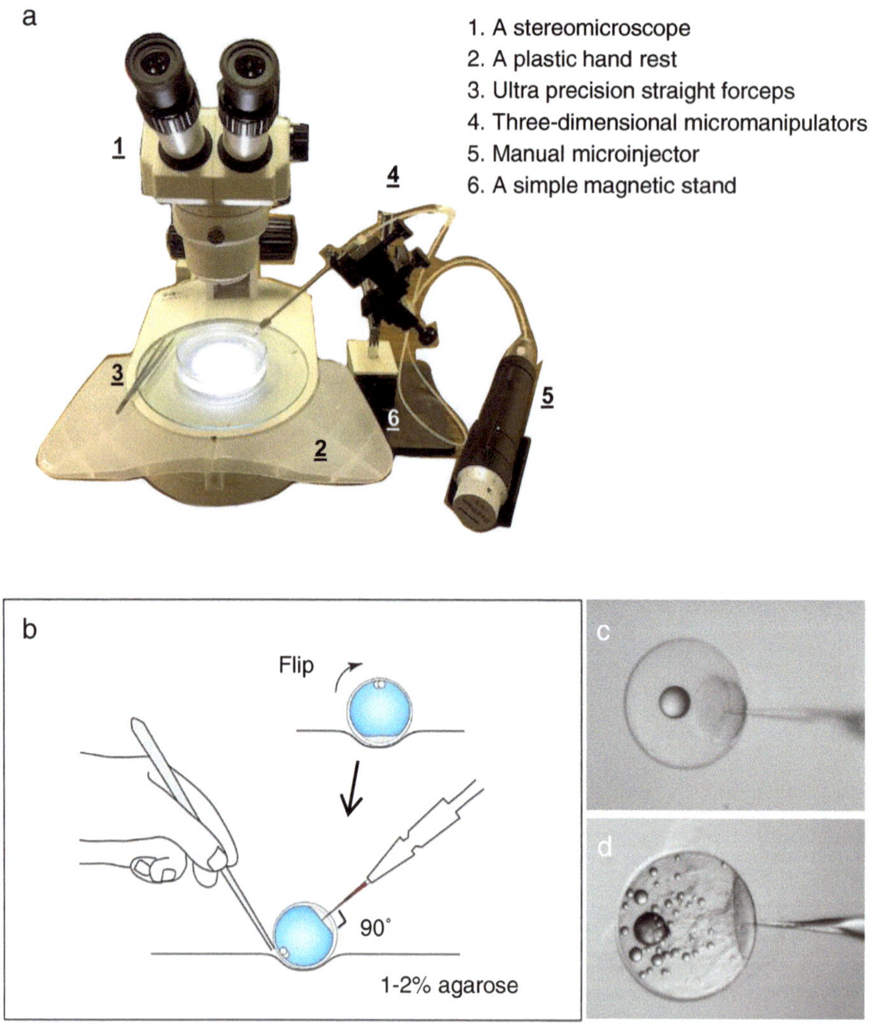

Fig. 5 Microinjection into pelagic eggs. (**a**) The stereomicroscope with peripheral equipment. (**b**) Positioning and angle of a pelagic egg, the injection needle and forceps for microinjection on the agar plate. (**c**) A microscopic image of microinjection into an egg of kawakawa, *Euthynnus affinis*, at 2-cell stage. (**d**) The microscopic image of microinjection into an egg of amberjack, *Seriola quinqueradiata*, at the 1-cell stage

4. Set one egg into the hole and orient it using forceps toward the microinjection needle (Fig. 5b–d) (*see* **Note 19**).
5. The injectant is microinjected into the base or center of the embryonic blastodisc. Microinjection should be performed from the animal pole if the egg has a single yolk mass (Fig. 1a). If the egg contains multiple yolk granules, injection can be performed from other positions as well as from the blastodisc (Fig. 1d).
6. After microinjection, the eggs are transferred to 100% seawater supplemented with antibiotics.

3.9 Culture the Eggs After Microinjection

1. Transfer each egg into a well of a multi-well plate filled with 100% filtered seawater supplemented with antibiotics.
2. Incubate the eggs at the optimal temperature for the species.
3. Under the microscope, transfer the egg or embryo into a dish filled with 2% methyl cellulose in 50% seawater and allow it to settle on the bottom of the dish (*see* **Note 20**). If the embryo needs anesthesia, use 0.025% tricaine in 50% seawater before the transfer to 2% methyl cellulose.
4. After observation, wash the specimen by gently pipetting with 50% seawater to remove methyl cellulose from its surface.

4 Notes

1. Kawakawa spawn not only at the water surface but also at the mid-level or bottom of a net cage. These eggs are the last to be found on the water surface; nevertheless, it is important to collect the eggs as soon as possible for microinjection. Therefore, a fish net with a long rope is useful for collecting eggs from any depth in the net cage.
2. Transmitted illumination through the stage base of the stereomicroscope enables visualization of transparent eggs and an adjustable mirror allows the optimal light position to be used. A compact base is better for stable microinjection since the magnetic holder can be set closer to the center of the stage.
3. Forceps needs to be long and fine enough to manipulate eggs but not sharp in order to prevent puncturing the egg chorion with the tip.
4. A stepwise supply voltage cannot be used for this instrument since heating the looped platinum wire needs fine control.
5. The three-dimensional manipulators should be fixed firmly on each side of the binocular compound microscope to control the position of the supporting glass capillary and the microinjection needle to add a "brake." In our laboratory, we added wing-like poles to the microscope and fixed the manipulator to them (Fig. 2a).
6. The pipette holder used for fixing the looped platinum wire needs to have a small plate at the front to set the alligator clips in place (Fig. 2a, b).
7. Cut 3–4 cm lengths of platinum wire, and make a smooth loop 1.0–1.5 mm size in diameter on each length by winding the wire onto a glass capillary tube.
8. Cut a 5 cm length of glass capillary tube and bend it to 90° by heating; glue the bent glass capillary to the surface of the glass slide (Figs. 2a and 3a).

9. If poor electrical conduction occurs, sand both ends of the looped platinum wire with copper foil to enable a better electric current flow.
10. Take care not to break the tip of the microinjection needle before injection. The microinjection needle should have a sharp, short, and rigid tip, especially when injecting into eggs <0.6 mm in diameter.
11. The "brake" enables the operator to control the pressure required to inject into the egg and also prevents back flow of cytoplasm from the egg.
12. The microinjection needle tends to bend after heating to form the brake. If the bend is small, then there is no problem in using the needle for microinjection. In such a case, place the needle so that it bends downward.
13. A disposable plastic pipette can also be used to handle eggs.
14. Concentration of Platinum TALEN RNA (50–150 ng/μL) has been validated in the kawakawa (unpublished data). A concentration of 50–100 μM morpholino oligos against the ***dead end*** gene is effective to ablate primordial germ cells in marine species including kawakawa, Japanese anchovy, and thread-sail filefish (unpublished data).
15. Select dyes depending on the application. We use 0.04% Brilliant Blue FCF to check the location of the injected solution and this dye disperses as development progress.
16. Centrifugation of the injection solution just before loading to the microinjection needle is important to prevent the brake in the injection needle from becoming clogged with undissolved material or dust.
17. Avoid using a fish net to collect Japanese anchovy eggs after spawning.
18. This step is useful if there are many different organisms such as *Noctiluca* spp. and copepods etc. present in the water. Gentle rotation of the dish arranges the organisms in a concentric pattern (Fig. 4).
19. Do not overfill the egg since pressure from the agar can induce cytoplasmic back flow after the removal of the needle.
20. The optimal dilution rate with seawater may have to be determined according to the target species. Storing the egg and the embryo in dilute seawater should be minimized to prevent any disturbance in development.

References

1. Rosen JN, Sweeney MF, Mably JD (2009) Microinjection of zebrafish embryos to analyze gene function. J Vis Exp (25):e1115. https://doi.org/10.3791/1115
2. Kinoshita M, Murata K, Naruse K, Tanaka M (2009) Microinjection technique for medaka eggs. In: Medaka: biology, management, and experimental protocols. Wiley-Blackwell, Ames, IA
3. Yamagami K, Hamazaki TS, Yasumasu S, Masuda K, Iuchi I (1992) Molecular and cellular basis of formation, hardening, and breakdown of the egg envelope in fish. Int Rev Cytol 136:51–92
4. Saito T, Psenicka M, Goto R, Adachi S, Inoue K, Arai K, Yamaha E (2014) The origin and migration of primordial germ cells in sturgeons. PLoS One 9(2):e86861
5. Takahashi E, Shimizu Y, Urushibara H, Kawakami Y, Arai K, Yamaha E (2017) Migration behavior of PGCs and asymmetrical gonad formation in pond smelt *Hypomesus nipponensis*. Int J Dev Biol 61:397–405
6. Kishimoto T (1986) Microinjection and cytoplasmic transfer in starfish oocytes. Methods Cell Biol 27:379–394
7. Saito T, Fujimoto T, Maegawa S, Inoue K, Tanaka M, Arai K, Yamaha E (2006) Visualization of primordial germ cells in vivo using GFP-nos1 3′UTR mRNA. Int J Dev Biol 50:691–699
8. Goto R, Saito T, Kawakami Y, Kitauchi T, Takagi M, Todo T, Arai K, Yamaha E (2015) Visualization of primordial germ cells in the fertilized pelagic eggs of the barfin flounder *Verasper moseri*. Int J Dev Biol 59:465–470
9. Goto R, Saito T, Takeda T, Fujimoto T, Takagi M, Arai K, Yamaha E (2012) Germ cells are not the primary factor for sexual sex determination in goldfish. Dev Biol 370:98–109

Chapter 28

Generating Gene Knockout *Oryzias latipes* and Rice Field Eel Using TALENs Method

Daji Luo, Ke Feng, Zuoyan Zhu, and Wei Hu

Abstract

Transcription activator-like effector nucleases (TALENs) methods based on engineered nucleases enable precise manipulations with genomic DNA. Within the field, genetic manipulation has surpassed the proof of principle stage and is now utilized in both applied strategies and performing in economic organism. The generation and study of gene modified (GM) fish using TALENs provides a novel and essential tool for fishes molecular breeding. The medaka, *Oryzias latipes*, is a small, freshwater fish and an idea model organism to study reproductive mechanism. Rice Field eel is a valuable model organism of high economic importance in PR China. In this chapter, we describe microinjection of *Oryzias latipes* and Rice Field eel embryos, in the context of preparing, evaluating, performing gene knockout and generate gene knockout fish with TALENs in detail, which are easy to prepare and proved to be efficient in targeting of a wide range of cleavage sites. Our procedure will facilitate broader applications of TALENs in freshwater aquaculture organism.

Key words *Oryzias latipes*, Rice Field eel, Gene knockout, Golden Gate assembly, TALEN method

1 Introduction

1.1 TALENs and Assembly

Modifications of genomes have laid the foundation of functional studies in modern biology and have led to significant discoveries [1]. Recently, Zinc finger nucleases (ZFNs) [2, 3], transcription activator-like effector nucleases (TALENs) [4, 5], and clusters of regularly interspaced short palindromic repeats (CRISPR) [6, 7] have been shown to edit genomic DNA in a variety of cell types and different model organisms at stable efficiency and specificity. Currently, a number of TALEN assembly platforms have been established such as Golden Gate assembly [8], PCR-based modular assembly [9], FLASH assembly [10], as well as some commercially available approaches including GeneArt ® Precision TALs (Life Technologies) and Cellectis bioresearch. We have chosen the Golden Gate assembly in our studies, which is one of easy, fast, cost effective, and therefore suitable for most laboratories. In

Chengyu Liu and Yubin Du (eds.), *Microinjection: Methods and Protocols*, Methods in Molecular Biology, vol. 1874,
https://doi.org/10.1007/978-1-4939-8831-0_28,

addition, we modified the Golden Gate method [8] using two constructs, pCS2-ELD and pCS2-KKR, which are feasible for microinjection into both zebrafish and other fish embryos with high gene disruption efficiency [11, 12]. Using our method, we performed a high efficiency gene disruption in zebrafish [12–14].

Many studies and our studies demonstrated that TALENs provide an efficient method to achieve targeted gene (containing noncoding RNAs) and genome modifications in model organisms. It is now utilized TALENs method in both applied strategies and performing in economic organism. Using TALENs, we perform highly efficient gene targeting both transgenic gene and endogenous gene in *Oryzias latipes* [15–17] and Rice Field eel *Monopterus albus* [18]. The generation and study of gene modified (GM) fish using TALENs provides a novel and essential tool for fishes molecular breeding. Here, we provide protocols for generating and screening TALENs induced gene disruption mutants of *Oryzias latipes* and Rice Field eel *Monopterus albus* in detail.

1.2 Germline Transmission of TALENs

For economic animals, on the one hand, it is important to successfully perform TALENs to generate mutants with excellent traits; and on the other hand, it is more important to efficiently transmit the knockout genotypes to next generation to easily generate homozygous gene knockout animals line. Germline transmission of the knockout genotypes is critical to obtain homozygous gene knockout animals. It is more difficult to select an ideal targeted TALENs site that there are incomplete and inaccurate genomic sequence and genetics information in economic animals. Generally, mutation rate of the F0 in economic animals may not be ideal. As a complement, this problem can be solved by modifying elements of TALEN mRNAs. To improve the germline integration efficiency, we incorporated the zebrafish nanos-3′ UTR into the TALEN construct, which was reported to protect mRNA from degradation in primordial germ cells and improve the germ cell targeting efficiency [12, 19]. Our results provide that DMY-nanos3UTR-TALEN induced higher proportion of mutant gametes than DMY-TALEN induced that in the F0 medaka [16]. This method may also be useful for aquaculture breeding. Thus, the incorporation of the specific elements into the TALEN construct improves the germ cell targeting efficiency in aquaculture production, permitting the generation knockout lines. We also introduce this in this chapter.

1.3 Off Targeted Sites Evaluation of TALENs

Specificity is essential to establish precisely targeted gene knockout lines. TALENs have become an accepted tool for targeted mutagenesis, but undesired off-targets, in addition to the targeted genomic region, remain an important issue [4, 11, 20, 21]. To test for potential nonspecific mutations induced by TALENs, we also developed a simple and reliable off-target prediction program for confirming TALEN-induced mutation. The criteria of the novel program for determining off-target sites were that the 0 position of the EBEs must be T, from 1 to 10 mismatch bases occur in the pairs of EBEs, and the spacer between the two putative EBE regions is less than 100 bp, because it has been suggested that longer spacers interfere with Fok I dimerization. In our studies, our program was more efficient than ePCR, Primer3 and Blast to predict TALENs off-target sites [15, 16]. Recently, tools for predicting TALEN off-targets have been developed, such as idTALE (http://idtale.kaust.edu.sa), Paired Target Finder (https://tale-nt.cac.cornell.edu), and TALENoffer [22]. However, a more detailed investigation on the possible off-target effects of TALENs and more accurate program will be needed in the future.

In this chapter, we provide the detailed protocols that focused on the generation of and screening for targeted *Oryzias latipes* and Rice Field eel mutants generated by microinjection of TALEN mRNAs into their embryos. We believe that Our procedure will facilitate broader applications of TALENs in freshwater aquaculture organisms.

2 Materials

All water in molecular operation required for double-distilled. All plasmids for microinjection must be prepared using an endotoxin-free kit.

2.1 Instruments

1. Stereomicroscope.
2. Pressure Microinjector.
3. PN-30 Puller.
4. Thermocycler.
5. Nanodrop.
6. Centrifuge.
7. Themomixer.
8. Constant temperature Incubator.
9. Thermostatic Shaker.

2.2 TALENs Assembly

1. Golden Gate vectors, Addgene Cat #1000000016.
2. pCS2-TALEN-ELD and pCS2-TALEN-KKR (*see* **Note 1**).
3. DH5α.
4. LB Broth medium.
5. LB Agar medium.
6. Spectinomycin.
7. Tetracycline.

8. Ampicillin.
9. QIAprep Spin Miniprep Kit.
10. Plasmid-Safe™ ATP-Dependent Dnase.
11. T4 DNA ligase (*see* **Note 2**).
12. BsaI.
13. BsmbI.
14. XbaI.
15. NotI.
16. RNeasy MiniElute Cleanup Kit.
17. mMessage mMachine SP6 Kit.
18. RNase-free water, tips, and Eppendorf tubes.
19. Primers:
    pCR8_F: 5′-TTGATGCCTGGCAGTTCCCT-3′
    pCR8_R: 5′-CGAACCGAACAGGCTTATGT-3′
    NTalF: 5′-GATGACAAGGGTACCGTG-3′
    CTalR: 5′-CTAGTTGGGATCCGGCAAC-3′

2.3 Manipulation of Oryzias latipes and Rice Field Eel Embryos

1. Orange strains of *Oryzias latipes* or wild-type of Rice Field eel.
2. Embryos Buffer (1 × PBS with 1 × Penicillin-streptomycin solution) (*see* **Note 3**).
3. Penicillin-streptomycin solution.
4. Microinjection molds to hold embryos for microinjection.
5. Phenol red solution.
6. Dumont #5 Tweezer to break injection needles.
7. Disposable capillaries.
8. Glass capillaries.
9. Microloader tips.
10. Nitrogen gas, dishes, and tank.

2.4 Detection of Mutations and Germline Transmission

1. Taq DNA polymerase.
2. 10 mM dNTP.
3. pMD18-T vector.
4. LB Broth medium.
5. LB Agar medium.
6. Ampicillin.
7. Gene-specific primers.

3 Methods

3.1 TALENs Assembly

3.1.1 TALENs-Targeted Site Selection

1. A full TALEN target site is composed of two half-sites separated by a spacer. Most of the indel-mutations will occur in the spacer region. To disrupt a protein-coding gene, the targeted region should preferably be placed in the exons at the end or the exon encoding the functional domain. According to gene sequence, choose the ideal TALEN target site. For each target, a pair of TALEN monomers should be constructed to bind the left and the right half-site (Fig. 1).
2. Each half-site sequence of TALEN is chosen as three rules. First, position 0 ought to be T. Second, the targeted half-site length of TALEN site (left and right, respectively) is strongly recommended to be 16–18 bp. Third, there is less than five G in each half site of TALEN (*see* **Note 4**).
3. The length of the spacer between the left and the right half-site of TALEN is around 16–18 bp (*see* **Note 5**).

3.1.2 First-Step Golden Gate Assembly

1. For each obtained clone (Fig. 2), culture the bacteria in 5 mL LB with appropriate antibiotics (tetracycline, spectinomycin, or ampicillin) overnight. Extract plasmid DNA using the Plasmid Miniprep Kit. Adjust the concentration of these plasmids (especially, 40 RVD modular plasmids) to 100 ng/μL (*see* **Note 6**).
2. Label the 5′-T of TALEN half-site as position 0 and the remaining nucleotides with serial numbers (e.g., 1–10 and X). Pick up pFUS_A1 and RVDs plasmids recognizing positions 1–10. Pick up pFUS_Bn (e.g., if pFUS_B is to contain 5 RVDs, pick up pFUS_B5 as backbone) and RVDs plasmids recognizing positions 11 to (X – 1–10) (*see* **Note 7**).

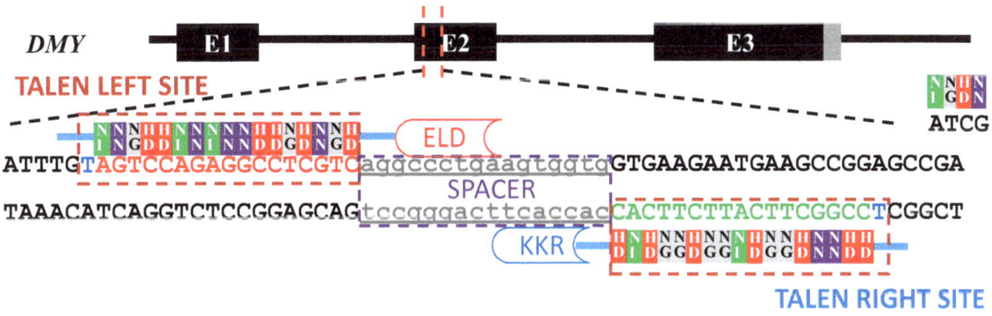

Fig. 1 The design of TALEN sites. Full TALEN site contains TALEN left site (blue and red capital letters, labeled into red dashed box), TALEN right site (blue and green capital letters, labeled into blue dashed box) and Spacer (gray lowercase, labeled into purple dashed box). Module unit NI recognize A, NG recognize T, HD recognize C, NN recognize G

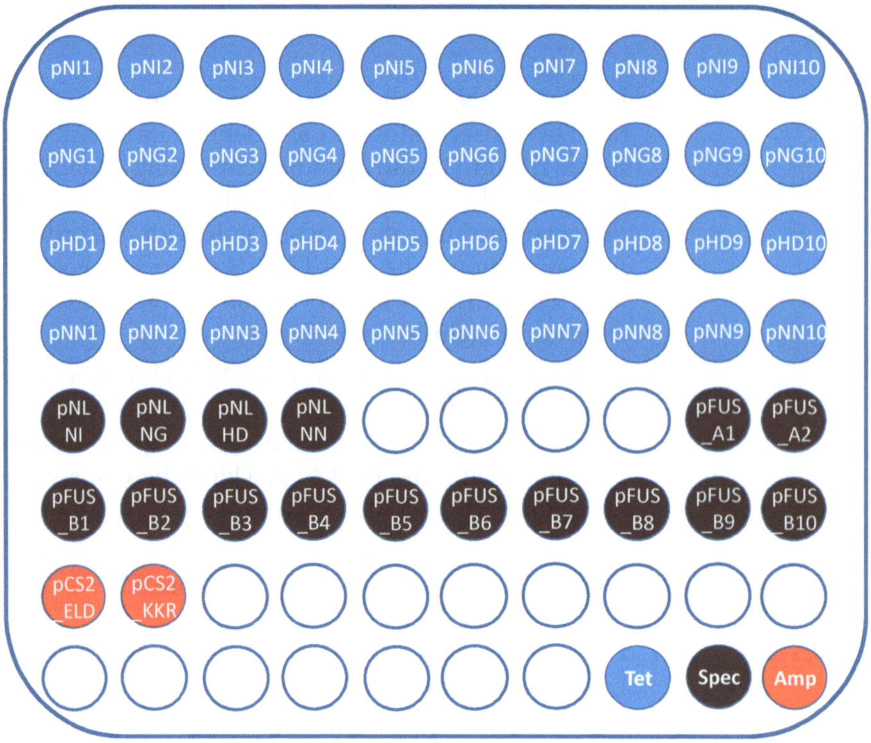

Fig. 2 The contents of the Golden Gate TALEN kit. The toolkit includes 40 RVD modular plasmids (blue), 12 middle array plasmids (black), four last repeat plasmids (pNL_NI, NG, HD, NN, black), and two TALEN expression plasmids (pCS2-ELD and pCS2-KKR). The 40 modular plasmids recognize four genetic bases at ten different positions. The middle array plasmid pFUS_A1 serves as the vector backbone for insertion of the RVDs recognizing nucleotides at positions 1–10. If the length of TALEN recognized site is over 21 bp; pFUS_A2 serves as that of first step assembly for the insertion of the RVDs recognizing nucleotides at positions 11–20. Then, pFUS_B1 to pFUS_B10 serve as the vector backbones after pFUS_A2 step assembly for insertion of RVDs recognizing nucleotide at positions 21–30. If the length of TALEN recognized less than 20, pFUS_B1 to pFUS_B10 serve as the vector backbones of first step assembly for insertion of RVDs recognizing nucleotide at positions 11–20. The last-repeat plasmids recognize the nucleotide in the last position. The two TALEN expression plasmids serve as the vector backbones for insertion of the assembled RVDs to express functional TALENs. The antibiotic property of each plasmid is coded by different colors. Tet, tetracycline resistance, show in blue; Spec: spectinomycin resistance, show in black; Amp: ampicillin resistance, show in red

3. Set up the first step Golden Gate assembly reaction as tabulated in Tables 1 and 2, then run the reaction in Thermocycler as following program:

   Six cycles of 37 °C for 20 min, 16 °C for 20 min;

   Hold at 37 °C for 15 min;

   Finally hold at 80 °C for 5 min.

4. Add 1 μL Plasmid-Safe nuclease to the reaction mixture, then incubate it at 37 °C for 30 min.
5. Add all mixture into 100 μL HD5a competent cells. Incubate on ice for 10 min. Heat-shock at 42 °C for 90 s. Then, hold on

Table 1
Components of first step Golden Gate assembly for pFUS_A

Components	Amount (μL)
10_ NEB Buffer	1
25 mM ATP	0.4
RVD Plasmids 1–10 (100 ng/μL)	0.6 (for each)
pFUS_A (100 ng/μL)	1.0
BsaI	0.8
T4 DNA Ligase H.C.(NEB)	0.8

Table 2
Components of first step Golden Gate assembly for pFUS_Bn

Components	Amount (μL)
10_ NEB Buffer	1
25 mM ATP	0.4
RVD Plasmids 1–*n* (100 ng/μL)	0.6 (for each)
pFUS_Bn (100 ng/μL)	1.0
BsaI	0.5
T4 DNA Ligase H.C. (NEB)	0.5
ddH_2O	Adjust to 10

ice for 10 min. Add 200 μL LB medium into it and culture at 37 °C with shaking at 200 rpm for 30 min. Centrifuge it at 4,000 g for 3 min. Remove the supernatant and reconstitute it. Smear it on the Spec-plate (supplemented with 50 μg/mL spectinomycin). Hold the plate at 37 °C, overnight.

6. Randomly pick up six clones from pFUS_A plate and two clones of pFUS_Bn into 50 μL LB medium (spec+), culture it at 37 °C for 2 h, shaking at 250 rpm (*see* **Note 8**).
7. Prepare PCR reaction in Table 3 and run the reaction in Thermocycler as following program:

   Denature template at 95 °C for 5 min;

   32–35 cycles of 95 °C for 30 s, 52 °C for 30 s, and 72 °C for 60 s;

   Hold at 72 °C for 5 min.
8. Run the PCR products on 1.5% agarose gel. Smearing ladder bands will appear in the correct clones (Fig. 3).

Table 3
Components of final TALEN assembly for TALEN left and right sites

Components	Amount (μL)
10× Taq DNA polymerase buffer	2.0
10 mM ATP	0.4
pCR8_F1 (10 μM)	1.0
pCR8_R1 (10 μM)	1.0
pFUS_A or pFUS_Bn plasmid template	1.0
Taq DNA polymerase	0.2
ddH_2O	Adjust to 20

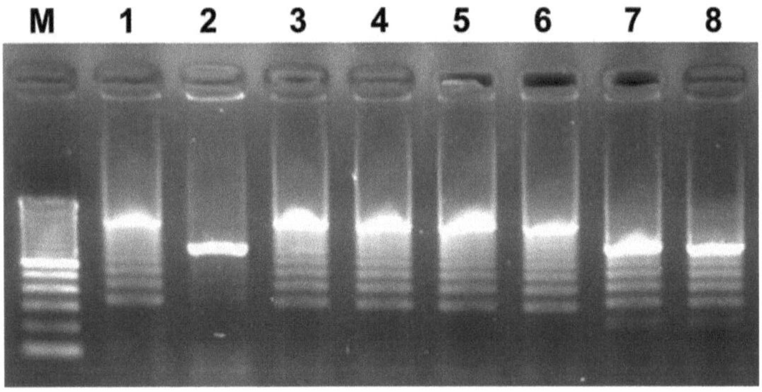

Fig. 3 Representative agarose gels of first-step Golden Gate assembly. M: Marker, Invitrogen 100 bp ladder. Lanes 1–6: pFUS_A; Lane 1, 3–6: the correct pFUS_A array containing 10 RVDs; Lane 2 the typical failure of pFUS_A assembly; Lanes 7 and 8: the correct pFUS_B containing 5 RVDs

9. Culture correct pFUS_A and pFUS_Bn clone with 1.5 mL LB medium (spec+) overnight.
10. Extract pFUS_A and pFUS_Bn plasmid DNA with QIAprep Spin Miniprep Kit. Elute the plasmid DNA with 30 μL ddH_2O. Measure plasmid DNA concentrations with Nanodrop. Adjust the concentration of these plasmids to 100 ng/μL.

3.1.3 The Final TALEN Plasmid Assembly

1. Set up the reaction as described in Table 4. TALE repeats recognizing the left half-site were cloned into pCS2-ELD and those recognizing the right half-site were cloned into pCS2-KKR (*see* **Note 9**).
2. Run the reaction in Thermocycler as following program:

Table 4
Components of first step Golden Gate assembly for pFUS_Bn

Components	Amount (μL)
10_ NEB Buffer	1.0
25 mM ATP	0.4
pFUS_A (100 ng/μL)	1.0
pFUS_Bn (100 ng/μL)	1.0
pCS2-ELD or pCS2-KKR (100 ng/μL)	1.0
pLR (100 ng/μL)	1.0
Esp3I	0.4
T4 DNA Ligase H.C. (NEB)	0.4
ddH_2O	Adjust to 10

Table 5
Components of final TALEN assembly for TALEN left and right sites

Components	Amount (μL)
10× Taq DNA polymerase buffer	2.0
10 mM ATP	0.4
NTalF (10 μM)	1.0
CTalR (10 μM)	1.0
Final ELD or KKR plasmid template	1.0
Taq DNA polymerase	0.2
ddH_2O	Adjust to 20

Six cycles of 37 °C for 15 min, 16 °C for 15 min;

Hold at 37 °C for 15 min;

Hold at 80 °C for 5 min.

3. Add all mixture into 100 μL HD5a competent cells. Incubate on ice for 10 min. Heat-shock at 42 °C for 90 s. Then, hold on ice for 10 min. Add 200 μL LB medium into it and culture at 37 °C with shaking at 200 rpm for 30 min. Centrifuge it at 4,000 g for 3 min. Remove the supernatant and reconstitute it. Smear it on the Amp-plate (supplemented with 50 μg/mL ampicillin). Hold the plate at 37 °C, overnight.
4. Randomly pick up three clones from the plate into 50 μL LB medium (Amp+), culture it at 37 °C for 2 h, shaking at 250 rpm.

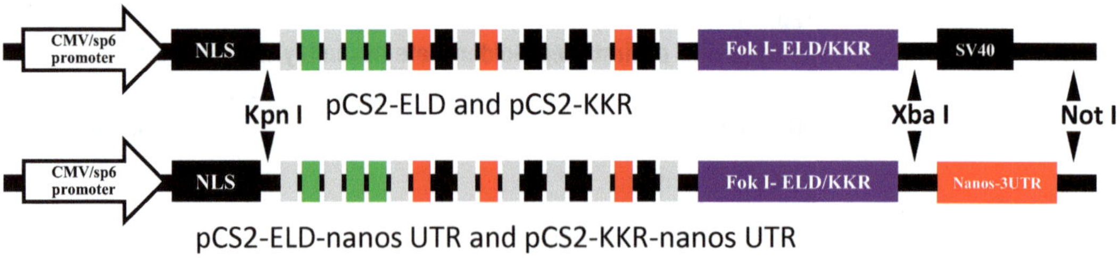

Fig. 4 Reconstructions of pCS2-ELD-nanos3UTR and pCS2-KKR-nanos3UTR plasmids. The zebrafish nanos-3UTR was cloned into the end of the pCS2-ELD and pCS2-KKR coding sequence between XbaI and NotI, replacing SV40

5. Prepare PCR reaction in Table 5 and run the reaction in Thermocycler as following program:

   Denature template at 95 °C for 5 min;

   30 cycles of 95 °C for 30 s, 52 °C for 30 s, and 72 °C for 90 s;

   Hold at 72 °C for 5 min.

6. Run the PCR products on 1.5% agarose gel. Strong band range between 1700 and 2000 bp will appear in the correct clones. Sequence the clone for double check.
7. Culture correct clone with 5 mL LB medium (Amp+) overnight.
8. Extract final TALEN left and right plasmid DNA with QIAprep Spin Miniprep Kit. Elute the plasmid DNA with 30 μL ddH_2O. Measure plasmid DNA concentrations with Nanodrop (*see* **Note 10**).
9. If necessary, the zebrafish nanos-3UTR was cloned into the end of the pCS2-ELD and pCS2-KKR coding sequence between XbaI and NotI, replacing SV40. pCS2-ELD-nanos3UTR and pCS2-KKR-nanos3UTR plasmids are prepared for microinjection (Fig. 4) (*see* **Note 11**).

3.1.4 TALEN mRNA Preparation

1. Linearize pCS2-ELD and pCS2-KKR plasmids (pCS2-ELD-nanos3UTR and pCS2-KKR-nanos3UTR plasmids) with NotI as described in Table 6. Purify the digested plasmid using the PCR Purification Kit. Elute the plasmid DNA with 30 μL RNase-free water (*see* **Note 12**).
2. TALEN mRNAs for microinjection were prepared using the mMESSAGE mMACHINE SP6 kit. Set up in vitro transcription reaction as shown in Table 7. Incubate the mixture at 37 °C for 2 h. Add 1 μL Tubo DNase and incubate the mix at 37 °C for 15 min.
3. Purify the mRNA using RNeasy Mini Kit referring to the manufacturer's instructions in detail. Elute the mRNA with 30 μL RNase-free water.

Table 6
Components of final TALEN plasmid linearization

Components	Amount (μL)
10× NEB buffer 3	2.0
Not I	1.0
TALEN_L/TALEN_R (10 μg)	Variable
ddH_2O	Adjust to 20

Table 7
Components of in vitro transcription of final TALEN mRNA (total volume is 20 μL)

Components	Amount (μL)
NotI-linearized plasmid DNA	6.0
2XNTP/ATCA	10.0
10× SP6 reaction buffer	2.0
SP6 enzyme mix	2.0

4. Check the quality and integrity of RNAs on a denatured agarose gel and measure the concentration of RNAs on the NanoDrop (*see* **Note 13**).

3.2 Manipulation of Oryzias latipes and Rice Field Eel Embryos

1. The Orange strain of medaka was used in our studies (Originally from National University of Singapore). Medakas were maintained in aquaria under an artificial 14 h light and 10 h dark photoperiod at 26 °C in the Institute of Hydrobiology, Chinese Academy of Science.
2. In the evening before microinjection, pick up several pairs of medaka into individual breeding tanks. Separate the male and female fish. In the morning of microinjection, let male and female fish mate. Collect eggs at the bottom of the breeding tank.
3. Prepare the injection solution, the pair of TALEN mRNAs should be mixed equally (*see* **Note 14**).
4. Microinjection could be performed as medaka handbook (Fig. 5a) (*see* **Note 15**).
5. Calculate the number of the dead/deformed/normal embryos at 72 h post fertilization. Remove the dead and severely deformed embryos. Sample several embryos to evaluate mutation induced by TALENs and cultivate others into adulthood. Then, evaluate germline transmission of mutation.

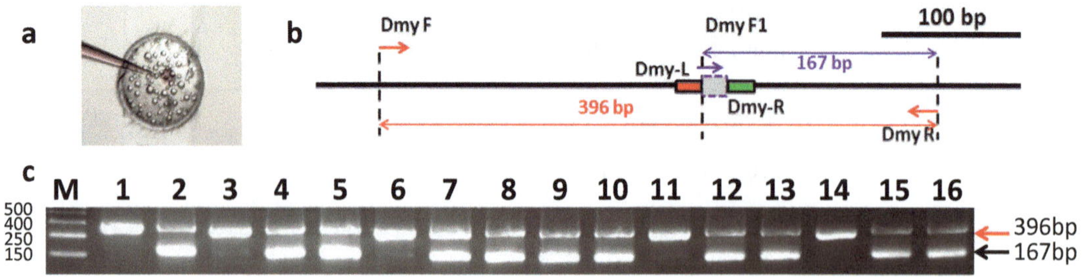

d

Medaka *DMY*-TALENs F0 mutations

```
WT      TAGTCCAGAGGCCTCGTCAGGCCCTGAAGTGGTGGTGAAGAATGAAGCCGGA   WT
No.1    TAGTCCAGAGGCCTCGTCAGGCC------TGGTGGTGAAGAATGAAGCCGGA   Δ6
No.3    TAGTCCAGAGGCCTCGTCAG---------TGGTGGTGAAGAATGAAGCCGGA   Δ9
No.6    TAGTCCAGAGGCCTCGTCAcc--------TGGTGGTGAAGAATGAAGCCGGA   Δ8   (Δ10,+2)
No.11   TAGTCCAGAGGCCTCGTCAGGCC------TGGTGGTGAAGAATGAAGCCGGA   Δ6
No.14   TAGTCCAGAGGCCT---------CTGAAGTGGTGGTGAAGAATGAAGCCGGA   Δ9
```

Fig. 5 Microinjection and efficiency evaluation of *dmy* TALEN mRNA. (**a**) Microinjection of *dmy* TALEN mRNA into medaka embryos. (**b**) The design of mutations detection in TALEN targeted medaka embryos. Primers *dmy* F, *dmy* F1 and *dmy* R were used to amply the *dmy* gene. If primer *dmy* F and *dmy* R generated a PCR fragment, while primer *dmy* F1 and *dmy* R failed to do so, this suggested that the targeted gene is disrupted by the *dmy* TALEN. (**c**) Electrophoretic detection of mutations in TALEN-injected medaka embryos. Lines 1–16, TALENs injected embryos. 1, 3, 6, 11, 14 show mutated embryos. (**d**) Sequencing detection of mutations in TALEN-induced medaka embryos. -, deleted nucleotide; lowercase letter, added nucleotide. +, insertions; Δ, deletions

3.3 Detection of Mutations and Germline Transmission

3.3.1 Detection of Mutations in F0 Generation

1. At 72 h post fertilization, randomly collect and pool 8–10 microinjected medaka embryos for genomic DNA isolation.
2. Detection of mutations in TALEN-targeted medaka embryos.

   e.g., *dmy* TALEN mRNA induced *dmy* gene disruption. Primers *dmy* F, *dmy* F1 and *dmy* R were used to amply the *dmy* gene. Primers *dmy* F and *dmy* R bridge both EBE regions, while primers *dmy* F1 and *dmy* R link the spacer region and the downstream EBE. *dmy* F and *dmy* R generated the 396 bp PCR fragment. *dmy* F1 and *dmy* R generated the 167 bp PCR fragment. If primer *dmy* F and *dmy* R generated a PCR fragment, while primer *dmy* F1 and *dmy* R failed to do so, this suggested that the targeted gene is disrupted by the *dmy*-TALEN (Fig. 5b).
3. Generate PCR reaction of detecting mutations as described in Table 8, and run it as following program in thermocycler:

   Denature template at 95 °C for 5 min;

   30 cycles of 95 °C for 30 s, 55 °C for 30 s, and 72 °C for 90 s;

   Hold at 72 °C for 10 min.
4. Gel purification of target bands. Elute DNA with 30 μL ddH_2O.

Table 8
Components of gene-specific PCR reaction of genomic DNA

Components	Amount (μL)
10× Taq DNA polymerase buffer	2.0
10 mM ATP	0.4
Gene specific forward primer (10 μM)	1.0
Gene specific reverse primer (10 μM)	1.0
Genomic DNA	1.0
Taq DNA polymerase	0.1
ddH_2O	Adjust to 20

Table 9
Components of TA clone

Components	Amount (μL)
pMD18-T vector	1.0
Insert DNA (>0.3 pmol)	4.0
Solution I	5.0

5. Ligate the purified PCR product into pMD18-T vector as tabulated in Table 9.
6. Transform the ligation product into 100 μL HD5a competent cells. Incubate on ice for 10 min. Heat-shock at 42 °C for 90 s. Then, hold on ice for 10 min. Add 200 μL LB medium into it and culture at 37 °C with shaking at 200 rpm for 30 min. Centrifuge it at 4,000 g for 3 min. Remove the supernatant and reconstitute it. Smear it on the Amp-plate (supplemented with 50 μg/mL ampicillin). Hold the plate at 37 °C, overnight.
7. Pick up 32 single clones and culture these clones in a 50 μL amp+ medium on a 96-well plate with shaking at 37 °C, 250 rpm for 3–4 h.
8. Set up competitive PCR as described in Table 10 and run it as following program in Thermocycler:

   Denature template at 95 °C for 5 min;

   32 cycles of 95 °C for 30 s, 55 °C for 15 s, and 72 °C for 30 s;

   Hold at 72 °C for 5 min.
9. Run 1.5% agarose gel to detect mutations (Fig. 5c).
10. Sequence clones to double check, blast the sequencing results at Ensembl database to identify the mutated genotypes (Fig. 5d).

Table 10
Components of PCR reaction of detecting mutations

Components	Amount (μL)
10× Taq DNA polymerase buffer	2.0
10 mM ATP	0.4
Gene specific forward primer (10 μM)	0.5
Gene specific reverse primer (10 μM)	1.0
Spacer region forward primer (10 μM)	0.5
Subclone template	1.0
Taq DNA polymerase	0.1
ddH_2O	Adjust to 20

3.3.2 Evaluation of Germline Transmission

1. Raise the TALEN-injected embryos to adulthood.
2. Randomly pick out several mature F0 medaka and cross with wild-type medaka, individually.
3. Collect and pool 8 to 10 F1 medaka embryos from each cross, for genomic DNA isolation, genomic PCR, colony assay and sequencing as described in Subheading 3.3.1, then evaluate the efficiency of germline transmitted indel-mutations.
4. Cut a small piece of the tail fin for genomic DNA isolation for genotyping the F1 adults with indel-mutations.

3.4 Off-Target Analysis

1. Use e-PCR to perform BLAST searches to identify potential off-target sites.
2. Or use idTALE (http://idtale.kaust.edu.sa), Paired Target Finder (https://tale-nt.cac.cornell.edu), and TALENoffer.
3. To test for potential nonspecific mutations induced by TALENs, we designed a program (Please forward such request to Prof. Wei Hu or Prof. Xiaoqin Xia) (*see* **Note 16**).

   E.g., we use this program to scan the medaka genomic sequence (http://www.ensembl.org/Oryzias_latipes) to identify potential off-target sites potentially targeted by *dmy* TALENs.

3.5 Screen and Establishment of Mutant Line

1. Rapid mode to obtain mutants (Fig. 6a)

   TALEN-injected medaka embryos were raised to sexual maturity. F0 mutated males were crossed with F0 mutated females. After 90 dpf, screen and find heterozygous genotype homozygous mutants in F1 generation (*see* **Note 17**).
2. Typical genetic mode to generate homologs mutant line (Fig. 6b).

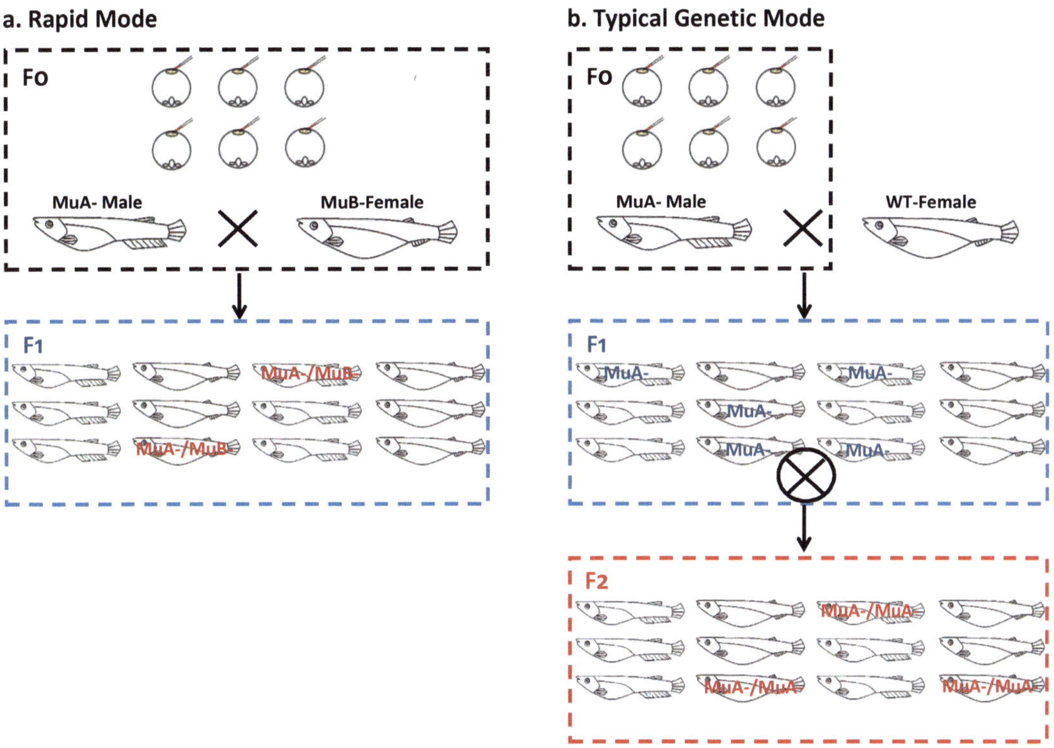

Fig. 6 Rapid mode and typical genetic mode to generate homologs mutants. (**a**) Flowchart to generate heterozygous genotype homozygous mutants in rapid mode; (**b**) Flowchart to generate typical genetic homozygous mutants in typical genetic mode. F0: F0 generation of TALEN mRNA microinjection; F1: F1 generation; F2: F2 generation. MuA-: genotype A of specific gene mutation; MuB-: genotype B of specific gene mutation

TALEN-injected medaka embryos were raised to sexual maturity. F0 mutated males were crossed with wild-type females; and F1 embryos were collected at 72 h post fertilization. Genomic DNA was extracted from 8 to 10 randomly pooled embryos to assess mutagenesis at the TALEN-targeted site by PCR and sequencing. F1 embryos were individually collected at 7 days post fertilization (dpf), and genomic DNA was extracted from each individual embryo to assess mutagenesis at the TALEN-targeted site by PCR and sequencing. F1 mutated males were crossed with the same mutated genotype females (*see* **Note 18**). About 25% of F2 medaka will be homozygous mutants. The genotype of F2 could be checked by genomic PCR and sequencing.

4 Notes

1. The recombinant TALEN expression plasmids, pCS2-TALEN-ELD and pCS2-TALEN-ELD, could be obtained from Cheng's laboratory in CUHK (Please forward such request to

Prof. Christopher H.K. Cheng) and Hu's laboratory in IHB (Please forward such request to Prof. Wei Hu).

2. The success rate of first step Golden Gate assembly is greatly affected by the high quality of T4 DNA ligase. We have tried T4 DNA ligase (Promega Cat. No. M1801, Invitrogen Cat. No. 15224041 and NEB, Cat. No. M0202L, M0202M), and we strongly recommend to use high concentration of T4 DNA ligase (NEB, Cat. No. M0202M).
3. In our experience, whatever buffer you chose for embryos culture, we strongly recommend to add penicillin and streptomycin add in to protect embryos after microinjection.
4. According to original and our Golden Gate Assembly platform of TALENs, assembled half-site of TALEN is up to recognize 31 bp target site in gene sequences. Theoretically, longer targeted sites can enhance the specificity of TALENs, however, in fact, efficiency of gene disruption is too low to generate mutant line.
5. If the sequence of targeted gene is allowed, we suggest that there is less than five G in your designed half site of TALEN.
6. The success rate of Golden Gate assembly is greatly affected by the quality of the plasmid DNA. The concentration of 40 RVD modular plasmids makes sure to adjust to 100 ng/μL for each one. These plasmids should be avoided repeated thaw/frozen. Store at 4 °C for up to 3 months.
7. The full length TALEN site ($16 < X < 19$) divides into three part to assembly, which is 1–10, 11 to $X - 1$, and last repeats. For recognizing positions 1–10, the A at position 1 of TALEN site is recognized by pNI1, the C at position 2 of that is recognized by pHD2, etc. For recognizing positions 11 to ($X - 1$), the A at position 11 of TALEN site is recognized by pNI1, the C at position 12 of that is recognized by pHD2, etc. The plasmid recognizing the last nucleotide will be assembled later.
8. In our experience, about beyond 80% of the pFUS_A clones and nearly 100% of the pFUS_Bn contain the correct RVDs. Randomly pick up six clones from pFUS_A plate and two clones of pFUS_Bn is enough for selecting correct single clones for next step.
9. According to the last base pair of left and right TALEN sites, four pLR plasmids are picked up. For example, dmy TALEN L is 5′-TAGTCCAGAGGCCTCGTC-3′, last repeat of TALEN left site is C, pLR HD is selected to construct the final TALEN plasmid assembly.
10. Sequence the final TALEN plasmid using primers NtalF and CtalR. To analyze the sequencing result, translate the DNA sequence into amino acid using translate tool (http://web.expasy.org/translate/). Comparing with the sequence of

designed TALENs, double check the RVDs correction from sequenced final TALEN plasmid.

11. In our experience, the zebrafish nanos-3UTR, which was replaced SV40 into the end of the pCS2-ELD and pCS2-KKR coding sequence between XbaI and NotI, could significantly increase germline transmission.
12. UV illumination of the linearized plasmid DNA template for in vitro transcription should be avoided.
13. RNA should be aliquoted and stored at 80 °C. All procedures should be carefully done to avoid possible RNase contamination and refreezing-thawing.
14. Prepare different dosage of TALEN mRNA, evaluate the dosage effects and efficiency of TALEN mRNA at the first time. TALEN mRNAs induce embryonic cell toxicity. If majority of the injected embryos (beyond 50%) are deformed or dead, it means that the injection amount of TALEN mRNA should be lower.
15. After collecting medaka eggs, microinject as quickly as you can. Medaka eggs harden in 30 min. After that, it is too hard to perform microinjection. In order to ensure high efficiency of TALEN-induced gene disruption, it is best to microinject into animal pole of embryos.
16. The core criteria of the novel program for determining off-target sites were that the 0 position of the EBEs must be T, from 1 to 10 mismatch bases occur in the pairs of EBEs, and the spacer between the two putative EBE regions is less than 100 bp.
17. According to high efficiency of this TALEN platform, generally, mutated males and mutated females (maybe the genotype of mutation in male and female are different) are easy to find in F0 generation. In farmer fish, if you would like to faster inverstigate this gene knockout effects, you could cross these male and female mutants and find heterozygous genotype homozygous mutants in this F1 generation. The technique described in this protocols can greatly accelerate your research.
18. In our experience, F0 mutants cross with wild type, there may be more than single mutated genotype in F1 generation. For generating strict genetic homozygous, we strongly recommend to check genotype in F1 generation before F1 self-cross.

Acknowledgment

This work was supported by the National Science Foundation of China (31325026, 31461163006), Natural Science Foundation of Hubei (2015CFA007) to W. Hu and the National Science Foundation of China (31472263) to D. Luo.

References

1. Esvelt KM, Wang HH (2013) Genome-scale engineering for systems and synthetic biology. Mol Syst Biol 9:641
2. Bibikova M, Golic M, Golic KG, Carroll D (2002) Targeted chromosomal cleavage and mutagenesis in Drosophila using zinc-finger nucleases. Genetics 161:1169–1175
3. Kim S, Lee MJ, Kim H, Kang M, Kim JS (2011) Preassembled zinc-finger arrays for rapid construction of ZFNs. Nat Methods 8:7
4. Hockemeyer D, Wang H, Kiani S, Lai CS, Gao Q, Cassady JP, Cost GJ, Zhang L, Santiago Y, Miller JC et al (2011) Genetic engineering of human pluripotent cells using TALE nucleases. Nat Biotechnol 29:731–734
5. Huang P, Xiao A, Zhou M, Zhu Z, Lin S, Zhang B (2011) Heritable gene targeting in zebrafish using customized TALENs. Nat Biotechnol 29:699–700
6. Cong L, Ran FA, Cox D, Lin S, Barretto R, Habib N, Hsu PD, Wu X, Jiang W, Marraffini LA et al (2013) Multiplex genome engineering using CRISPR/Cas systems. Science 339:819–823
7. Hwang WY, Fu Y, Reyon D, Maeder ML, Tsai SQ, Sander JD, Peterson RT, Yeh JR, Joung JK (2013) Efficient genome editing in zebrafish using a CRISPR-Cas system. Nat Biotechnol 31:227–229
8. Cermak T, Doyle EL, Christian M, Wang L, Zhang Y, Schmidt C, Baller JA, Somia NV, Bogdanove AJ, Voytas DF (2011) Efficient design and assembly of custom TALEN and other TAL effector-based constructs for DNA targeting. Nucleic Acids Res 39:e82
9. Zhang F, Cong L, Lodato S, Kosuri S, Church GM, Arlotta P (2011) Efficient construction of sequence-specific TAL effectors for modulating mammalian transcription. Nat Biotechnol 29:149–U190
10. Reyon D, Tsai SQ, Khayter C, Foden JA, Sander JD, Joung JK (2012) FLASH assembly of TALENs for high-throughput genome editing. Nat Biotechnol 30:460
11. Lei Y, Guo X, Liu Y, Cao Y, Deng Y, Chen X, Cheng CH, Dawid IB, Chen Y, Zhao H (2012) Efficient targeted gene disruption in Xenopus embryos using engineered transcription activator-like effector nucleases (TALENs). Proc Natl Acad Sci U S A 109:17484–17489
12. Liu Y, Luo D, Zhao H, Zhu Z, Hu W, Cheng CH (2013) Inheritable and precise large genomic deletions of non-coding RNA genes in zebrafish using TALENs. PLoS One 8: e76387
13. Liu Y, Luo D, Lei Y, Hu W, Zhao H, Cheng CH (2014) A highly effective TALEN-mediated approach for targeted gene disruption in Xenopus tropicalis and zebrafish. Methods 69:58–66
14. Tang H, Liu Y, Luo D, Ogawa S, Yin Y, Li S, Zhang Y, Hu W, Parhar IS, Lin H et al (2015) The kiss/kissr systems are dispensable for zebrafish reproduction: evidence from gene knockout studies. Endocrinology 156:589–599
15. Chen J, Cui X, Jia S, Luo D, Cao M, Zhang Y, Hu H, Huang K, Zhu Z, Hu W (2016) Disruption of dmc1 produces abnormal sperm in medaka (Oryzias latipes). Sci Rep 6:30912
16. Luo D, Liu Y, Chen J, Xia X, Cao M, Cheng B, Wang X, Gong W, Qiu C, Zhang Y et al (2015) Direct production of XY (DMY-) sex reversal female medaka (Oryzias latipes) by embryo microinjection of TALENs. Sci Rep 5:14057
17. Qiu C, Cheng B, Zhang Y, Huang R, Liao L, Li Y, Luo D, Hu W, Wang Y (2014) Efficient knockout of transplanted green fluorescent protein gene in medaka using TALENs. Mar Biotechnol 16:674–683
18. Feng K, Luo HR, Li YM, Chen J, Wang YP, Sun YH, Zhu ZY, Hu W (2017) High efficient gene targeting in rice field eel Monopterus albus by transcription activator-like effector nucleases. Sci Bull 62:162–164
19. Koprunner M, Thisse C, Thisse B, Raz E (2001) A zebrafish nanos-related gene is essential for the development of primordial germ cells. Genes Dev 15:2877–2885
20. Mussolino C, Cathomen T (2011) On target? Tracing zinc-finger-nuclease specificity. Nat Methods 8:725–726
21. Tesson L, Usal C, Menoret S, Leung E, Niles BJ, Remy S, Santiago Y, Vincent AI, Meng XD, Zhang L et al (2011) Knockout rats generated by embryo microinjection of TALENs. Nat Biotechnol 29:695–696
22. Grau J, Boch J, Posch S (2013) TALENoffer: genome-wide TALEN off-target prediction. Bioinformatics 29:2931–2932

Chapter 29

Functional Studies of Transcriptional Cofactors via Microinjection-Mediated Gene Editing in *Xenopus*

Yuki Shibata, Lingyu Bao, Liezhen Fu, Bingyin Shi, and Yun-Bo Shi

Abstract

The anuran *Xenopus laevis* has been studied for decades as a model for vertebrate cell and developmental biology. More recently, the highly related species *Xenopus tropicalis* has offered the opportunity to carry out genetic studies due to its diploid genome as compared to the pseudo-tetraploid *Xenopus laevis*. Amphibians undergo a biphasic development: embryogenesis to produce a free-living tadpoles and subsequent metamorphosis to transform the tadpole to a frog. This second phase mimics the so-called postembryonic development in mammals when many organs/tissues mature into their adult form in the presence of high levels of plasma thyroid hormone (T3). The total dependence of amphibian metamorphosis on T3 offers a unique opportunity to study postembryonic development in vertebrates, especially with the recent development gene editing technologies that function in amphibians. Here, we first review the basic molecular understanding of the regulation of *Xenopus* metamorphosis by T3 and T3 receptors (TRs), and then describe a detailed method to use CRISPR to knock out the TR-coactivator SRC3 (steroid receptor coactivator 3), a histone acetyltransferase, in order to study its involvement in gene regulation by T3 in vivo and *Xenopus* development.

Key words *Xenopus laevis*, *Xenopus tropicalis*, Thyroid hormone receptor, Metamorphosis, Postembryonic development, Coactivator, Histone acetyltransferase, Histone modification, CRISPR

1 Introduction

The anuran amphibian *Xenopus laevis* has long been used as a model for vertebrate development. For decades, its external development and the ability to manipulate gene function in vivo through microinjection of mRNAs, proteins, and antisense oligonucleotides, etc., has made possible many seminal discoveries on early embryogenesis in vertebrates. In addition, after a finite period, the free-living tadpole formed at the end of embryogenesis undergoes a second phase of development: metamorphosis, during which essentially every organ/tissue changes drastically into its adult form (Fig. 1) [1]. This process is controlled by thyroid hormone (T3) and resembled postembryonic development in mammals, a period around birth when many organs/tissues mature into the

Chengyu Liu and Yubin Du (eds.), *Microinjection: Methods and Protocols*, Methods in Molecular Biology, vol. 1874, https://doi.org/10.1007/978-1-4939-8831-0_29,

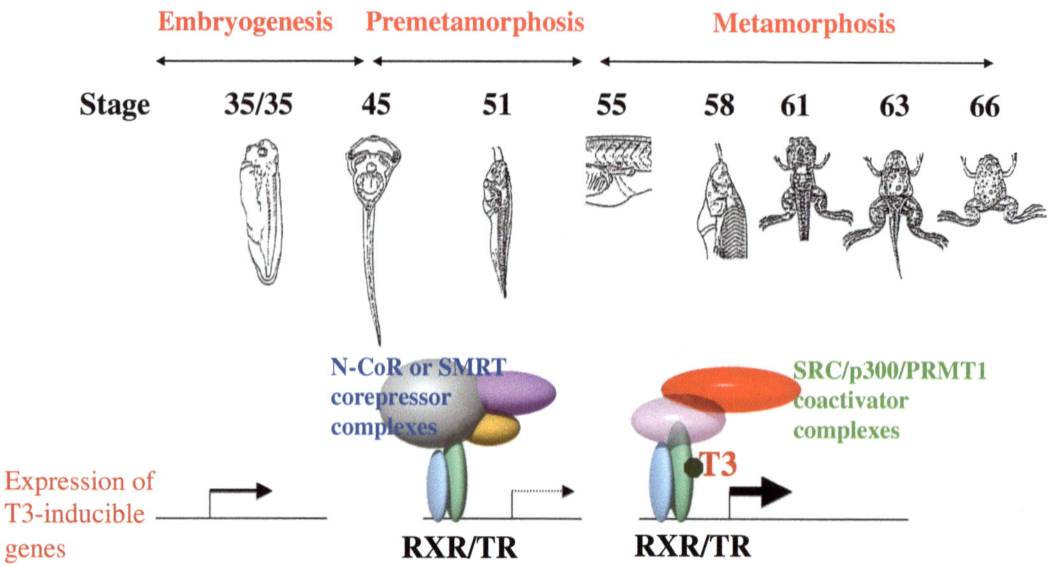

Fig. 1 A dual function model for TR during frog development. Top: The embryos hatch around stage 35 and tadpole feeding begins around stage 45 [71]. Subsequently, the tadpole enters the premetamorphic phase (till stage 54) when it undergoes mostly growth with little morphological changes as the plasma concentration of T3 remains very low [30]. Metamorphosis begins after stage 54 (prometamorphosis: stage 54–58) and most dramatic changes occur at climax stages 58–66 and end at stage 66 when the tail is completely resorbed. Bottom: A molecular mechanism for gene regulation by TR during development. T3-inducible genes are expressed at the basal levels during embryogenesis when there are little T3 and TRs. After the onset of tadpole feeding (stage 45), the expression of TR and RXR leads to the formation of TR/RXR heterodimers that bind to the T3 response elements (TREs) in the target genes and recruit corepressor complexes to the promoter region. This represses the expression of the genes prior to the synthesis of endogenous T3 around stage 54. Subsequently, T3 binds to TR and causes the release of the corepressor complexes and recruitment of coactivator complexes to the genes, leading to the activation of T3-inducible genes and metamorphosis. Note that exemplary cofactor complexes are shown here

adult form [1, 2]. Amphibian metamorphosis offers a number of advantages for studying postembryonic development in vertebrates. First, it occurs externally, independent of the maternal influence as in mammals. Second, it is totally dependent on T3 and can be easily manipulated by altering T3 levels available to the tadpoles. Blocking the synthesis of endogenous T3 prevents metamorphosis, thus allowing animals remaining as tadpoles for at least a few years (for comparison, normal development leads to the completion of metamorphosis at about 2 months after fertilization in *Xenopus laevis*). On the other hand, tadpoles as young as the onset of feeding (stage 45, about 1 week old for *Xenopus laevis*) can be induced to undergo precocious metamorphosis with physiological concentrations of exogenous T3 in rearing water. Third, most, if not all, tissue transformations are mostly, if not all, organ-autonomous, enabling studies of this process in organ or primary cell cultures. Finally, while embryonic lethality due to gene alternations makes it impossible to study gene function during postembryonic development in mammals, the ability to microinject mRNAs/proteins into fertilized eggs coupled with the external development can rescue the embryonic defect due to gene alteration, leading to the formation of free-feeding tadpoles that allows functional studies of the gene alteration during tadpole development and metamorphosis. Taking advantage of these properties, we and others have been studying the molecular mechanisms by which T3 regulates *Xenopus* development. Here, we focus on epigenetic enzymes that are involved in gene regulation by T3.

1.1 Gene Regulation by Thyroid Hormone Receptors (TRs)

T3 can regulate target gene transcription through its nuclear receptors, TRs. T3 can both activate and repress gene expression. While the mechanism for gene repression by T3 is less clear, for genes induced by T3, TRs generally bind to thyroid hormone response elements (TREs) present in the target genes as heterodimers with 9-*cis* retinoic acid receptors (RXRs). TR regulates gene regulation by recruiting corepressor or coactivator complexes. In the absence of T3, unliganded TR binds to corepressors. Two best-studied corepressors are the highly related proteins N-CoR (nuclear corepressor) and SMRT (silencing mediator of retinoid and thyroid hormone receptors), both of which can form large complexes containing histone deacetylase 3 (HDAC3) [3–13]. Thus, gene repression by unliganded TR is often accompanied by deacetylation of the histones around the TRE. When T3 is present, TR releases corepressors and recruits coactivators. Many TR-coactivators have been identified. Among them include ATP-dependent chromatin remodeling proteins, histone acetyltransferases such as steroid receptor coactivators (SRC1, 2, 3) and p300, and histone methyltransferases PRMT1 (protein arginine methyltransferase 1) and CARM1 (coactivator-associated arginine methyltransferase 1, or PRMT4) [12, 14–24]. Consistently, gene activation by T3 is associated with increased local acetylation and methylation.

1.2 TR Function During Xenopus laevis Development

Based on the molecular properties of TRs above and expression profiles of TRs during *Xenopus laevis* development, we have previously proposed a dual function model for TRs during *Xenopus* development [25, 26]. That is, during embryogenesis, there is little T3 or TRs and T3-inducible genes are expressed at basal level for embryonic development. At the end of embryogenesis, a free-living tadpole is formed and the animal begins to feed (stage 45). By this stage, high levels of expression of TRα and RXRα lead to the binding of TR/RXR heterodimers to the target genes (Fig. 1) [27–29]. As there is little T3 present [30], the unliganded TR/RXR recruits corepressor complexes to repress these genes. The repression of these genes prevents precocious metamorphosis and ensures proper tadpole growth and development. After stage 54, the rising concentration of endogenous T3 enables T3-binding to TR, thus releasing the corepressor complexes and recruiting coactivator complexes. This results in the activation of the target genes and metamorphosis.

Over the years, several laboratories have provided strong molecular and transgenic evidence to support the dual functions of TRs during *Xenopus* development [31–38]. More recently, with the development of TALEN and CRISPR-nuclease-mediated gene editing technologies for both the pseudo-tetraploid *Xenopus laevis* and diploid *Xenopus tropicalis*, one can study endogenous genes during *Xenopus* development [39–45]. Using the TALEN technology, two groups independently knocked down the endogenous TRα during *Xenopus tropicalis* development and observed that the TRα knockdown animals developed faster and initiate metamorphosis at younger age than the wild-type siblings, accompanied by the upregulation (derepression) of T3-target genes [46–49]. Similar findings were made in total knockout animals [50, 51]. These results indicate that TRα is important for metamorphic timing and the rate of metamorphosis progression, further supporting the dual function model for TR in *Xenopus* development.

1.3 Roles of Cofactors in TR Action and Xenopus Development

As indicated above, TRs repress or activated T3-inducible gene by recruiting corepressors or coactivators in the absence or presence of T3, respectively. The dependence of amphibian metamorphosis on T3 and the ability to manipulate this process make it possible to study the involvement of such cofactors in gene regulation by T3 during vertebrate development in vivo. Early experiments by using chromatin immunoprecipitation (ChIP) assays on tadpole tissues have shown that TRs indeed recruit corepressors or coactivators to endogenous T3-inducible genes in premetamorphic tadpoles when there is little endogenous T3, or during metamorphosis when T3 is present (Fig. 1) [19, 36, 52–58]. Furthermore, transgenic studies have revealed that interferes with the interaction of endogenous corepressor complexes with unliganded TR leads to premature derepression of T3-inducible genes and precocious metamorphosis [59]. On the other hand, disrupting coactivator function with TR inhibits gene activation by T3 and metamorphosis (Fig. 2) [53, 54] while transgenic overexpression of the wild-type coactivator PRMT1 (protein arginine methyltransferase 1) enhances gene activation induced by T3 and accelerates metamorphic progression [19].

TR-interacting corepressor complexes are known to contain histone deacetylases while coactivator complexes contain histone acetyltransferases and/or methyltransferases and/or chromatin remodeling activity. Consistently, ChIP assays with antibodies again core histones H2B and H3 have recently shown that T3 induces the removal of core histones, equivalent to about two nucleosomes, from the TRE regions of T3 target genes during

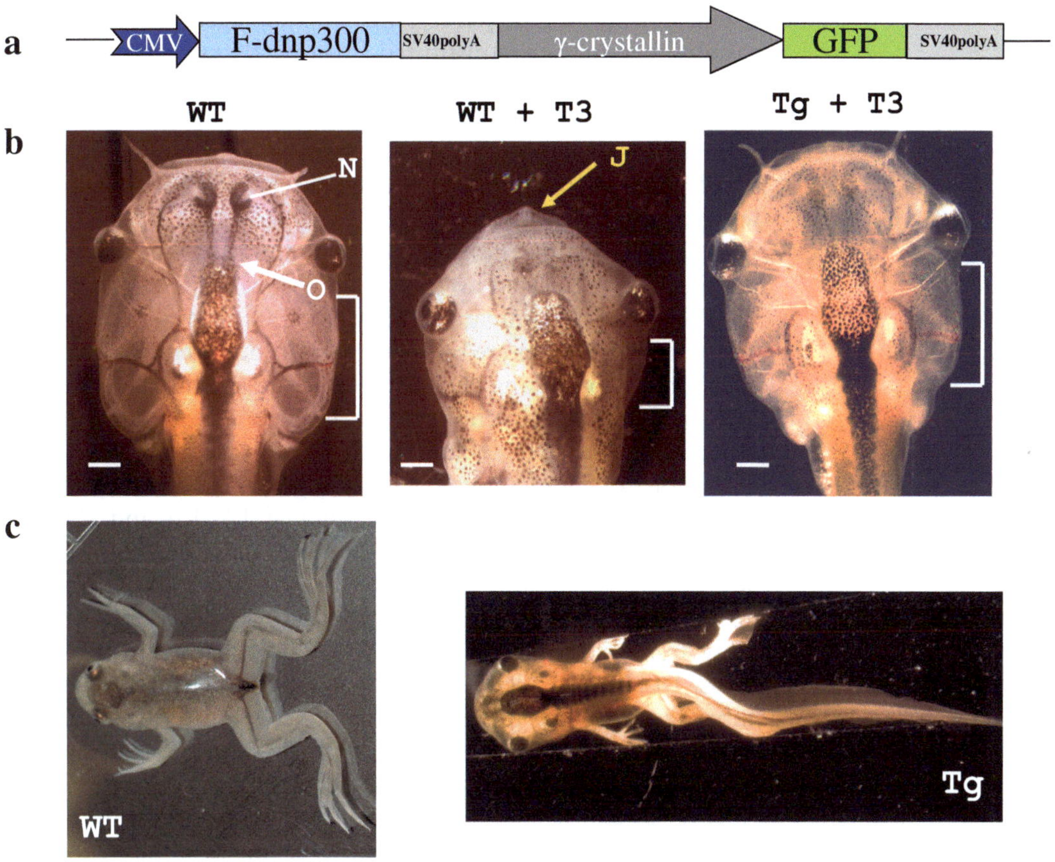

Fig. 2 Transgenic overexpressing a Flag-tagged dominant negative (F-dn) p300 inhibits T3-induced and natural metamorphosis. (**a**) Schematic representation of the double promoter construct used for generation of transgenic animals. The constitutively active cytomegalovirus (CMV) promoter drives the expression of the transgene, F-dnp300, and is followed by the SV40 polyadenylation signal. The construct also harbors GFP under the control of the eye-specific *γ-crystallin* promoter as a marker to identify transgenic animals. (**b**) F-dnp300 inhibits T3-induced metamorphic changes. Wild-type (WT) and transgenic (Tg) tadpoles at pre-metamorphic stage 54 were treated with 5 nM T3 for 3 days in the same container. At the end of the treatment, transgenic tadpoles were identified by the green fluorescence in the eye lens. The wild-type tadpoles which were treated with T3 underwent the typical T3-induced changes such as resorption of gills (bracketed) and development of the Meckel's cartilage leading to protrusion of the jaw (marked as J). Furthermore, the anterior end of the olfactory lobe (marked as O in the wild-type animal without T3 treatment) is closer to the nostrils (marked as N) in wild-type animals treated with T3 (WT + T3) as compared to untreated wild-type siblings (WT). These changes were not observed or greatly reduced in F-dnp300 transgenic animals treated with T3 (Tg + T3). (**c**) The overexpression of F-dnp300 arrests or delays natural metamorphosis. The progression of natural metamorphosis of F-dnp300 transgenic animals (Tg) and their wild-type siblings (WT) was monitored. Note that a tadpole which displayed arrested metamorphosis, remained at stage 61 for more than a month and then died, possibly due to partial metamorphosis, compared to a wild-type sibling which completed metamorphosis at about 2 months of age. For more details, *see* ref. 54

metamorphosis in *Xenopus tropicalis* [60], in agreement with what was observed in the reconstituted frog oocyte transcription system [28, 61, 62]. They have further revealed T3-dependent histone modifications at T3 target genes [60, 63, 64]. While direct evidence for a role of such epigenetic modifications in gene regulation by T3 is still lacking, treating premetamorphic tadpoles with a histone deacetylase inhibitor increases the expression of T3 target genes [65, 66], suggesting that epigenetic modifications may be important for gene regulation by TR.

These studies summarized above indicate that corepressor and coactivator complexes are both involved in coordinating amphibian development by altering chromatin structure and histone modifications to affect TR function in the absence and presence of T3, respectively. However, so far, there have been no reported analyses on the roles of endogenous cofactors in gene regulation by TR and metamorphosis. With the newly developed TALEN and CRISPR gene editing technologies [39–45], we can now begin to investigate the function of endogenous genes such as SRC3 (Fig. 3) during *Xenopus* development, particularly during metamorphosis. Many transcription cofactors, such as the histone methyltransferases Dot1L and PRMT1, are known to be essential for mouse embryogenesis. This makes it impossible to study the roles of such genes during postembryonic development (around birth) in mammals. Anuran metamorphosis offers a unique opportunity to address this gap since metamorphosis can be induced with exogenous T3 as early as the onset of tadpole feeding (around stage 45 or 3 days old in *Xenopus tropicalis*). This is possible because if the knockout of a transcription cofactor leads to embryonic lethality, one can rescue this development by injecting mRNA encoding

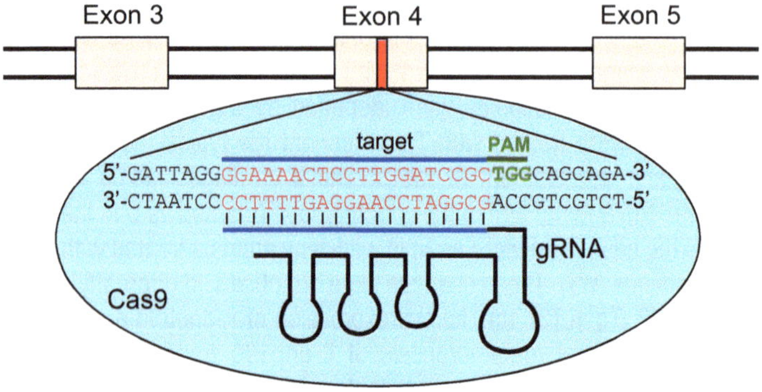

Fig. 3 Schematic diagram for targeting SRC3 with CRISPR/Cas9 in *Xenopus tropicalis*. The protein coding exons are shown as numbered, colored boxes. The CRISPR/Cas9-recognized sequence is shown as red boxes (top) or red letters and blue line (below). The PAM sequences that are required for gRNA guided target-recognition and cleavage by Cas9 protein (light blue shape) are shown in green letters

wild-type cofactor into fertilized egg. Since the mRNA and the resulting translated protein can persist for a few days in the embryos, this should rescue the knockout effect to allow animal development into feeding stage tadpoles (3 days or older) for studying the role of the cofactor during T3-induced metamorphosis. Thus, by combining the advantage of the anuran development and the gene-editing technologies, it will be possible to investigate cofactor function during postembryonic development regardless of its role during embryogenesis. Furthermore, if a CRISPR is highly efficient (e.g., >80% mutation rate) in causing mutations after injecting the corresponding RNAs into fertilized eggs, it is even possible to study the function of the cofactor in the resulting F0 generation animals, allowing one to know the outcome within days after micro-injecting of the RNAs.

2 Materials

2.1 Design and Synthesize a CRISPR Guide RNA for the Target Site

Gene knockdown/knockout methods by using CRISPR/Cas9 system in frog were as described by [67].

1. RNA transcription kit, Ambion mMessage® SP6 mMachine.
2. MEGAscript® T7 Transcription Kit.
3. pCS2 + hSpCas9 (Cat: Plasmid #51815, Addgene, MA).
4. Zero Blunt TOPO + mCherry (a gift from Dr. Rudolf Jaenisch (Addgene plasmid # 59995).
5. QIAGEN QIAquick® PCR Purification Kit (250).
6. Takara PrimeSTAR Max.
7. Primers: The primers used for the target gene [67]:

   Forward;

   5′-TAATACGACTCACTATAG(N)16-19GTTTTAGAGCTA-GAAATAGCAAG-3′,

   Reverse;

   5′-AAAAGCACCGACTCGGTGCCACTTTTTCAAGTTGAT AACGGACTAGCCT TATTTTAACTTGCTATTTCTAG CTCTAAAAC-3′.
8. *Online tools for CRISPR guide RNA selection*

   CHOPCHOP (https://chopchop.cbu.uip.no) [68].

   CRISPR/Cas9 target predictor (http://crispr.cos.uni-heidelberg.de) [69].

   CRISPR direct (http://crispr.dbcls.jp/) [70].

2.2 Generate Knockdown/Knockout Animals Through mRNA Injection into Fertilized Eggs

Adults of the West African clawed frog *Xenopus tropicalis* were obtained from Nasco (Wisconsin). Embryos were prepared by in vitro fertilization as described [31]. Developmental stages were determined according to [71].

1. Mark's Modified Ringer's (MMR): 0.1 M NaCl, 2.0 mM KCl, 1 mM $MgSO_4$, 2 mM $CaCl_2$, 5 mM HEPES (pH 7.8), 0.1 mM EDTA.
2. Merck Human Chorionic Gonadotropin (HCG).
3. Adult *Xenopus tropicalis.*
4. Sutter Instruments P-97 Needle puller (*see* **Note 1**).
5. Drummond Microcaps Needles 1-000-0300 (*see* **Note 1**).
6. World Precision Instruments PV820 Injector (*see* **Note 1**).

2.3 Identify Out-Frame Mutants via a Color Assay

1. QIAGEN DNeasy Blood and Tissue kit.
2. QIAGEN PCR Purification Kit.
3. QIAGEN Gel Extraction Kit.
4. NEB Qiuck Ligation kit.
5. Invitrogen *E. coli* TOP10 competent cells.
6. SalI-HF.
7. EcoRI-HF.
8. NEB CutSmart Buffer.
9. IPTG solution: 10 mM IPTG.
10. Xgal solution: 20 mg/mL Xgal in DMSO.
11. Takara PrimeSTAR GXL DNA Polymerase.
12. Primers for PCR-amplification of the SRC3 gRNA target region:

    SRC3 Forward: 5′-GGCTGCATCTTTGTCGCTAA-3′.

    SRC3 Reverse: 5′-GAGAGCCAATCACAGCCTTG-3′.

    SRC3 NestedF: 5′-CCGCATCGTCGACAATGCGCCTTCAGCTGCCTTTTTG-3′ (SalI site in underline).

    SRC3 NestedR: 5′-GCCAGTTGAATTCCTACTGTTTGCCATGCCTC-3′ (EcoRI site in underline).

    SRC3 gRNA forward: 5′-TAATACGACTCACTATAGGAAAACTCCTTGGATCCGCGTTT-3′.

    5′-TAGAGCTAGAAATAGCAAG-3′ [67].

    SRC3 gRNA reverse: 5′-AAAAGCACCGACTCGGTGCCACTTTTTCAAGTTGATAACGGACTAGCCTTATTTTAACTTGCTATTTCTAGCTCTAAAAC-3′ [67].
13. pLacZα-GFP [72] (Note that it is available through Addgene distribution at https://www.addgene.org/Yun-Bo_Shi/).

14. Nano Drop 2000.
15. Fluorescent dissection microscope.
16. MicroPublisher 5.0 digital camera attached to the microscope.

3 Methods

3.1 Design and Synthesize a CRISPR Guide RNA for the Target Site

3.1.1 Design CRISPR Guide RNA

The first step to generate gene knockout animals is to design a CRISPR guide RNA that contains a sequence 16–20 bp of the target site. The target site should have a G residue at the +1 position and should be followed by a PAM sequence in the target gene [*S. pyogenes* Cas9 (SpCas9): 3 bp NGG], which is recognized by Cas9 to make a double strand break (Fig. 3) (*see* **Note 2**). One can use online tools to search the optimal target site and putative off-target site based on the *Xenopus tropicalis* genome for designing CRISPR guide RNA [68–70]. We designed some CRISPR guide RNAs for SRC3 (Fig. 3) as described [67] (*see* **Note 3**).

3.1.2 In Vitro Transcription for the CRISPR Guide RNA

1. The 5′ Primer, 5′-<u>TAATACGACTCACTATAG</u>$(N)_{16-19}$ GTTTTAGAGCTAGAAATAGCAAG-3′, consists of T7 promoter sequence, under lined and the target specific sequence, G$(N)_{16-19}$ which is designed as describe above. On the other hands, the 3′ primer is used for all CRISPR guide RNA and as shown below 5′-AAAAGCACCGACTCGGT GCCACTTTTTCAAGTTGATAACGGACTAGCCTTATTTTAACTTGCTATTTCTAGCTCTAAAAC-3′ (*see* **Note 4**).
2. The primeSTAR Max DNA polymerase was used for template assembly by PCR. Assembly reactions were performed in a final volume of 100 μL and carried out with the PCR program: 98 °C for 10 s, 60 °C for 5 s, 72 °C for 15 s repeat 15 cycles.
3. After electrophoresis to identify the specific band, the PCR product was purified by QIAGEN PCR purification kit.
4. One μg of the DNA was transcribed in vitro with T7 RNA polymerase (*see* **Note 5**).
5. Remove template DNA, purify RNA and quantitate it (*see* **Note 6**).

RNA can be aliquoted and stored at −80°C for months.

3.1.3 Prepare Cas9 and mCherry mRNAs by In Vitro Transcription

1. Linearize the plasmid DNA with an appropriate restriction enzyme such that it can be used for in vitro transcription to produce sense mRNA. Prepare enough linearized DNA (e.g., digest about 20 μg) for at least several transcription reactions at 1–1.5 μg per reaction.
2. Purify and quantitate the DNA, and confirm its linearization, purity, and quantification by running a small aliquot on an agarose gel (*see* **Note 7**).

3. Transcribe RNA by following the directions of Ambion's mMessage mMachine kit (*see* **Note 5**). In our studies, pCS2+hSpCas9 plasmid and Zero Blunt TOPO + mCherry were linearized with Not I, twice phenol/chloroform-extracted, and the ethanol-precipitated. One μg of the DNA was transcribed in vitro with SP6 RNA polymerase.
4. Remove template DNA, purify RNA, and quantitate it (*see* **Note 6**).

   RNA can be aliquoted and stored at −80 °C for months.

3.2 Generate Knockdown/Knockout Animals Through mRNA Injection into Fertilized Eggs

3.2.1 Fertilize Eggs In Vitro

1. *X. tropicalis* females are primed with 20 U of human chorionic gonadotropin (HCG) 1 day before ovulation.
2. The primed frogs are injected with 200 U of HCG next morning.
3. A *X. tropicalis* male frog is sacrificed. The testes are removed for in vitro fertilization (*see* **Note 8**) and if needed, the remainder can be stored in 1× MMR at 4 °C for up to several days.
4. With gentle squeezing, eggs are expelled from ovulating females onto petri dishes containing 1 mL 1× MMR into which a small fragment of testes has recently been macerated.
5. After gently splaying out of eggs into the sperm solution and 2 min incubation, the dishes are filled with 0.1× MMR and incubated for an additional 15 min.
6. To remove the jelly coat, the fertilized eggs are washed with 0.1× MMR-3.5% cysteine, pH 8.0, until they started to touch each other (about 3–5 min).
7. Then wash the fertilized eggs/embryos 4–6 times with 0.1× MMR before being transferred into fresh 0.1× MMR (*see* **Note 8**).
8. Embryos are immediately transferred to 0.1× MMR–6% Ficoll.

3.2.2 Inject and Culture the Embryos

1. The embryos above are injected with RNAs (guide RNA for SRC3, mRNAs for Cas9 and mCherry at 400 pg, 400 pg, 200 pg per embryo, respectively) in a 2 nL total volume per embryo (*see* **Note 9**).
2. Control and injected embryos are kept in 0.1× MMR–6% Ficoll for 4–6 h after injection and then transferred to 0.1× MMR for rearing.
3. Embryos are incubated for 3 days at 25 °C in incubator. The culture medium is changed daily.
4. At 48, 72 h after fertilization, the embryos are phenotypically examined and sorted based on mCherry signal under the microscope.

5. Five embryos are randomly selected for mutation analysis, and the rest of the F0 generation animals are kept in a 3-L plastic container for further rearing.

3.3 Identify Out-of-Frame Mutants via a Color Assay

3.3.1 Extract and Purify Genomic DNA

1. Collect five embryos 3–5 days after fertilization and isolate genomic DNA by using DNeasy Blood and Tissue kit.
2. Design two pairs of primers for detection and screening of CRISPR/Cas9-induced out-of-frame mutations in SRC3 gene: F (GGCTGCATCTTTGTCGCTAA) and R (GAGAGCCAATCACAGCCTTG), and NestedF (CCGCATCGTCGACAATGCGCCTTCAGCTGCCTTTTTG) and NestedR (GCCAGTTGAATTCCTACTGTTTGCCATGCCTC), respectively (*see* **Note 10**).
3. Perform the first round PCR amplification by using primers F and R for 15–20 cycles on the genomic DNA (*see* **Note 11**).
4. Dilute the PCR products at 1:100 to 1:1000, respectively, to serve as templates in the second round PCR amplification with the nested primers (Nested F and Nested R) for 25–30 cycles (*see* **Note 11**).
5. Check the PCR specificity by analyzing the PCR products on an agarose gel.
6. Proceed to purify the nested PCR products by using PCR Purification Kit.
7. Double digest the purified PCR products with restriction enzymes SalI-HF and EcoRI-HF, in CutSmart Buffer, at 37 °C for at least 2 h (*see* **Note 12**).
8. Add calf intestinal alkaline phosphatase (CIP) to the reactions (1 μL CIP per 100 μL reactions) and incubate for 10 min to dephosphorylate the DNA ends (*see* **Note 13**).
9. Add 10 μL 5 M NaCl and 80 μL isopropanol to 100 μL reaction, centrifuge at 15,000 × g at 4 °C for 10 min.
10. Carefully remove the supernatant by using a pipette. Air dry the DNA and add 10–20 μL TE buffer to dissolve the DNA.
11. Use agarose gel electrophoresis to resolve the dephosphorylated PCR product bearing an EcoRI and a SalI end and excise out the agarose gel slice containing the DNA band.
12. Proceed to purify the DNA fragment by using Gel Extraction Kit.
13. Measure the concentration of the DNA fragments by using Nano Drop 2000 and store it at −20 °C.

3.3.2 Prepare the Color Assay Vector

1. Double-digest the assay vector pLacZα-GFP with the same restriction enzymes used for digesting the PCR product above (*see* Subheading 3.3.1, **step** 7), and perform single digestions

with each enzyme separately as controls to make sure that both enzymes digest the vector completely (*see* **Note 14**).

2. Check the efficiency of the enzyme digestions by using agarose-gel electrophoresis. Note that the digested vector in all three reactions would be seen as a single band in the agarose gel because the small fragment released by double digestion from the multiple cloning sites within the vector is too small to be visible in the gel.
3. Add 10 μL 5 M NaCl and 70 μL isopropanol per 100 μL reaction, centrifuge at 15,000 × *g* for 10 min at 4 °C.
4. Carefully remove the supernatant by using a pipette. Air dry and add 10–20 μL TE buffer to dissolve the linear pLacZα-GFP.
5. Purify the linear plasmid by using agarose gel electrophoresis and Gel Extraction Kit.
6. Measure the concentration of the linear pLacZα-GFP, and store it at −20 °C.

3.3.3 Clone the Genomic DNA and Screen for Out-of-Frame Mutations

1. Ligate the double-digested PCR products (get from Subheading 3.3.1) into pLacZα-GFP predigested with EcoRI and SalI (get from Subheading 3.3.2) by using Quick Ligation kit for 5–10 min at room temperature (*see* **Note 15**).
2. Prepare IPTG/Xgal/Amp plates by spreading evenly 50 μL Xgal and 100 μL IPTG each on the surface of ampicillin plates.
3. Transform the ligated DNA above into *E. coli* TOP10 competent cells and use 300 μL SOC medium to activate the transformed bacteria.
4. Add 30–300 μL transformed *E. coli* TOP10 competent cells per IPTG/Xgal/Amp plates, and then spread the solution evenly (*see* **Note 16**).
5. Invert the plates and incubate them at 37 °C overnight.
6. Observe the plates under a fluorescent dissection microscope (Model MZ10F) and photograph with a MicroPublisher 5.0 digital camera attached to the microscope under bright or fluorescent field (*see* **Note 17**).
7. Process the images with Adobe PhotoShop.
8. Count GFP-negative colonies under florescent conditions and the total number of blue or greenish blue colonies on the same plate. GFP-negative blue colonies contain out-of-frame mutations in the SRC3 fragment.
9. Determine the out-of-frame mutation rate by dividing the numbers of GFP-negative colonies by the total number of blue/greenish blue colonies on the same plate.

4 Notes

1. A variety of instruments and approaches can be used to pull needles and inject RNA or DNA samples. They include simple hand pulled needles using a Bunsen Burner, and forced air injectors using "house air" found in most laboratories. The crucial criteria are choosing a needle and a method that allow for a small needle diameter (about 10 μm) to be pulled and a way to calibrate or control injection volumes (about 2–5 nL per injection).
2. When designing the CRISPR guide RNA, it is preferable to choose sequences near the 5′-end encoding a region important for the function of the target protein. If there are multiple in frame initiation codons, the guide RNA should target the last one. If there are multiple spliced forms, the guide RNA should target a common region.
3. The online tools often provide a list of potential CRISPR guide RNA sequences and putative off targets. We choose the highly ranked guide RNA with fewer off targets. To control for specificity, we would recommend analyzing possible mutations at 3–5 off target sites in the list, especially if one uses F0 or F1 generation animals for phenotypic studies.
4. According to [67], two types of templates can be used for in vitro transcription to prepare CRISPR guide RNA. One is plasmid-based and the other PCR-based. We used the PCR-based strategy for our CRISPR guide RNA because it allows prompt synthesis of CRISPR guide RNA without cloning. Alternative strategy may also be used to produce gRNA through CSY4 nuclease processing which generates gRNA starts from any nucleotides at its 5′-end and expands the choices of CRISPR targets on the gene of interest [73].
5. Though we used a commercial RNA transcription kit, in vitro transcription can be performed with transcription components being purchased individually at cheaper costs.
6. Once the transcription reaction is complete (following the removal of the DNA template with DNase), the RNA must be purified prior to injection. LiCl precipitation is a good and quick method of removing unincorporated nucleotides and proteins. Phenol/chloroform extraction followed by ethanol precipitation is a more stringent method of purification, though care must be taken to remove all traces of phenol and chloroform prior to injection.
7. Use gel electrophoresis to confirm that the DNA is pure and completely linearized. Circular plasmid templates will generate

extremely long, heterogeneous RNA transcripts because RNA polymerases are very processive.

8. In vitro fertilization for *Xenopus tropicalis* is essentially the same as that for *Xenopus laevis*. For details, see *Xenopus* fertilization protocols [74].
9. It would be better to complete the injection before the first cell division, i.e., within about 90 min after fertilization.
10. pLacZα-GFP vector [72] is used for the color assay here. Primers F and R are designed for amplifying an extended region encompassing the CRISPR/Cas target (the PCR product should be around 500–1000 bp). The nested PCR primers (NestedF and NestedR) should each contain a unique restriction enzyme site present in the multi-cloning sites (MCS) of the pLacZα-GFP vector but absent in the rest of the resulting PCR product. The 6–8 extra bp at the 5′-end beyond the restrict site is necessary to ensure that the nested PCR product can be digested efficiently by the restriction enzymes. It is preferable that the two chosen enzymes can digest their substrate DNA efficiently in a single reaction buffer at the same incubation temperature. Restriction enzyme star activity may lead to false outcomes from the assay, thus high fidelity enzymes are preferred.
11. To ensure accuracy during PCR amplification, high fidelity DNA polymerases such as Takara PrimeSTAR GXL DNA Polymerase are preferred.
12. Set up a 100 μL double enzyme digestion reaction as below: DNA 1 μg; 10× CutSmart Buffer 10 μL; SalI-HF 1 μL; EcoRI-HF 1 μL; plus water to 100 μL.
13. Dephosphorylation of the double restriction enzyme-digested PCR products is necessary to ensure that only a single insert is ligated into the color assay vector. Multiple inserts would produce false positive results, thus should be avoided.
14. Another color assay vector, pmCherry-GFP, can also be used for determining the out-of-frame mutations by using two fluorescent proteins, i.e., mCherry and GFP [72]. Complete digestion of the vector DNA by the chosen two restriction enzymes is critical for accurate evaluation of the out-of-frame mutation rate from CRISPR/Cas-mediated mutagenesis (linear DNA from single-enzyme digestion would be self-ligated very efficiently, leading to false negative colonies).
15. For ligation, the insert to vector molar ratio should be between 2 and 6 to obtain high ligation efficiency.
16. When plating transformed *E. coli* TOP10 competent cells, it is better to plate two different volumes of *E. coli* per plate for each transformation sample in order to obtain optimal numbers of

colonies per plate for quantification purpose, such as 50 and 150 μL.

17. The double positive pLacZα-GFP colonies look greenish in the bright field and bright green under fluorescent condition for GFP. This may be due to relatively weaker blue color (in frame GFP expression may reduce the color from LacZα expression) of the X-gal staining plus weak green color due to the GFP under visible light [72]. Restriction enzyme star activity may lead to false results from the assay, and low quality of the nested PCR primers, i.e., having wrong nucleotides inserted during DNA synthesis, can also contribute to false positive results. If a high background mutation rate is observed, please check if the primer sequences and the sequences at the cloning sites are correct.

Acknowledgments

This work in the laboratory was supported by the Intramural Research Program of National Institute of Child Health and Human Development, National Institutes of Health. Y. Shibata was supported in part by a Japan Society for the Promotion of Science Research Fellowship for Japanese Biomedical and Behavioral Researchers at the National Institutes of Health.

References

1. Shi Y-B (1999) Amphibian metamorphosis: from morphology to molecular biology. Wiley, New York
2. Tata JR (1993) Gene expression during metamorphosis: an ideal model for post-embryonic development. BioEssays 15(4):239–248
3. Burke LJ, Baniahmad A (2000) Co-repressors 2000. FASEB J 14(13):1876–1888
4. Jones PL, Shi Y-B (2003) N-CoR-HDAC corepressor complexes: roles in transcriptional regulation by nuclear hormone receptors. In: Workman JL (ed) Current topics in microbiology and immunology: protein complexes that modify chromatin, vol 274. Springer, Berlin, pp 237–268
5. Glass CK, Rosenfeld MG (2000) The coregulator exchange in transcriptional functions of nuclear receptors. Genes Dev 14(2):121–141
6. Zhang J, Lazar MA (2000) The mechanism of action of thyroid hormones. Annu Rev Physiol 62:439–466
7. Yoon H-G, Chan DW, Huang ZQ, Li J, Fondell JD, Qin J, Wong J (2003) Purification and functional characterization of the human N-CoR complex: the roles of HDAC3, TBL1 and TBLR1. EMBO J 22:1336–1346
8. Zhang J, Kalkum M, Chait BT, Roeder RG (2002) The N-CoR-HDAC3 nuclear receptor corepressor complex inhibits the JNK pathway through the integral subunit GPS2. Mol Cell 9:611 623
9. Ishizuka T, Lazar MA (2003) The N-CoR/histone deacetylase 3 complex is required for repression by thyroid hormone receptor. Mol Cell Biol 23:5122–5131
10. Guenther MG, Lane WS, Fischle W, Verdin E, Lazar MA, Shiekhattar R (2000) A core SMRT corepressor complex containing HDAC3 and TBL1, a WD40-repeat protein linked to deafness. Genes Dev 14:1048–1057
11. Li J, Wang J, Wang J, Nawaz Z, Liu JM, Qin J, Wong J (2000) Both corepressor proteins SMRT and N-CoR exist in large protein complexes containing HDAC3. EMBO J 19:4342–4350
12. Shi YB, Matsuura K, Fujimoto K, Wen L, Fu L (2012) Thyroid hormone receptor actions on transcription in amphibia: the roles of histone

modification and chromatin disruption. Cell Biosci 2(1):42. https://doi.org/10.1186/2045-3701-2-42 2045-3701-2-42 [pii]
13. Jones PL, Sachs LM, Rouse N, Wade PA, Shi YB (2001) Multiple N-CoR complexes contain distinct histone deacetylases. J Biol Chem 276 (12):8807–8811
14. Heimeier RA, Hsia VS-C, Shi Y-B (2008) Participation of BAF57 and BRG1-containing chromatin remodeling complexes in thyroid hormone-dependent gene activation during vertebrate development. Mol Endocrinol 22:1065–1077
15. Huang Z-Q, Li J, Sachs LM, Cole PA, Wong J (2003) A role for cofactor–cofactor and cofactor–histone interactions in targeting p300, SWI/SNF and Mediator for transcription. EMBO J 22:2146–2155
16. Demarest SJ, Martinez-Yamout M, Chung J, Chen H, Xu W, Dyson HJ, Evans RM, Wright PE (2002) Mutual synergistic folding in recruitment of CBP/p300 by p160 nuclear receptor coactivators. Nature 415:549–553
17. Chen D, Ma H, Hong H, Koh SS, Huang SM, Schurter BT, Aswad DW, Stallcup MR (1999) Regulation of transcription by a protein methyltransferase. Science 284:2174–2177
18. Koh SS, Chen DG, Lee YH, Stallcup MR (2001) Synergistic enhancement of nuclear receptor function by p160 coactivators and two coactivators with protein methyltransferase activities. J Biol Chem 276:1089–1098
19. Matsuda H, Paul BD, Choi CY, Hasebe T, Shi Y-B (2009) Novel functions of protein arginine methyltransferase 1 in thyroid hormone receptor-mediated transcription and in the regulation of metamorphic rate in Xenopus laevis. Mol Cell Biol 29:745–757
20. Matsuda H, Paul BD, Choi CY, Shi Y-B (2007) Contrasting effects of two alternative splicing forms of coactivator-associated arginine methyltransferase 1 on thyroid hormone receptor-mediated transcription in Xenopus laevis. Mol Endocrinol 21(5):1082–1094
21. Matsuura K, Fujimoto K, Das B, Fu L, Lu CD, Shi YB (2012) Histone H3K79 methyltransferase Dot1L is directly activated by thyroid hormone receptor during Xenopus metamorphosis. Cell Biosci 2(1):25. https://doi.org/10.1186/2045-3701-2-25 2045-3701-2-25 [pii]
22. Yen PM (2001) Physiological and molecular basis of thyroid hormone action. Physiol Rev 81(3):1097–1142
23. McKenna NJ, O'Malley BW (2001) Nuclear receptors, coregulators, ligands, and selective receptor modulators: making sense of the patchwork quilt. Ann N Y Acad Sci 949:3–5
24. Li J, O'Malley BW, Wong J (2000) p300 requires its histone acetyltransferase activity and SRC-1 interaction domain to facilitate thyroid hormone receptor activation in chromatin. Mol Cell Biol 20(6):2031–2042
25. Shi Y-B, Wong J, Puzianowska-Kuznicka M, Stolow M (1996) Tadpole competence and tissue-specific temporal regulation of amphibian metamorphosis: roles of thyroid hormone and its receptors. BioEssays 18:391–399
26. Sachs LM, Damjanovski S, Jones PL, Li Q, Amano T, Ueda S, Shi YB, Ishizuya-Oka A (2000) Dual functions of thyroid hormone receptors during Xenopus development. Comp Biochem Physiol B Biochem Mol Biol 126(2):199–211
27. Yaoita Y, Brown DD (1990) A correlation of thyroid hormone receptor gene expression with amphibian metamorphosis. Genes Dev 4 (11):1917–1924
28. Wong J, Shi Y-B (1995) Coordinated regulation of and transcriptional activation by Xenopus thyroid hormone and retinoid X receptors. J Biol Chem 270:18479–18483
29. Wang X, Matsuda H, Shi Y-B (2008) Developmental regulation and function of thyroid hormone receptors and 9-cis retinoic acid receptors during Xenopus tropicalis metamorphosis. Endocrinology 149:5610–5618
30. Leloup J, Buscaglia M (1977) La triiodothyronine: hormone de la métamorphose des amphibiens. CR Acad Sci 284:2261–2263
31. Puzianowska-Kuznicka M, Damjanovski S, Shi Y-B (1997) Both thyroid hormone and 9-cis retinoic acid receptors are required to efficiently mediate the effects of thyroid hormone on embryonic development and specific gene regulation in Xenopus laevis. Mol Cell Biol 17:4738–4749
32. Nakajima K, Yaoita Y (2003) Dual mechanisms governing muscle cell death in tadpole tail during amphibian metamorphosis. Dev Dyn 227:246–255
33. Schreiber AM, Brown DD (2003) Tadpole skin dies autonomously in response to thyroid hormone at metamorphosis. Proc Natl Acad Sci U S A 100:1769–1774
34. Das B, Schreiber AM, Huang H, Brown DD (2002) Multiple thyroid hormone-induced muscle growth and death programs during metamorphosis in Xenopus laevis. Proc Natl Acad Sci U S A 99:12230–12235
35. Schreiber AM, Das B, Huang H, Marsh-Armstrong N, Brown DD (2001) Diverse developmental programs of Xenopus laevis

metamorphosis are inhibited by a dominant negative thyroid hormone receptor. Proc Natl Acad Sci U S A 98:10739–10744
36. Buchholz DR, Hsia VS-C, Fu L, Shi Y-B (2003) A dominant negative thyroid hormone receptor blocks amphibian metamorphosis by retaining corepressors at target genes. Mol Cell Biol 23:6750–6758
37. Buchholz DR, Tomita A, Fu L, Paul BD, Shi Y-B (2004) Transgenic analysis reveals that thyroid hormone receptor is sufficient to mediate the thyroid hormone signal in frog metamorphosis. Mol Cell Biol 24:9026–9037
38. Hasebe T, Buchholz DR, Shi YB, Ishizuya-Oka A (2011) Epithelial-connective tissue interactions induced by thyroid hormone receptor are essential for adult stem cell development in the Xenopus laevis intestine. Stem Cells 29 (1):154–161. https://doi.org/10.1002/stem.560
39. Wen L, Fu L, Guo X, Chen Y, Shi YB (2015) Histone methyltransferase Dot1L plays a role in postembryonic development in Xenopus tropicalis. FASEB J 29:385–393. https://doi.org/10.1096/fj.14-252171
40. Guo X, Zhang T, Hu Z, Zhang Y, Shi Z, Wang Q, Cui Y, Wang F, Zhao H, Chen Y (2014) Efficient RNA/Cas9-mediated genome editing in Xenopus tropicalis. Development 141(3):707–714. https://doi.org/10.1242/dev.099853
41. Lei Y, Guo X, Deng Y, Chen Y, Zhao H (2013) Generation of gene disruptions by transcription activator-like effector nucleases (TALENs) in Xenopus tropicalis embryos. Cell Biosci 3 (1):21. https://doi.org/10.1186/2045-3701-3-21 2045-3701-3-21 [pii]
42. Lei Y, Guo X, Liu Y, Cao Y, Deng Y, Chen X, Cheng CH, Dawid IB, Chen Y, Zhao H (2012) Efficient targeted gene disruption in Xenopus embryos using engineered transcription activator-like effector nucleases (TALENs). Proc Natl Acad Sci U S A 109:17484–17489. https://doi.org/10.1073/pnas.1215421109 1215421109 [pii]
43. Nakayama T, Fish MB, Fisher M, Oomen-Hajagos J, Thomsen GH, Grainger RM (2013) Simple and efficient CRISPR/Cas9-mediated targeted mutagenesis in Xenopus tropicalis. Genesis 51(12):835–843. https://doi.org/10.1002/dvg.22720
44. Blitz IL, Biesinger J, Xie X, Cho KW (2013) Biallelic genome modification in F(0) Xenopus tropicalis embryos using the CRISPR/Cas system. Genesis 51(12):827–834. https://doi.org/10.1002/dvg.22719
45. Wang F, Shi Z, Cui Y, Guo X, Shi YB, Chen Y (2015) Targeted gene disruption in Xenopus laevis using CRISPR/Cas9. Cell Biosci 5:15. https://doi.org/10.1186/s13578-015-0006-1
46. Yen PM (2015) Unliganded TRs regulate growth and developmental timing during early embryogenesis: evidence for a dual function mechanism of TR action. Cell Biosci 5:8. https://doi.org/10.1186/2045-3701-5-8
47. Wen L, Shi YB (2015) Unliganded thyroid hormone receptor alpha controls developmental timing in Xenopus tropicalis. Endocrinology 156:721–734. https://doi.org/10.1210/en.2014-1439
48. Choi J, Suzuki KI, Sakuma T, Shewade L, Yamamoto T, Buchholz DR (2015) Unliganded thyroid hormone receptor alpha regulates developmental timing via gene repression as revealed by gene disruption in Xenopus tropicalis. Endocrinology 156:735–744. https://doi.org/10.1210/en.2014-1554
49. Sachs LM (2015) Unliganded thyroid hormone receptor function: amphibian metamorphosis got TALENs. Endocrinology 156 (2):409–410. https://doi.org/10.1210/en.2014-2016
50. Choi J, Ishizuya-Oka A, Buchholz DR (2017) Growth, development, and intestinal remodeling occurs in the absence of thyroid hormone receptor alpha in tadpoles of Xenopus tropicalis. Endocrinology 158(6):1623–1633
51. Wen L, Shibata Y, Su D, Fu L, Luu N, Shi Y-B (2017) Thyroid hormone receptor α controls developmental timing and regulates the rate and coordination of tissue specific metamorphosis in Xenopus tropicalis. Endocrinology 158(6):1985–1998
52. Paul BD, Buchholz DR, Fu L, Shi Y-B (2005) Tissue- and gene-specific recruitment of steroid receptor coactivator-3 by thyroid hormone receptor during development. J Biol Chem 280:27165–27172
53. Paul BD, Fu L, Buchholz DR, Shi Y-B (2005) Coactivator recruitment is essential for liganded thyroid hormone receptor to initiate amphibian metamorphosis. Mol Cell Biol 25:5712–5724
54. Paul BD, Buchholz DR, Fu L, Shi Y-B (2007) SRC-p300 coactivator complex is required for thyroid hormone induced amphibian metamorphosis. J Biol Chem 282:7472–7481
55. Havis E, Sachs LM, Demeneix BA (2003) Metamorphic T3-response genes have specific co-regulator requirements. EMBO Rep 4:883–888

56. Paul BD, Shi Y-B (2003) Distinct expression profiles of transcriptional coactivators for thyroid hormone receptors during Xenopus laevis metamorphosis. Cell Res 13:459–464
57. Tomita A, Buchholz DR, Shi Y-B (2004) Recruitment of N-CoR/SMRT-TBLR1 corepressor complex by unliganded thyroid hormone receptor for gene repression during frog development. Mol Cell Biol 24:3337–3346
58. Sachs LM, Jones PL, Havis E, Rouse N, Demeneix BA, Shi Y-B (2002) N-CoR recruitment by unliganded thyroid hormone receptor in gene repression during Xenopus laevis development. Mol Cell Biol 22:8527–8538
59. Sato Y, Buchholz DR, Paul BD, Shi Y-B (2007) A role of unliganded thyroid hormone receptor in postembryonic development in Xenopus laevis. Mech Dev 124:476–488
60. Matsuura K, Fujimoto K, Fu L, Shi Y-B (2012) Liganded thyroid hormone receptor induces nucleosome removal and histone modifications to activate transcription during larval intestinal cell death and adult stem cell development. Endocrinology 153:961–972
61. Wong J, Shi Y-B, Wolffe AP (1997) Determinants of chromatin disruption and transcriptional regulation instigated by the thyroid hormone receptor: hormone-regulated chromatin disruption is not sufficient for transcriptinal activation. EMBO J 16:3158–3171
62. Wong J, Shi YB, Wolffe AP (1995) A role for nucleosome assembly in both silencing and activation of the Xenopus TR beta A gene by the thyroid hormone receptor. Genes Dev 9 (21):2696–2711
63. Bilesimo P, Jolivet P, Alfama G, Buisine N, Le Mevel S, Havis E, Demeneix BA, Sachs LM (2011) Specific histone lysine 4 methylation patterns define TR-binding capacity and differentiate direct T3 responses. Mol Endocrinol 25:225–237
64. Grimaldi A, Buisine N, Miller T, Shi YB, Sachs LM (2013) Mechanisms of thyroid hormone receptor action during development: lessons from amphibian studies. Biochim Biophys Acta 1830(7):3882–3892. https://doi.org/10.1016/j.bbagen.2012.04.020 S0304-4165(12)00125-0 [pii]
65. Sachs LM, Amano T, Shi YB (2001) An essential role of histone deacetylases in postembryonic organ transformations in Xenopus laevis. Int J Mol Med 8(6):595–601
66. Sachs LM, Amano T, Rouse N, Shi YB (2001) Involvement of histone deacetylase at two distinct steps in gene regulation during intestinal development in Xenopus laevis. Dev Dyn 222 (2):280–291
67. Nakayama T, Blitz IL, Fish MB, Odeleye AO, Manohar S, Cho KW, Grainger RM (2014) Cas9-based genome editing in Xenopus tropicalis. Methods Enzymol 546:355–375
68. Montague TG, Cruz JM, Gagnon JA, Church GM, Eivind Valen E (2014) CHOPCHOP: a CRISPR/Cas9 and TALEN web tool for genome editing. Nucleic Acids Res 42: W401–W407
69. Stemmer M, Thumberger T, Del Sol Keyer M, Wittbrodt J, Mateo JL (2015) CCTop: an intuitive, flexible and reliable CRISPR/Cas9 target prediction tool. PLoS One 10(4):e0124633. https://doi.org/10.1371/journal.pone.0124633
70. Naito Y, Hino K, Bono H, Ui-Tei K (2015) CRISPRdirect: software for designing CRISPR/Cas guide RNA with reduced off-target sites. Bioinformatics 31 (7):1120–1123. https://doi.org/10.1093/bioinformatics/btu743
71. Nieuwkoop PD, Faber J (1965) Normal table of Xenopus laevis. North Holland Publishing, Amsterdam
72. Fu L, Wen L, Luu N, Shi YB (2016) A simple and efficient method to visualize and quantify the efficiency of chromosomal mutations from genome editing. Sci Rep 6:35488. https://doi.org/10.1038/srep35488
73. Tsai SQ, Wyvekens N, Khayter C, Foden JA, Thapar V, Reyon D, Goodwin MJ, Aryee MJ, Joung JK (2014) Dimeric CRISPR RNA-guided FokI nucleases for highly specific genome editing. Nat Biotechnol 32 (6):569–576. https://doi.org/10.1038/nbt.2908
74. Wu M, Gerhart J (1991) Raising Xenopus in the laboratory. Methods Cell Biol 36:3–18

Chapter 30

Microinjection of Live Mammalian Cells: A Delivery Method that Provides Added Versatility to the Study of Cellular Function

Robert F. Gahl

Abstract

Microinjection is a technique that allows for the delivery of a diverse array of compounds and biomolecules into live mammalian cells. As a result, the behavior of injected biomolecules and their resulting effects can be observed in the native environments of mammalian cells in real time. This capability allows for more accurate observations and conclusions about biological systems of interest. This chapter discusses the protocol and guidelines to successfully microinject live mammalian cells to maintain their viability. This chapter outlines considerations into the preparation of samples and the microscopy setup as well as a description of a stepwise presentation of the methods for proper execution of a microinjector. The presentation of the technique serves as a starting point to understanding how to apply microinjection to a cellular system compatible with a wide array of live-cell imaging techniques.

Key words Microinjection, Mammalian cells, Live-cell imaging

1 Introduction

The successful understanding of the functions of cellular components relies on the observation of these components in their native environment. The execution of experiments to make these observations presents a challenge in introducing probes or modified components to aid in understanding biological function. Microinjection is a technique that allows for additional freedom for circumventing this challenge. A wide array of compounds can be introduced to live cells with minimal invasiveness that maintains their viability. This chapter outlines some important procedures and considerations to maximize the successful microinjection of mammalian cells.

Microinjection has been employed to study a variety of biological functions in mammalian cells. The orientation of SNARE proteins has been studied by microinjecting derivatized variants into live cells and imaged using confocal microscopy

Chengyu Liu and Yubin Du (eds.), *Microinjection: Methods and Protocols*, Methods in Molecular Biology, vol. 1874,
https://doi.org/10.1007/978-1-4939-8831-0_30,

[1]. The mechanisms of gap-junctions during cellular communication can be studied by microinjecting dye into cells and observing the diffusion in real time [2]. Antibodies targeting histones can be microinjected to determine multiple binding partners and complex interactions [3]. In addition, microinjection can also be utilized to introduce components that are difficult to produce or are only present in small quantities.

One valuable insight that microinjection can bring to research is observations under environments that can only be recreated in live cells and not in artificial systems. While there are many instances of recreating cellular function outside of cells, there are aspects of the native environment of the cell that cannot be recreated in artificial systems. Therefore, one distinct advantage of the ability to introduce compounds capable of customization is that targeted research questions can be answered in native cellular environments. The implementation of microinjection resulted in better insight to understanding the structure and role of proteins of the Bcl-2 family as part of the regulation of apoptosis.

One of the principle questions about the regulation of apoptosis is how the Bcl-2 protein, Bax, undergoes conformational changes as it translocates from the cytosol to the mitochondria, where it oligomerizes to permeabilize the mitochondria. This step is known as the "point of no return" for the cell and a cascade of events ensues to continue the process of apoptosis [4, 5]. An important observation about how the Bcl-2 family regulates apoptosis is that another Bcl-2 protein Bcl-xL is responsible for keeping Bax from entering the mitochondria [6, 7] by "retrotranslocating" Bax. The importance of this observation is that Bax is "poised" to enter the mitochondria in the native environment of live cells. In artificial systems, Bax was not observed to enter membranes and only done so when additional proteins were present [8]. In another study that wanted to determine what the structure of Bax was in live cells, microinjection was employed to deliver Bax conjugated with FRET probes to measure structural changes [9, 10]. Not only did this study also observe behavior that was consistent with retrotranslocation of Bax, this study shows that the helix responsible for Bax entering mitochondria was continually exposed to the cellular environment. This structural feature allowed Bax to be "poised" to enter the mitochondria. Previous studies performed outside a cellular context proposed that this helix is tucked into the structure of Bax [11] and additional work proposed that additional interactions are required for Bax to insert into the mitochondria [12]. Later studies using gene editing technology confirmed that Bax is indeed "poised" to enter the mitochondria because Bax inserts into mitochondria in the absence of Bcl-xL or any other Bcl-2 protein [13, 14]. These studies illustrate that more accurate conclusions about biological systems of interest can be reached by

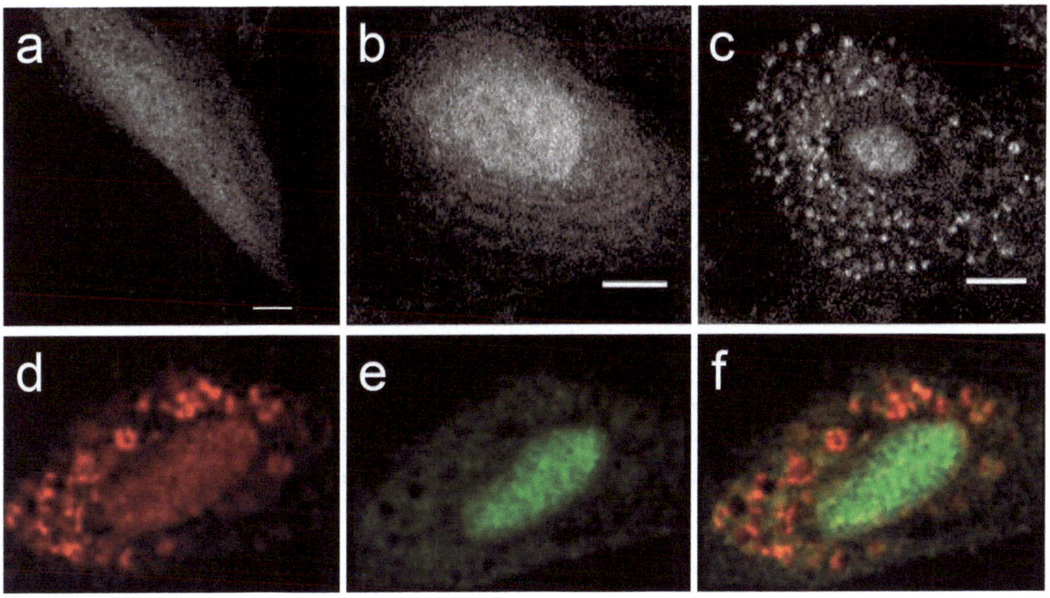

Fig. 1 Examples of successful microinjections of MEF cells with biologically active material. The scale bars represent 10 μm. (**a**) Microinjected GFP. The outline of the cell is visible, but the location of the injection is not. (**b**) Biologically active Bax labeled with fluorescent probes injected into MEF cells. In healthy cells, Bax is diffuse throughout the cytosol, as observed in the image. (**c**) After the addition of staurosporin (STS), Bax translocates to mitochondria and forms a punctate pattern, which is visible in the image. (**d**) The protein Bid gets cleaved after it associates with the mitochondria. This image shows the part that associates the mitochondria. (**e**) This image shows the other portion of Bid that diffuses throughout the cytosol. (**f**) An overlay of the images in **d** and **e** shows no appearance of the characteristic yellow color that indicates translocation between regions of two images. Adapted from [16]

observing how relevant biomolecules function in live cells. Microinjection proved to be a versatile tool to make these observations.

The most important task to ensure success when utilizing microinjection is the confirmation of viability of the microinjected cells and the bio-activity of the microinjected compounds. Viable cells should not appear to have been microinjected. Figure 1a shows the microinjection of GFP [10]. The outline of the cell is visible and the location of the microinjection is not visible. Derivatized biomolecules should exhibit the same bio-activity as has been previously demonstrated. The same translocation of Bax that has been reported previously [10] is also observed by the change from a diffuse distribution, Fig. 1b, to a punctate distribution, Fig. 1c. Another example of biological activity that was verified in microinjected samples was performed on another protein, Bid. Bid is another Bcl-2 protein that undergoes cleavage when it associates with the mitochondria [15]. The cleavage can be observed by labeling Bid with two fluorescent tags and observing the signal from the two proteins in different parts of the cell [16]. Figure 1d shows the signal from the portion of Bid that associates with the mitochondria while Fig. 1e shows that signal from the portion of Bid that diffuses away after the cleavage. An overlay of the two signals, Fig. 1f, does not show any correlation. Therefore, it is critically important to have tests for the viability of cells and the bioactivity of the microinjected samples. Since there is increased variability in what you can microinject, other assays that utilize staining of organelles can be used in conjunction with microinjection to verify biological activity. Mitochondria labeling using covalent agents [17] or potentiometric stains [18] can be used effectively to determine cell viability and biological activity.

The method presented is not meant to be applied to every experiment that utilizes microinjection, but rather should be a starting point to understanding the pitfalls that could arise whenever this technique is employed. The "Notes" contained in this chapter are intended to maximize the success of microinjection to study viable cells. The suggested parameters for the execution of microinjection are meant for mouse embryonic fibroblast (MEF) cells. However, other cell types can be studied and different variations of this protocol can also be successful in studying a variety of cellular functions. There is no substitute for practice. Familiarity with the cellular system and imaging equipment are necessary for successful experiments. When employed successfully, microinjected cells can be studied with a wide array of imaging techniques and analysis.

2 Materials

2.1 Microscope Setup

1. Inverted Microscope with moveable stage and access to medium.
2. Stage should connect to pressure-powered microinjector with access to sample medium.
3. Fluorescence Imaging as well as Visible Light Imaging should be accessible through eye-piece.
4. 43× or similar objective installed for imaging (*see* **Note 1**).
5. Microscope should be ready for live-cell imaging technique of choice (wide-field, confocal, TIRF, two-photon, etc.) once cells are microinjected (*see* **Note 2**).
6. Oil for objective.

2.2 Cell Culture

1. Dishes with coverslips on the bottom specified to imaging objective.
2. Attaching agent for cells, fibronectin.
3. Phosphate-Buffered Saline (PBS) solution.

4. Medium that does not contain optical artifacts and is sufficient for studying cells. The medium should contain 10% fetal bovine serum (FBS), penicillin, and streptomycin (*see* **Note 3**).
5. Cell line of interest (*see* **Note 4**).

2.3 Sample Preparation

1. Sample in solution that matches the ionic strength of the cellular medium to be used for microinjection (*see* **Note 5**).
2. Sample in solution that maintains the bio-activity of the molecule of interest for storage in future experiments (*see* **Note 6**).
3. PBS (sterile) or diluting solvent to adjust the concentration of the sample for microinjection.

2.4 Needles

1. Premade or produced using a puller right before use.
2. Syringe for loading sample into needle.
3. 20% ethanol solution.
4. Ultra-purified water.

3 Methods

3.1 Growing Cells on Coverslips for Imaging (Previous Day of Microinjection and Imaging)

1. Dilute 1 mg/mL solution of fibronectin 20× into PBS. A total volume of 200 μL is sufficient to coat a cover-slip in 35 mm glass bottom dishes.
2. Coat the cover-slip area for each dish with diluted solution of fibronectin.
3. Incubate the dishes for 4 h at 37 °C.
4. As the dishes are being incubated, the cells can be passaged.
5. Aspirate cellular media from dishes containing the cells of interest. The cells should have been growing and are approaching confluency.
6. Rinse cells 3× with PBS.
7. Add enough 5% trypsin solution to coat the cells. Typically, ¼ of volume of media used to grow the cells is sufficient.
8. After 5 min at room temperature, add media that will be used for the imaging experiments to stop the trypsinization process. The total volume of the medium should be the same as the volume used to grow the cells of interest.
9. Gently mix the trypsinized cells, trypsin solution, and cellular medium, then transfer to a large enough falcon tube, and centrifuge the mixture for 5 min at 1,200 × *g*.
10. After centrifugation, aspirate the medium off the pelleted cells, taking care not to aspirate the cell pellet itself.

11. Gently reconstitute the cell pellet in the medium used for imaging to a total volume used to grow the cells.
12. Dilute cells into a new falcon tube to a desired concentration (*see* **Note 7**).
13. Remove the dishes with cover-slips that have been incubating at 37 °C.
14. Rinse off the fibronectin suspended in solution with PBS.
15. Pipette the cells at the desired concentration in the imaging media into the 35 mm dishes with cover-slips coated with fibronectin. 2 mL is sufficient for 35 mm dishes.
16. Incubate 35 mm dishes overnight at 37 °C.

3.2 Setting Up the Microscope for Microinjection (Day of Microinjection and Imaging)

1. Select 43× objective and imaging settings.
2. Ensure that fluorescent light and visible light can easily be interchanged while viewing through the eye-piece.
3. Add oil to objective.
4. Place dish with cells onto stage on top of objective.
5. Move stage to get cells in focus.

3.3 Loading the Sample into the Needle

1. Pull needle or remove from sterile packaging (*see* **Note 8**).
2. Disinfect syringe with 3× wash with 20% ethanol solution.
3. Wash syringe with 3× wash with ultra-purified water (*see* **Note 9**).
4. Use syringe to load 4 μL of sample into needle and eliminate air-bubbles (*see* **Note 10**). If the sample is already in a solvent matching the ionic strength of the media, that can be directly loaded into the syringe. Otherwise, an aliquot of the sample solution can be diluted with PBS to a concentration compatible with imaging and cellular function before it is loaded into the syringe.

3.4 Loading the Needle onto Microinjector

1. Turn on Microinjector to charge pressure needed for injections.
2. Apply a "CLEAN" burst of air to ensure that pressure is properly applied through tubing and through opening (*see* **Note 11**).
3. Attach needle to microinjector manipulator taking care not to break the needle.
4. Apply another "CLEAN" burst of air to make sure all pressure is applied to needle and sample (*see* **Note 12**).
5. Make sure there is sufficient pressure for a constant flow of sample to medium (*see* **Note 13**).

3.5 Microinjecting Cells

1. Make sure the cells are in focus.
2. The manipulator should be in a "COARSE" setting.
3. Slowly bring needle down to just below the level of medium so that reflected light is visible.
4. Find this reflected light in the eye-piece by moving the position of the needle from the manipulator, Fig. 2a.

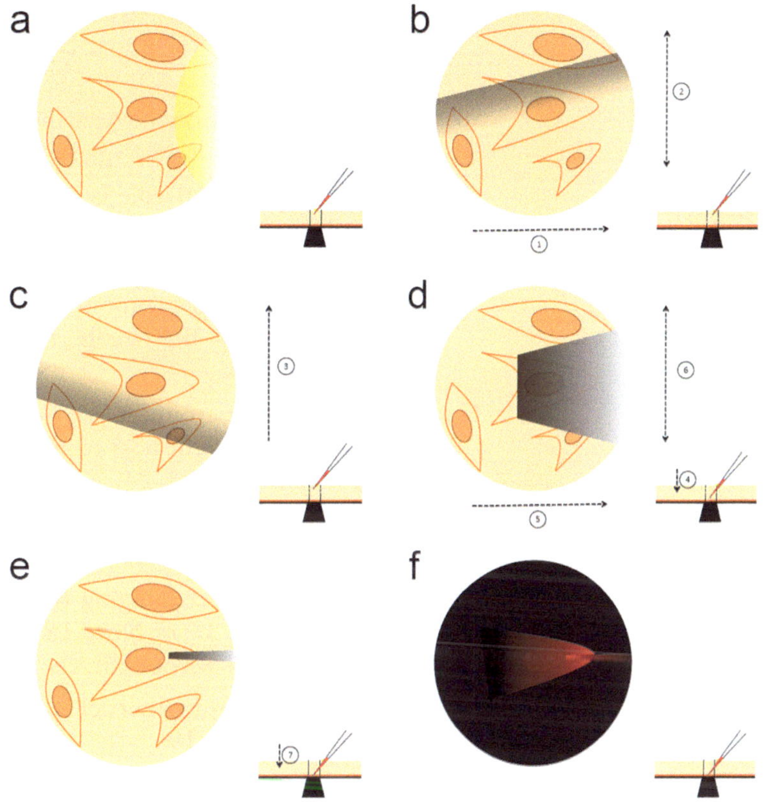

Fig. 2 Protocol to find a needle while looking through the eye-piece of a microscope. (**a**) The yellow color represents brighter light that is observed when the meniscus created by a needle in solution comes into view. The manipulator should be in a "COARSE" movement mode. (**b**) After moving the needle to the right, ①, and forwards and back, ②, an edge of the needle comes into view. (**c**) The needle can be moved forward and back again, ③, to find the other edge of the needle. (**d**) To get the tip of the needle into view, incremental movements of bringing the needle closer to the cells, ④, moving the needle to the right, ⑤, and moving the needle back and forth to make sure the ends of the needle are in view, ⑥, gradually bring the end into view, but is still unfocused. When the end of the needle is in view, change the movement setting of the manipulator to "FINE". (**e**) Move the Needle to a region next to the nucleus of the cell, ⑦, and puncture the cell. Bring the needle out of the cell when other parts of the cell move. This can take less than a second. (**f**) Switch the viewing setting to fluorescent light and the outline of the cell should be visible

5. Slowly move the needle to the right, ①, while making adjustments forward and backward, ②, to find a dark shadow of the needle, Fig. 2b. The needle is out of focus but you have found an edge of the needle.
6. Move the needle forward and back, ③, until you can see either side of the out-of-focus needle, Fig. 2c.
7. Bring the needle closer to the cells by incrementally lowering the needle, ④, and moving it to the right, ⑤, where at each step you can still see the two sides of the out-of-focus needle, ⑥. After enough increments, both sides of the needle come into view (Fig. 2d).
8. If the needle is lost from the viewing area, bring the needle back up to the surface and start again (*see* **Note 14**).
9. Once the tip of the needle is in view, change the movement setting of the "joystick" to "FINE."
10. Apply another burst of pressure by pressing "clean" and watching fluorescent sample come through the needle. If no signal is coming through the needle, then there is a blockage (*see* **Note 15**).
11. Choose a cell to microinject (*see* **Note 16**) and bring the needle down to the cell, right next to the nucleus, and puncture the cellular membrane, Fig. 2e, ⑦.
12. When different regions of the cell move, bring the needle back out of the cell (*see* **Note 17**).
13. Turn on fluorescence imaging to make sure the shape of the cell is visible, Fig. 2f (*see* **Note 18**). If the outline of the cell is visible under fluorescent light, the cell has been microinjected.
14. Move the needle to another cell, refocus the cell, and repeat the injections (*see* **Note 19**).
15. Since the needle is already in view, the cells just need to be slightly refocused before the needle can approach and inject another cell (*see* **Note 20**).

4 Notes

1. The choice for an objective for imaging is a balance between detail and number of cells per viewing area. There should be enough detail in the images to investigate a particular question, but enough microinjected cells need to be viewed to acquire a sufficient sample size. The best magnification would be 43×, which allows you to view up to 15–20 cells at a time in the viewing area. However, at higher magnifications, such as 73×, fewer cells can be imaged.

2. The goal is to image in the same setup as microinjection. Ideally, you should not remove the dish from the stage after microinjection. It is very difficult to microinject cells on one microscope then move the dish to another microscope and find the same cells again although, with some practice it is possible. You would only want to make fine adjustments to center microinjected cells for imaging.
3. In previous work [9], OPTIMEM worked well for reducing imaging artifacts. 10% FBS was present in the medium and some should be present to maintain cell viability. Another customization for the cellular medium could include removing components that contribute the cellular growth if imaging is required for long amounts of time. There are many types of medium that are compatible for using microinjection.
4. Mouse embryonic fibroblasts (MEF) work well for microinjection because their cell shape is relatively flat which makes observing the instance of an injection readily visible. This leads to shorter, more innocuous injections that improve the chances of viable microinjected cells. Other cell lines that have been compatible with microinjection include HeLa and HCT116 cells, but these are more difficult to work with compared to MEFs. These cell lines can be injected and imaged at room temperature. If a cell line is known to be highly dependent on a temperature of 37 °C, then other cell lines should be considered.
5. The sample should be at a sufficient concentration for assays and detection and must match the ionic strength of the cytosol of the cell. This improves the success rate for viable cells after microinjection. 0.1 M NaCl would be sufficient to avoid and drastic changes in the cellular environment, which would increase the success of microinjection. Since the amount injected is roughly 10% of the cellular volume [19] the pH of the sample should be within a unit of the cellular pH to not perturb the cellular pH. If quantity of sample is an issue, it would be beneficial to set up the last step in the prep to be a buffer that would be compatible with microinjection.
6. Great care should be taken to make sure no detrimental consequences such as precipitation or aggregation occur in the sample for microinjection. These side effects can be harmful to cells or prevent the sample from flowing through the needle.
7. The density of the cells should maximize the cells in the viewing area to get a sufficient sample size but not induce confluence during imaging. This largely depends on the cell type used. Another consideration for passaging is preparing dishes at varying dilutions (1/4, 1/8, 1/16) so that in successive days, a similar density of cells can be studied and more cells can be

microinjected from a single passage. This would reduce the variability in cells when making multiple measurements.

8. For a Sutter Instrument P-97 Flaming/Brown Micropipette Puller, one method that worked with Sutter Instruments BF100-78-10 Needles is a two-stage method with the following settings: P = 500, Line 1: Heat = 640, Pull=, Vel = 50, Time = 200; Line 2: Heat = 640, Pull = 25, Vel = 65, Time = 200. Additional recommendations for pullers and needles can be found through manufacturers such as Sutter.
9. These two steps are very important to avoid contaminating your sample with bacteria, which can be seen in images and can rupture the cells during experiments. Bacteria can be identified by their rapid movements and mostly show up as small particles. If washing the needle does not solve the problem, then the medium that contains the sample is contaminated and must be remade.
10. Careful introduction of the sample to the needle can ensure no air bubbles are present in addition to vigorous shaking of the needle. Air bubbles should always be removed because they impede the flow of sample during microinjection.
11. A "clean" burst of air is the application of an intense amount of pressure put through the tubing about several thousand hBar. Other microinjectors may call this burst of air by a different name, but "CLEAN" is used for an Eppendorf Femtojet. A "CLEAN" application should not be applied for extended periods of time.
12. Another application of the "CLEAN" function should not produce any sound from the needle. If any sounds are heard, then the tubing in the needle is broken or the needle itself broke while it was attached to the manipulator.
13. Having a constant flow of sample from the needle can be accomplished with only 20–24 hbar of pressure if injecting a buffered aqueous solution with 0.1 M NaCl. In the Eppendorf Femtojet, this pressure setting is called *pc*.
14. Breaking the needle not only causes a loss of sample, but contaminates the dish with excess fluorescent signal from the sample that could contaminate other experiments in the same dish. When a needle breaks, the needle as well as the dish need to be replaced.
15. A burst of pressure should introduce a small amount of fluorescent sample to the medium. Not the entire contents form the needle. If there is no flow coming out, then there could be an air bubble, insufficient back pressure, or a pressure leak.
16. The ideal location to inject a cell is right next to the nucleus because it is next to the largest part of the cell. This is the best

part of the cell to handle a temporary expansion of the cellular membrane. Also, you can see how much other parts of the cell respond to a microinjection.

17. You can also judge if the flow of sample is too much for the cell if the nucleus begins to compress. The reaction time would be approximately 1 s, so quick response is required. The nucleus is the largest and most dense organelle so if there is enough flow to move the nucleus, then other organelles are being stressed as well. Over the course of imaging, these cells might begin to undergo morphologies of being stressed and may not be viable during the duration of the experiment.
18. A good rule of thumb is that if you cannot see the cell with the fluorescent lamp looking through the eyepiece, then there is a slim chance you can view the cell with other types of imaging.
19. If you want to inject multiple dishes, the Eppendorf microinjector has a "home" feature that can remove the needle from the medium then return it to the exact place after a dish is replaced. This feature saves a lot of time. The only thing to remember is bring the needle sufficiently above the cells before activating the home function. Otherwise, the needle would come back to close to the layer of cells and break the needle.
20. There are functions in microinjectors that have an "INJECT" setting. However, the microinjection level needs to be constant to inject cells in a sequential fashion without breaking the needle or completely missing the cell. Therefore, the plane of the cells can differ by microns, which could be the difference between a successful or unsuccessful microinjection. To maximize the probability of viable, microinjected cells, the cell chosen for microinjection should be refocused before the needle is brought down manually for each microinjection.

References

1. Sakon JJ, Weninger KR (2010) Detecting the conformation of individual proteins in live cells. Nat Methods 7(3):203–205
2. Alberto AV, Bonavita AG, Fidalgo-Neto AA, Bercot F, Alves LA (2017) Single-cell microinjection for cell communication analysis. J Vis Exp. https://doi.org/10.3791/50836
3. Dixon CR, Platani M, Makarov AA, Schirmer EC (2017) Microinjection of antibodies targeting the lamin A/C histone-binding site blocks mitotic entry and reveals separate chromatin interactions with HP1, CenpB and PML. Cell 6(2):E9
4. Tait SW, Green DR (2013) Mitochondrial regulation of cell death. Cold Spring Harb Perspect Biol 5(9):a008706
5. Moldoveanu T, Follis AV, Kriwacki RW, Green DR (2014) Many players in BCL-2 family affairs. Trends Biochem Sci 39(3):101–111
6. Edlich F, Banerjee S, Suzuki M, Cleland MM, Arnoult D, Wang C, Neutzner A, Tjandra N, Youle RJ (2011) Bcl-x(L) retrotranslocates Bax from the mitochondria into the cytosol. Cell 145(1):104–116
7. Todt F, Cakir Z, Reichenbach F, Youle RJ, Edlich F (2013) The C-terminal helix of Bcl-x (L) mediates Bax retrotranslocation from the mitochondria. Cell Death Differ 20 (2):333–342
8. Lovell JF, Billen LP, Bindner S, Shamas-Din A, Fradin C, Leber B, Andrews DW (2008) Membrane binding by tBid initiates an ordered

series of events culminating in membrane permeabilization by Bax. Cell 135(6):1074–1084
9. Gahl RF, Tekle E, Tjandra N (2014) Single color FRET based measurements of conformational changes of proteins resulting from translocation inside cells. Methods 66(2):180–187
10. Gahl RF, He Y, Yu S, Tjandra N (2014) Conformational rearrangements in the pro-apoptotic protein, Bax, as it inserts into mitochondria: a cellular death switch. J Biol Chem 289(47):32871–32882
11. Westphal D, Kluck RM, Dewson G (2014) Building blocks of the apoptotic pore: how Bax and Bak are activated and oligomerize during apoptosis. Cell Death Differ 21 (2):196–205
12. Birkinshaw RW, Czabotar PE (2017) The BCL-2 family of proteins and mitochondrial outer membrane permeabilisation. Semin Cell Dev Biol 72:152–162
13. O'Neill KL, Huang K, Zhang J, Chen Y, Luo X (2016) Inactivation of prosurvival Bcl-2 proteins activates Bax/Bak through the outer mitochondrial membrane. Genes Dev 30 (8):973–988
14. Zhang J, Huang K, O'Neill KL, Pang X, Luo X (2016) Bax/Bak activation in the absence of Bid, Bim, Puma, and p53. Cell Death Dis 7: e2266
15. Billen LP, Shamas-Din A, Andrews DW (2008) Bid: a Bax-like BH3 protein. Oncogene 27 (Suppl 1):S93–S104
16. Gahl RF, Dwivedi P, Tjandra N (2016) Bcl-2 proteins bid and bax form a network to permeabilize the mitochondria at the onset of apoptosis. Cell Death Dis 7(10):e2424
17. Warnes G (2015) Flow cytometric assays for the study of autophagy. Methods 82:21–28
18. Iannetti EF, Willems PH, Pellegrini M, Beyrath J, Smeitink JA, Blanchet L, Koopman WJ (2015) Toward high-content screening of mitochondrial morphology and membrane potential in living cells. Int J Biochem Cell Biol 63:66–70
19. Pepperkok R, Scheel J, Horstmann H, Hauri HP, Griffiths G, Kreis TE (1993) Beta-COP is essential for biosynthetic membrane transport from the endoplasmic reticulum to the Golgi complex in vivo. Cell 74(1):71–82

Index

A

Acepromazine 376, 385
Acidic Tyrode's solution 182–185, 187
American Large White boars 362
Ampicillin 47, 385, 492–494, 497, 501, 518
Anesthesia 20, 27, 28, 33, 36, 94, 96, 155, 165, 239, 240, 259, 275, 276, 286, 292, 322, 331, 338, 343, 484
AttB 71–73, 76, 81, 83, 418
AttP 71–74, 79, 81–83, 418
Avertin 36

B

Bacterial artificial chromosome (BAC) 43, 52, 58, 60–62
  concentration, typical 53
  library 64
  ordering 50
  purification 52
  quantitation 53
  restriction mapping 55
Blastocyst .... v, 5, 6, 36, 99–113, 140, 165, 170, 176, 189, 192, 207, 212, 223, 224, 239, 268, 287, 295, 313–325, 336, 337, 357, 358, 367, 368, 370
Blastodisc 475, 477, 484
Buffalo rat liver (BRL) cells 404–412, 414–417, 419, 421, 422, 428
Bupivacaine 104, 110, 144, 155, 195, 201
Buprenorphine 20, 28, 33

C

Caenorhabditis elegans 374, 431–455
CARD Hyperova 230, 248, 252, 254
Carprofen 332, 338, 340
Cauda epididymides 237, 238, 249, 250
Center for Animal Resources Development (CARD) method 229–241, 243–254
Cesarean section 20, 32, 36, 223, 224, 386
Chloramphenicol 47, 52, 53
Clustered regularly interspaced short palindromic repeats (CRISPR) 7, 17, 83–84, 88, 101, 115, 140, 169, 179, 192, 257, 274, 297, 314, 328, 347, 373–389, 418, 432, 459, 475, 489, 510
Cre recombinase 116, 141, 300, 305, 420
CRISPR-associated protein 9 (Cas9) 11, 83, 95, 117, 118, 122, 130, 136, 140, 141, 143, 144, 146, 149, 153, 162–164, 171, 179–183, 185, 189, 193, 196, 198, 200, 207, 257–270, 274, 279, 281, 297, 314, 329, 332, 333, 336, 341, 348, 349, 352, 358, 374, 420, 432, 436, 439, 440, 449, 451–453, 459, 465, 472, 515, 516
CRISPR RNA (crRNA) 83, 141, 146, 163, 179–181, 348–350, 432, 433, 439, 440, 451, 452
Cryoprotectant agents (CPAs) 243, 245, 248, 249, 254
Cumulus-oocyte complexes (COCs) 234, 235, 240, 252–254, 354, 362, 365, 382, 398
Cybb 141, 143, 144, 146, 148, 153, 156–158, 160–162
Cycloheximide (CHX) 351, 377, 385
Cytochalasin B (CB) 170, 171, 173, 174, 176, 212, 213, 215, 220, 222, 223, 279, 282, 283, 291, 351, 355, 361, 377, 384, 385, 393, 398, 399
Cytoplasm-cell complexes (CCCs) 356, 357
CZB 21, 28–31, 36, 38, 39, 214–218, 221–223, 225, 227
CZB/H 21, 23, 28–31, 36, 38

D

DAP213 .... 231, 233, 236, 237, 241
DH10B *E. coli* 52, 56–59
Dimethyl sulfoxide (DMSO) 105, 112, 117, 171, 213, 214, 231–233, 236, 237, 241, 279, 310, 351, 375, 378, 380, 394, 405, 512
DNASTAR Lasergene 15

E

EF medium 215, 220
Electrofusion 212, 361, 368, 369
Electroporation .... v, 11, 21, 46, 60, 90, 92–93, 145, 164, 170, 179, 183, 258, 297, 352, 394, 396, 397, 419, 427
Embryo culture-tested mineral oil 21, 104, 144, 195
Embryo transfer, oviduct 31–33, 79, 239, 240, 338, 339

Chengyu Liu and Yubin Du (eds.), *Microinjection: Methods and Protocols*, Methods in Molecular Biology, vol. 1874, https://doi.org/10.1007/978-1-4939-8831-0,

Embryo transfer pipette 24, 26, 31, 32, 37, 38, 103, 106, 108, 109, 142, 153, 155, 200, 223, 224, 277, 287, 288, 292, 320
Embryo transfer, uterine 320, 322, 343
Ensembl Genome Browser 64
Enucleation media 215, 216, 220
Euthasol (Euthanize solution) 332, 335

F

Femtojet 12, 142, 154, 194, 200, 275, 442, 534
FERTIUP® (PM) 232, 237, 238
Flunixin meglumine 376, 385
Fluorinert FC-770 (mercury substitute) 9, 171
Fusion medium 356, 362, 365, 384

G

γ-irradiator 105, 112
Gelatin 88, 102, 105, 107, 112, 261, 406, 438
Geneservice Limited 65
Germline stem (GS) cells 87–89, 92–97
Glutathione (GSH) 187, 244, 248, 252, 253
Goat mammary gland cell culture medium 393–395
Golden Gate vectors 491
Gonocytes 403–430

H

Handmade cloning (HMC) 348, 349
Hepatocyte 257–270
Histone deacetylase inhibitor (HDACi) trichostatin A (TSA) 212
H11 locus 72–74, 81, 82
Holding pipette 22–24, 26, 30, 103, 107, 108, 113, 142, 153, 154, 172, 174–177, 195, 200, 220, 221, 224, 275, 277, 283, 284, 320, 322, 324, 335–337, 384
Holtfreiter's buffer 461, 464, 467, 470
Human chorionic gonadotropin (hCG) 20, 28, 170, 172, 183, 184, 187, 195, 199, 214, 217, 230, 231, 234, 248, 252, 254, 261, 279, 280, 290, 332, 333, 335, 338, 362, 512, 516
Hyaluronidase 21, 28, 29, 153, 171, 172, 183, 184, 195, 200, 214, 218, 219, 231, 235, 279–281, 290, 341, 351, 355, 363, 375, 384, 393, 398
Hydrodynamics-mediated gene delivery 258

I

ICR (CD-1) mice 19, 35, 213
ImaGenes GmbH 65
Intratubular injection 87
In vitro fertilization (IVF) .... v, 8, 187, 189, 208, 230, 238, 241, 243–245, 252, 260, 262
Isoflurane 20, 27, 32, 36, 275, 276, 286–288, 292, 317, 322, 332

J

JK1 .... 88, 91–93, 95

K

Kanamycin (kan) 47, 60, 61, 143, 149, 317, 323
Kawakawa 477, 483, 484, 486
Ketamine 36, 104, 108, 111, 144, 153, 160, 195, 200, 205, 376, 385
K-RVCL-50 medium 184, 185
KSOM 164, 176, 239, 240, 279–281, 283, 285, 287, 291, 292

L

Laparotomy 264
Latrunculin A (Lat A) 212–215, 218
Lipofectamine 2000 .... 117, 123, 129, 130, 136
Low melting agarose (LMA) 21, 36
LoxP 66, 73, 191, 192, 194, 198, 199, 201, 203, 205, 299, 302, 305, 309, 310, 420, 433
Luteinizing hormone (LH) 279, 363, 364

M

M2 .... 21, 104, 106, 108, 144, 152, 153, 171–173, 175, 176, 183–186, 188, 195, 200, 240, 281, 282, 287, 291, 393, 398
M16 .... 21, 144, 154, 164, 195, 200, 393, 398, 399
Mammalian cells .... v, 9, 71, 72, 348, 373, 525–535
Mammary gland bioreactors 391–400
Marine fish eggs 475–486
Mark's Modified Ringer's (MMR) 512, 516
Maturation media (MM) 350
Medaka 475, 490, 499–501, 503, 505
Mouse embryonic fibroblast (MEF) feeders 105, 107, 112, 133, 136, 319
Meganuclease (MN) 9, 296
MEGAshortscript T7 .... 144, 149, 261, 263
Meloxicam 104, 110, 144, 155, 195, 201, 279, 288
Methyl-β-cyclodextrin (MBCD) 244, 247
Mice
  B6CBAF1 34
  B6C3F1 213
  B6DBAF1 34
  B6D2F1 .... 194, 199, 213, 226
  C57BL/6 .... 34, 109, 110, 244, 262
  CF-1 112

C3H/He ... 213
FVB/N ... 34
NSG ... 139
129/Sv ... 34
Swiss Webster ... 19, 35
Microforge ... 22, 26, 103, 212, 215, 316, 324, 329, 335
Microhomology-mediated end joining (MMEJ) ... 140, 146, 158, 161, 162, 167
Microinjection
blastocyst ... 99–113
buffer ... 36, 47, 153, 164, 171, 200, 290, 298
chamber ... 23, 26, 28, 30, 103, 277, 282, 291
cytoplasmic ... 152, 274, 275, 283–285
pipette ... 24, 31, 176, 260
pronuclear ... 5, 6, 17–40, 74, 79, 169, 262, 284, 291, 314, 328
Microinjector ... 3, 4, 9, 12, 22, 23, 30, 38, 79, 171, 172, 174, 177, 277, 283, 329, 336, 448, 461, 470, 478, 480, 482, 491, 528, 530, 534, 535
Microinsemination ... 87, 96, 97
Micromanipulator ... 2, 12, 22, 23, 103, 107, 108, 142, 171, 172, 174, 176, 194, 200, 212, 215, 220, 221, 226, 275, 284, 285, 316, 320, 329, 335, 336, 348, 394, 436, 437, 442, 444, 453, 461, 479
Mitomycin C ... 88, 91, 105, 112, 317
Modified human tubal fluid (mHTF) ... 231–236, 239, 240, 245, 246, 248–250, 252–254
Modified Tris-buffered medium (mTBM) ... 364, 365
Mos1-mediated single-copy transgene insertions (MosSCI) ... 432
Mouse Genome Informatics ... 64

N

NanoDrop spectrophotometer ... 53
NCBI Map Viewer ... 64
Nuclear isolation medium (NIM) ... 215
Nucleobond ... 46, 52, 53

O

Oocyte activation medium ... 212
Oocyte manipulation medium ... 364, 365
Oryzias latipes ... 489–505

P

Parthenogenetic activation ... 361
PB1 ... 231–233
Phenylmethanesulfonyl fluoride (PMSF) ... 214, 215
PhiC31 integrase ... 72, 79, 83, 84, 418, 420
Phytohemagglutinin (PHA) ... 351, 356, 359
Piezo ... 8, 9, 169, 211, 213, 215, 220, 221, 225–227, 316, 320, 322
Pipette puller ... 12, 22, 26, 37, 103, 142, 194, 200, 212, 215, 277, 316, 461, 470
PMM Operation Liquid ... 215
Polyamine (PA) ... 47
Polyflex ... 376, 385
Polyvinylpyrrolidone (PVP) ... 170, 173, 174, 216, 219, 220, 223, 225, 226, 317, 320, 324
Porcine ... 88, 347–359, 361
Porcine zygote medium-3 (PZM-3) ... 350, 351, 357, 365, 367
Pregnant mare's serum gonadotropin (PMSG) ... 20, 195
Primordial germ cells (PGCs) ... 314, 403–430, 482, 490
Pronase ... 351, 355
Pseudopregnant ... 18, 19, 29, 31, 35, 79, 101, 113, 142, 153, 164, 169, 176, 184, 186, 188, 213, 224, 239, 253, 287
Pulsed field gel ... 55
Pulsed field gel electrophoresis (PFGE) ... 45, 55
Puromycin ... 90, 93, 104, 405, 411, 418, 422

R

Rabbit ... v, 6, 296, 327–343, 393, 397
Rat ... v, 5–7, 10, 44, 45, 62, 73, 88, 170, 273–292, 296, 298, 299, 308, 309, 313–325, 404, 408
Rat C6 cells ... 297, 298, 303
Recombineering ... 43–67
Rice Field eel *Monopterus albus* ... 490
Rosa26 knockin ... 73

S

SeqBuilder ... 49–51
Sertoli cells ... 218
Somatic cell nuclear transfer (SCNT) ... v, 2, 7, 8, 211, 212, 214, 216, 218, 328, 347, 373 389, 392, 398, 400
Spermatogonial stem cell (SSC) ... 87–97, 274
Sperm preincubation medium: TYH with MBCD ... 245, 250–253
Sprague Dawley, Fisher 344 and Brown Norway ... 274
Streptococcus pyogenes (SpCas9) ... 118, 122, 130, 141, 144, 146, 198, 515
Strontium chloride ($SrCl_2$) ... 213, 215
Superovulation ... 20, 28, 34, 112, 152, 230, 252, 262, 274, 290, 329, 333, 336, 341, 398
Surveyor assay ... 122–124, 134

Synthetic oviduct fuid (SOF) medium 376, 377, 385
Systemic Cas9 expressed Tg mouse (sCAT mouse) 258, 259, 262–264, 269

T

TARGATT^{TM} technology 71–85
Tetracycline (tet) 47, 57, 130, 491, 493, 494
Tetraploid 6, 101, 361, 367–370, 510
Thoracotomy 264
T7 in vitro transcription (T7-IVT) 116, 118–121, 143
Trans-activating crRNA (tracrRNA) 83, 118, 119, 141, 146, 163, 179, 180, 348, 432, 438, 440, 449, 451
Transcription activator-like effector nucleases (TALEN) .... v, 10, 17, 88, 101, 296, 297, 314, 328, 329, 373, 459, 465, 475, 482, 486, 489–505, 510, 512
TransferMan 22, 103, 194, 275
2-2-2-Tribromoethanol (TBE) 36
Tricaine 461, 479, 484
Trichostatin A (TSA) 214, 215, 218
Trypan blue solution 91, 94

U

UCSC Genome Browser 64

V

Vasectomy 21, 27, 28, 35, 102, 142, 274
Vitrification, oocyte 229–241
Vitrification, sperm 230

W

WBB6F1 W/W^{v} 91

X

Xenopus laevis 507–510, 520
Xenopus tropicalis 510, 512, 515, 520
Xylazine 36, 104, 108, 144, 153, 195, 200, 376, 385

Y

Yangtze River Delta white goats 398
Yeast artificial chromosome (YAC) 17, 44

Z

Zebrafish 10, 374, 459–473, 475, 490, 498, 505
Zinc finger nucleases (ZFN) .... v, 10, 101, 274, 295–311, 314, 328, 329, 373, 459, 465, 489

MIX
Papier aus verantwortungsvollen Quellen
Paper from responsible sources
FSC® C105338

FSC
www.fsc.org

If you have any concerns about our products,
you can contact us on
ProductSafety@springernature.com

In case Publisher is established outside the EU,
the EU authorized representative is:
Springer Nature Customer Service Center GmbH
Europaplatz 3, 69115 Heidelberg, Germany

Printed by Libri Plureos GmbH
in Hamburg, Germany